Natural Enemies of Insect Pests in Neotropical Agroecosystems

Brígida Souza • Luis L. Vázquez
Rosangela C. Marucci
Editors

Natural Enemies of Insect Pests in Neotropical Agroecosystems

Biological Control and Functional Biodiversity

Springer

Editors
Brígida Souza
Federal University of Lavras (UFLA)
Department of Entomology
Lavras, MG, Brazil

Luis L. Vázquez
Instituto de Investigaciones de Sanidad Vegetal (INISAV)
Miramar, La Habana, Cuba

Rosangela C. Marucci
Federal University of Lavras (UFLA)
Department of Entomology
Lavras, MG, Brazil

ISBN 978-3-030-24732-4 ISBN 978-3-030-24733-1 (eBook)
https://doi.org/10.1007/978-3-030-24733-1

This Springer imprint is published by the registered company Springer Nature Switzerland AG
The registered company address is: Gewerbestrasse 11, 6330 Cham, Switzerland

We dedicate this book to our colleague, César Freire Carvalho (1949–2018), one of the authors of this book, a great friend, and a professional at the Entomology Department of Lavras Federal University, Lavras, MG, Brazil. Always very active, participatory, and enthusiastic, he dedicated more than 40 years of his life to the entomology studies. César left us the memory of a unique dedication to teaching, research, and extension, besides a great legacy of research, mainly the lacewing group.

Foreword

Since the first pesticide products, growers have commonly relied on them to control insects and maintain crop production. Awareness of the negative impacts of pesticides has increased over the years including examples of when pest populations develop pesticide resistance, bee mortality, worker safety, runoff to waterways, and loss of market due to food surpassing the insecticide residue limits for export. Biological control through the use of predators, parasitoids, and pathogens represent an attractive alternative. However, biological control can be a challenge to implement since these living organisms have their needs and behave optimally under certain conditions, and therefore require more forethought than conventional pest products.

Therefore, this book addresses ways to optimize natural enemies via conservation, understanding their ecology, rearing, a specific focus in crop systems, and integration into IPM. While the focus is in Neotropical regions, a reader from any part of the world will find information in this book valuable. The ecology section provides background on various natural enemies and is useful for introducing new users or students to the topic. As a reader from a non-tropical region, I was encouraged by the popularity of augmentative biological control, with one entire section dedicated to mass rearing. Several chapters provide the reader a handbook to rearing, with photos, diagrams, and an explanation of the common problems encountered.

Also, this book is timely discussing current topics of interest. How does climate change impact biological control? What is the recent evidence on the impact of transgenic plants on natural enemies? Given the increase in biopesticide products, the authors discuss their incorporation in IPM, and selectivity for compatibility with biological control. The use of silicon has long been used for enhancing plant vigor and can protect plants from pests, and the volatiles that attract natural enemies to infested plants have been increasingly used in the field for pest control.

Jana Lee
Research Entomologist, USDA ARS
Oregon, USA

Preface

The use of biological agents for phytosanitary management has increased considerably in recent years, which is due to the awareness of the benefits derived from its use. Several factors contribute to this new perception, among which are the selection of populations of resistant organisms caused by indiscriminate use of phytosanitary products, the problems related to environmental contamination, and exposure of the applicators to the products used. Besides that, producers have recognized this form of control as a highly efficient technique and responsible for guaranteeing quality and free of chemical residue products, which meets the demands of a consumer market increasingly aware of the need for healthy and safe health foods.

These arguments have encouraged the scientific community to develop more sustainable control strategies with the consequent formation of research groups on biological control of pest arthropods, weeds, and plant diseases in order to contribute to the expansion and success of this control method. In addition, regulatory requirements imposed by the United States and European countries, which determine acceptable levels of residues and banned active ingredients, have raised expectations of growth in the number of companies producing biological control agents in the world.

The megadiversity of species present in the Neotropical region, which is represented by the countries of Latin America, and the Caribbean islands, gives this region high exploration potential for bioprospecting new control agents. However, although there are research projects and published studies, pest management with the use of natural enemies is still incipient, and the knowledge generated is dispersed, not contextualizing the reality about the use of biological control in diverse environments in that part of the continent.

The reader will find in *Natural Enemies of Insect Pests in Neotropical Agroecosystems: Biological Control and Functional Biodiversity* a set of information related to the functionality of plant diversification, focusing on agricultural production and biological control (conservative and large-scale production) and the use of these bioagents in large crops, pastures, forests, ornamental and horticultural plants, weeds, and plant diseases. Finally, examples of biological control integration with other pest management strategies resulting from research carried out in Latin America will be addressed.

The book has the participation of 92 researchers from 33 research institutions and universities from 8 Latin American countries. Many of the results presented in the book are the fruit of research projects of undergraduate students and the graduate program in Entomology (PPGEN) of Lavras Federal University (UFLA), as well as of former PPGEN graduates who currently act as researchers in Brazilian institutions and other Latin American countries. Thus, 35 PPGEN graduates contributed to the preparation of the book, which will be a vehicle for disseminating the knowledge generated in the area of CB in the last 30 years of PPGEN/UFLA.

Lavras, MG, Brazil — Brígida Souza
Miramar, La Habana, Cuba — Luis L. Vázquez
Lavras, MG, Brazil — Rosangela C. Marucci

Acknowledgments

We are grateful to the state and federal institutions, Fundação de Amparo à Pesquisa do Estado de Minas Gerais (FAPEMIG) and Conselho Nacional de Desenvolvimento Científico e Tecnológico (CNPq), for the financial support given to our research. We are especially grateful to Coordenação de Aperfeiçoamento de Pessoal de Nível Superior (CAPES), for the financial support obtained through the "Accord of Technical Cooperation and Academic CAPES/FAPEMIG," Process CAG 30042/12, which made possible the revision of the English text and the normalization of the bibliographic references.

We would like to thank the authors of the chapters for accepting the challenge and taking their time to complete the book; trainees, scientific initiation fellows, and master's and PhD students, who have dedicated themselves to projects in the area of biological control and integrated pest management; entomologists of the Neotropical region who believe in biological control as a sustainable strategy for the regulation of pest population in agroecosystems and who, through their research, have contributed to the expansion of areas under the use of this control method in Latin America.

Contents

Part IV Biological Control in Major Crops, Forests, Pasture, Weeds and Plant Diseases in the Neotropical Region

Contributors

Juan Pablo Molina Acevedo Colombian Agricultural Research Corporation (Agrosavia)/Embrapa CPATC, Rio Largo, AL, Brazil

Janet Alfonso-Simonetti Plant Health Research Institute (INISAV), Zoology and Herbology Department, Laboratory of Entomology, La Habana, Cuba

Miguel A. Altieri University of California, Berkeley, CA, USA

Dejane Santos Alves Coordenação do Curso de Licenciatura em Ciências Biológicas – Universidade Tecnológica Federal do Paraná, Santa Helena, PR, Brazil

Dany Silvio Souza Leite Amaral Prefeitura Municipal de Belo Horizonte, Belo Horizonte, Minas Gerais, Brazil

Vanessa Andaló Federal University of Uberlândia, Monte Carmelo Campus, Institute of Agrarian Sciences, Monte Carmelo, MG, Brazil

Jociani Ascari Coordenação do Curso de Licenciatura em Ciências Biológicas – Universidade Tecnológica Federal do Paraná, Santa Helena, PR, Brazil

Franscinely Aparecida Assis University Center of Goiatuba (UNICERRADO), Goiatuba, Goiás, Brazil

Gleice Aparecida Assis Federal University of Uberlândia (UFU), Monte Carmelo Campus, Institute of Agrarian Sciences (ICIAG), Monte Carmelo, Minas Gerais, Brazil

Alexander Machado Auad Embrapa Gado de Leite/Entomologia Department, Juiz de Fora, MG, Brazil

Leonardo Rodrigues Barbosa Empresa Brasileira de Pesquisa Agropecuária - Embrapa Florestas, Colombo, Paraná, Brazil

Lorena Barra-Bucarei Instituto de Investigaciones Agropecuarias, INIA, Banco de Recursos Genéticos Microbianos Department, Quilamapu, Chillán, Chile

Antonio Batista Filho Instituto Biológico, Agência Paulista de Tecnologia dos Agronegócios (APTA), Campinas, SP, Brazil

Carlos Eduardo Souza Bezerra Department of Agricultural, Biological and Engineering Sciences, Várzea Grande University Center, Várzea Grande, MT, Brazil

Yaril Matienzo Brito Institute of Plant Health Research (INISAV), La Havana, Cuba

Juliana Mendonça Campos Departamento Fitotecnia, Universidade Federal de Viçosa, Viçosa, Minas Gerais, Brazil

Ernesto Oliveira Canedo-Júnior Departamento de Educação e Ciências Humanas, Universidade do Estado de Minas Gerais - Unidade Poços de Caldas, Poços de Caldas, Minas Gerais, Brazil

Deise María Fontana Capalbo Embrapa Environment, Jaguariúna, SP, Brazil

César Freire Carvalho †Deceased

Geraldo Andrade Carvalho Federal University of Lavras (UFLA), Department of Entomology, Lavras, MG, Brazil

Lívia Mendes Carvalho Empresa de Pesquisa Agropecuária de Minas Gerais (EPAMIG), Campo Experimental Risoleta Neves, São João del Rei, MG, Brazil

Stephan Malfitano Carvalho Department of Entomology, Federal University of Lavras (UFLA), Lavras, MG, Brazil

Julie Giovanna Chacon-Orozco Instituto Biológico, Campinas, SP, Brazil

Daniela Resende Chaves Consultor in Risk Assessment, Sete Lagoas, MG, Brazil

Juliana Andrea Martinez Chiguachi Doutoranda Entomologia/Universidade Federal de Viçosa (UFV), Viçosa, Minas Gerais, Brazil

Francisco Jorge Cividanes Instituto Biológico, Laboratório de Parasitologia Vegetal, Campinas, SP, Brazil

Aloisio Coelho Jr Escola Superior de Agricultura "Luiz de Queiroz"/Universidade de São Paulo (USP/ESALQ), Piracicaba, SP, Brazil

Cecilia Czepak Universidade Federal de Goiás (UFG) – Escola de Agronomia, Goiânia, GO, Brazil

Tiago Cardoso da Costa-Lima Empresa Brasileira de Pesquisa Agropecuária (Embrapa Semiárido), Petrolina, PE, Brazil

Viviane Araujo Dalbon Federal University of Alagoas, Laboratory of Research in Chemistry of Natural Products – LPQPN, Maceió, AL, Brazil

Bernardo de Almeida Halfeld-Vieira Embrapa Meio Ambiente, Jaguariúna, SP, Brazil

José Eduardo Marcondes de Almeida Instituto Biológico, Agência Paulista de Tecnologia dos Agronegócios (APTA), Campinas, SP, Brazil

Carolina Rodrigues de Araujo Empresa Brasileira de Pesquisa Agropecuária - EMBRAPA, Embrapa Agrobiologia, Seropédica, Brazil

Alessandra de Carvalho Silva Empresa Brasileira de Pesquisa Agropecuária - EMBRAPA, Embrapa Agrobiologia, Seropédica, Brazil

Regiane Cristina Oliveira de Freitas Bueno Faculdade de Ciências Agrárias, UNESP, Botucatu, SP, Brazil

Kátia de Lima Nechet Embrapa Meio Ambiente, Jaguariúna, SP, Brazil

Flávio Henrique Vasconcelos de Medeiros Department of Plant Pathology, Universidade Federal de Lavras, Lavras, MG, Brazil

Regina de Oliveira Arruda Probiom Technology Ltda, Campinas, Brazil;

Universidade Guarulhos, Research and Development Department, Guarulhos, SP, Brazil

Leopoldo Ferreira de Oliveira Bernardi Federal University of Lavras (UFLA), Department of Entomology, Lavras, MG, Brazil

Rodrigo Lopes de Oliveira Federal University of Lavras (UFLA), Department of Entomology, Lavras, MG, Brazil

Antônio César Medeiros de Queiroz Laboratório de Ecologia de Formigas, Department of Biology - Setor Ecologia, Universidade Federal de Lavras, Lavras, Minas Gerais, Brazil

Alexandre de Sene Pinto Centro Universitário Moura Lacerda, Ribeirão Preto, SP, Brazil

Clara Yalexy Delgado TicoFrut, Alajuela, Costa Rica

Alexandre José Ferreira Diniz Faculdade de Ensino Superior Santa Bárbara (FAESP), Tatuí, SP, Brazil

Karina Cordeiro Albernaz Godinho Universidade Federal de Goiás (UFG) – Escola de Agronomia, Goiânia, GO, Brazil

Pablo da Costa Gontijo Universidade Federal de Goiás (UFG) – Escola de Agronomia, Goiânia, GO, Brazil

Anderson Dionei Grützmacher Federal University of Pelotas (UFPel), Department of Phytosanitary, Pelotas, RS, Brazil

María Elena Márquez Gutiérrez Havana University, Havana, Cuba

Andrés France Iglesias Instituto de Investigaciones Agropecuarias, INIA, Banco de Recursos Genéticos Microbianos Department, Quilamapu, Chillán, Chile

Gabriel C. Jacques Instituto Federal de Minas Gerais, Campus Bambuí, Bambuí, MG, Brazil

Sóstenes Eduardo Leal Trujillo Department of Entomology, Lavras, MG, Brazil

Aldomário Santo Negrisoli Júnior Embrapa Coastal Tablelands – CPATC, Research and Development Execution Unit, Rio Largo, AL, Brazil

Luís Garrigós Leite Instituto Biológico, Agência Paulista de Tecnologia dos Agronegócios (APTA), Campinas, SP, Brazil

Rosangela C. Marucci Federal University of Lavras (UFLA), Department of Entomology, Lavras, MG, Brazil

Simone Martins Mendes Lab de Entomologia, Embrapa Milho e Sorgo, Sete Lagoas, MG, Brazil

Heisler Alexsander Gomez Mendez Protección de Plantas Department, Universidad de San Carlos de Guatemala, Guatemala, Guatemala

Ângelo Barbosa Monteiro †Deceased

Jair Campos Moraes Federal University of Lavras (UFLA), Department of Entomology, Lavras, Minas Gerais, Brazil

Clara I. Nicholls University of California, Berkeley, Berkeley, CA, USA

Rodrigo de Oliveira Moraes Probiom Tecnologia - Pesquisa e Desenvolvimento Experimental em Ciências Físicas e Naturais Ltda., Research and Development Department, Campinas, SP, Brazil

Luis Clepf Passos Federal University of Lavras (UFLA), Department of Entomology, Lavras, MG, Brazil

Maria Fernanda Gomes Villalba Peñaflor Federal University of Lavras (UFLA), Department of Entomology, Lavras, MG, Brazil

Ernesto Prado EPAMIG Sul, Lavras, MG, Brazil

Eliane Dias Quintela Empresa Brasileira de Pesquisa Agropecuária (Embrapa Arroz e Feijão), Goiânia, GO, Brazil

Paulo Rebelles Reis Empresa de Pesquisa Agropecuária de Minas Gerais – EPAMIG Sul/EcoCentro, Lavras, Minas Gerais, Brazil

Janayne Maria Rezende Agbitech Controles Biológicos Ltda, São Paulo, SP, Brazil

Claudio Salas Instituto de Investigaciones Agropecuarias INIA, Centro Regional de Investigación Intihuasi, La Serena, Coquimbo Region, Chile

Marcus Vinicius Sampaio Federal University of Uberlândia, Agricultural Sciences Institute, Uberlândia, MG, Brazil

Lenira V. C. Santa-Cecília IMA/EPAMIG Sul, Lavras, MG, Brazil

Terezinha Monteiro dos Santos-Cividanes Laboratory of Plant Parasitology, Biological Institute, Ribeirão Preto, SP, Brazil

Graziele Santiago Silva Laboratório de Ecologia de Formigas, Department of Biology - Setor Ecologia, Universidade Federal de Lavras, Lavras, Minas Gerais, Brazil

Julio Carlos Pereira da Silva Department of Plant Pathology, Universidade Federal de Lavras, Lavras, MG, Brazil

Kethullyn H. Silva EPAMIG Sul, Lavras, MG, Brazil

Sandra Elisa Barbosa da Silva Federal University of Lavras (UFLA), Department of Entomology, Lavras, MG, Brazil

Érika Carla da Silveira Federal University of Lavras (UFLA), Department of Entomology, Lavras, MG, Brazil

Luis Cláudio Paterno Silveira Federal University of Lavras (UFLA), Department of Entomology, Lavras, MG, Brazil

Ana Luiza Viana de Sousa Department of Entomology, Rutgers University, New Brunswick, NJ, USA

Rutgers University, New Brunswick, NJ, USA

Brígida Souza Federal University of Lavras (UFLA), Department of Entomology, Lavras, MG, Brazil

Camila S. F. Souza Federal University of Lavras (UFLA), Department of Entomology, Lavras, MG, Brazil

Ivana Lemos Souza Federal University of Lavras (UFLA), Department of Entomology, Lavras, MG, Brazil

Pedro Henrique Brum Togni Universidade de Brasília (UnB), Ecology Department, Brasília, Distrito Federal, Brazil

Vitor Barrile Tomazella Federal University of Lavras (UFLA), Department of Entomology, Lavras, MG, Brazil

Carlos Pino Torres Centro I+D Agroecología, Curicó, Maule, Chile

Fernando Hercos Valicente Embrapa Maize and Sorghum, Sete Lagoas, MG, Brazil

Luis L. Vázquez Instituto de Investigaciones de Sanidad Vegetal (INISAV), Miramar, La Habana, Cuba

Madelaine Venzon Empresa de Pesquisa Agropecuária de Minas Gerais (EPAMIG), Viçosa, Minas Gerais, Brazil

Bruno Sérgio Vieira Universidade Federal de Uberlândia, Monte Carmelo, MG, Brazil

Marcelo Diniz Vitorino Fundação Universidade Regional de Blumenau, Blumenau, SC, Brazil

Carlos Frederico Wilcken Departamento de Proteção Vegetal, Faculdade de Ciências Agronômicas, Botucatu, SP, Brazil

Maurício Sergio Zacarias Embrapa Coffee, Brazilian Agricultural Research Corporation, PqEB, W/3 norte (at the end), Brasília, Federal District, Brazil

Bruno Zachrisson Biological Control Laboratory, Institute of Agricultural Research of Panama (IDIAP), Chepo, Panama

Jose Cola Zanuncio Departamento de Entomologia/BIOAGRO, Universidade Federal de Viçosa, Viçosa, Minas Gerais, Brazil

Part I
Conservation of Natural Enemies and Functional Biodiversity in Neotropical Agroecosystems

Chapter 1
Vegetational Designs to Enhance Biological Control of Insect Pests in Agroecosystems

Miguel A. Altieri and Clara I. Nicholls

1.1 Introduction

About 80% of the 1.5 billion hectares devoted to agriculture in the planet are occupied by industrial large-scale monocultures of maize, soybean, rice, and wheat. Due to their genetic homogeneity, these systems have proven to be highly vulnerable to insect invasions and disease epidemics (Heinemann et al. 2013). To keep pests at bay, about 2.3 billion kg of pesticides are applied worldwide per year, and today there are about 450 insecticide-resistant insects, indicating that the technology is reaching its limits. Moreover, secondary pest outbreaks are common in pesticide-loaded crops due to elimination of natural enemies (Hayes and Hansen 2017). Total removal of pesticides can restore natural enemy diversity and lead to renewed biological control of specific pests. Within 2 years, virtually, all banana insect pests in Golfito, Costa Rica, dropped to below economic threshold levels due to enhanced parasitization and predation after stopping insecticide (dieldrin and carbaryl) sprays. Similarly, in walnut orchards of California, natural biological control of the frosted scale and the calico scale was soon achieved by encyrtid parasitoids after removal of dichlorodiphenyltrichloroethyne (DDT) sprays (Croft 1990).

In insecticide-free agroecosystems, such as organic farms, communities of predators tend to be more even exerting stronger pest control pressure (Crowder et al. 2010). Despite the fact that organic farming potentially offers a means of augmenting natural pest control, most commercial organic farms practice monoculture under input-substitution management, which does not offer optimal environmental conditions (habitat, refuge, alternative food sources, etc.) to natural enemies. Many researchers have shown that by adding plant diversity to monocultures, it is possible to exert changes in habitat complexity which in turn favors natural enemy abundance and effectiveness due to enhanced availability of alternate prey, nectar sources, and suitable microhabitats (Altieri and Nicholls 2004).

M. A. Altieri (✉) · C. I. Nicholls
University of California, Berkeley, Berkeley, CA, USA
e-mail: agroeco3@berkeley.edu

B. Souza et al. (eds.), *Natural Enemies of Insect Pests in Neotropical Agroecosystems*, https://doi.org/10.1007/978-3-030-24733-1_1

A key strategy in agroecology is to restore functional biodiversity at field and landscape levels. Biodiversity performs key ecological services and if correctly assembled in time and space can lead to agroecosystems capable of sponsoring their own soil fertility, crop protection, and productivity (Altieri 1999). Diversity can be enhanced in crop fields with cover crops, intercropping, agroforestry, crop/livestock mixtures, and in the surrounding landscapes via shelterbelts, hedgerows, corridors, etc. Generally, correct agroecosystem diversification strategies result in pest regulation through restoration of natural control of insect pests (Altieri and Nicholls 2014).

1.2 Manipulating Biodiversity at the Landscape Level

One key characteristic of modern agricultural systems is the large size and homogeneity of crop monocultures that fragment the natural landscape, directly affecting abundance and diversity of natural enemies. Such simplification can have serious ecological implications for biological control, such as in the case of four US Midwest states where recent biofuel-driven growth in maize and soybean planting resulted in lower landscape diversity, decreasing the supply of pest natural enemies to maize and soybean fields and reducing biocontrol services by 24%. This loss of biocontrol services cost soybean and maize producers in these states an estimated $58 million per year in reduced yield and increased pesticide use (Landis et al. 2008).

One way to reintroduce biodiversity into large-scale monocultures is by establishing vegetationally diverse field margins and/or hedgerows that may serve as biological corridors. There is wide acceptance of the importance of field margins as reservoirs of the natural enemies of crop pests, as these habitats provide overwintering sites, increased resources such as alternative prey/hosts, pollen, and nectar for parasitoids and predators (Gurr et al. 1998).

1.2.1 Effects of Hedges and Surrounding Natural Vegetation

Many studies have demonstrated increased abundance of natural enemies and more effective biological control where crops are bordered by wild vegetation from which natural enemies colonize adjacent crop fields. Parasitism of armyworm, *Pseudaletia unipunctata* (Haworth) (Noctuidae), was significantly higher in maize fields embedded in a complex landscape than in maize fields surrounded by simpler habitats. In a 2-year study, researchers found higher parasitism of *Ostrinia nubilalis* (Hübner) (Crambidae) larvae by the parasitoid *Eriborus terebrans* (Gravenhorst) (Ichneumonidae) in edges of maize fields adjacent to wooded areas than in interior fields (Landis et al. 2000).

Similarly, in Germany, parasitism of rape pollen beetle was about 50% at the edge of the fields, while at the center of the fields, parasitism dropped significantly to 20% (Thies and Tscharntke 1999). In Hawaii, the presence of nectar-source plants in sugarcane field margins allowed population levels to rise and increased the efficiency of the sugarcane weevil parasite, *Lixophaga sphenophori* (Villeneuve) (Tachinidae) (Thopham and Beardsley 1975). The authors suggest that the effective range of the parasite within cane fields is limited to about 45–60 meters from nectar sources present in the field margins. Continuous herbicidal elimination of field margin nectar source plants had a detrimental effect on populations of *Lixophaga* and, therefore, led to a decrease in the efficiency of the parasite as a biocontrol agent of the weevil. Maier (1981) observed higher parasitization rates of apple maggot [*Rhagoletis pomonella* (Walsh) (Tephritidae)] by braconids in apple and hawthorn orchards in northern Connecticut, where plants such as blueberry (*Vaccinium* spp.), dogwood (*Cornus* spp.), and winterberry (*Ilex ciliata*) commonly grew nearby. These plants support populations of several frugivorous Tephritidae that serve as alternate hosts to braconids.

In many cases, weeds and other natural vegetation around crop fields harbor alternate hosts/prey for natural enemies, thus providing seasonal resources to bridge the gaps in the life cycles of entomophagous insects and crop pests. A classic case is that of the egg parasitoid wasp *Anagrus epos* Girault (Mymaridae), whose effectiveness in regulating the grape leafhopper, *Erythroneura elegantula* Osborn (Cicadellidae), was increased greatly in vineyards near areas invaded by wild blackberry (*Rubus* sp.). This plant supports an alternative host leafhopper [*Dikrella cruentata* (Gillette)], which breeds in its leaves in winter (Doutt and Nakata 1973). Recent studies have shown that prune trees planted next to vineyards also allowed early-season buildup of *A. epos*. After surviving the winter on an alternate host, the prune leafhopper, *Anagrus*, wasps move into the vineyard in the spring, providing grape leafhopper control up to a month earlier than in vineyards not near prune tree refuges (Pickett and Bugg 1998).

Murphy et al. (1998) completed a rigorous evaluation of the effectiveness of French prune trees in increasing control of the grape leafhopper. Results from this study indicated that there was a consistent and significant pattern of higher parasitism in grape vineyards with adjacent prune tree refuges than in vineyards lacking refuges. Researchers now recommend that trees should always be planted upwind from the vineyard, but otherwise can be managed as a typical commercial prune orchard; and to plant as many trees as is economically feasible, since the more trees there are, the more productive the refuge is likely to be (Corbett and Rosenheim 1996) (Fig. 1.1).

1.2.2 *Corridors*

One way to introduce the beneficial biodiversity from surrounding landscapes into large-scale monocultures is by establishing vegetationally diverse corridors that allow the movement and distribution of useful arthropod biodiversity into the

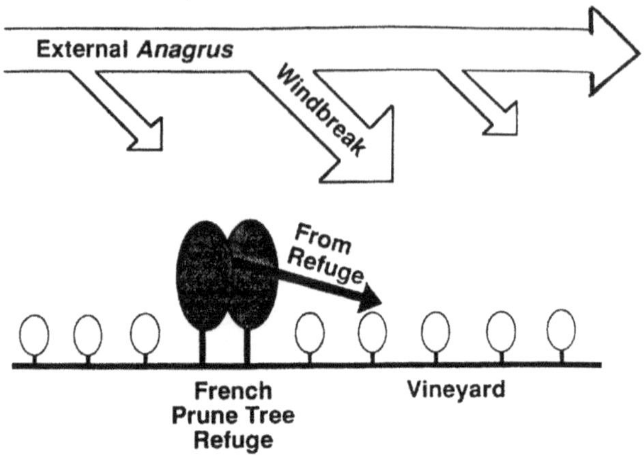

Fig. 1.1 Hypothesized sources of *Anagrus* colonizing vineyards early in the season. *Anagrus* colonize vineyards from adjacent French prune tree refuges. *Anagrus* also colonize from external overwintering sites. The windbreak effect generated by prune trees causes increased colonization by external *Anagrus* immediately downwind of refuges. (Corbett and Rosenheim 1996)

center of monocultures. Nicholls et al. (2001) established a vegetational corridor that connected to a riparian forest and cut across a vineyard monoculture. The corridor allowed natural enemies emerging from the riparian forest to disperse over large areas of otherwise monoculture vineyard systems. The corridor provided a constant supply of alternative food for predators effectively decoupling predators from a strict dependence on grape herbivores and avoiding a delayed colonization of the vineyard.

This complex of predators continuously circulated into the vineyard interstices, establishing a set of trophic interactions leading to a natural enemy enrichment, which also led to lower numbers of leafhoppers and thrips on vines. Generalist predators in the families Coccinellidae, Chrysopidae, Nabidae, and Syrphidae exhibited a density gradient in the vineyard, indicating that the abundance and spatial distribution of these insects were influenced by the presence of the corridor that channeled dispersal of the insects into adjacent vines. Adult leafhoppers exhibited a clear density gradient reaching lowest numbers in vine rows near the corridor and forest and increasing in numbers toward the center of the field, away from the adjacent vegetation. The highest concentration of leafhoppers occurred after the first 20–25 rows (30–40 m) downwind from the corridor.

Researchers in Switzerland introduced successional strips of annual flowering plants into cereal fields (Lys and Nentwig 1992). Significantly higher predator activity was found in the strip-managed fields than in the control monocultures, especially for carabid beetles such as *Poecilus cupreus* L., *Carabus granulates* L., and *Pterostichus melanarius* Illiger. Several observations led to the conclusion that this higher activity was generally due to a prolongation of the reproductive period in the strip-managed area. Besides the marked increase in activity and density, a large increase in the diversity of ground beetle species was observed. After 3 years of research, the authors concluded that flower strips offer not only higher food availability but also more suitable overwintering sites. Flower strips increase the chance of survival of many carabid species in arable ecosystems thus counteracting the faunal impoverishment trends promoted by monocultures. Nentwig (1988) found similar effects with 3–9-m-wide sown weed strips dividing large fields in small parts so that the distance between strips does not exceed 50–100 m. A favorite plant to be used as strips within or around fields is *Phacelia tanacetifolia*.

1.3 Manipulating Plant Biodiversity at the Field Level

1.3.1 Ecological Theory

It is accepted by many entomologists that inter-species plant diversity reduces crop vulnerability to insect pests. There is a large body of literature documenting that diversification of cropping systems (variety mixtures, polycultures, agroforestry systems, etc.) often lead to reduced herbivore populations (Risch et al. 1983; Altieri and Nicholls 2004). Two hypotheses have been offered to explain such reductions (Andow 1991).

The *natural enemy hypothesis* predicts that there will be a greater abundance and diversity of natural enemies of pest insects in polycultures than in monocultures. Predators tend to be polyphagous and have broad habitat requirements, so they would be expected to encounter a greater array of alternative prey and microhabitats in a heterogeneous environment (Russell 1989).

A study that supports the enemy hypothesis was conducted in tropical corn/bean/squash systems where Letourneau (1987) studied the importance of parasitic wasps in mediating the differences in pest abundance between simple and complex crop arrangements. A squash-feeding caterpillar, *Diaphania hyalinata* L. (Crambidae), occurred at low densities on intercropped squash in tropical Mexico (Letourneau 1987). Part of the effect of the associated maize and bean plants may have been to render the squash plants less apparent to ovipositing moths. Polyculture fields also harbored greater numbers of parasitic wasps than did squash monocultures. Malaise trap captures of parasitic wasps in monoculture consisted of one-half the number of individuals caught in mixed culture. The parasitoids of the target caterpillars were also represented by higher numbers in polycultures throughout the season. Not only were parasitoids more common in the vegetationally diverse traditional system; also the parasitization rates of *D. hyalinata* eggs and larvae on squash were higher in polycultures. Approximately 33% of the eggs in polyculture samples over the season were parasitized and only 11% of eggs in monocultures. Larval samples from polycultures showed an incidence of 59% parasitization for *D. hyalinata* larvae, whereas samples from monoculture larval specimens were 29% parasitized.

The *resource concentration hypothesis* is based on the fact that insect populations can be influenced directly by the concentration and spatial dispersion of their food plants. Many herbivores, particularly those with narrow host ranges, are more likely to find and remain on hosts that are growing in dense or nearly pure stands and which are thus providing concentrated resources and monotonous physical conditions (Andow 1991).

One study that supports this hypothesis (Risch 1981) looked at the population dynamics of six chrysomelid beetles in monocultures and polycultures of maize/bean/squash (*Cucurbita pepo*). In polycultures containing at least one nonhost plant (maize), the number of beetles per unit was significantly lower relative to the numbers of beetles on host plants in monocultures. Measurement of beetle movements in the field showed that beetles tended to emigrate more from polycultures than from host monocultures.

Apparently, this was due to several factors: (a) beetles avoided host plants shaded by maize, (b) maize stalks interfered with flight movements of beetles, and (c) as beetles moved through polycultures, they remained on nonhost plants for a significantly shorter time than on host plants. There were no differences in rates of parasitism or predation of beetles between systems (Risch 1981).

A second study (Bach 1980) examined the effects of plant diversity on the cucumber beetle, *Acalymma vittata* (Fabricius). Population densities were significantly greater in cucumber (*Cucumis sativus*) monocultures than in polycultures containing cucumber and two nonhost species. Bach also found greater tenure time of beetles in monocultures than in polycultures. She also determined that these differences were caused by plant diversity per se and not by differences in host plant density or size. Nevertheless, she did not reveal if differences in numbers of herbivores between monocultures and polycultures are due to diversity or rather to the interrelated and confounding effects of plant diversity, plant density, and host plant patch size.

1.3.2 Research Evidence

Over the last 40 years, many studies have evaluated the effects of crop diversity on densities of herbivore pests and have tried to prove one or both ecological hypothesis. An early review by Risch et al. (1983) summarized 150 published studies on the effect of diversifying an agroecosystem on insect pest abundance; 198 total herbivore species were examined in these studies. Fifty-three percent of these species were found to be less abundant in the more diversified system, 18% were more abundant in the diversified system, 9% showed no difference, and 20% showed a variable response.

Eight years later, Andow (1991) analyzed results from 209 studies involving 287 pest species and found that compared with monocultures, the population of pest insects was lower in 52% of the studies, i.e., 149 species and higher in 15% of the studies, i.e., 44 species. Of the 149 pest species with lower populations in intercrops,

60% were monophagous and 28% polyphagous. The population of natural enemies of pests was higher in the intercrop in 53% of the studies and lower in 9%. The reduction in pest numbers was almost twice for monophagous insects (53.5% of the case studies showed lowered numbers in polycultures) than for polyphagous insects (33.3% of the cases).

In a meta-analysis of 21 studies comparing pest suppression in polyculture versus monoculture, Tonhasca Jr. and Byrne (1994) found that polycultures significantly reduced pest densities by 64%. In a later meta-analysis, Letourneau et al. (2011) found a 44% increase in abundance of natural enemies (148 comparisons), a 54% increase in herbivore mortality, and a 23% reduction in crop damage on farms with species-rich vegetational diversification systems than on farms with species-poor systems.

Unequivocally, earlier reviews and recent meta-analyses suggest that diversification schemes generally achieve significant positive outcomes including natural enemy enhancement, reduction of herbivore abundance, and reduction of crop damage from a combination of bottom-up and top-down effects.

A study conducted in Kenya at the International Center of Insect Physiology and Ecology (ICIPE) added a new dimension to the above studies by showing that interactions in polycultures are mediated by the relationship between chemical ecology and agrobiodiversity. Scientists developed a habitat management system to control the stem borer, which uses two kinds of crops that are planted together with maize: a plant that repels these borers (the push) and another that attracts (pulls) them (Khan et al. 2000).

The plant chemistry responsible for stemborer control involves release of attractive volatiles from the trap plants and repellent volatiles from the intercrops. Two of the most useful trap crops that pull in the natural enemies of borers, such as the parasitic wasp [*Cotesia sesamiae* (Cameron) (Braconidae)], are Napier grass and Sudan grass, both important fodder plants; these are planted in a border around the maize.

Two excellent borer-repelling crops, which are planted between the rows of maize, are molasses grass, which also repels ticks, and the leguminous silverleaf (*Desmodium*), which in addition can suppress the parasitic weed *Striga* by a factor of 40 compared to maize monocrop. The N-fixing ability of *Desmodium* increases soil fertility, leading to a 15–20 percent increase in maize yield (Kahn et al. 1998). The push-pull strategy was adopted by more than 10,000 households in 19 districts in Kenya, 5 districts in Uganda, and 2 districts in Tanzania helping participating farmers to increase their maize yields by an average of 20% in areas where only stemborers are present and by more than 50% in areas where both stemborers and *Striga* weed are problems.

Participating farmers in the breadbasket of Trans-Nzoia reported a 15–20% increase in maize yield. In the semiarid Suba district – plagued by both stemborers and striga – a substantial increase in milk yield has occurred in the last 4 years, with farmers now being able to support grade cows on the fodder produced by *Desmodium* and other plants. When farmers plant maize, napier, and desmodium together, a return of US$ 2.30 for every dollar invested is made.

1.4 Conclusions

A community of organisms in an agroecosystem becomes more complex when a larger number of different kinds of plants are included, leading to more interactions among arthropods and microorganisms, components of above and below ground food webs. As diversity increases, so do opportunities for coexistence and beneficial interference between species that can enhance agroecosystem sustainability.

Diverse systems encourage complex food webs, which entail more potential connections and interactions among members, creating many alternative paths for energy and material flow. For this reason, a more complex community exhibits more stable production and less fluctuations in the numbers of undesirable organisms. By enhancing functional biodiversity, a major goal of the agroecological conversion process is achieved: strengthening the weak ecological functions in the agroecosystem, allowing farmers to gradually eliminate inputs altogether by relying instead on ecosystem functions (Nicholls et al. 2016).

Hundreds of studies show that complementary interactions occur between crops grown in polycultures and between adjacent cultivated and uncultivated vegetational components of agroecosystems. These interactions can have positive or negative, direct or indirect effects on the biological control of specific crop pests.

The exploitation of these interactions in real situations involves agroecosystem design and management and requires an understanding of the numerous relationships among plants, herbivores, and natural enemies (Altieri and Nicholls 2014). One of the major problems has been predicting which plant biodiversity spatial arrangements will lead to reduced pest abundance, since not all combinations of crops will produce the desired effect and blind adherence to the principle that a more diversified system will reduce pest infestation is clearly inadequate (Gurr et al. 1998).

Therefore, there is a need for greater understanding of the mechanisms involved to explain how, where, and when pest reduction occur and of identifying the type of biodiversity that is desirable to maintain and/or enhance in order to carry out ecological services and then to determine the best practices that will encourage the desired biodiversity components. Regardless of the need for more research in this area, several farmers and Integrated Pest Management (IPM) practitioners have identified various combinations of crops, trees, and natural vegetation that bolster biological control, and many use them commercially, such as introducing flowering cover crops in vineyards or deploying corridors of alyssum in vegetable crops, phacelia strips and beetle banks in cereal crops, etc.

There are many agricultural practices and designs that have the potential to enhance functional biodiversity, and others that negatively affect it. The idea is to apply the best management practices in order to enhance or regenerate the kind of biodiversity that can best subsidize the sustainability of agroecosystems by providing ecological services such as biological pest control, nutrient cycling, water and soil conservation, etc. (Fig. 1.2).

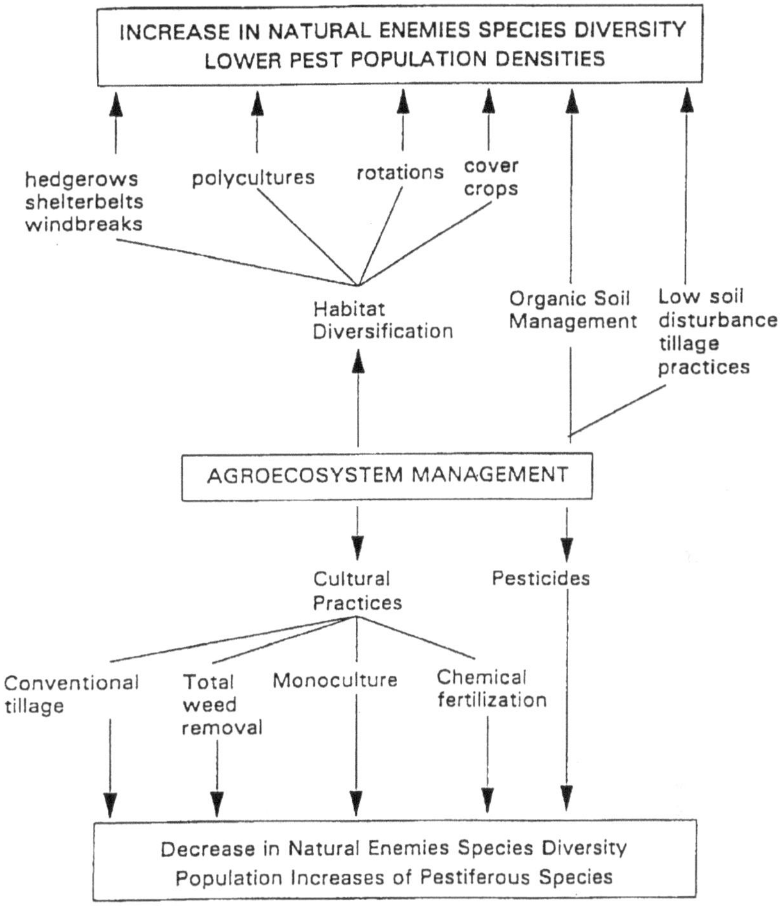

Fig. 1.2 The effects of agroecosystem management and associated cultural practices on the biodiversity of natural enemies and the abundance of insect pests. (Altieri and Nicholls 2004)

The role of agroecologists should be to encourage those agricultural practices that increase the abundance and diversity of above- and below-ground organisms, which in turn provide key ecological services to agroecosystems (Altieri and Nicholls 1999). In order for this diversification strategy to be more rapidly implemented, a much better understanding of the ecology of parasitoids and predators within and outside of the cultivated habitat and identifying those resources that are necessary for their survivorship and reproduction (Altieri and Letourneau 1982). It is also important to determine to what extent populations within the crop contribute to the overall natural enemy metapopulation in subsequent years. If these contributions are minor, then investments in habitat management should be oriented specifically to increasing the source populations outside the crop to ensure a greater number of immigrants each year, an action parallel to increasing the dosage of a chemical biocide. However, if the subpopulations within the cropping system contribute significantly to the year-to-year metapopulation dynamics, then habitat modifications should not only consider tactics fostering immigration into the crop but also those augmenting the probability of successional emigration when this habitat becomes unsuitable. Such actions could include the addition of plant species to provide alternate hosts and/or food sources, habitats as suitable overwintering sites, or the provision of corridors within the cropping system to facilitate movement between the subcomponents of the metapopulation (Pickett and Bugg 1998).

From a management perspective, there are four key issues to consider when implementing habitat management: (a) the selection of the most appropriate plant species and their spatial/temporal deployment; (b) identify the natural enemy complexes associated with such plant arrangements, the predator/parasitoid behavioral mechanisms that are influenced by the manipulation; (c) the spatial scale over which the habitat enhancement operates; and (d) the potential negative aspects associated with adding new plants to the agroecosystem, as obviously proposed habitat management techniques must fit existing cropping systems and adapt to the needs and circumstances of farmers.

References

Altieri MA (1999) The ecological role of biodiversity in agroecosystems. Agric Ecosyst Environ 74(1/3):19–31

Altieri MA, Letourneau DK (1982) Vegetation management and biological control in agroecosystems. Crop Prot 1(4):405–430

Altieri MA, Nicholls CI (1999) Biodiversity, ecosystem function and insect pest management in agricultural systems. In: Collins WW, Qualset CO (eds) Biodiversity in agroecosystems. CRC Press, Boca Raton, pp 69–84

Altieri MA, Nicholls CI (2004) Biodiversity and pest management in agroecosystems, 2nd edn. Waworth Press, New York

Altieri MA, Nicholls CI (2014) Manage insects in your farm: a guide to ecological strategies, sustainable agriculture research and education handbook series, vol 7. Sustainable Agriculture Network, College Park

Andow DA (1991) Vegetational diversity and arthropod population response. Annu Rev Entomol 36:561–586

Bach CE (1980) Effects of plant density and diversity on the population dynamics of a specialist herbivore, the striped cucumber beetle, *Acalymma vittata* (Fab.). Ecology 61(6):1515–1530

Corbett A, Rosenheim JA (1996) Impact of a natural enemy overwintering refuge and its interaction with the surrounding landscape. Ecol Entomol 21(2):155–164

Croft BA (1990) Arthropod biological control agents and pesticides. Wiley, New York

Crowder DW, Northfield TD, Strand MR et al (2010) Organic agriculture promotes evenness and natural pest control. Nature 466(7302):109–112

Doutt RL, Nakata J (1973) The rubus leafhopper and its egg parasitoid: an endemic biotic system useful in grape-pest management. Environ Entomol 2(3):381–386

Gurr GM, Van Emden HF, Wratten SD (1998) Habitat manipulation and natural enemy efficiency: implications for the control of pests. In: Barbosa P (ed) Conservation biological control. Academic Press, New York, pp 155–183

Hayes TB, Hansen M (2017) From silent spring to silent night: agrochemicals and the anthropocene. J Elementa Sci Anthrop. https://doi.org/10.1525/elementa.246. Accessed 2 Nov 2017

Heinemann JA, Massaro M, Coray DS et al (2013) Sustainability and innovation in staple crop production in the US Midwest. Int J Agr Sustain 12(1):71–88
Kahn ZR, Ampong-Nyarko K, Hassanali A et al (1998) Intercropping increases parasitism of pests. Nature 388:631–632
Khan ZR, Pickett JA, Van den Berg J et al (2000) Exploiting chemical ecology and species diversity: stemborer and striga control for maize and sorghum in Africa. Pest Manag Sci 56:957–962
Landis DA, Gardiner MM, Van der Werf W et al (2008) Increasing corn for biofuel production reduces biocontrol services in agricultural landscapes. Proc Natl Acad Sci USA 105:20552–20557
Landis DA, Wratten SD, Gurr GA (2000) Habitat management to conserve natural enemies of arthropod pests in agriculture. Annu Rev Entomol 45:175–201
Letourneau D (1987) The enemies hypothesis: tritrophic interactions and vegetational diversity in tropical agroecosystems. Ecology 68(6):1616–1622
Letourneau DK, Armbrecht I, Rivera BS et al (2011) Does plant diversity benefit agroecosystems?: a synthetic review. Ecol Appl 21(1):9–21
Lys JA, Nentwig W (1992) Augmentation of beneficial arthropods by strip-management: 4. surface activity, movements and density of abundant carabid beetles in a cereal field. Oecologia 92(3):373–382
Maier CT (1981) Parasitoids emerging from puparia of *Rhagoletis pomonella* (Diptera: Tephritidae) infesting hawthorn and apple in Connecticut. Can Entomol 113(9):867–870
Murphy BC, Rosenheim JA, Dowell RV et al (1998) Habitat diversification tactic for improving biological control: parasitism of the western grape leafhopper. Entomol Exp Appl 87(3):225–235
Nentwig W (1988) Augmentation of beneficial arthropods by strip management: 1. Succession of predaceous arthropods and long-term change in the ratio of phytophagous and predaceous species in a meadow. Oecologia 76(4):597–606
Nicholls CI, Altieri MA, Vazquez L (2016) Agroecology: principles for the conversion and redesign of farming systems. J Ecosys Ecograph S5:10
Nicholls CI, Parrella MP, Altieri MA (2001) The effects of a vegetational corridor on the abundance and dispersal of insect biodiversity within a northern California organic vineyard. Landsc Ecol 16(2):133–146
Pickett CH, Bugg RL (1998) Enhancing biological control: habitat management to promote natural enemies of agricultural pests. University of California Press, Berkeley
Risch SJ (1981) Insect herbivore abundance in tropical monocultures and polycultures: an experimental test of two hypotheses. Ecology 62(5):1325–1340
Risch SJ, Andow D, Altieri MA (1983) Agroecosystem diversity and pest control: data, tentative conclusions, and new research directions. Environ Entomol 12(3):625–629
Russell EP (1989) Enemies hypothesis: a review of the effect of vegetational diversity on predatory insects and parasitoids. Environ Entomol 18(4):590–599
Thies C, Tscharntke T (1999) Landscape structure and biological control in agroecosystems. Science 285(5429):893–895
Thopham M, Beardsley JW (1975) An influence of nectar source plants on the New Guinea sugarcane weevil parasite, *Lixophaga sphenophori* (Villeneuve). Proc Hawaii Entomol Soc 22:145–155
Tonhasca A Jr, Byrne DN (1994) The effects of crop diversification on herbivorous insects: a meta-analysis approach. Ecol Entomol 19(3):239–244

Chapter 2
Interactions of Natural Enemies with Non-cultivated Plants

Madelaine Venzon, Dany Silvio Souza Leite Amaral, Pedro Henrique Brum Togni, and Juliana Andrea Martinez Chiguachi

2.1 Introduction

Conservation of natural enemies in agroecosystems is directly associated with farm management practices that provide more suitable resources and conditions to biological control agents. Using such strategy, it is expected that natural enemies provide the ecosystem the service of biological control, thereby reducing the need for external outputs (UN 2017). In this sense, strategies related with natural enemy diversity conservation are important tools to reduce negative trends in modern agriculture, including pesticide resistance among insect pests, environmental and human health impacts, biodiversity loss, and introduction of invasive exotic pests (Tscharntke et al. 2007; Bianchi et al. 2008). This perspective demands the improvement of ecosystem services via management strategies that increase and conserve the biodiversity of flora and fauna within and around crop fields (Gurr et al. 2003). Typically, management strategies to enhance vegetational diversity, which, subsequently, increase the biodiversity of associated beneficial arthropods, vary with crop type (perennial or annual) and region (temperate or tropical) at both local and landscape scales (Thies and Tscharntke 1999). Importantly, efforts to conserve the biodiversity may positively affect natural enemies that provide biological control of insect pests (Landis et al. 2000; Norris and Kogan 2000, 2005; Letourneau et al. 2011).

M. Venzon (✉)
Empresa de Pesquisa Agropecuária de Minas Gerais (EPAMIG), Viçosa, Minas Gerais, Brazil
e-mail: madelaine@epamig.br

D. S. S. L. Amaral
Prefeitura Municipal de Belo Horizonte, Belo Horizonte, Minas Gerais, Brazil

P. H. B. Togni
Universidade de Brasília (UnB), Ecology Department, Brasília, Distrito Federal, Brazil

J. A. M. Chiguachi
Doutoranda Entomologia/Universidade Federal de Viçosa (UFV),
Viçosa, Minas Gerais, Brasil

B. Souza et al. (eds.), *Natural Enemies of Insect Pests in Neotropical Agroecosystems*, https://doi.org/10.1007/978-3-030-24733-1_2

A practice that has been widely adopted for a long time for habitat management in agroecosystems is the management of non-crop strips, which can increase beneficial arthropod populations (Altieri and Whitcomb 1978; Nentwig 1998; Liljesthröm et al. 2002; Gurr et al. 2003; Norris and Kogan 2000, 2005). Farmers can easily adopt the management of non-crop plants due to characteristics such as rapid plant growth and low investment (Amaral et al. 2013). However, non-crop plants are traditionally called and treated as weeds by some researchers, farmers, and citizens. Using this terminology implies that these plants have no obvious function in the agroecosystem and are indeed only deleterious to agriculture. Non-crop plants in agroecosystems provide resources and conditions to allow natural enemy survival, growth, and reproduction, even when their prey is scarce or absent. Thus, understanding the complex multitrophic interactions between natural enemies and non-crop plants provides a critical framework for the implementation of conservation biological control strategies.

In this chapter, we reviewed the direct and indirect effects of non-crop plant management on natural enemies. We addressed the main mechanism mediating the interactions among plants, herbivores, and natural enemies, focusing on examples from Latin America, mainly Brazil. We emphasized arthropod predators as model of action of non-crop vegetation on biological control. Thus, our specific objectives were firstly to examine how non-crop management improves the abundance and diversity of natural enemies, by the quantification of the specific mechanisms that mediated the interaction between non-crop plants and natural enemies. Secondly, we focused on possible applications on agroecosystems and effects upon evaluating plant characteristic or functional traits that may be chosen to improve key natural enemies.

2.2 Multitrophic Interactions Mediated by Non-crop Plants

The classical question "Why is the world green?" proposed by Hairston, Smith, and Slobodkin (HSS), in their influential 1960 study, was important to understand the role of natural enemies to regulate herbivore populations. In the HSS theory, or the *green world hypothesis*, Hairston and colleagues argued that plant biomass is not completely limited by herbivores, because natural enemies (third trophic level) regulate the populations of herbivores (second trophic level), reducing their impact on plants (first trophic level) (Hairston et al. 1960). This concept was previously defined as top-down regulation, where the third trophic level indirectly affects the populations from the first trophic level in a trophic cascade. Complimentary, Root (1973) demonstrated that brassica plots containing non-crop vegetation usually harbor a more diverse community of predatory and parasitoid arthropods than monoculture plots. The author also observed that the diversified plots presented less herbivore and associated damage, suggesting that plants can favor the diversity of natural enemies, which, in turn, reduces herbivore populations. After that, Price et al. (1980)

discussed that considering the role of natural enemies is imperative to understand the plant-herbivore interactions and the predator-prey dynamics due to direct and indirect effects of plants on natural enemies and vice versa. There are evidences that showed the importance of predators and other natural enemies' action against herbivores, reducing their negative effects on plants via trophic cascades (Dyer and Letourneau 1999; Costamagna and Landis 2011; Wilkinson and Sherratt 2016). Top-down forces, such as predation and parasitism, directly influence agricultural communities and can be managed to reduce pest outbreaks (Stireman et al. 2005; Macfadyen et al. 2009).

The top-down control is one of the most important theoretical bases to justify the implementation of conservation biological control in agroecosystems via non-crop plants. Beyond abundance of predator and other natural enemies, the effectiveness of biological control may increase with predator species abundance and diversity (Root 1973; Ives et al. 2005; Snyder et al. 2005). Specifically, modifications on leaf litter or crop residues, vegetation structure, and plant diversity (e.g., associated with polyculture, mulching techniques, or weedy cropping systems) all affect the diversity of beneficial arthropods (Uetz 1991; Denno et al. 2005) and, subsequently, the biological control service provided (Dornelas et al. 2009). However, simply adding plants and natural enemies to agroecosystems is not a guarantee of biological control (Letourneau et al. 2011) because the vegetational diversification by different techniques must be functional to attract and favor the performance of natural enemies in the farm (Venzon et al. 2015).

In this perspective, to attract and keep natural enemies in agroecosystems, practitioners often adopt habitat management strategies to conserve non-crop vegetation strips within or surrounding crop fields (Altieri and Whitcomb 1978; Nentwig 1998; Landis 2017). The manipulation of non-crop plants may be associated with a wide array of mechanisms that explain the attraction and maintenance of natural enemy populations as follows: (i) source of preferential prey species (Weyman and Jepson 1994); (ii) presence of patches used as refuge against competitors or as alternative for oviposition (Finke and Denno 2002; Langellotto and Denno 2006); (iii) provision of complementary and supplementary food resources (alternative prey and plat-provided food) (Jonsson et al. 2008; Lundgren 2009); (iv) access to favorable micro-climate conditions (Alderweireldt 1994; Chen et al. 2011); (v) reduction of the impact of negative interactions, such as cannibalism or intraguild predation, favoring synergistic or additive predation (Losey and Denno 1998; Finke and Denno 2006; Robinson et al. 2008); and (vi) the improvement of spatial structure that facilitates the web building, resulting in increase in prey capture (Robinson 1981; Mcnett and Rypstra 2000).

Agroecosystems are managed by human labor, which means that provisioning of biological control as an ecosystem service depends on cooperation between human and nature (Bengtsson 2015). The more the ecological theory is used as a management tool for pest control using ecosystem services besides artificial techniques (i.e., insecticide spraying), the less human counterpart is needed. In this perspective,

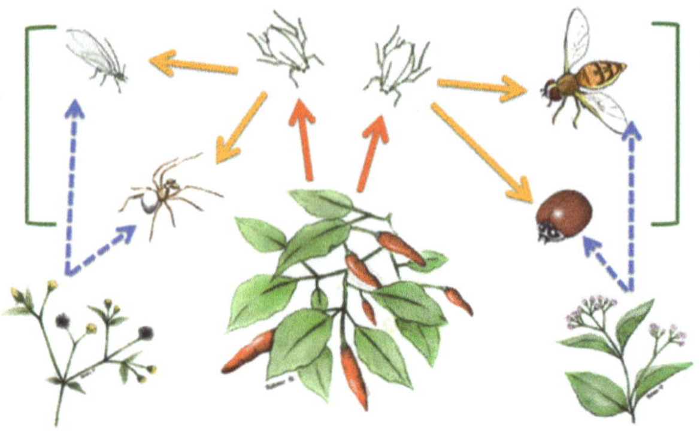

Fig. 2.1 *Next-to* effect (blue arrows) of non-crop plants contributing to top-down pest regulation (yellow arrows)

the strength of top-down control relies on the manipulation of the associated plant to the main crop, such as non-crop plants. Associated with top-down control, we predicted the *next-to hypothesis,* when the non-crop or another associated plant acts as a "scaffold" that helps to improve the coexistence of diverse natural enemy populations and promotes biological control (Fig. 2.1). *Why is the world green?* Because there is top-down control of herbivores, but complementary there are *next-to* traits on non-crop or other associated plants that build and reinforce the actions of natural enemies.

2.3 Non-crop Plants Affecting Natural Enemies

One of the main advantages of using non-crop plants to habitat manipulation is that plants are naturally present on agroecosystems; they grow rapidly and spontaneously, and farmers know them well. However, the effectiveness of this strategy depends on finding the functional role of each plant to specific biological control agents. To achieve this and successfully integrate non-crop plants into pest management decision-making, it is necessary to quantify and understand the ecological mechanisms that influence the response of natural enemies to such plants (Andow 1988; Snyder et al. 2005).

In temperate agroecosystems, there are several reported beneficial effects of non-crop plants on the distribution and abundance of beneficial arthropods (e.g., Wyss 1995; Nentwig 1998; Leather et al. 1999; Norris and Kogan 2000; Showler and Greenberg 2003; Silva et al. 2010). However, in tropical agroecosystems, there is a paucity of information pertaining to the resources provided by non-crop plants and their interactions with natural enemies. In the tropics, crops are grown all over the year with almost no interruption, and non-crop interactions with natural enemies are continuous in space and time. Herein, we sought to find and examine possible mechanisms associated with interactions between non-crop plants and natural enemies in tropical agroecosystems to clarify the role of such plants in cropped areas, mainly from South America.

2.4 Non-crop Plants as Resource to Natural Enemies

Results obtained from research carried out in chili pepper agroecosystems in Brazil revealed the role of non-crop plants as a resource and source of natural enemies. Amaral et al. (2013) observed that the presence of non-crop plants within or surrounding chili pepper fields differentially affected the abundance of different groups of aphidophagous predators by providing alternative prey, refuge, nectar, and pollen as a complementary resource. Coccinellidae were more abundant when aphids were present on non-crop plants, but the ladybirds were also observed foraging on flowers and extrafloral nectaries and using plants as refuge. Adults of Syrphidae were more frequently recorded feeding on nectar and pollen from flowers when compared to any other resource. Anthocoridae, Neuroptera, and Araneae were observed equally exploiting the resources from non-crop plants (Amaral et al. 2013).

Cycloneda sanguinea (Linnaeus) (Coccinellidae) was observed frequently preying on aphids commonly found on annual sow thistle (*Sonchus oleraceus*) and on American black nightshade (*Solanum americanum*) (Amaral et al. 2013). It was also observed on flowers of tropic ageratum (*Ageratum conyzoides)* and of beggar-ticks (*Bidens pilosa).* Feeding on flowers of *B. pilosa* increased predator survival in the absence of prey (Fonseca et al. 2017). In Central Mexico, *Coccinella nugatoria* Mulsant was more frequently observed interacting with aphids in maize crops with non-crop plants than in monocultures (Trujillo-Arriaga and Altieri 1990).

Complementary, non-crop plants may contribute to the diversification of plant communities, which may aggregate generalist arthropods with complementary foraging strategies, promoting functional diversity of natural enemies. Spiders exhibit a diversity of foraging characteristics and behavior (Uetz et al. 1999) that, acting together, may increase top-down control of pests. Studies on chili pepper agroecosystems in Brazil provided evidence for associations between spider families and non-crop plants (Amaral et al. 2016). The composition of non-crop plant communities altered the abundance and guild structure of spiders; the greatest number of spiders was found on Asteraceae plants (Amaral et al. 2016). These plants are attractive to spiders, particularly to the family Thomisidae, potentially due to complex inflorescences that act as sites to capture prey attracted to their flower resources (Nyffeler et al. 1994; Peterson et al. 2010). Additionally, spiders were also observed using non-crop plants as substrate to build webs, mainly on taller and ramified plants (e.g., *A. conyzoides*, *Leonurus sibiricus*) (Amaral et al. 2016).

Non-crop plants are important reservoirs of natural enemies for many crop systems due to the characteristics discussed above. For example, *B. pilosa*, *Amaranthus* sp., *Parthenium hysterophorus*, and *Alternanthera ficoidea* are important sources of many *Orius* species for crops such as maize, beans, and lettuce (Silveira et al. 2003). In soybean, stink bugs can oviposit on non-crop plants, such as *A. ficoidea*, adjacent to the crop field. These eggs can be used by Platygastridae parasitoids (e.g., *Telenomus podisi* Ashmead, 1881) and favor the biological control of stink bugs in soybean fields (Maruyama et al. 2002). Pollen of *Peltaea riedelii* (Malvaceae) can be used as supplementary food by the predatory mites *Euseius concordis* (Chant) and *Iphiseiodes zuluagai* Denmark & Muma (Acari: Phytoseiidae) on physic nut crops (Marques et al. 2014). Non-crop plants *S. americanum* and *Salpichroa origanifolia* harbor *Phytoseiulus longipes* Evans (Acari: Phytoseiidae), a potential effective candidate for augmentative biological control of *Tetranychus evansi* Baker & Pritchard in tomatoes (Furtado et al. 2006). Supporting these findings, traditional cornfields with a high non-crop plant diversity in the Colombian Andes presented a more diverse community of predators that could potentially act as biological control agents (Martínez et al. 2015). Moreover, the presence of non-crop plants can provide a great amount of non-pest species that can compete with pest species and serve as supplementary or alternative prey to natural enemies. In this sense, Sánchez-Monge et al. (2011) found 19 thrips species (mostly non-pests) associated with non-crop plants in Costa Rica.

2.5 Survival of Natural Enemies on Non-crop Plants

The survival parameters of arthropod predators may be directly affected by specific non-crop plants. The survival of *C. sanguinea* (native from America), in the absence of prey, differed between plant species, with significantly greater survival on *A. conyzoides* and *B. pilosa* than on *S. oleraceus* (Amaral et al. 2013). In the same study, any evidence was gathered to suggest that non-crop floral resources provided by these plants offered nutritional benefits to the exotic Coccinellidae *Harmonia axyridis* (Pallas). The non-crop plant species studied are native from the Americas, and *H. axyridis* is native from the Palearctic region, demonstrating that life history traits are important factors to be considered when selecting non-crop plants in conservation biological control strategies (Amaral et al. 2013).

Chrysopidae larvae can benefit by feeding on flower resources of non-crop plants during periods of prey scarcity. Salgado (2014) reported that these benefits vary with predator and non-crop plant species. Larvae of *Chrysoperla externa* (Hagen, 1861) had higher survival when *A. conyzoides* flowers were offered. *Ceraeochrysa cubana* (Hagen, 1861) larvae had higher survival with *A. conyzoides* and with *B. pilosa* flowers. Adults of both species did not benefit from the presence of these flower species. Larvae of *C. externa* completed their development and turned to adults when fed exclusively on pollen of elephant grass *Pennisetum purpureum* (Oliveira et al. 2010).

2.6 Spatial Distribution

The spatial distribution should be also considered when evaluating possible benefits of management of non-crop plants, by determining the natural enemy cover span from the source of the plant diversity (Holland et al. 2004; Thomson and Hoffmann 2013). More than intrinsic characteristic among species or life stages, the movement behavior may be influenced by environmental heterogeneity (Winder et al. 2001;

Park and Obrycki 2004). The strategy of manipulation of non-crop plant may be affected by the pattern of movement and spatial distribution of natural enemies, contributing to biological control. On a spatiotemporal experiment in chili pepper, the presence of non-crop plants affected positively the distribution of coccinellids and spiders, promoting more densities of arthropods in non-crop plant patches (Amaral 2014; Amaral et al. 2016). Non-crop plant strips influenced the aggregation tendencies of coccinellids and spiders on season crop, mainly from the middle to the end, when the non-crop plants were established. Other studies have also shown that the presence of vegetation surrounding crop fields increases predator colonization and abundance (Sunderland and Samu 2000). In Colombia, the presence of non-crop plants affected the predatory species turnover among habitat types, contributing to a higher diversity of potential biological control agents and to rare arthropod predatory species conservation in tropical landscapes (Martínez et al. 2015). Similarly, conventional soybean monoculture expansion across the Rolling Pampa, Argentina, threatened the occurrence of non-crop plants, and, consequently, non-crop plant removal imposed a risk to beneficial arthropod species and functional groups occurrence in the landscape (De La Fuente et al. 2010).

2.7 Behavior

Beyond the provision of food, non-crop plants may affect indirectly the survival of predators by reducing negative effect of cannibalism and intraguild predation. The diversification of vegetation mixing squash (*Cucurbita pepo*) and non-crop mugwort *Artemisia vulgaris* affected life-stage specific interactions between the American coccinellid, *Hippodamia convergens* (Guerin-Méneville, 1842), and the exotic *H. axyridis,* which overlap in spatial distribution in many crop systems (Amaral et al. 2015) (Fig. 2.2). In a small scale, the association of squash and non-crop plants promotes spatial heterogeneity and prey availability that reduce larval intraguild predation and cannibalism. The reproductive output of *H. convergens* was improved by reducing intra- and interspecific egg predation. The mugwort leaf morphology has complex architecture and provides better shelter for coccinellid eggs when compared to squash plants. Similarly, competition between several arthropods sharing similar resources and prey can be alleviated by the presence of non-crop plants that provides more resources besides the more complex spatial structure within the habitat (Tixier et al. 2013).

Aphidophagous coccinellids such as *Eriopis connexa* (Germar, 1824), *C. sanguinea*, *H. axyridis*, and *H. convergens* share many prey types and, consequently, forage and oviposit in similar habitats. Sicsú et al. (2015) observed that, although adults from these species are found in many non-crop plants, each species oviposit on specific plants to avoid interspecific competition. Consequently, larval distribution is strongly related to the oviposition site on a preferred plant. Each plant also harbors different species of aphids. Therefore, there is an interaction between non-crop plant species and the availability of aphids, which can affect the ovipositional behavior of predatory coccinellids and the distribution of eggs and larvae.

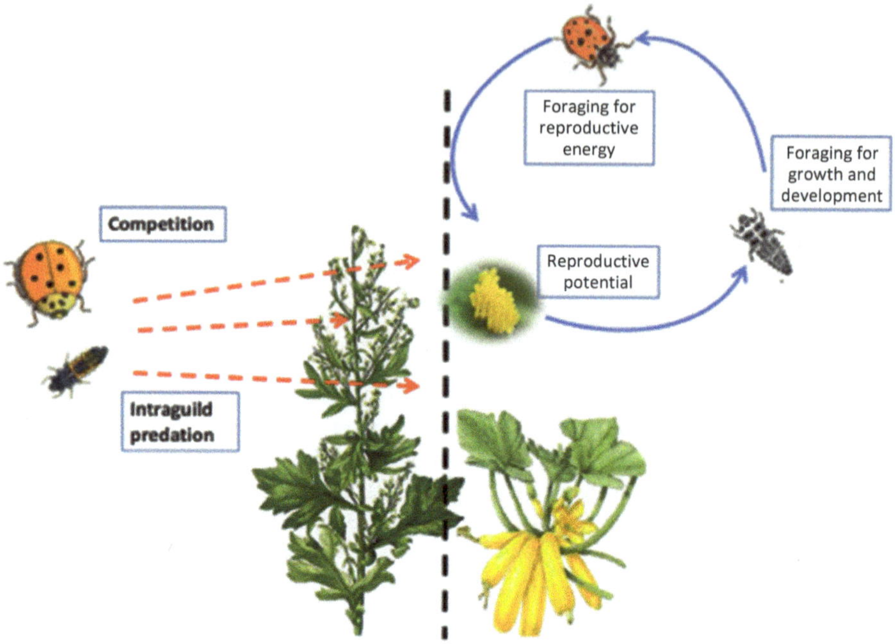

Fig. 2.2 Spatial guild division created by the management of non-crop plant (left), reducing negative effects of competition and intraguild predation between two species of coccinellids

In Mexico, Penagos et al. (2003) observed that the presence of non-crop plants favored the colonization of maize by natural enemies. The most common natural species associated with the reduction on *Spodoptera frugiperda* (J.E. Smith) (Noctuidae) and the presence of non-crop plants were the generalist predators *Doru taeniatum* (Dohrn) (Forficulidae) and *Chrysoperla* spp. (Chrysopidae) and predatory beetles such as carabid beetles. The authors also observed a significantly reduction on *S. frugiperda* eggs, on the number of aphid colonies, and on the number of sap beetles compared to maize plots without non-crop plants. However, the presence of non-crop plants reduced the parasitism rate of *S. frugiperda* eggs by *Chelonus insularis* (Cresson, 1865) (Braconidae) (Penagos et al. 2003). Similar results were found in the Peruvian Andes, where maize, bean, and non-crop plants treatments reduced the abundance of key herbivores similar to the conventional monoculture plots (Gianoli et al. 2006). These findings demonstrated that it is necessary to investigate specific traits of a given plant and its effects on target natural enemies.

2.8 Applications and Future Directions

Some growers can argue that non-crop vegetation can interfere on crop yield by the competition or the allelopathic effect. In fact, non-crop plants can be viewed as weedy, when they are not properly managed in agroecosystems. To avoid such problems, there is the need to encourage growers and researchers to search for the best design options when using non-crop plants as a diversification tool for pest management. For example, this could be achieved by maintaining strips of non-crop plants between crop rows or around the main crop. A selective removal of some species of non-crop plants can be also done in order to maintain only species that benefit natural enemies. Therefore, it is necessary to understand the ecological interactions among plants, herbivores, and natural enemies. Other complementary strategies, such as habitat manipulation at different spatial scales (e.g., using agroforestry systems, fallow areas, maintenance of natural areas), may broaden the positive effects of non-crop plants.

In a recent field study, Chiguachi et al. (2017) compared different pest management strategies for chili pepper pest control. Although pest population was lower in all sampled areas, the abundance and diversity of Coccinellidae were higher on cropped areas with non-crop plants without pesticides, compared with areas with and without pesticides but in absence of non-crop plants. Thus, the presence of non-crop plants contributes to the maintenance of biodiversity in cultivated areas.

In order to implement and recommend efficient strategies using non-crop plants and other habitat manipulation strategies, it is necessary to study deeply the biology and ecology of different natural enemies to identify factors disrupting biological control. This is an important challenge in tropical agroecosystems because there are much more species and possibilities of interactions with non-crop plants than in temperate regions. Non-crop plants are well known by growers, and it is also necessary that researchers focus their attention in their needs to produce applied research in accordance with real agricultural demands.

References

Alderweireldt M (1994) Habitat manipulations increasing spider densities in agroecosystems: possibilities for biological control. J Appl Entomol 118(1/5):10–16

Altieri MA, Whitcomb WH (1978) The potential use of weeds in manipulation of beneficial insects. HortScience 14:12–18

Amaral DSSL (2014) More than weed: non-crop plants, arthropod predators and conservative biological control. Universidade Federal de Viçosa, Tese

Amaral DSSL, Venzon M, Duarte MVA et al (2013) Non-crop vegetation associated with chili pepper agroecosystems promote the abundance and survival of aphid predators. Biol Control 64(3):338–346

Amaral DSSL, Venzon M, Perez AL et al (2015) Coccinellid interactions mediated by vegetation heterogeneity. Entomol Exp Appl 156(2):160–169

Amaral DSSL, Venzon M, Santos HH et al (2016) Non-crop plant communities conserve spider populations in chili pepper agroecosystems. Biol Control 103:69–77

Andow DA (1988) Management of weeds for insect manipulation in agroecosystems. In: Altieri MA, Liebman M (eds) Weed management in agroecosystems: ecological approaches. CRC Press, Florida, pp 265–301

Bengtsson J (2015) Biological control as an ecosystem service: portioning contributions of nature and human inputs to yield. Ecol Entomol 40(S1):45–55

Bianchi F, Booij C, Tscharntke T (2008) Sustainable pest regulation in agricultural landscapes: a review on landscape composition, biodiversity and natural pest control. Proc Biol Sci 273(1595):1751–1727

Chen LL, Chen SB, You MS (2011) Effects of cover crops on spider communities in tea plantations. Biol Control 59(3):326–335
Chiguachi JAM, Martins EF, Amaral DSSL et al (2017) Abundância e diversidade de coccinelídeos em cultivos de pimenta-malagueta com diferentes práticas de manejo de pragas. In: 6° Congresso latino-americano de agroecologia, 10° congresso brasileiro de agroecologia, 6° seminário de agroecologia do Distrito Federal e entorno, ABA, Brasília, 2013
Costamagna AC, Landis DA (2011) Lack of strong refuges allows top-down control of soybean aphid by generalist natural enemies. Biol Control 57(3):184–192
De La Fuente EB, Perelman S, Ghersa CM (2010) Weed and arthropod communities in soyabean as related to crop productivity and land use in the Rolling Pampa, Argentina. Weed Res 50(6):561–571
Denno RF, Lewis D, Gratton C (2005) Spatial variation in the relative strength of top-down and bottom-up forces: causes and consequences for phytophagous insect populations. Ann Zool Fenn 42(4):295–311
Dornelas M, Moonen AC, Magurran AE et al (2009) Species abundance distributions reveal environmental hetero-geneity in modified landscapes. J Appl Ecol 46(3):666–672
Dyer LA, Letourneau DK (1999) Trophic cascades in a complex, terrestrial community. Proc Natl Acad Sci U S A 96(9):5072–5076
Finke DL, Denno RF (2002) Intraguild predation diminished in complex-structured vegetation: implications for prey suppression. Ecology 83(3):643–652
Finke DL, Denno RF (2006) Spatial refuge from intraguild predation: implications for prey suppression and trophic cascades. Oecologia 149(2):265–275
Fonseca MM, Lima E, Lemos F et al (2017) Non-crop plant to attract and conserve an aphid predator (Coleoptera: Coccinellidae) in tomato. Biol Control 115:129–134
Furtado IP, Moraes GJ, Kreiter S et al (2006) Search for effective natural enemies of *Tetranychus evansi* in south and southeast Brazil. Exp Appl Acarol 40(3/4):157–174
Gianoli E, Ramos I, Alfaro-Tapia A et al (2006) Benefits of a maize–bean–weeds mixed cropping system in Urubamba Valley, Peruvian Andes. Int J Pest Manag 10(4):11–19
Gurr GM, Wratten SD, Luna JM (2003) Multi-function agricultural biodiversity: pest management and other benefits. Basic Appl Ecol 4(2):107–116
Hairston NG, Smith FE, Slobodkin LB (1960) Community structure, population control, and competition. Am Nat 94(879):421–425
Holland J, Winder L, Woolley C et al (2004) The spatial dynamics of crop and ground active predatory arthropods and their aphid prey in winter wheat. Bull Entomol Res 94(5):419–431
Ives AR, Cardinale BJ, Snyder WE (2005) A synthesis of subdisciplines: predator-prey interactions, and biodiversity and ecosystem functioning. Ecol Lett 8(1):102–116
Jonsson M, Wratten SD, Landis DA et al (2008) Recent advances in conservation biological control of arthropods by arthropods. Biol Control 45(2):172–175
Landis D, Wratten S, Gurr G (2000) Habitat management to conserve natural en-emies of arthropod pests in agriculture. Annu Rev Entomol 45:175–201
Landis DA (2017) Designing agricultural landscapes for biodiversity-based ecosystem services. Basic Appl Ecol 18:1–12
Langellotto GA, Denno RF (2006) Refuge from cannibalism in complex- structured habitats: implications for the accumulation of invertebrate predators. Ecol Entomol 31(6):575–581
Leather S, Cooke R, Fellowes M et al (1999) Distribution and abundance of ladybirds (Coleoptera: Coccinellidae) in non-crop habitats. Eur J Entomol 96(1):23–27
Letourneau DK, Armbrecht I, Rivera BS et al (2011) Does plant diversity benefit agroecosystems? A synthetic review. Ecol Appl 21(1):9–21
Liljesthröm G, Minervino E, Castro D et al (2002) The spider community in soybean cultures in the Buenos Aires province, Argentina. Neotrop Entomol 31(2):197–209
Losey JE, Denno RF (1998) Positive predator-predator interactions: enhanced predation rates and synergistic suppression of aphid populations. Ecology 79(6):2143–2152
Lundgren JG (2009) Relationships of natural enemies and non-prey foods. Springer, Dordrecht
Macfadyen S, Gibson R, Polaszek A et al (2009) Do differences in food web structure between

organic and conventional farms affect the ecosystem service of pest control? Ecol Lett 12(3):229–238

Marques RV, Sarmento RA, Ferreira VA et al (2014) Alternative food sources to predatory mites (Acari) in a *Jatropha curcas* (Euphorbiaceae) crops. Rev Colomb Entomol 40(1):74–79

Martínez E, Rös M, Bonilla MA et al (2015) Habitat heterogeneity affects plant and arthropod species diversity and turnover in traditional cornfields. PLoS One 10:e0128950

Maruyama WI, Pinto AS, Gravena S (2002) Parasitoides de ovos de percevejos (Hemiptera: Heteroptera) em plantas daninhas. Ceres 49(284):453–459

Mcnett BJ, Rypstra AL (2000) Habitat selection in a large orb-weaving spider: vegetational complexity determines site selection and distribution. Ecol Entomol 25(4):423–432

Nentwig W (1998) Weedy plant species and their beneficial arthropods: potential for manipulation in field crops. In: Pickett CH, Bugg RL (eds) Enhancing biological control: habitat management to promote natural enemies of agricultural pests. UC Press, Berkeley, pp 49–72

Norris R, Kogan M (2005) Ecology of interactions between weeds and arthropods. Annu Rev Entomol 50:479–503

Norris RF, Kogan M (2000) Interactions between weeds, arthropod pests, and their natural enemies in managed ecosystems. Weed Sci 48(1):94–158

Nyffeler M, Sterling WL, Dean DA (1994) How spiders make a living. Environ Entomol 23(6):1357–1367

Oliveira SA, Souza B, Auad AM et al (2010) Can larval lacewings *Chrysoperla externa* (Hagen): (Neuroptera, Chrysopidae) be reared on pollen? Rev Bras Entomol 54(4):697–700

Park Y, Obrycki J (2004) Spatio-temporal distribution of corn leaf aphids (Homoptera: Aphididae) and lady beetles (Coleoptera: Coccinellidae) in Iowa cornfields. Biol Contro 131(2):210–217

Penagos DI, Magallanes R, Valle J et al (2003) Effect of weeds on insect pests of maize and their natural enemies in Southern Mexico. Int J Pest Manag 49(2):155–161

Peterson JA, Romero SA, Harwood JD (2010) Pollen interception by linyphiid spiders in a corn agroecosystem: implications for dietary diversification and risk-assessment. Arthropod Plant Interact 4(4):207–217

Price PW, Bouton CE, Gross P et al (1980) Interactions among three trophic levels: influence of plants on interactions between insect herbivores and natural enemies. Annu Rev Ecol Syst 11:41–65

Robinson JV (1981) The effect of architectural variation in habitat on a spider community:an experimental field study. Ecology 62(1):73–80

Robinson KA, Jonsson M, Wratten SD et al (2008) Implications of floral resources for predation by an omnivorous lacewing. Basic Appl Ecol 9(2):172–181

Root RB (1973) Organization of a plant-arthropod association in simple and diversehabitats: the fauna of collards (*Brassica oleracea*). Ecol Monogr 43(1):95–124

Salgado DN (2014) Plantas espontâneas favorecem crisopídeos em plantio de pimenta malagueta. Dissertação, Universidade Federal de Viçosa

Sánchez-Monge A, Retana-Salazar A, Brenes S et al (2011) A contribution to thrips-plant associations records (Insecta: Thysanoptera) in Costa Rica and Central America. Fla Entomol 94(2):330–339

Showler A, Greenberg S (2003) Effects of weeds on selected arthropod herbivore and natural enemy populations, and on cotton growth and yield. Environ Entomol 32(1):39–50

Sicsú PR, Macedo RH, Sujii ER (2015) Oviposition site selection structures niche portioning among coccinellid species in a tropical ecosystem. Neotrop Entomol 44(5):430–438

Silva EB, Franco JC, Vasconcelos T et al (2010) Effect of ground cover vegetation on the abundance and diversity of beneficial arthropods in citrus orchards. Bull Entomol Res 100(4):489–499

Silveira LCP, Bueno VHP, Pierre LSR et al (2003) Plantas cultivadas e invasoras como habitat para predadores do gênero *Orius* (Wolff) (Heteroptera: Anthocoridae). Bragantia 62(2):261–265

Snyder WE, Chang GC, Prasad RP (2005) Conservation biological control: biodiversity influences the effectiveness of predators. In: Barbosa P, Castellanos I (eds) Ecology of predator-prey interactions. Oxford University Press, New York, pp 211–239

Stireman JOIII, Dyer LA, Matlock RB (2005) Top-down forces in managed and unmanaged habi-

tats. In: Barbosa P, Castellanos I (eds) Ecology of predator-prey interactions. Oxford University Press, New York, pp 303–322
Sunderland K, Samu F (2000) Effects of agricultural diversification on the abundance, distribution, and pest control potential of spiders: a review. Entomol Exp Appl 95(1):1–13
Thies C, Tscharntke T (1999) Landscape structure and biological control in agroecosystems. Science 285(5429):893–895
Thomson LJ, Hoffmann AA (2013) Spatial scale of benefits from adjacent woody vegetation on natural enemies within vineyards. Biol Control 64(1):57–65
Tixier P, Dagneaux D, Mollot G et al (2013) Weeds mediate the level of intraguild predation in arthropod food webs. J Appl Entomol 137(9):702–710
Trujillo-Arriaga J, Altieri MA (1990) A comparison of aphidophagous arthropods on maize polycultures and monocultures, in Central Mexico. Agric Ecosyst Environ 31(4):337–349
Tscharntke T, Bommarco R, Clough Y et al (2007) Conservation biological control and enemy diversity on a landscape scale. Biol Control 43(3):294–309
Uetz GW (1991) Habitat structure and spider foraging. In: Bell SS, McCoy ED, Mushinsky HR (eds) Habitat structure: the physical arrangement of objects in space. Chapman & Hall, London, pp 325–348
Uetz GW, Halaj J, Cady AB (1999) Guild structure of spiders in major crops. J Arachnol 27:270–280
United Nations (2017) Report of the special rapporteur on the right for food. UN, Geneva
Venzon M, Togni PHB, Amaral DSSL et al (2015) Manejo agroecológico de pragas. Inf Agropecu 36:19–30
Weyman GS, Jepson PC (1994) The effect of food supply on the colonization of barley by aerially dispersing spiders (Araneae). Oecologia 100(4):386–390
Wilkinson DM, Sherratt TN (2016) Why is the world green? The interactions of top–down and bottom–up processes in terrestrial vegetation ecology. Plant Ecol Divers 9:127–140
Winder L, Alexander C, Holland J et al (2001) Modelling the dynamic spatio-temporal response of predators to transient prey patches in the field. Ecol Lett 4(6):568–576
Wyss E (1995) The effects of weed strips on aphids and aphidophagous predators in an apple orchard. Entomol Exp Appl 75(1):43–49

Chapter 3
Quality of Agroecosystems as Habitats to Natural Enemies and Biological Control Agents

Yaril Matienzo Brito, Luis L. Vázquez, and Janet Alfonso-Simonetti

3.1 Introduction

Biological control should be considered a much more complex practice than multiplying in mass and introduce into the fields an effective agent to control pests. As a living organism, the characteristics of the cultivation and production system and the area where it will be released or applied must be known and improved.

Habitat management is a way of conserving biological control agents and an ecological tactic that benefits and enhances the activity of natural enemies of phytophagous insects in agricultural systems. In addition, this strategy increases alternative availability of food resources such as nectar and pollen. It also favors refuge and moderate microclimate, protecting biological control agents from environmental factors and pesticides as well as promoting a habitat for their prey and alternative hosts (Landis et al. 2000).

In their coevolutionary process with their guests or prey, mainly in their area of origin, biological control agents reach different trophic relation degrees. These relationships involve crops, weeds growing in fields and surrounding areas, phytophagous insects that can be hosts or preys, soil and climatic characteristics, crop technology, and production system management. Thus, there is a complex system and the biological characteristics of the natural enemies that determines its regulatory activity (Vázquez et al. 2008).

There is a need to pay greater attention to the effects of diversity on stability (Landis et al. 2000) and the occurrence of harmful organisms and their natural enemies in the agroecosystem (Nicholls et al. 2001). Likewise, interactions that con-

Y. Matienzo Brito (✉) · L. L. Vázquez
Institute of Plant Health Research (INISAV), La Havana, Cuba
e-mail: ymatienzo@inisav.cu

J. Alfonso-Simonetti
Plant Health Research Institute (INISAV), Zoology and Herbology Department, Laboratory of Entomology, La Habana, Cuba

B. Souza et al. (eds.), *Natural Enemies of Insect Pests in Neotropical Agroecosystems*, https://doi.org/10.1007/978-3-030-24733-1_3

tribute to ecological services of functional biodiversity should be encouraged (Nicholls 2010), including the connections between production systems and natural ecosystems (Perfecto et al. 2009).

In this chapter, we will contribute to the base knowledge on agroecosystem reformulation, aiming to achieve higher-quality habitats for the natural enemies that occupy such environments, as well as the biological control agents that are incorporated.

3.2 Interactions in Agroecosystems

There are similarities among these kinds of species that are related to each other in different ways, giving rise to complex networks of interaction. Networks may occur depending on the type of interaction, competing, trophic, mutualists, and facilitation, among others. The structure of ecological networks determines many of the ecosystems functions they represent. Therefore, when the architecture of these networks is lost, many other functions are changed (Schulze and Mooney 1994; Levin 1999; Montoya et al. 2001).

Several scientific hypotheses support the contribution of diversified systems to the arthropod richness and biota interactions (Altieri and Nicholls 2007). In particular, these hypotheses highlight the phytophagous reduction, and the diversity and activity of their natural enemies increase. Other factors such as visual clues, microclimatic changes, food preferences, or direct effects on the host plant vigor could affect the habitat or location-seeking behavior of both herbivores and natural enemies. Despite all the described experimental studies, it is still not possible to develop a predictive theory that allows the determination of which specific elements of biodiversity we should maintain, add, or remove to improve natural pest control (Altieri and Nicholls 2007).

Although natural enemies vary widely in response to crop distribution, density, and cultures dispersion, evidence indicates that certain agroecosystems structural attributes (plant diversity, entry levels, etc.) strongly influence the dynamics and predators and parasitoids diversity. Most of these attributes relate to biodiversity and are management subject, such as crop rotation and associations, presence of weed blooms, and genetic diversity, among others (Altieri 1994; Matienzo et al. 2015).

It is important to consider the observations made by McCann (2000), expressing that greater biodiversity increases the probability that an ecosystem has (a) species that can respond differently under different environmental conditions and disturbances and (b) functional redundancy, that is, species that are capable of replacing an extinct species function. In addition, higher levels of biodiversity in a non-fragmented network can maintain ecosystem functions. However, in the case of a fragmented network with few species, many of these functions can be changed.

The procedures performed in agroecosystems (crop rotation, new harvests, mechanized works, and chemical products applications, among others) and the simplicity of established designs, especially in conventional farming systems, can cause a dynamic and multicausal fragmentation of biodiversity interactions. This fact affects the natural enemy species that inhabit or those ones introduced as agents of biological control, as well as interactions with their own natural enemies, preys, or hosts and the plants where they cohabit. The foregoing requires research to study and propose the functional integration of seminatural vegetative structures.

3.3 Agroecosystem Quality as Habitat

Habitat is an area that combines resources (food, water, and cover) and environmental conditions (temperature, rainfall, predators, and competitors) and that promotes occupation by individuals of a specific species and allows them to survive and reproduce.

Characteristics, attributes, or habitat elements are contained in several habitat units in a specific geophysical region named landscape (or landscape matrix). These regions can get together and offer variety vegetation units in combination with the land physical aspects (or habitats). They can be used with more efficiency by one species (Morrison et al. 2008). All landscapes are composed by a set of universal elements that are represented by tiles (patches or fragments), corridors, and the matrix (Aguilera 2010).

Generally, the habitat is specific for each species (regardless of its definition as a term in real or potential habitat). It depends on the preference of organisms to the particular characteristics of their environment, which can be shared by one or more species, and is characterized by a certain uniformity of biotic and abiotic conditions (Morrison et al. 1999; Storch 2003). Other authors stated that, in the first step for the characterization and evaluation of habitat, one must take into account the minimum biotic and abiotic requirements that a species needs (Morrison et al. 1999).

In addition, it has been considered that biological conservation should have an appropriate amount of efficient protected areas (Turner 1996), alternating with high-quality ecological land use matrices to facilitate the withdrawal of wild flora and fauna (Perfecto and Vandermeer 2002). In other words, it is possible to stabilize the populations of several arthropods in agroecosystems if vegetation architectures that increase the natural enemy diversity are designed. The landscape structure also modifies the control practiced, which reflects in a parasitism and/or predation pressure increase (Marino and Landis 1996; Baggen et al. 1999).

Habitat management consists on modifying the agroecosystems biodiversity, substantially improving interactions at different trophic levels. A direct consequence of this action is the regulation of the harmful abundance of organisms by their natural enemies (Altieri 1995; Landis et al. 2000).

In tropical areas, these approaches include adaptive management as a strategy for agroforestry in the landscapes designs (Beever et al. 2006) and the structural and functional ecosystems restoration that allows their resilience. In other words, adaptive management seeks to recover or reorganize the agricultural systems after periods of turmoil caused by humans and natural agents (Chazdon et al. 2008).

Thus, several studies related to the conservation of natural enemies study or with the introduction of biological control agents and their integration into pests'

Table 3.1 Summary of the elements, design, and management of agricultural production systems that contribute to the habitat quality improvement for entomophagous (DeBach 1965; Matienzo 2005; Santos 2006; Vázquez et al. 2007; Martínez and Vázquez 2013; Wyckhuys et al. 2013)

Elements, designs, and management	Functions[a]
Phytophagous host plants	a
Plants with nectar and flowers with exposed pollen	b
Systems of cultivation of different species and structure in the same field (shrub, tree). Polyculture, polyhedra	c, f
Integration of cover crops and profuse root system	c, d
Dynamics of planting at different times, relay, and crop rotation	a, c, f
Integration of shrubs and trees (agroforestry design)	c, d, e
Adjacent cultures	c, d, e, f
Structure of the production system internal matrix. Subdivided into management units, separated by internal live fences	c, d, e, f
Size and shape of fields. Several, preferably with smaller dimensions	d
Culture practices (soil and cultivation). Rational realization	d
Vegetation around the production system	c, d
Promotion of live barriers on the fields sides with entomophilic plants (corn, sorghum, sunflower, tithonia)	b, c, d, e, f
Tolerance of weeds in and around the fields	b, c, d, e
Seminatural environments of auxiliary vegetation in production system sites	b, c, d, e, f
Living perimeter fences with mixed design and entomophilic species	b, c, d, e, f
Promotion and management of plants with functions to maintain populations	e
Integration of plants for entomophagous refuge	c
Reduction or elimination of degrading practices (over-tillage, monoculture, agrochemicals, burning of crop residues, gravity irrigation, flooding, etc.)	g

[a]*Functions*: (a) Provide hosts or preys; (b) adult feeding of entomophagous; (c) shelter for adverse factors and time factor; (d) up and down connectivity; (e) populations reservoir; (f) microclimate regulation; (g) reduce physical and chemical effects

management recommend certain conditions or characteristics. They are desired in the crop and production systems, to contribute to a better efficacy in the regulatory capacity of these organisms (Table 3.1).

3.4 Multifunction in the Agroecosystems Design and Management

The agroecosystems diversification increases environmental opportunities for natural enemies and, consequently, influences the pest biological control. The large variety of vegetation arrangements available as polyculture, diversified systems and cutting weed systems, or hedge plants conserve natural enemies that guarantee these species a series of ecological requirements appropriate for their populations development (Altieri and Nicholls 2007).

It is important to understand that the design and management carried out in diversified systems requires an understanding of the relationships between plants and their functions (harvesting, barriers, weeds, etc.) and the associated biota (phytophagous, entomophagous, or others) (Martínez and Vázquez 2013).

Agroecological designs of mixed cropping systems can achieve several functions. A study demonstrated it in different regions of Cuba (Vázquez et al. 2015), which showed that the highest functional coefficient (CF) reached the design that joins cassava-corn-beans (CF: 86.7%), followed by the designs of sweet potato-corn and cassava-corn (CF: 76.7%), banana-cassava (CF: 73.3%), beans-corn and banana-beans-corn (CF: 70%), and avocado-mamey-coffee (CF: 66.7%). In these systems, the functions predatory arthropods reservoir and parasitoids and feeding source (nectar) of adult arthropods (pollinators and entomophagous) obtained the highest values.

To introduce as a criterion the functions performed by productive plants integrated in the designs of complex crop systems and that are obtained because of the design and spatial-temporal management interactions reinforces the hypothesis that it is not enough to achieve complexity in agroecological designs but that multifunctionality is required. As expressed by Vandermeer (1995), polycultures are systems in which two or more cultures are, at the same time, established and close enough to produce interspecific competition or complementarity.

Studies have shown beneficial arthropod movements from the edges to the harvest, and there has been a greater biological control in the crop lines near the edges than in the middle of the plantations (Thies and Tscharntk 1999). For this reason, in recent years, the functional role of seminatural environments coupled with crops or inserted into the landscape matrix as vegetation patches has been reassessed. It occurs mainly by the services of pest regulation services provided (Marasas et al. 2012; Bianchi et al. 2013). The ideal management of these environments can favor several groups that manifest as predators, parasitoids, and pollinators (Altieri et al. 2014).

In a study using vines, Nicholls and Altieri (2002) concluded that a habitat agroecological management with adequate biodiversity leads to the establishment of the necessary infrastructure that provides the resources (pollen, nectar, alternative preys, shelter, etc.) for an ideal diversity and abundance of natural enemies. These resources should be integrated into the agricultural landscape in a spatial and temporal manner, favoring natural enemies and, of course, easily implemented by farmers. Success depends on (a) the selection of the most appropriate plant species, (b) the entomofauna associated with plant biodiversity, (c) how natural enemies answer to diversification, and (d) the spatial scale in which the regulatory effects of habitat manipulation operate.

The purpose of the agroecological projects of the cropping systems and of the whole production system is to achieve several functional attributes that, besides contributing to the productive efficiency and soil conservation, also reduce the arrival, establishment, and increase of pest populations. In turn, the occurrence of natural enemies is favored (Fig. 3.1), contributing with their ability to self-regulation of pests (Vázquez 2015).

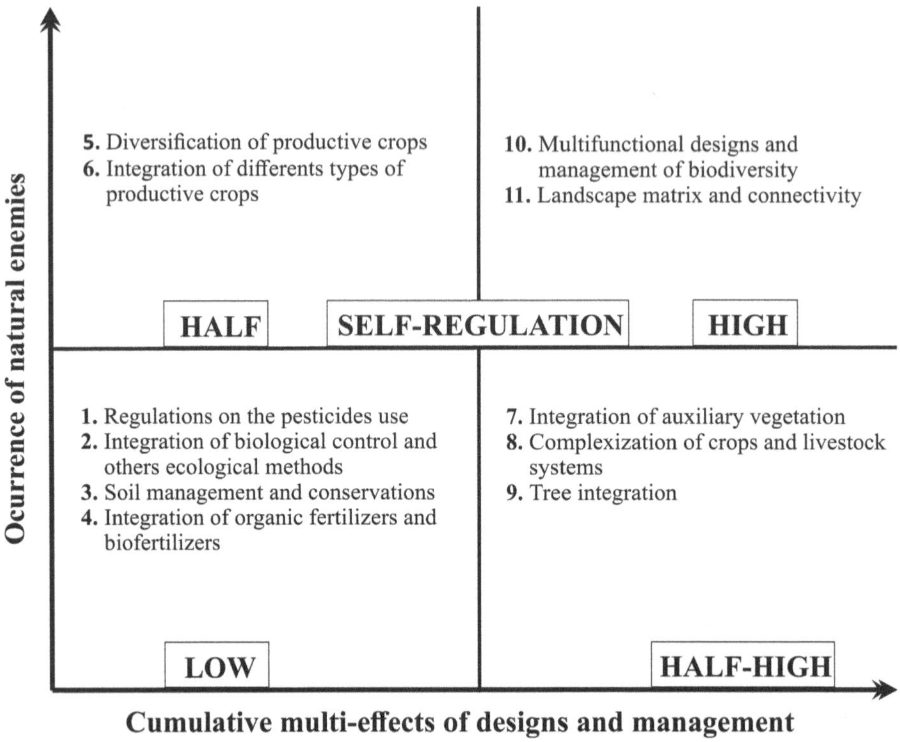

Fig. 3.1 Transition to the greater capacity of self-regulation of pests in agroecosystems. (Figure: Authors)

References

Aguilera F (2010) Application of landscape ecology metrics for the analysis of urban occupation patterns in the Metropolitan Area of Granada. An Geogr Univ Complut 30:9–29

Altieri MA (1994) Biodiversity and pest management in agroecosystems. Haworth Press, New York

Altieri MA (1995) Agroecology: the science of sustainable agriculture. Westview Press, Boulder

Altieri MA, Nicholls CI (2007) Biodiversidad y manejo de plagas en agroecosistemas:perspectivas agroecológicas. Icaria, Barcelona

Altieri MA, Nicholls CI, Montalba R (2014) El papel de la biodiversidad en la agricultura campesina en América Latina. LEISA Rev Agroecol 30(1):5–8

Baggen LR, Gurr GM, Meats A (1999) Flowers in tri-trophic systems: mechanisms allowing selective exploitation by insect natural enemies for conservation biological control. Entomol Exp Appl 91(1):155–161

Beever EA, Swihart RK, Bestelmeyer BT (2006) Linking the concept of scale to studies of biological diversity: evolving approaches and tools. Divers Distrib 12(3):229–235

Bianchi JJA, Schellhorn N, Cunningham SA (2013) Habitat functionality for the ecosystem service of pest control: reproduction and feeding sites of pests and natural enemies. Agr Forest Entomol 15(1):12–23

Chazdon RL, Harvey C, Komar O et al (2008) Beyond reserves: a research agenda for conserving biodiversity in tropical human-modified landscapes. Biotropica 41(2):142–153

DeBach P (1965) Biological control of insect pests and weeds. ICL, La Habana

Landis DA, Wratten SD, Gurr GA (2000) Habitat management to conserve natural enemies of arthropod pests in agriculture. Annu Rev Entomol 45:175–201

Levin S (1999) Fragile dominion. Perseus Books, Massachusetts

Marasas ME, Cap G, Luca L et al (2012) El camino de la transición agroecológica. INTA, Buenos Aires

Marino PC, Landis DL (1996) Effect of landscape structure on parasitoid diversity and parasitism in agroecosystems. Ecol Appl 6(1):276–284

Martínez A, Vázquez LL (2013) Características de la colindancia de cultivos en tres sistemas agrícolas convencionales y su relación con la incidencia de insectos nocivos y reguladores naturales. Fitosanidad 17(2):65–74

Matienzo Y (2005) Conservación de artrópodos benéficos en un sistema de producción agrícola urbano. Tesis, Universidad Agraria de La Habana

Matienzo Y, Vázquez LL, Alfonso-Simonetti J et al (2015) Manejo del hábitat para la conservación de reguladores naturales: experiencia cubana en agricultura urbana. Rev Inter Nos 20:8–11

McCann KS (2000) The diversity-stability debate. Nature 405:228–233

Montoya JM, Solé RV, Rodríguez MA (2001) La arquitectura de la naturaleza: complejidad y fragilidad en redes ecológicas. Ecosistemas 10:1–14

Morrison ML, Block WM, Strickland MD et al (2008) Wildlife study design,. 2nd. Springer, New York

Morrison ML, Marcot BG, Mannan RW (1999) Wildlife habitat relationships: concepts and applications. J Mammal 80:1382–1385

Nicholls CI (2010) Bases agroecológicas para diseñar e implementar una estrategia de manejo del hábitat para control biológico de plagas. In: Leon T, Altieri MA (eds) IDEAS 21. Instituto de Estudios Ambientales, Bogotá, pp 203–225

Nicholls CI, Altieri MA (2002) Biodiversidad y diseño agroecológico: un estudio de caso de manejo de plagas en viñedos. Manejo Integr Plagas Agroecol 65:50–64

Nicholls CI, Parrella MP, Altieri MA (2001) Effects of a vegetational corridor on the abundance and dispersal of insect biodiversity within a northern Californian organic vineyard. Landsc Ecol 16(2):133–146

Perfecto I, Vandermeer J (2002) Quality of agroecological matrix in a tropical montane landscape: ants in coffee plantations in southern of Mexico. Conserv Biol 16(1):174–182

Perfecto I, Vandermeer J, Wright A (2009) Nature's matrix: linking agriculture, conservation and food sovereignty, Routledge, Londres

Santos TM dos (2006) Manejo do habitat–uma tática de controle biológico conservativo. Pesqui Tecnol 3(1):1–6

Schulze ED, Mooney HA (1994) Ecosystem function of biodiversity: a summary. In: Shultze ED, Mooney HA (eds) Biodiversity and ecosystem function. Springer-Verlag, Berlin, pp 497–510

Storch I (2003) Linking a multiscale habitat concept to species conservation. In: Bissonette JA, Storch I (eds) Landscape ecology and resource management: linking theory with practice. Island Press, Washington, pp 303–320

Thies C, Tscharntk T (1999) Landscape structure and biological control in agroecosystems. Science 285(5429):893–895

Turner IM (1996) Species loss in fragments of tropical rain forest: a review of the evidence. J Appl Ecol 33(2):200–209

Vandermeer JH (1995) The ecological basis of alternative agriculture. Annu Rev Ecol Syst 26:201–224

Vázquez LL (2015) Diseño y manejo agroecológico de sistemas de producción agropecuaria. En: Unión Europea y la Agencia Española de Cooperación al Desarrollo. Sembrando en tierra viva: manual de agroecología. La Habana, CERAI-ANAP, pp 133–160

Vázquez LL, Fernández E, Alfonso-Simonetti J (2007) Desarrollo de prácticas de manejo de la diversidad de plantas en sistemas de producción de la agricultura urbana. Agricultura Agric Org 3:39–41

Vázquez LL, Matienzo Y, Veitía M et al (2008) Manejo y conservación de enemigos naturales de insectos fitófagos en los sistemas agrícolas de Cuba. CIDISAV, La Habana
Vázquez LL, Porras A, Alfonso-Simonetti J (2015) Tipos funcionales de plantas productivas integradas en diseños de sistemas de cultivos complejos innovados por agricultores. En: Anais do 5° Congreso Latinoamericano de Agroecología, Socla, La Plata, 7–9 Oct 2015
Wyckhuys K, Li AG, Morales H et al (2013) Current status and potential of conservation biological control for agriculture in the developing world. Biol Control 65(1):152–167

Chapter 4
Plants as Food for Adult Natural Enemies

Claudio Salas

4.1 Introduction

From the time man started to develop monoculture agriculture, an ecological imbalance in agricultural systems occurred, resulting in increased arthropod pest species. However, synthetic pesticides were discovered in the mid-twentieth century, which effectively decreased certain pest infestations. Due to the indiscriminate use of these pesticides, a series of problems arose such as the selection of resistant populations, contamination of natural, cultivated, and aquatic environments, death of useful organisms, imbalances, and resurgence of secondary pests. Nowadays, agriculture is oriented toward sustainable production worldwide; this reaffirms the importance of developing integrated technologies that promote the survival and effectiveness of pest bioregulator agents.

Small Latin-American farmers face economic rationality and a relative scarcity of natural and capital resources (Benencia 1997). The decrease in pesticide use is therefore a priority, particularly in peasant family farming that cannot deal with the costs associated with pesticide use. Arthropod pest management strategies based on agroecological principles would allow peasant farmers to redesign their systems and reduce the use of these inputs. However, to advance in the implementation of agroecological strategies in these systems, it is necessary to understand how they work and the potentialities that can be developed, in order to improve the self-regulatory mechanisms that lead to greater stability and resilience to pests (Paleologos et al. 2008).

A negative trend of modern agriculture is the expansion of monocultures at the expense of surrounding natural vegetation that maintains biodiversity at the landscape level and plays a very important role in the conservation of auxiliary insects

C. Salas (✉)
Instituto de Investigaciones Agropecuarias INIA, Centro Regional de Investigación Intihuasi, La Serena, Coquimbo Region, Chile
e-mail: claudio.salas@inia.cl

B. Souza et al. (eds.), *Natural Enemies of Insect Pests in Neotropical Agroecosystems*, https://doi.org/10.1007/978-3-030-24733-1_4

(Altieri and Nicholls 2002). Insect plants [plants with flowers that attract and maintain a population of natural enemies with their nectar and pollen resources (Landis et al. 2000; Jonsson et al. 2008; Parolin et al. 2012)] offer refuge for prey and alternative hosts, especially when the latter are scarce in the fields; insect plants also provide food in the form of nectar (floral or extrafloral), pollen, seeds, or plant juices (Alomar and Albajes 2005).

Many insect species visit flowers to obtain food in the form of nectar and/or pollen. These floral sources can provide the necessary energy for survival, maintenance, ovule maturation, and dispersal through adult flight. This suggests that food supplements from floral resources can increase the effectiveness of biological controllers in the field (van Lenteren et al. 1987; van Lenteren 1999; Gurr et al. 2005). The use of corridors or flower hedges in a crop, either outdoors or in greenhouses, can be a good option to improve the availability of food resources other than their prey for insects present in agroecosystems, thus promoting conservative biological control (CBC) (Fig. 4.1).

Agricultural habitat management contributes to biological pest management in crops through the incorporation of so-called insect plants. In these plants, natural enemies not only find protein sources but also refuge when faced with adverse climatic conditions and/or predators.

Fig. 4.1 *Lobularia maritima* flower beds as food sources to attract Syrphidae as aphid biocontrollers associated with commercial lettuce crops, Pan de Azúcar, Coquimbo Region, Chile. (a) Cultivation of lettuce in association with *Lobularia maritima.* (b) Adult of Syrphidae in flower of *Lobularia maritima.* (c) Syrphidae egg next to colony of aphids on lettuce. (d) Syrphidae larva preying on aphid on lettuce. (Photos: Claudio Salas)

The aim of this chapter is to present studies related to plants as a natural food source for natural enemies with a greater emphasis on those developed in Latin America.

4.2 Biodiversity and Its Influence on Natural Enemies: Field Experiments

Many comparative studies that show the impact of sustainable and conventional agricultural systems verify that habitat management affects the richness and abundance of arthropod communities (Pfiffner and Balmer 2011). This is due to the availability of a wide variety of vegetation structures such as polycultures, diversified crop-weed systems, and cover crops that conserve natural enemies by ensuring them a series of ecological requirements (Altieri and Nicholls 1994, 2004). It is very important to maintain and promote the increase in natural enemy populations through practices such as the availability of floral resources, because it is currently estimated that natural enemies are responsible for controlling 50–90% of arthropod pests in fields (Pimentel et al. 2005).

Some factors related to pest regulation in diversified agroecosystems include the increase of the parasitoid and predator population given greater alternative food availability and habitat, decrease in pest colonization and reproduction, food inhibition of plants unattractive to pests by chemical repellents, and prevention of movement and optimal synchronization between natural enemies and the pest. When the agroecosystem is more diverse and the diversity is less disturbed, trophic interactions increase, developing synergisms that promote population stability (Nicholls 2006).

4.3 Plants as a Food Source for Natural Enemies

Many natural enemies require carbohydrates and proteins for growth, basic metabolism, and reproduction. If the crop does not provide these compounds, natural enemies will search outside the crop for nectar, pollen, or host plants to feed on. These emigrants might not find their way back to the crop, thus reducing pest control. Small natural enemies, such as wasps of the genus *Trichogramma,* can die rapidly if the crop or agroecosystem does not offer resources such as nectar. On the contrary, crops or agroecosystems that provide nectar, pollen, or alternative prey will retain larger, better nourished, and more fertile natural enemy populations (Van Driesche et al. 2007). Therefore, the diversity of plant species in natural communities increases these requirements and prolongs their availability.

Monocultures with synchronized development can concentrate flowering in short periods of overabundance or eliminate them completely in certain crops. Crops that are adequate as habitats for parasitoids will depend on the particular needs of each species as well as field size compared with uncultivated neighboring areas and the vegetational composition in adjacent areas. Extensive monocultures without nectar and pollen sources and with few prey or alternative hosts offer few resources for natural enemies. Although mechanized crops need to be simplified, the availability of some functional plant species can be possible in well-studied systems and it is common in less intensive agricultural systems (Gurr et al. 2005).

Vegetation accompanying the crops contributes in maintaining beneficial entomofauna, among which are neutral phytophages, predators, and parasitoids. To reach normal fertility and longevity, these arthropods feed on nectar secretions (intra- and extrafloral), wound exudates, pollen, and alternative prey found in the vegetation (Mexzón and Chinchilla 2003).

Nectar secretions contain carbohydrates such as glucose, saccharose, and fructose as well as some essential amino acids (Baker and Baker 1973). Pollen is also rich in amino acids and it is a fundamental part of the diet of some insects (Leius 1967; Altieri and Whitcomb 1979).

On the other hand, plants emit chemical signals (kairomones) that are perceived by entomophagous insects, encouraging them to migrate from the surroundings to locate the habitat of their prey or hosts (Mexzón and Chinchilla 2003).

In Latin America and in the rest of the world, some studies in this area of biological control have been conducted to determine which requirements must be satisfied by some plants to provide refuge and food to natural enemies in fields or greenhouses with monocultures. These types of vegetation can be herbaceous or woody species, depending on the type of main crop. All of this is under the premise of making a CBC model. It is necessary to take into account certain criteria to determine which types of plants are best suited to meet the objectives of setting up hedges or strips of insect crops.

According to Rodríguez and González (2014), selection of the plants that constitute these ecological infrastructures is of vital importance. Some basic parameters to consider in habitat management aimed at promoting functional diversity are the use of native plants adapted to the edaphoclimatic conditions of the location that are not fungal or viral disease reservoirs, offer both food resources, such as nectar (floral and extrafloral) and pollen, and refuge. Another important factor to consider is the establishment of plant species that provide the above resources on a regular basis over time.

Jervis and Heimpel (2005) explain that at some point in their life cycle, parasitoids and predators feed on different plant parts, which are essential for their growth, development, survival, and reproduction. These authors classify them according to the location where they feed:

- *Directly* on plants, consuming floral and extrafloral nectar, pollen, seeds (either whole seeds or specific tissues) and, less frequently, materials such as sap (including fruit juices), plant exudates, epidermis, and trichomes (Majerus 1994 on Coccinellidae, Canard 2002 on Chrysopidae, Gilber and Jervis 1998 on parasite flies, Jervis 1998 on flies).

- *Indirectly* consuming the honeydew produced by Hemiptera-Sternorrhyncha such as aphids, mealybugs, and whiteflies (Evans 1993).

For a long time, most of the research about the feeding behavior of predators and parasitoids and population dynamics referred to the interaction of a natural enemy with its prey/hosts and ignored or overlooked its interaction with nonprey/nonhost food sources (Jervis and Heimpel 2005).

The phenomenon of phytophagy by natural enemies is one of the most important aspects of the biology of natural enemies; however, it is now the least studied and detailed concept in scientific literature if compared with studies about types of prey for each predator and parasitoid (Jervis and Heimpel 2005). However, the world's trend to develop sustainable agriculture that promotes CBC has produced great interest in understanding the importance of plant-derived foods on the behavior and ecology of parasitoids and predators (Evans 1993; Cisneros and Rosenheim 1998; Gilber and Jervis 1998; Heimpel and Jervis 2004).

4.4 Natural Enemies and Their Relationship with Host Plants in the Context of Food: Field Experiments

The range of plant materials under field conditions that are used as food by most predators and almost all parasitoids is not well known. An obvious potential food source is flowers. However, there are few scientific reports dedicated to a conclusive study of natural enemy behavior, whether parasitoid or predator, to generate a register of plant species that provide food that allows them to complete their life cycle in the field.

Since the early 1990s in Canada, as well as in European countries and the United States, some experiments have been conducted to implement flower hedges in monoculture fields and greenhouses to supply pollen to natural enemies for their sexual maturity to occur as well as to improve their longevity. For example, experiments in Canada with the parasitoid of the pine shoot moth *Rhyacionia buoliana* (Denis and Schiffmüller) (Tortricidae), which incorporated umbelliferous plants (e.g., carrot) in the surrounding area, increased pest control from 15% to 90%.

In Chile an apple export orchards, which maintained weed-free plots, strong foci of woolly aphid *Eriosoma lanigerum* (Hausmann) (Aphididae) and other insect pests were detected, while aphids were hard to find in other neighboring plots with flowering plants (Zuñiga 1987). The author indicates the importance of maintaining certain flower-producing vegetation on the edges of crops and plots or reproducing or promoting others such as fennel, carrot, radish, and wild herbs that are highly attractive and visited by entomophages to consume pollen and nectar.

Venzon et al. (2006) conducted studies with *Chrysoperla externa* (Hagen) (Chrysopidae), females fed with pollen from Indian hemp, pea, and castor bean in single diets or diets combined with honey to determine the preoviposition period, oviposition rates, and longevity. They also analyzed proteins present in each of the plant species under study and concluded that female *C. externa* did not oviposit when fed castor bean and honey diets as their only food; however, the oviposition rate increased 1.5 times with the Indian hemp and honey diet and 1.2 times with the pea and honey diet. It is inferred that a diet is efficient when it exhibits a correlation between protein and carbohydrate content present in each type of pollen. Thus, castor bean had higher protein content but lower carbohydrate content, followed by Indian hemp and pea. This shows that legumes have a poor carbohydrate/protein correlation.

In other studies conducted by Venzon and Carvalho (1992) in Brazil, it was determined that there was a higher fecundity rate in lacewings when these were subjected to maize pollen diets; this phenomenon is associated with the high percentage of carbohydrates in maize pollen due to its high starch content. Patt et al. (2003) indicated that vitamins, mineral nutrients, and sterols are important for digestion and other metabolic processes in addition to the nutritional value of the macronutrients found in pollen (Stanley and Linskens 1974; Waldbauer and Friedman 1991). It is expected that differences in nutrient digestibility and assimilation of different pollen species influence biocontroller fecundity.

Another Brazilian study by Morales and Köhler (2008) collected 1187 individuals of the Syrphidae family (74 species) visiting 51 plant species distributed in 23 botanical families; it was determined that the most visited families were Apiaceae and Asteraceae, which demonstrated their greater importance as foragers (75%) when compared with the other families. These authors also concluded that the frequency and abundance of visits by syrphids in the flowers of the different plant families are related to the type (shape and coloring), quantity, and ease of access to available resources. Therefore, dark-colored flowers that are scarce and have limited access to floral resources tend to exhibit fewer visitors than light-colored (yellow, white) flowers that are abundant and have easy access to floral resources.

Inflorescences of the Poaceae species are particularly important to provide pollen to small-sized Syrphidae (Proctor et al. 1996); this has been observed with the *Allograpta* sp. (OstenSacken) (Syrphidae) and *Pseudodorus clavatus* (Fabricius) (Syrphidae) species. However, some larger-sized species were collected, such as *Palpada* sp. (Mcquart) (Syrphidae), which visited species of this family.

Proctor et al. (1996) indicate that the elongated proboscis of syrphids is clearly associated with the foraging habit, but the great majority does not exhibit any specific adaptations because they usually feed on open flowers. Therefore, it was established in the study that the Syrphidae community is regulated by local interactions between species, mainly between environmental conditions and food resource availability. The Apiaceae and Asteraceae species are important to maintain the Syrphidae community because they have a large number of flowers and availability of abundant food resources.

Carrillo et al. (2006) conducted an assay with Brassicaceae in Colombia. Although these crops are attacked by a group of pests, such as aphids, Lepidoptera, and mollusks, they face the diamondback moth *Plutella xylostella* (L.) (Plutellidae), which is a more aggressive pest. Biological controllers associated with this pest are cited worldwide as being mainly or the Braconidae and Ichneumonidae species. In the latter family, *Diadegma insulare* (Cresson) (Ichneumonidae) is highlighted due to its good searching ability, high fecundity, ability to avoid superparasitism, and to regulate the development of its host (Idris and Grafius 1996).

Diadegma insulare is the main parasitoid of *P. xylostella* in several countries (Johanowicz and Mitchell 2000; Xu et al. 2001). In some cases, field parasitism exceeds 90%; however, the success of biological control has been reduced in many other cases because the food requirements of the parasitoid were not considered (Muckenfuss et al. 1990; Wratten et al. 2003). Understanding the relative importance of flowers as a food source for *D. insulare* is very important to improve its efficiency as a biological controller of *P. xylostella*. A floristic inventory was conducted in 13 vegetable-producing farms in several municipalities of the Bogota savanna to determine the effect of several nectariferous plants normally encountered in the Brassicaceae crops of this area, on the reproductive ability of *D. insulare* as well as the degree of acceptance and access to floral structures where food is found. Four species were selected for these studies as nectar sources for *D. Insulare*: turnip *Brassica rapa* (L.) (Brassicaceae), marigold *Calendula officinalis* (L.) (Asteraceae), rue *Ruta graveolens* (L.) (Rutaceae), and borage *Borrago officinalis* (L.) (Boraginaceae). Results of this assay indicated that the reproductive success of *D. insulare* measured in terms of its longevity and fecundity is influenced by the type of floral resources it can feed on.

Flower architecture and insect morphological characteristics influence the degree of accessibility of nectar sources. Among the evaluated plants, those with the widest and longest corolla prolonged longevity and increased parasitoid fecundity. Borage, *B. officinalis* was the best food source because it increased the longevity of *D. insulare* fourfold and fecundity fivefold compared with a honey solution. These results evidence the need to incorporate nectariferous plants to increase the effect of *D. insulare* on the management of *P. xylostella* in an integrated Brassicaceae crop management program in the Bogota savanna.

Studies conducted by Diehl et al. (2012) in Brazil with predatory mites and plants present in vineyards indicated that plants such as *Plantago tomentosa* (L.) (Plantaginaceae), *Sonchus oleraceus* (L.) (Asteraceae), and *Chromolaena laevigata* (R. M. King and H. Rob) (Asteraceae) exhibited a wealth of predatory mite species from the Phytoseiidae, Stigmaeidae, and Iolinidae families. This can be related to the morphophysiological characteristics such as the presence of pilosities, trichomes, and nerves along the leaf, which can provide shelter for these predators. In addition, these species constantly provide pollen for different species of predatory mites because they bloom all year-round (Lorenzi 2000).

Similar results were obtained by Collier et al. (2001) in apples *Malus domestica* (Borkham) (Rosaceae), demonstrating that the associated plants serve as a food source for the predatory mite *Neoseiulus californicus* (McGregor) (Phytoseiidae) by offering pollen or hosting secondary prey at an unfavorable time for the main prey.

Galaz and Navarro (2018) conducted a study in Chile to evaluate the effect of incorporating floral resources on the presence and abundance of arthropods in a commercial avocado orchard compared with the presence of these insects in spontaneous vegetation. The study was implemented in a commercial "Hass" avocado orchard located in the commune of Quillota, Valparaíso region. In a sector of the property, flower mixes or patches were established in 90 m^2 areas. The floral species were the following: *Papaver rhoeas* (L.) (Papaveraceae), *Calendula officinalis* (L.) (Papaveraceae), *Centaurea* sp. (L.) (Asteraceae), *Gypsophila elegans* (M. Bieb) (Caryopyllaceae), *Linaria maroccana* (Hook. F) (Plantaginaceae), *Lobularia maritima* (L. Desv) (Brassicaceae), *Phacelia campanularia* (A. Gray) (Boraginaceae), and *Zinnia elegans* (L.) (Asteraceae). Monitoring of the avocado trees adjacent to the floral mix and spontaneous vegetation led to the observation of two families of natural enemies, which are Coccinellidae and Syrphidae.

In the case of these biological control agents, the established floral resource was an important contribution thanks to the great attractiveness of the floral mixes, which can be used as refuge and/or food by the insects. As for the abundance of pests in avocado trees adjacent to the floral mix, it was possible to identify a decrease in mealy bugs *Pseudococcus viburni* (Signoret) (Pseudococcidae) and red spider *Tetranychus urticae* (C.L. Kock) (Tetranychidae).

Field studies conducted by Arias Roda (2012) in Honduras indicated that insect crops should be plants rich in nectaries and foliage. As for nectaries, they should not only be abundant but also accessible to the insects they maintain (Nicholls and Altieri 2012). In addition, this functional flora should not be attractive for arthropod pest species, not be prone to viruses and other fungal diseases, or inclined to be converted into weeds. Therefore, the objective of the study was to evaluate the adaptability of 30 plant species under the conditions found in El Zamorano, Honduras, and their ability to serve as refuge for natural enemies. Among the plants that demonstrated better behavior compared with the rest are carrot *Daucus carota* (L.) (Apiaceae), coriander *Eryngium foetidum* (L.) (Apiaceae), and dill *Anethum graveolens* (L.) (Apiaceae) to attract parasitoid wasps and Coleoptera of the Coccinellidae and Cantharidae families because their small flowers provide accessible nectaries for these insects. The most frequently encountered predators were the bedbugs of the Reduviidae family, Coleoptera of the Coccinellidae, Cantharidae, and Carabidae families, and flies of the Syrphidae family. Parasitoids of the Ichneumonoidea and Chalcidoidea superfamilies were also found. Among the families considered as phytophagous, individuals of the Miridae and Coreidae (Hemiptera) and Chrysomelidae (Coleoptera) families were detected.

Díaz et al. (2018), in a region of Argentina, evaluated the ability of buckwheat *Fagopirumesculentum* (Moench) (Caryophyllales: Polygonaceae) as a nectary plant in agroecological horticulture; they focused on a holistic approach to pest management and considered biodiversity promotion and/or conservation using floral plant resources. The objective of the study was to determine the entomofauna associated with buckwheat in order to consider its contribution to biological control in horticultural agroecosystems. Individuals were classified by order, family, and functional group (pests, pollinators, and natural enemies). The trophic structure revealed that pollinators were more abundant than natural enemies and phytophages. Natural enemies were represented by the Diptera order with the Dolichopodidae and Syrphidae (genera *Toxomerus* and *Allograpta*) families as predators and the Tachinidae family as parasitoids. In the Hemiptera order, the Reduviidae and

Nabidae families are highlighted as predators. Phytophages were in the Coleoptera and Lepidoptera orders. Traps were used to capture 424 individuals from the Diptera (44.3%), Hymenoptera (36.0%), Hemiptera (9.5%), and other (10.2%) orders. The authors concluded that buckwheat as an "insect plant" can attract insects that contribute to biological control and pollination of different agroecological horticultural crops.

4.5 Applications and Future Guidelines

A greater perception of risk by consumers regarding pesticide use is obliging farmers to reengineer agricultural landscapes and make them more environmentally friendly. This is achieved by conducting and implementing integrated pest management (IPM) programs, including CBC, focused on incorporating plants in field or greenhouse monocultures that, in addition to being a reservoir for natural enemies, are an essential food source to complete part of their life cycles.

However, the mere presence of flowering plants in an agroecosystem is not enough to guarantee improvements in agroecosystems. Several factors should be considered such as the type of crop to be established and the most suitable companion plants or insect crops to prevent agronomic interference. On the other hand, all insects have highly specific requirements and the selection of different floral resources is determined by several factors related to flower color, morphology, smell, and size as well as nectar and pollen quality, abundance and age, and markings of previous visitors (Ambrosino et al. 2006).

A plant can contain a lot of nectar that is not necessarily accessible to natural enemies; therefore, it is not useful for CBC. As well as nectar production, plant floral architecture and nectary location within the flower (hidden, shallow, or exposed) should also be considered.

There are certain aspects of plant morphology such as extrafloral nectaries or the presence of leaf trichomes that can greatly influence phytophagous insects, the natural enemies they attract, and their interactions. They can be located in different plant parts such as leaf edges, petioles, bracts, rachis, stipules, pedicels, and fruits; their size, shape, and secretions vary with plant taxa. However, extrafloral nectar is an important food source, independently of flowering, to provide food resources to adults of different natural enemy species that are essential for pest biological control such as chrysopids, phytoseids, parasitoids, or predatory flies. It has been ascertained that the presence of plants with extrafloral nectaries in fields can be advantageous for biological control (Koptur 2013).

An indispensable condition when designing a plantation for CBC is that it consists of plant species that bloom sequentially throughout the year. This is intended to ensure continuity in food resources and keep natural enemies close to the crops (Long 1995). It is also important to select plants that bloom during the winter when there is a shortage of flowering plants in the fields (Rodríguez and González 2014).

Another factor to consider is to determine whether the target natural enemy population belongs to the parasitoid or predator group because pest control performance of these biocontrollers is variable. Sabelis (1992) explains that many predators must consume several preys before they reproduce. Delayed reproduction often results in a slow numerical response to increase prey populations and reduces opportunities for acceptable control by some predators. The predator functional responses also stabilize faster than those of many parasitoids because predators are satiated with food, resulting in lower attack rates per unit of time spent searching for and handling prey.

In summary, increased biodiversity in agroecosystems can lead to a hierarchy of pest management benefits. Although there is a wide range of flowering plants that have been used for this purpose in agroecosystems, relatively little research has been done on the most appropriate plants. Plants of different species can have different effects on herbivores compared with their natural enemies, while many that enhance the activity of the parasitoid also increase that of the herbivore. This occurs as a result of the improved aptness of the herbivore by enhancing trophic processes or by masking odors that the host plant induces by affecting the parasitoid. This is the masking of odors (which parasitoids use to find hosts) by the odors of associated plants, such as those that can be achieved by the added floral resources (Lavandero et al. 2006).

It is inferred that the increased effectiveness of natural enemies in biological control programs can be achieved by providing selective floral resources. This can ensure the effectiveness of natural enemies without increasing the density or activity of the herbivore. For example, if flowering plant species improve the condition of herbivores, then there will be no beneficial effect of the floral subsidy in the natural enemy population because there is a positive effect on the herbivore; therefore, the action of the natural enemy will be masked or reduced (Lavandero et al. 2006).

Further studies are required to better focus on the use of these insect crops that provide refuge and food to these insects with the aim of reducing the indiscriminate use of insecticides and contribute to quality agriculture with safe products obtained in an environmentally friendly way. The key is to consider the above guidelines because each country, region, and even each field has a different ecosystem and the plants to be selected must completely fulfill their function.

4.6 Final Considerations

Knowledge about the crop, its pests and natural enemies, and where and what their preferences are in order to select the best floral mix is crucial. Thus, it is essential to continue conducting field and laboratory assays and correlate which are the best plant families that provide refuge and food and their sowing dates in the crop. This is to obtain the best benefits, decrease chemical pesticide use, and increase conservation crop management in the fields of Latin America to obtain and to offer attractive and safe products to buyers worldwide given the comparative advantages of this continent.

References

Alomar O, Albajes R (2005) Control biológico de plagas: biodiversidad funcional y gestión del agroecosistema. Biojournal Net 1:1–10. https://www.recercat.cat/bitstream/handle/2072/4643/Biodiversidad_Funcional.pdf?sequence=1

Altieri M, Nicholls C (2002) Biodiversidad y diseño agroecológico: un estudio de caso de manejo de plagas en viñedos. ManejoIntegrado de Plagas y Agroecología 65:50–64

Altieri M, Nicholls C (2004) Biodiversity and pest management in agroecosystems. CRC Press, Boca Raton

Altieri M, Nicholls CI (1994) Biodiversity and pest management in agroecosystems. Haworth Press, New York

Altieri MA, Whitcomb WH (1979) The potential us e of weeds in the manipulation of beneficial insects. HortScience 14(1):12–17

Ambrosino MD, Luna JM, Jepson PC et al (2006) Relative frequencies of visits to selected insectary plants by predatory hoverflies (Diptera: Syrphidae), other beneficial insects, and herbivores. Environ Entomol 35(2):394–400

Arias Roda F (2012). Refugios para enemigos naturales de plagas insectiles: Selección inicial de plantas para condiciones de El Zamorano. Proyecto especial de graduación del programa de Ingeniería Agronómica, Escuela Agrícola Panamericana, Zamorano

Baker HG, Baker I (1973) Aminoacids in nectar and their evolutionary significance. Nature 241:543–545

Benencia R (1997) Los productores hortícolas. En: Benencia R (coord) Área hortícolas Bonaerense. Cambios en la producción y su incidencia en los sectores sociales. La Colmena, Buenos Aires, pp 123–150

Canard M (2002) Natural food and feeding habits of lacewings. In: McEwen PK, New TR, Whittington AE (eds) Lacewings in the crop environment. Cambridge University Press, Cambridge, pp 116–129

Carrillo D, Serrano MS, Torrado-León E (2006) Effect of nectar producing on the reproduction of *Diadegma aff insulare* Cresson (Hymenoptera: Ichneumonidae), parasitoid of *Plutella xylostella* L. (Lepidoptera: Plutellidae). Rev Colomb Entomol 32(1):18–23

Cisneros JJ, Rosenheim JA (1998) Changes in foraging behaviour, within-plant vertical distribution, and microhabitat selection of a generalist insect predator: an age analysis. Environ Entomol 27(4):949–957

Collier KFS, Albuquerque GS, Eiras AE et al (2001) Estímulos olfativos envolvidos na localização de presas pelo ácaro predador *Neoseiulus californicus* (McGregor) (Acari: Phytoseiidae) em macieiras e plantas hospedeiras alternativas. Neotrop Entomol 30(4):631–634

Díaz BM, Martínez MA, Cavigliasso P (2018) Evaluación del trigo sarraceno (*Fagopirum esculen tum*) como "planta insectario" en la horticultura agroecológica. In: Poster en el 20° Congreso Argentino de Entomología, [s.n.], Mendoza, 21–24 May 2018

Diehl M, Ferla NJ, Johann L (2012) Plantas associadas à videiras: uma estratégia para o controle biológico no Rio Grande do Sul. Arq Inst Biol 79(4):579–586

Evans EW (1993) Indirect interactions among phytophagous insects: aphids, honeydew and natural enemies, in Individuals, Populations and Patterns. In: Watt AD, Leather S, Mills NJ et al (eds) Ecology. Intercept, Andover, pp 287–298

Galaz F, Navarro D (2018) Efecto de la incorporación de recursos florales en la abundancia de enemigos naturales en huertos de palto en la localidad de Quillota, región de Valparaíso. Magíster en Producción Vegetal, Escuela de Agronomía de Santiago

Gilber FS, Jervis MA (1998) Functional, evolutionary and ecological aspects of feeding-related mouthpart specializations in parasitoid flies. Biol J Lin Soc 63(4):495–535

Gurr GM, Wratten SD, Stephen D et al (2005) Providing plant foods for natural enemies in farming systems: balancing practicalities and theory. In: Wäckers FL, Van Rijn PCJ, Bruin J (eds) Plant provided food for carnivorous insects: a protective mutualism and its applications. Cambridge University Press, Cambridge, pp 326–347

Heimpel GE, Jervis MA (2004) An evaluation of the hypothesis that floral nectar improves biological control by parasitoids. In: Wackers F, Van Rijn P, Bruin J (eds) Plant-provided food and plant-carnivore mutualisms. Cambridge University Press, Cambridge

Idris AB, Grafius E (1996) Effects of nectar-producing plants on *Diadegma insulare* (Cresson), a biological control agent of diamondback moth, *Plutella xylostella* (L.). The management of diamondback moth and other crucifer pests. In: Proceedings of the 3rd international workshop, Kuala Lumpur, Malaysia, 29 Oct-1 Nov 1996

Jervis MA (1998) Functional and evolutionary aspects of mouthpart structure in parasitoid wasps. Biol J Lin Soc 63(4):461–493

Jervis MA, Heimpel GE (2005) Phytopagy. In: Jervis MA (ed) Insects as natural enemies. A practical perspective. Springer, New York, pp 525–550

Johanowicz DL, Mitchell ER (2000) Effects of sweet alyssum flowers on the longevity of the parasitoid wasps *Cotesia marginiventris* (Hymenoptera: Braconidae) y *Diadegma insulare* (Hymenoptera: Ichneumonidae). Fla Entomol 83(1):41–47

Jonsson M, Wratten SD, Landis DA et al (2008) Recent advances in conservation biological control of arthropods by arthropods. Biol Control 45(2):172–175

Koptur S (2013) Nectar as fuel for plant protectors. In: Wäkers FL, Van Rijn PCJ, Bruin J (ed) Plant-provided food and herbivore–carnivore interactions. Cambridge University Press, Cambridge

Landis DA, Wratten SD, Gurr GM (2000) Habitat management to conserve natural enemies of arthropod pests in agriculture. Annu Rev Entomol 45:175–201

Lavandero B, Wratten D, Didham R et al (2006) Increasing floral diversity for selective enhancement of biological control agents: a double-edged sward? Basic Appl Ecol 7(3):236–243

Leius K (1967) Food sources and preferences of adults of a parasite, *Scambus buolianae* (Hym: Ichneumonidae) and their consequences. Can Entomol 99(8):865–887

Long R (1995) Insectary plants. Improving natural enemy activity. Small Farm News 4

Lorenzi H (2000) Plantas daninhas do Brasil, 3a ed. Instituto Plantarum, Nova Odessa

Majerus MEN (1994) Ladybirds. Harper Collins, London Management of diamondback moth and other crucifer pests. In: 3rd Proceedings of the International Workshop, [s.n.], Kuala Lumpur, 2018

Mexzón RG, Chinchilla CM (2003) Especies vegetales atraventes de la entomofauna benéfica en plantaciones de palma de aceite (*Elaeis guineensis* Jacq.) en Costa Rica. Revista Palmas 24(1):33–57

Morales MN, Köhler A (2008) Comunidade de Syrphidae (Diptera): diversidade e preferências florais no Cinturão Verde (Santa Cruz do Sul, RS, Brasil). Rev Bras Entomol 52(1):41–49

Muckenfuss AE, Sheppard BM, Ferrer ER (1990) Natural mortality of diamondback moth in coastal South Carolina. Diamondback moth and other crucifer pests. In: Proceedings of the 2nd international workshop, AVRDC Publication, Tainan, 10–14 Dec

Nicholls C (2006) Bases agroecológicas para diseñar e implementar una estrategia de manejo de hábitat para control biológico de plagas. Agroecología 1:37–48

Nicholls CI, Altieri MA (2012) Modelos ecológicos y resilientes de producción agrícola para el siglo XXI. Agroecología 6:29–37

Paleologos MF, Flores CC, Sarandon SJ et al (2008) Abundancia y diversidad de la entomofauna asociada a ambientes seminaturales en fincas hortícolas de La Plata, Buenos Aires, Argentina. Rev Bras Agroecol 3(1):28–40

Parolin P, Bresch C, Desneux N et al (2012) Secondary plants used in biological control: a review. Int J Pest Manag 58(2):91–100

Patt JM, Wainright SC, Hamilton GC et al (2003) Assimilation of carbon and nitrogen from pollen and nectar by a predaceous larva and its effects on growth and development. Ecol Entomol 28(6):717–728

Pfiffner L, Balmer O (2011) Factsheet organic agriculture and biodiversity. DepartamentD'Agricultura, Ramaderia. FichaTécnica PAE 21

Pimentel D, Hepperly P, Hanson J et al (2005) Environmental, energetic, and economic comparisons of organic and conventional farming systems. Bioscience 55(7):573–582

Proctor M, Yeo P, Lack A (1996) The natural history of pollination. Harper Collins Publishers, London

Rodríguez E, González M (2014) Vegetación autóctona y control biológico: diseñando una horticultura intensiva sostenible. Grupo Cooperativo Cajamar, São Paulo. Fichas de Transferencia 4

Sabelis MW (1992) Predatory arthropods. In: Crawley MJ (ed) Natural enemies: the population biology of predators, parasites and diseases. Blackwell Science, Oxford, pp 225–264

Stanley RG, Linskens HG (1974) Pollen, biology, biochemistry and management. Sringer-Verlag, Berlin

Van Driesche RG, Hoddle MS, Center TD (2007) Control de plagas y malezas por enemigos naturales. Forest Health Technology Enterprise Team, Morgantown

Van Lenteren JC (1999) Fundamental knowledge about insect reproduction: essential to develop sustainable pest management. Invertebr Reprod Dev 36:1):1–1)15

Van Lenteren JC, Van Vianen A, Kortenhoff HF (1987) The parasite host relationship between *Encarsia formosa* Gahan (Hymenoptera: Aphelinidae) and *Trialeurodes vaporariorum* (Westwood) (Homoptera: Aleyrodidae). J Appl Entomol 103(1/5):69–84

Venzon M, Carvalho CF (1992) Biologia da fase adulta de *Ceraeochrysa cubana* (Hagen, 1861) (Neuroptera, Chrysopidae) en diferentes dietas y temperaturas. Cienc Prát 16:315–320

Venzon M, Rosado MC, Euzébio DE et al (2006) Suitability of leguminous cover crop pollens as food source for the green lacewing *Chrysoperla externa* (Hagen) (Neuroptera: Chrysopidae). Neotrop Entomol 35(3):371–376

Waldbauer GP, Friedman S (1991) Self-selection of optimal diets by insects. Annu Rev Entomol 36:43–63

Wratten S, Berndt L, Gurr G et al (2003) Adding floral diversity to enhance parasitoid fitness and efficacy. In: 1st international symposium on biological control of arthropods, Honolulu, USDA Forest Service, 14–18 Jan 2003

Xu J, Shelton AM, Cheng X (2001) Comparison of *Diadegma insulare* (Hymenoptera: Ichneumonidae) and *Microplitis plutellae* (Hymenoptera: Braconidae) as biological control Agents of *Plutella xylostella* (Lepidoptera: Plutellidae): field parasitism, insecticide susceptibility, and host-searching. J Econ Entomol 94(1):14–20

Zuñiga E (1987) Cómo mejorar el efecto de los enemigos naturales en el campo: medidas prácticas para facilitar la acción de estos extraordinarios controladores de plagas. IPA La Platina 44:1924

Chapter 5
Dispersion and Increase of Natural Enemies in Agroecosystems

Luis L. Vázquez and Gabriel C. Jacques

5.1 Introduction

The dispersion capacity of natural enemies by biological control agents released or applied in cultivations fields as well as natural enemies, which inhabit the agroecosystem, is of great importance for the sustainability of pest management, because it facilitates their survival, establishment, and continuous activity, among other advantages. Therefore, the dispersion of entomopathogenic and entomophagous is an important factor that affects its establishment and its efficiency in pest suppression (McDougall and Mills 1997).

However, the expansion of an extensive and strongly industrialized agriculture has simplified the agricultural landscape to small fragments of the natural habitats. Habitat fragmentation can affect insect dispersal in the landscape, especially considering modern agricultural practices that modify the landscape significantly, eliminating natural or semi-natural vegetation where biological control agents may seek refuge (Grez et al. 2008).

The management of ecological corridors becomes possible to foment the biodiversity in the monoculture systems, improving the biological structure of the agroecosystems and facilitating the ecological control (Altieri and Nicholls 1994). The corridors structure is influenced by the management of the trees due to their utility in the conservation of the associated biodiversity, by providing habitats and preserving certain level of connectivity of the landscape (Harvey et al. 1999), that is, the continuity of a habitat or covering type through an area determined in the landscape (Turner et al. 2001). The connectivity implies the connection of habitats, species, communities, and ecological processes to multiple special scales and storms; it can

L. L. Vázquez (✉)
Instituto de Investigaciones de Sanidad Vegetal (INISAV), Miramar, La Habana, Cuba

G. C. Jacques
Instituto Federal de Minas Gerais, Campus Bambuí, Bambuí, MG, Brazil

B. Souza et al. (eds.), *Natural Enemies of Insect Pests in Neotropical Agroecosystems*, https://doi.org/10.1007/978-3-030-24733-1_5

be defined as the grade in that the landscape facilitates or impedes the movement among patches of resources (Noss 1990).

The objective of this chapter is to offer a panoramic on routes and factors of dispersion phytophagous of natural enemies of pest insects in agroecosystems, as a base for pest management programs.

5.2 Dispersion Routes

The dispersion of populations of natural enemies can occur for three main routes: biotic, abiotic, and anthropogenic as well as their combinations. The complexity of the interactions of natural enemies with their cultivations and other host plants and the characteristics of the agroecosystems constitute decisive factors in the necessity and success of the dispersion.

Pathogens can be dispersed in several ways: active discharge of infective spores (local dispersal), weather factors (wind and rain), or by host and non-host dispersal (long-distance dispersal) (Fuxa and Tanada 1987). The transmission of a fungal pathogen is dependent on a number of processes: conidia production, discharge, dispersal, survival, and germination (Hajek and Leger 1994). The large number of conidia produced by infected cadavers partially compensates for the high probability and that many conidia will not actually infect a host (Shimazu and Soper 1986). Indeed, the density and distribution of a pathogen population is one of the most important factors determining whether a disease becomes epizootic (Carruthers et al. 1991).

In an unmanaged host/pathogen system, epizootics are usually host-density dependent, developing as the host population increases. However, an epizootic may develop at low host densities if the pathogen is widely distributed within the host habitat. Pathogen populations are often distributed at low densities or discontinuously in the host habitat and, therefore, pathogen dispersal is essential (Fuxa and Tanada 1987).

Most pathogens have a limited capacity for active dispersal; however, entomophthoralean fungi (except *Massospora*) produce conidiophores that forcibly discharge primary conidia across the leaf boundary layer (Steinkraus et al. 1996).

Dispersal of conidia over a greater distance requires physical factors such as rain and wind or dispersal by hosts and non-hosts. Rain can remove pathogens in splash droplets or by vibration caused by the impact of raindrops. Many studies have shown the importance of rain in the dispersal of plant pathogens (Pedersen et al. 1994), though this may be less important for some entomopathogens such as *Pandora neoaphidis* (= *Erynia*) (Remaudière & Hennebert) (Pell et al. 1997).

Wind is undoubtedly important in the long-distance dispersal of many fungal entomopathogens (Steinkraus et al. 1996) with most fungal conidia being small. However, the dispersion of entomopathogens by weather factors is random in contrast to dispersal by hosts and non-host vectors, which are targeted.

Schellhorn et al. (2014) affirmed that insects use two mechanisms for entomophagous dispersion: auto-directed movements (to walk, to jump, to fall, and to fly) and passive movements (wind, foresis, and transport mediated for human). Among the passive movements, wind is considered a factor that influences the air dispersion of parasitoids (McManus 1988; Kristensen et al. 2013). The parasitoids dispersion is influenced by biotic and abiotic factors such as the size of the insect, the gender and state of mating of the parasitoid (Liu et al. 2015), the density of the host (Petit et al. 2008), the direction and speed of the wind (Kristensen et al. 2013), and the presence of chemical signs (Liu et al. 2015).

The insects pass a long part of their time moving among places that offer different resources such as food, refuge, couple, and hosts; this is fundamental for most of organisms and necessary for their survival (Schellhorn et al. 2014). The result of the process of parasitoids dispersion is the encounter and acceptance of hosts (Rehman and Powell 2010). Some insects can control the direction and the flight speed by combining active and passive mechanisms when entering, remaining, and leaving the current of the wind above the canopy (Schellhorn et al. 2014).

Studies carried out by Hernández and Manzano (2016) on the parasitoid *Amitus fuscipennis* MacGown & Nebeker (Platygastridae), a promissory biological control of the *Trialeurodes vaporariorum* (Westwood) (Aleyrodidae), showed that the parasitoid was dispersed up to 12 m and it was the biggest rate that happened in the east direction (15.08%), which suggested that the wind contributes to the passive dispersion once the insect begins the flight. *Amitus fuscipennis* measures less than 1 mm, which helps its passive initial wind dispersion (Hernández and Manzano 2016), as it has been reported for the micro wasps *Trichogramma* (Trichogrammatidae) that cannot fly up against the wind (Keller et al. 1985). On the contrary, a bigger parasitoid, *Gonatocerus ashmeadi* Girault (Mymaridae), with a longitude of the posterior tibia <0.7 mm is not affected by the wind in its dispersion pattern (Petit et al. 2008).

Winds, including those of tropical hurricanes, can directly or in organs of their host plants disperse populations of insects pests and their natural enemies. Two citrus pests and their natural enemies favored by the winds of hurricanes were introduced in Cuba: the citrus leaf miner (*Phyllocnistis citrella* Stainton) (Gracillariidae) with the parasitoid *Ageniaspis citricola* Logvinovskaya (Encyrtidae) (Vázquez and Pérez 1997) and the citrus psilid, *Diaphorina citri* Kuwayama (Liviidae) with *Tamarixia radiata* (Waterston) (Eulophidae) (Vázquez et al. 2009).

Parasitoids and predators of herbivores have evolved and they function within a multitrophic context. Consequently, their physiology and behavior are influenced by elements from other trophic levels such as their herbivore victim (second trophic level) and its plant food (first trophic level). Natural enemies base their foraging decisions on information from these different trophic levels, and chemical information plays an important role. The importance of infochemicals, a subcategory of semiochemicals, in foraging by parasitoids and predators has been well documented (Vet and Dicke 1992).

In a study of larval movement by using video techniques, Furlong and Pell (1996) determined that larvae foraged on by *Diadegma semiclausum* (Hellen) (Ichneumonidae) moved significantly further and into significantly more new areas of leaf than control larvae. Interestingly, larvae foraged on by a different parasitoid, *Cotesia plutellae* (Kurdjumov) (Braconidae), were intermediate in their movement and were not more likely to become infected by the entomopathogen *Zoophthora radicans* (Brefeld) Batkothan control larvae. This suggests that there is a threshold of movement necessary for the fungal transmission to be encouraged and that this threshold was exceeded by larvae foraged on by *D. semiclausum* but not by *C. plutellae* larvae foraged. These interactions are complex and difficult to predict because they vary between natural enemies. In a similar study, the transmission of *P. neoaphidis* was increased in the presence of the parasitoid *Aphidius rhopalosiphi* de Stefani-Perez (Braconidae) (Fuentes-Contreras et al. 1998).

Intraguild interactions are common among communities of biological control agents including interactions between parasitoids and entomopathogenic fungi (Rashki et al. 2009). Mohammed and Hatcher (2017) showed that combining *Aphidius colemani* Viereck (Braconidae) and *Lecanicillium muscarium* R. Zare & W. Gams (Mycotal®) to control *Myzus persicae* Sulzer (Aphididae) may be feasible for controlling the aphid in greenhouse situation as a result of environmental conditions, especially the temperature that favors fungal growth, which can be controlled and also parasitoids can be kept in. A more detailed knowledge of the effects of naturally occurring parasitoids on pest control and their interaction with other biological control agents will help to develop environmentally sound crop management strategies with reduced insecticide applications.

The release of predatory mites accompanied by spray applications of *Beauveria bassiana* (Bals.-Criv.) Vuill. may be a viable alternative in the management of *Tetranychus urticae* Koch (Tetranychidae) populations and possibly improve efficiencies in controlling *T. urticae* (Chandler et al. 2005). The potted plant investigations realized by Shengyong et al. (2018) indicated that *Phytoseiulus persimilis* Athias-Henriot (Phytoseiidae) showed significant aversion behavior to the initial fungal spray but gradually dispersed over the entire bean plants. The study suggests that fungal spray did not affect the predation capability of *P. persimilis* and poses a negligible risk to their behavior.

5.3 Dispersion Factors

Getting natural enemies to disperse and multiply in the environment where they are released is one of the greatest challenges of applied biological control. All elements that influence the establishment of these organisms ought to be predicted and avoided for the success of any biological control program. These elements are mainly related to three factors: climatic conditions during the release of the controlling agents, crop characteristics (e.g., architecture and plant age), and attributes that are intrinsic to the natural enemy, such as development stage, biology, and ecology. The study of these factors can provide information on the best technique to be applied and in the calculation of the number of specimens, release points, and the frequency and interval between releases.

Climate Influence Most of the natural enemies used in biological control programs (mites and insects) are ectothermic animals. These animals depend on the heat of the environment to activate their metabolism. Thus, a release above or below their optimum temperature can affect the survival and dispersion of these natural enemies in the field. Rain is also a very important climatic factor. Natural enemies in the egg and pupal stages can be washed away by rain and die on the ground before their emergence. The release of *Trichogramma* sp. (Trichogrammatidae) in the rainy season together with the predatory activity of their natural enemies eliminated nearly 80% of parasitized eggs (Smith 1994).

Crop Characteristics The architecture and the age of the plants can directly influence the dispersion and persistence of the natural enemies in the field. Characteristics such as leaf width and length, branching, presence of flowers, and structures on the leaf surface (e.g., trichomes) can modify the microclimate as well as food sources and shelter under adverse environmental conditions. *Trichogramma pretiosum* Riley (Trichogrammatidae) was less dispersed in a 90-day tomato crop when compared with the same culture at younger ages (Pratissoli et al. 2005). Larger leaf area, size, and thickness of older tomatoes caused variations in the microclimate and interfered on the search capacity of the parasitoid.

The presence of flowers in cultivated plants might increase populations of natural enemies such as syphidians, coccinellids, chrysopedes, and parasitoids. Parasitoids are mainly used in monoculture systems that present a poor environment during most of the phenological cycle, negatively affecting their dispersion and population increase. The lower availability of floral nectar reduces the search activity by the host and favors the migration to other food-rich areas. Adequate feeding of adults causes several benefits such as increased reproduction and better egg maturation.

Natural Enemy Characteristics Biological characteristics such as fecundity, fertility, longevity, and sexual ratio contribute to the reproductive capacity of the natural enemy and, consequently, to its effectiveness in biological control programs. The evaluation of these characteristics in laboratory studies can provide information on their ability to survive and multiply in field conditions.

The age and stage of development of the natural enemy is very important. However, studies are needed to verify the age or stage that provides greater dispersion and survival of these enemies in the field. For example, the stage and the best age for release of *Cotesia flavipes* (Cameron) (Braconidae) are adults with up to 24 hours due to lower mortality rates (Oliveira et al. 2012). Host specificity can also influence the establishment, dispersion, and effectiveness of natural enemies. Predators use more than one individual to complete their life cycle. However, the vast majority are generalists and end up preying on non-target prey. Although parasitoids use only one individual per cycle, they are usually more specialists, converging their parasitism in the target pest. Strains of the same species of parasitoid present differences in relation to host preference, plant recognition and acceptance, search behavior, and tolerance to environmental conditions. Differences in the parasitism rate of different strains of *T. pretiosum* were observed in the control of *Chrysodeixis includens* (Walker) (Noctuidae) (Bueno et al. 2009).

5.4 Dispersion of Entomopathogens by Their Phytophagues and Other Hosts

The development of a low-cost autodissemination technique for entomopathogenic fungi where the insect can spread the fungus via horizontal transmission to conspecifics (e.g., during mating) is necessary (Avery et al. 2009). Similar autodissemination techniques for controlling pests have been evaluated in other systems (Tsutsumi et al. 2003; Scholte et al. 2004).

Based on laboratory studies, Avery et al. (2009) proposed the potentiality of the autodissemination of *Isaria fumosorosea* Wize is inoculating yellow cards with the entomopathogenic fungus to the Asian citrus psyllid (*D. citri*) as a disseminate technique for managing the pest populations.

Dispersal flight, a well-known strategy for aphids to locate suitable plants, was studied by Feng et al. (2007) for its possible role in disseminating fungal pathogens and parasitoids as natural control agents of aphids by air captures in Hangzhou, China, during 2001–2005. Up to 3183 migratory-winged forms of the green peach aphid *M. persicae* were captured from air using a yellow-plus-plant trap on the top platform of a six-storey building in an urbanized area and individually reared in a laboratory for ≥7 days. Among the captured winged, 28.9% survived on average for 2.5 days and then died from mycoses attributed to 10 species of obligate or non-obligate aphid pathogens. These were predominated by *P. neoaphidis*, which caused 80% of the winged mycoses. Another 4.4% survived for an average of 3.7 days, followed by mummification by *Aphidius gifuensis* Ashmead (Braconidae) (52.9%) and *Diaeretiella rapae* (M'Intosh) (Braconidae) (47.1%).

An autoinoculative device was used to test the ability of sap beetles (Nitidulidae) to carry a specific strain of *B. bassiana* to over wintering sites in a multiyear field study. The device was baited with the pheromone and coattractants for the dusky sap beetle (*Carpophilus lugubris* Murray) and placed in the field in the fall of each year. The autoinoculating device provides selective contamination of sap beetles in overwintering sites when used in the fall. It may be useful in providing some control of sap beetles or other insects where limited numbers of mass overwintering sites (such as tree holes) occur (Down and Vega 2003).

Studies carried out by Gross et al. (1994) suggested that *Apis mellifera* L. (Apidae) can disseminate insect pathogens be applicable for pest control will depend on an improved understanding of the interrelationships among *A. mellifera*, flowering plants on which they forage, targeted phytophages that reside on the plants, and the availability of pathogens that attack the phytophages. The mean percentage of HNPV-induced mortality was significantly higher among *Helicoverpa zea* (Boddie) (Noctuidae) larvae that fed on clover heads from fields foraged by HNPV-contaminated *A. mellifera* and among *Helicoverpa* spp. larvae collected from those fields than among similarly exposed control larvae.

Pollen beetles, *Meligethes aeneus* Fabricius (Nitidulidae), are pests that feed and oviposit in the buds and flowers of oilseed rape. Honeybees foraging from a hive fitted with an inoculum dispenser at the entrance effectively delivered dry conidia of the entomogenous fungus, *Metarhizium anisopliae* (Metchnikoff) Sorokin to the flowers of oilseed rape in caged field plots. Mortality (61% on winter rape, 100% on spring rape) was higher during peak flowering, when the feeding activity of both bees and beetles from the flowers was maximal, providing optimal conditions for inoculum dissemination and infection. Conidial sporulation occurred on a significant proportion of the dead pollen beetles. There was no evidence of any adverse effect on the honeybee colonies (Butt et al. 2010).

5.5 Dispersion of Entomopathogens by Entomophagous

It has long been recognized that predators and parasitoids contribute to the dissemination of viral entomopathogens and the development of epizootics. However, only a few more recent studies have assessed the transmission of fungal entomopathogens by non-host vectors (Butt et al. 2010).

The presence of insect natural enemies may have an impact on local transmission of a fungal pathogen. The presence of a foraging adult coccinellid, for example, resulted in a substantial increase in the local transmission of the aphid pathogen *P. neoaphidis* within a population of pea aphids, *Acyrthosiphon pisum* Harris (Aphididae) on individual bean plants in the laboratory (Roy et al. 1998). Foraging ladybirds cause an increase in aphid movement, although the degree to which this occurs depends on the aphid species and host plant (Hajek and Dahlsten 1987). The increase in movement of aphids in the presence of foraging predators such as coccinellids would increase the probability of the aphid coming into contact with the sporulating cadaver, and therefore receiving more inoculum.

In laboratory conditions, studies carried out by Meyling et al. (2006) revealed that the aphid *Microlophium carnosum* Buckton (Aphididae) and its predator *Anthocoris nemorum* (L.) (Anthocoridae) dispersed conidia of *B. bassiana* from the soil toward the cultivation and from cadavers in the foliage of the plant.

A good example of natural autodispersion happens in the interaction between the coffee berry borer, *Hypothenemus hampei* (Ferrari) (Curculionidae) and *B. bassiana* in coffee plantations. According to Bustillo (2006), among the entomopathogenic fungi that naturally regulate the populations of this pest in Colombia, *B. bassiana* is the one that is observed in almost all the regions where this insect appears. Similarly, Vázquez et al. (2010) informed that in Cuba the epizooties caused by this fungus happened between 21 and 98 months after the detection of the pest in the different territories. A cadaver of *H. hampei* well sporuled with *B. bassiana* is able to produce about ten million spores that can be dispersed by a natural way to establish in the coffee plantations (Narváez et al. 1997). It is considered that approximately 49% of total coffee berry borer population in a coffee plantation dies as consequence of the regulatory activity of *B. bassiana* (Ruiz 1996).

A very interesting case was studied by Forschler and Young (1993), in Arkansas, USA, who observed adults of *Diabrotica undecimpunctata* Howardi (Chrysomelidae) to feed on cadavers of *H. zea* killed by *H. zea* nuclear polyhedrosis virus. Laboratory bioassays showed that excreta from adults of *D. undecimpunctata* fed on virus-killed cadavers collected up to 3 days after feeding resulted in virus mortality of *H. zea*.

5.6 Anthropogenic Dispersion

The releases and applications of entomophagous and entomopathogenic that are carried out for the pest population control be inundative or inoculative constitute the main route of anthropogenic dispersion of the biological control agents. Many techniques have been developed and are applied in the agricultural production with this purpose and for the dispersion of natural enemies that inhabit the agroecosystems.

Capture-Inoculation Trap with Entomopathogenic Releases Pozo et al. (2006) found that scarabs of the genera *Phyllophaga* and *Cyclocephala* (Scarabaeidae) that attack the cultivation of the pineapple, *Ananas comosus* (L.) Merrill, were captured by a system of traps that combines the attraction of adults to the light and their collection in a tray that contains a suspension in water of entomopathogenic nematode, in a dose of 3×10^6 juvenile for trays of 50 × 50 and a million for trays of 25 × 25 cm. The captured adults were infected with the juvenile of the nematode, which were transferred by the adults that escaped from the trap and disperse them toward the system radicular of the cultivation where they inhabit larvae of the pest.

Pozo (2013) referred an effectiveness of 74% in the biological control of *Cosmopolites sordidus* (Germar) (Curculionidae) and *Metamasius hemipterus sericeus* (Olivier) (Curculionidae) in banana plantation by using traps elaborated with pseudo stem of bananas cut as sandwich, where a suspension with entomopathogenic nematode was sprayed using a dose of 3000 immature juvenile of instar 3/trap.

Direct Collect-Artisanal Rearing-Released Another way of dispersion is the breeding of the natural enemy in its own agroecosystem to later disperse it in different places. In Cuba, Castiñeiras et al. (1982) created and introduced a methodology to collect and multiply the ant *Pheidole megacephala* (Fabricius) (Formicidae) by means of artificial reservoirs. From these reservoirs, workers and reign populations were moved to inoculate the colonies in recently planted fields of the cultivation of the sweet potato, *Ipomoea batatas* (L.) Lam. (Convolvulaceae), with the purpose of controlling populations of the sweet potato weevil, *Cylas formicarius* (Fabricius) (Brentidae). Castiñeiras and Ponce (1991) found that by placing nine colonies of ants/ha against sweet potato, there was a 22% damage reduction and 68% of the population of adult pests reduction.

A similar procedure was proposed by Roche and Abreu (1983) by developing artificial reservoirs of the ant *Tetramorium bicarinatum* (Nylander) (Formicidae) to transfer and inoculate colonies in banana fields (*Musa* spp.) that are infested with banana black weevil (*C. sordidus*), which is also a pest of importance in this cultivation. According to Bendicho and Gonzalez (1986), *T. bicarinatum* was able to control approximately 65% of the total population of the pest in fields with high infestation and that in low populations it elevated the control percentage until 83.5%.

Direct Collect-Transfer Social insects can be manipulated by the farmer with the objective of increasing their populations to regulate pest. Wasps from the genus *Polistes* are important biological control agents and stand out as predators of agricultural pests, especially caterpillars. Their colonies are easily manipulated and transported to artificial shelters because of the small nests with peduncle. This practice was carried out in Colombia, where the farmers made shelters to the wasps *Polistes* to make their nests; later these shelters are transferred to the fields of cultivations with more necessity (García 2000).

Colonies of social wasps are best sampled early in the evening, which should be carefully performed so the peduncle is not destroyed. After capture, the colonies are placed in the desired environment using cyanoacrylate ester-based glue. Wooden board shelters (15 × 20 × 1 cm) in the form of a roof are attached to a 1.8-m support (Prezoto and Machado 1999) (Fig. 5.1). Burned oil and grease are applied to the base of the support to prevent ant attacks. The wood structure can be replaced by plastic containers, which facilitate the translocation of the shelters (Elisei et al. 2012) (Fig. 5.2). Four colonies of *P. simillimus* arranged on each side of an experimental plot cultivated with corn (10 × 10 m) at a distance of 2 m from the crop limit caused a reduction of 77.2% in the occurrence of *S. frugiperda* when compared to areas without the presence of wasp colonies.

A single technique of natural enemies dispersion is practiced by farmers, which consist in collecting leaves with populations of pests, which are parasited for entomopathogenic fungus and transfer them to fields or parts of them where has not shown the epizooty (Vázquez and Elósegui 2011).

Experiences have also been documented in agroecosystems of Cuba on the collection and transfer of small populations of entomophagous, such as adults of the *Zelus longipes* L. (Reduviidae), nymphs and adults of *Orius insidiosus* (Say) (Anthocoridae), larvae and adults of Coccinellidae, eggs and larvae of Chrysopidae, and larvae of lepidopters infected for entomopathogenic virus. Cases of entomophagous transfer have been systematized from experiences of farmers (Caballero et al. 2003; Vázquez et al. 2008). For example, eggs of lepidopterous parasited *Trichogramma* and *Telenomus*, leaves with mines of leafminer parasited (Agromyzidae); leaves with colonies of aphids parasited *Lysiphlebus testaceipes* (Cresson) (Braconidae); leaves with colonies of whiteflies parasited *Encarsia* and *Eretmocerus*; leaves and branches of coffee with colonies of *Coccusviridis* (Green) (Coccidae) parasited *Lecanicillium lecanii* R. Zare & W. Gams; leaves with colonies of aphids parasited *Entomophthora*; leaves with colonies of whiteflies parasited *Aschersonia*; leaves with larvae of *Spodoptera* parasited *Euplectrus*; cottons of *Apanteles* in cassava, leaves with eggs of *Pachnaeuslitus* (Germar) (Curculionidae) in citrus parasited *Tetrastichus*; leaves with colonies of predators acari; and fruits of coffee infested by *H. hampei* parasited *B. bassiana*.

Fig. 5.1 Shelter for translocation of social wasp colonies: (a) wooden support of 1.80 m; (b) social wasp colony shelter. (Photo: Gabriel C. Jacques)

Fig. 5.2 *Polistes versicolor* colony translocated to an artificial shelter. (Photo: Gabriel C. Jacques)

5.7 Final Remarks

The cluster previously used for different ways of natural enemy dispersions does not seek to establish a system to classify them, but it is rather carried out to contribute to the understanding of the diversity of these processes and their practical importance. It was not sought to drain the topic because the available information is wide. Nevertheless, there is a need for demand of basic and applied investigation to integrate practical of natural enemy dispersion in pest management.

The complexity of the designs of systems of mixed cultivations (polycultures, polyfruits, others) and the integration of auxiliary vegetation structures (ecological corridors, alive barriers, among others) in the matrix of the production systems, among other practices of biological control for conservation, are facilitating the dispersion of natural enemies.

References

Altieri MA, Nicholls CI (1994) Biodiversidad y manejo de plagas en agroecosistemas. Icaria, Capellades

Avery PB, Hunter WB, Hall DG et al (2009) *Diaphorina citri* (Hemiptera: Psyllidae) infection and Dissemination of the entomopathogenic fungus *Isaria fumosorosea* (Hypocreales: Cordycipitaceae) under laboratory conditions. Fla Entomo l92(4):608–618

Bendicho A, Gonzalez N (1986) Comportamiento de poblaciones de Cosmopolitessordidus y Tetramoriumguineense en condiciones naturales. Cienc Agric 17:9–12

Bueno RCOF, Parra JRP, Bueno AF et al (2009) Desempenho de tricogramatídeos como potenciais agentes de controle de *Pseudoplusia includes*Walker (Lepidoptera: Noctuidae). Neotrop Entomol 38(3):389–394

Bustillo AE (2006) Una revisión sobre la broca del café, *Hypothenemus hampei* (Coleoptera: Curculionidae: Scolytinae). Rev Colomb Entomol 32(2):101–116

Butt TM, Carreck NL, Ibrahim L et al (2010) Honey-bee-mediated infection of pollen beetle (*Meligethes aeneus* Fab.) by the insect-pathogenic fungus, *Metarhizium anisopliae*. Biocontrol Sci Technol 8(4):533–538

Caballero S, Carr A, Vazquez LL (2003) Guía de medios biológicos. Instituto de Investigaciones de Sanidad Vegetal. La Habana, CD-ROOM

Carruthers RI, Sawyer AJ, Hural K (1991) Use of fungal pathogens for biological control of insect pests. In: Sustainable agriculture research and education in the field. NationalAcademyPress, Washington

Castiñeiras A, Caballero S, Rego G et al (1982) Efectividad técnico-económica del empleo de la Hormiga Leona *Pheidole megacephala* en el control del Tetúan del Boniato *Cylas formicarius elegantulus*. CiencTecAgric. Prot Plantas 5(supl):103–109

Castiñeiras A, Ponce E (1991) Efectividad de la utilización de *Pheidole megacephala* (Hymenoptera: Formicidae) en la lucha biológica contra *Cosmopolites sordidus* (Coleoptera: Curculionidae). Prot Plantas 1:15–21

Chandler D, Davidson G, Jacobson RJ (2005) Laboratory and glasshouse evaluation of entomopathogenic fungi against the two-spotted spider mite, *Tetranychus urticae* (Acari: Tetranychidae), on tomato *Lycopersicon esculentum*. Biocontrol Sci Technol 15(1):37–54

Down PF, Vega FE (2003) Autodissemination of Beauveriabassiana by Sap Beetles (Coleoptera: Nitidulidae) to overwintering sites. Biocontrol Sci Technol 13(1):65–75

Elisei T, RibeiroJúnior C, FernandesJúnior AJ et al (2012) Management of social wasp colonies in eucalyptus plantations (Hymenoptera: Vespidae). Sociobiology 59(4):1–8

Feng MG, Chen C, Suwei S (2007) Aphid dispersal flight disseminates fungal pathogens and parasitoids as natural control agents of aphids. Ecol Entomol 32(1):97–104

Forschler BT, Young SY (1993) Southern corn rootworm adults (Coleoptera: Chrysomelidae) may act as a non-host disseminator of nuclear polyhedrosis virus. J Invertebr Pathol 61:313–314

Fuentes-Contreras E, Pell JK, Niemeyer HM (1998) Influence of plant resistance at the third trophic level: interactions between parasitoids and entomopathogenic fungi of cereal aphids. Oecologia 117(3):426–432

Furlong MJ, Pell JK (1996) Interactions between the fungal entomopathogen *Zoophthora radicans* Brefeld (Entomophthorales) and two hymenopteran parasitoids attacking the diamondback moth, *Plutella xylostella* L. J Invertebr Pathol 68(1):15–21

Fuxa JR, Tanada Y (1987) Epizootiology of insect diseases. Wiley-Interscience, New York

García Roa FA (2000) Control biológico de plagas: manual ilustrado. MIP, Palmira

Grez AA, Zaviezo T, Díaz S et al (2008) Effects of habitat loss and fragmentation on the abundance and species richness of aphidophagous beetles and aphids in experimental alfalfa landscapes. Eur J Entomol 105(3):411–420

Gross HR, Hamm JJ, Carpenter JE (1994) Design and application of a hive-mounted device that uses honey bees (Hymenoptera: Apidae) to disseminate *heliothis* nuclear polyhedrosis virus. Environ Entomol 23(2):492–501

Hajek AE, Dahlsten DL (1987) Behavioural interactions between three birch aphid species and *Adalia bipunctata* larvae. Entomol Exp Appl 45(1):81–87

Hajek AE, Leger RJS (1994) Interactions between fungal pathogens and insect hosts. Annu Rev Entomol 39:293–322

Harvey CA, Haber WA, Mejías F et al (1999) Árboles remanentes en potreros de Costa Rica: herramientas para la conservación? Agrofor Am 6:19–22

Hernández LM, Manzano MR (2016) Efecto del viento en la dispersión a corta distancia del parasitoide *Amitus fuscipennis* MacGown y Nebeker (Hymenoptera: Platygasteridae) en cultivos de fríjol y habichuela. Acta Agron 65(1):80–86

Keller M, Lewis W, Stinner R (1985) Biological and practical significance of movement by *Trichogramma* species: a review. Southwestern Entomol 8:138–155

Kristensen NP, Schellhorn NA, Hulthen AD et al (2013) Wind-borne dispersal of a parasitoid: the process, the model, and its validation. Environ Entomol 42(6):1137–1148

Liu TX, Stansly PA, Gerling D (2015) Whitefly parasitoids: distribution, life history, bionomics, and utilization. Annu Rev Entomol 60:273–292

McDougall S, Mills N (1997) Dispersal of *Trichogramma platneri* Nagarkatti Hymenoptera: Trichogrammatidae from point-source releases in an apple orchard in California. J Appl Entomol 121(1/5):205–209

McManus ML (1988) Weather, behaviour and insect dispersal. Mem Entomol Soc Can 120(S146):71–94

Meyling NV, Pell JK, Eilenberg J (2006) Dispersal of Beauveria bassiana by the activity of nettle insects. J Invertebr Pathol 93(2):121–126

Mohammed AA, Hatcher PE (2017) Combining entomopathogenic fungi and parasitoids to control the green peach aphid *Myzus persicae*. Biol Control 110:44–55

Narváez M, González MT, Bustillo AE et al (1997) Producción de esporas de aislamientos de *Beauveria bassiana* y *Metarhizium anisopliae* en diferentes sustratos. Rev Colomb Entomol 23(2/3):125–132

Noss RF (1990) Indicators for monitoring biodiversity: a hierarchical approach. Conserv Bio 14(4):355–364

Oliveira HN, Bellon PP, Santana DRS (2012) Critérios para determinação da idade ideal de liberação de *Cotesia flavipes*. Cad Agroecol 7(2):60–64

Pedersen EA, Morrall RAA, McCartney HA et al (1994) Dispersal of conidia of *Ascochyta fabae* f. sp. *lentis* from infected lentil plants by simulated wind and rain. Plant Pathol 43(1):50–55

Pell JK, Tydeman C, McCartney HA (1997) Impact of rainfall on the persistence and transmission of *Erynia neoaphidis*. IOBC/WPRS Bull 21:49

Petit J, Hoddle M, Grandgirard J et al (2008) Short-distance dispersal behavior and establishment of the parasitoid *Gonatocerus ashmeadi* Hymenoptera: Mymaridae in Tahiti: implications for its use as a biological control agent against *Homalodisca vitripennis* Hemiptera: Cicadellidae. Biol Control 45(3):344–352

Pozo E (2013) Empleo de los nematodos entomopatógenos en el manejo de plagas. En: Vázquez LL Manual para la adopción del manejo agroecológico de plagas en la agricultura urbana. INISAV-INIFAT, La Habana, pp 59–74

Pozo E, Sisne ML, Rodríguez U et al (2006) Susceptibilidad de escarabajos (Coleoptera: Scarabaeidae) presentes en la piña (*Ananas comosus*) en Ciego de Ávila a nematodos entomopatógenos. Parte II. Géneros *Phyllophaga* y *Cyclocephala*. Cent Agríc 33(3):83–86

Pratissoli D, Thuler RT, Andrade GS et al (2005) Estimate of *Trichogramma pretiosum* to control Tuta absolutain stalked tomato. Pesqui Agropecu Bras 40:715–718

Prezoto F, Machado VLL (1999) Ação de *Polistes (Aphanilopterus) simillimus* Zikán (Hymenoptera, Vespidae) no controle de *Spodoptera frugiperda* (Smith) (Lepidoptera, Noctuidae). Rev Bras Zool 16(3):841–850

Rashki M, Kharazi-Pakdel A, Allahyari H et al (2009) Interactions among the entomopathogenic fungus, *Beauveria bassiana* (Ascomycota: Hypocreales), the parasitoid, *Aphidius matricariae* (Hymenoptera: Braconidae), and its host, *Myzus persicae* (Homoptera: Aphididae). Biol Control 50:324–328

Rehman A, Powell W (2010) Host selection behaviour of aphid parasitoids Aphidiidae: Hymenoptera. J Plant Breed Crop Sci 2(10):299–311

Roche R, Abreu S (1983) Control del Picudo Negro del Plátano (*Cosmopolites sordidus*) por la hormiga *Tetramorium guineense*. Cienc Agric 17:41–49

Roy HE, Pell JK, Clark SJ et al (1998) Implications of predator foraging on aphid pathogen dynamics. J Invertebr Pathol 71(3):236–247

Ruiz R (1996) Efecto de la fenología del fruto del café sobre los parámetros de la tabla de vida de la broca del café; *Hypothenemus hampei* (Ferrari). Universidad de Caldas, Tesis

Schellhorn N, Bianchi F, Hsu C (2014) Movement of entomophagous arthropods in agricultural landscapes: links to pest suppression. Annu Rev Entomol 59:559–581

Scholte EJ, Knols BGJ, Takken W (2004) Autodissemination of the entomopathogenic fungus *Metarhizium anisopliae* amongst adults of the malaria vector *Anopheles gambiae*. Malar J 3(1):45

Shengyong W, Zhenlong X, Weinan S et al (2018) Effects of *Beauveria bassiana* on predation and behavior of the predatory mite *Phytoseiulus persimilis*. J Invertebr Pathol 153:51–56

Shimazu MS, Soper RS (1986) Pathogenicity and sporulation of *Entomophthora maimaiga* (Entomophthorales: Entomophthoraceae) on larvae of the gypsy moth, *Lymantria dispar* L. (Lepidoptera: Lymantriidae). Appl Entomol Zool 21:589–596

Smith SM (1994) Methods and timing of releases of *Trichogramma* to control lepidopterous pests. In: Wajnberg E, Hassan SA (eds) Biological control with egg parasitoids. CAB International, Wallingford, pp 113–144

Steinkraus DC, Hollingsworth RG, Boys GO (1996) Aerial spores of *Neozygites fresenii* (Entomophthorales: Neozygitaceae): density, periodicity and potential role in cotton aphid (Homoptera: Aphididae) epizootics. Environ Entomol 25(1):48–57

Tsutsumi T, Teshiba M, Yamanaka M et al (2003) Anautodissemination system for the control of brown winged green bug, *Plautia crossota stali* Scott (Heteroptera: Pentatomidae) by an entomopathogenic fungus, *Beauveria bassiana* E- 9102 combined with aggregation pheromone. Japanese J Appl Entomol Zool 47(4):159–163

Turner MG, Gardner R, O'Neill R (2001) Landscape ecology in theory and practice: pattern and process. Spring, New York

Vázquez LL, Elósegui O (2011) Manejo de epizootias por hongos entomopatógenos. En:Vázquez LL Manual para la adopción del manejo agroecológico de plagas en fincas de la agricultura suburbana. Inisav, La Habana, pp 145–162

Vázquez LL, Elósegui O, Leyva L et al (2010) Ocurrencia de epizootias causadas por *Beauveria bassiana* (Bals.) Vuill. en poblaciones de la broca del café (*Hypothenemus hampei* Ferrari) en las zonas cafetaleras de Cuba. Fitosanidad 14(2):111–116

Vázquez LL, Matienzo Y, Veitía M et al (2008) Manejo y conservación de enemigos naturales de insectos fitófagos en los sistemas agrícolas de Cuba. Cidisav, La Habana

Vázquez LL, Pérez I (1997) Introducción y dispersión del minador de la hoja de los cítricos (*Phyllocnistis citrella* Stainton) en la región Neotropical. Levante Agrícola 36(338):4–7

Vázquez LL, Veitía M, Fernández E et al (2009) Diagnóstico rápido de la ocurrencia de plagas en sistemas agrícolas de Cuba por eventos extremos de cambios en el Clima. Rev Bras Agroecol 4(2):2149–2152

Vet LEM, Dicke M (1992) Ecologyofinfochemicalusebynaturalenemiesinatritrophic context. Annu Rev Entomol 37:141–172

Chapter 6
Climate Change and Biological Control of Pests in Agriculture

Luis L. Vázquez

6.1 Introduction

What we know today about climatic changes and variability is that they manifest in the intensity, frequency and duration of extreme events, such as heat waves and cold fronts, severe droughts, hurricanes, penetrations of the sea, among others (Grupo Intergubernamental de Expertos sobre el Cambio Climático 2013). Furthermore, they generally unchain natural disasters (fires, water scarcity, soil salinization, among others) and unbalance biodiversity, causing complex socio-ecological repercussions.

There is scientific evidence on the impact of climatic changes in agriculture, based on their effects on outstanding related variables: temperature, precipitation, carbon dioxide concentration, and soil humidity (Grupo Intergubernamental de Expertos sobre el Cambio Climático 2007). However, management practices, technological changes, market dynamics, and public politicians affect the sector and create uncertainties, due to the difficulty of an isolate evaluation of the influence of the climate, based on other factors (Seo 2011). Therefore, deepening in the studies to demonstrate cause-effect relationships is necessary (Grupo Intergubernamental de Expertos sobre el Cambio Climático 2007).

The effects of climatic changes have special importance on the phenology, distribution, diversity, and composition of species in communities of plants (Morecroft and Keith 2009), birds (Fiedler 2009), mammals (Humphries 2009), and insects (Pelini et al. 2009) in ecosystems.

The influence of climate changes on plants health is evidenced by its direct effects on the development and survival of herbivores and pathogens, by the physiologic changes in defense mechanisms of the plants and, indirectly, by the abundance of natural enemies (e.g., parasitoids of herbivore insects), mutualist insects

L. L. Vázquez (✉)
Instituto de Investigaciones de Sanidad Vegetal (INISAV), Miramar, La Habana, Cuba

B. Souza et al. (eds.), *Natural Enemies of Insect Pests in Neotropical Agroecosystems*, https://doi.org/10.1007/978-3-030-24733-1_6

(e.g., pathogen vectors), and competitors (Ayres and Lombardero 2000). This situation increases the risk for the sustainability of pest management through the biological control (Ladányi and Horváth 2010).

Generally, for the biological control of pests, temperature increases are expected to rebound negatively (Guzmán 2014). Most natural enemies involved are arthropods, which are ectotherms, that is, without the capacity to regulate their body temperature; therefore, these pests are highly dependent on and susceptible to environmental changes. This factor influences the rate of conversion of food into biomass, since it determines the kinetics of biochemical reactions in the physiology of ectotherms (Sabelis 1985). In particular, entomopathogenic microorganisms are significantly depressed under these conditions because they are very sensitive to the high temperatures and dry environments (Walstad et al. 1970; Barbercheck 1992; Smits 1996).

Because of the importance that biological control has acquired in some countries, it is necessary to pay attention to the effects of the climatic change on the effectiveness of the agents that are used in augmentative programs, which will be discussed in the next chapter.

6.2 Exposure to Physical Stress

The stress caused by temperature and relative humidity of the environment, particularly the lingering drought, significantly affects the development and the survival of diverse groups of organisms, contributing to an unbalance of populations of natural enemies – phytophagues. The same happens to biological control agents that are released or applied in augmentative forms, be it entomophagues or entomopathogens (Stireman et al. 2005; Vázquez 2007).

The physical effects of hurricanes are the result of the combination of winds and rainfalls that, due to their high magnitude, affect not only plants and the soil but also pests and their natural enemies, be it those directly exposed, or those that are inside the organs of the plant, or even in the lower levels of the soil (Vázquez 2011).

Scenarios in which the organisms will be more exposed to high temperatures will be outstanding for the future of the biological control of pests, since most of the natural enemies used as biocontrol agents are arthropodal; therefore, many of the processes related to their survival, reproduction, and other parameters of biological effectiveness depend on abiotic conditions (Beveridge et al. 2010).

As a consequence, these species should adjust their tolerance to the abiotic environment where they are. However, if an organism alters its way of life as a direct consequence of the global change, any other organism with which it maintains some type of ecological relationship (depredation, competition, mutualism) will make adjustments as an indirect consequence of the changes of its partner, as it makes changes in the abiotic mark (Chapin et al. 2001).

Regarding the natural enemies of noxious insect populations that inhabit the agroecosystems, several effects have been documented in areas affected by drought and tropical hurricanes, which are also expressed in biological control agents (entomophagues and entomopathogens) that are applied or released in crops. Some of these effects are as follows: (a) dehydration of the phases due to direct exposition to solar radiation and due to the effects of heating the plant organs; (b) difficulties for ovoposition due to the temperature of the leaves; (c) dehydration by superficial currents of hot air; (d) dehydration by low humidity of the soil and the plant; (e) limitations to movement in the host's search for a non-existing sheet of water; (f) mechanic damage by the direct effect of air and water currents; (g) action of secondary microorganisms due to the effect of excessive humidity; (h) dehydration combined with physical effects of strong air currents; and (i) death due to excess of water from soil flooding (Vázquez 2007, 2011).

6.3 Tolerance

The sensitivity to temperature is intensified when ascending of trophic level (Voigt et al. 2003), an amplifying "bottom-up" effect that usually gives in trophic nets subjected to some interference type (Cagnolo et al. 2002). Therefore, if the global heating has a negative impact on the effectiveness or on the parameters of the cycle of the natural enemies' life, the cascade trophics could be muffled, which would harm the biological control of pests. In fact, recent studies have established a direct relationship between a detention in the control of herbivore populations and the severe abiotic conditions that occur in some agroecosystems (Stavrinides et al. 2010; Montserrat et al. 2013).

Natural enemies depend on the abiotic conditions of the environment for their metabolism, their development, their movement capacity and speed, and their predator rate (Guzmán 2014). As a consequence, the intensity at which the individuals interact can be modified if the abiotic conditions of the environment vary, that is, within a context of pests and natural enemies interacting in a cultivation, it could stop harming the biological control (Montserrat et al. 2013).

The combined effects of temperature increase, direct solar radiations, and heating of the surface of the plant organs diminish the rate of mycellal growth and viability of conidia of *Beauveria* and *Metarhizium* (Moore et al. 1993; Inglis et al. 1995). In the case of *Beauveria bassiana*, their conidia alone remain viable for 48 hours at direct exposition and up to 72 hours at indirect exposition (Estrada and Guelmes 2004). This is very negative when the product is not formulated to protect the structures of the microorganism, and it is applied at hours of the day before or during the greatest incidence of solar radiations.

6.4 Effects on the Interaction

Based on a revision of 688 publications, Tylianakis et al. (2008) evidence the effects the climatic change have on the intensity and/or directions of the biotic interactions, such as competition (plant-plant, animal-animal), parasitism, mutualism (plant-pollinator), depredator, etc.

At least for insects, with which the populational effects of this sort of interaction can be better analyzed, the first studies, such as those of Butler (1985), already evidence a reduction of the populations of herbivores and of predators that feed of them, among others that confirm these results (Hunter 2001; Holton et al. 2003). Thus, the set of changes in phenology and in the quality of the plant used as food by the herbivore will give place to an entirely new situation that will affect the security of the rest of the trophic chain (Holton et al. 2003).

As Wallner (1987) pointed out, temperature can have a variety of direct effects on the populations, for example, synchrony between phytophagues and hosts (phenology and other), predator and the relationship predator-prey, mortality during hibernation, selective mortality (the population genetics), modification of refuge areas (survival).

6.5 Resilience Capacity

In general, it is possible to state that the extinction of species, both at global and local levels, can be avoided if the populations migrate to more favorable habitats or if the organisms overcome the stress conditions through phenotypic plasticity or even if the populations undergo processes of adaptative evolution (Berg et al. 2010; Hoffmann and Sgrò 2011). However, due to the speed and magnitude of the climatic change, the response of many populations to resisting negative effects of temperature increases is not likely to be appropriate; under this point of view, the future success of biological control strategies will be limited by the capacity of natural enemies to have adaptative responses to quick changes in the climate (Hoffmann and Sgrò 2011).

Since abiotic conditions mediate many parameters related to the biological effectiveness of species (survival, longevity, reproduction) and how individuals interact to each other (predator rate, competitive ability) (Dunson and Travis 1991; Beveridge et al. 2010; Gilman et al. 2010), these changes in climate can cause alterations to the dynamics and structure of communities. Therefore, currently, one of the most important challenges in ecology is to predict the effect that climatic changes will have on the species and the communities (Hughes 2000; Voigt et al. 2003; Gilman et al. 2010).

The resilience of biological control to the effects of climatic changes, mainly of the augmentative programs, depends on several capacities, such as (a) resistance of the species used as biological control agents, whether entomophagues or entomopathogens; (b) the characteristic of these agents as products, specifically of the liberation technique for the entomophagues and of the protection of infectious structures or stages for the entomopathogens; (c) monitoring to decide moment and covering of the application or releases; (d) characteristic of the habitat where they will be release or applied, mainly the complex cultivation systems and the integration of the auxiliary vegetation in the scale of the production system; and (e) capacity to guarantee productions in biolaboratories or insectaries, in spite of being exposed to extreme events.

According to Hódar et al. (2004), the reality is that there are enough uncertainty and ignorance on the magnitude and the address of the effects of the global change. The difficulty when clarifying the processes implied in the global change is that the changes are multiple and interactive, and many of them are simultaneously a consequence of a change and the cause of another change.

6.6 Final Remarks

The appropriate climatic conditions for the massive multiplication of biological control agents have been relatively studied; however, the effects of the climatic change, such as the increase in the half temperature, the periods of drought, the tropical hurricanes, defrosting and other extreme events, are scarcely documented.

These investigations require contextual studies during periods of time that allow comparing the behavior of the pest and the effectiveness of the biological control agents, in their interactions with the management of the cultivations, including their effects on the natural enemies that live in agroecosystems. Thus, the biological control programs will be able to be adapted to the climatic change.

References

Ayres MP, Lombardero MJ (2000) Assessing the consequences of global change for forest disturbance from herbivores and pathogens. Sci Total Environ 262(3):263–286

Barbercheck ME (1992) Effect of soil physical factors on biological control agents of soil insect pests. Fla Entomol 75(4):539–548

Berg MP, Kiers ET, Driessen G et al (2010) Adapt or disperse: understanding species persistence in a changing world. Global Change Biol 16(2):587–598

Beveridge OS, Humphries S, Petchey OL (2010) The interacting effects of temperature and food chain length on trophic abundance and ecosystem function. J Anim Ecol 79(3):693–700

Butler GD (1985) Populations of several insects on cotton in open-top carbon dioxide enrichment chambers. Southwest Entomo 110:264–266

Cagnolo L, Molina SI, Valladares GR (2002) Diversity and guild structure of insect assemblages under grazing and exclusion regimes in a montane grassland from Central Argentina. Biodivers Conserv 11(3):407–420

Chapin IIIFS, Sala OE, Huber-Sannwald E et al (2001) The future of biodiversity in a changing world. In: Chapin IIIFS, Sala OE, Huber-Sannwald E (eds) Global biodiversity in a changing environment: scenarios for the 21st century. Springer, New York, pp 1–4

Dunson WA, Travis J (1991) The role of abiotic factors in community organization. Am Nat 138(5):1067–1091

Estrada ME, Guelmes J (2004) Persistencia de *Beauveria bassiana* (Bálsamo) Vuillemin en hojas de caña de azúcar (*Saccharum* sp. híbrido). Fitosanidad 8(4):53–56

Fiedler W (2009) Bird ecology as an indicator of climate and global change. In: Letcher T (ed) Climate change: observed impacts on planet Earth. Elsevier, Amsterdam, pp 181–195

Gilman SE, Urban MC, Tewksbury J et al (2010) A framework for community interactions under climate change. Trends Ecol Evol 25(6):325–331
Grupo Intergubernamental de Expertos sobre el Cambio Climático (2007) Cambioclimático 2007: informe de síntesis. IPPC, Ginebra
Grupo Intergubernamental de Expertos sobre el Cambio Climático (2013) Climate change 2013: the physical science basis. IPPC, Ginebra
Guzmán C (2014) Cambio climático y control biológico de plagas: efecto de las condiciones abióticas en las interacciones entre enemigos naturales presentes en el agro-ecosistema del aguacate. Tesis, Universidad de Málaga
Hódar JA, Zamora R, Peñuelas J (2004) El efecto del cambio global en las interacciones planta-animal. In: Valladares F (ed) Ecología del bosque mediterráneo en un mundo cambiante. Ministeriode Medio Ambiente, Madrid, pp 461–478
Hoffmann AA, Sgrò CM (2011) Climate change and evolutionary adaptation. Nature 470:479–485
Holton MK, Lindroth RL, Nordheim EV (2003) Foliage quality influences tree-herbivore-parasitoid interactions: effects of elevated CO2, O3, and plant genotype. Oecologia 137(2):233–244
Hughes L (2000) Biological consequences of global warming: is the signal already apparent? Trends Ecol Evol 15(2):56–62
Humphries MM (2009) Mammal ecology as an indicator of climate change. In: Letcher T (ed) Climate change: observed impacts on planet earth. Elsevier, Amsterdam, pp 197–214
Hunter MD (2001) Effects of elevated atmospheric carbon dioxide on insect-plant interactions. Agric For Entomol 3(3):153–159
Inglis GD, Jonson DL, Goettel MS (1995) Influence of ultraviolet light protectans on persistence of the entomopathogenic fungus *Beauveria bassiana*. Biol Control 5(4):581–590
Ladányi M, Horváth L (2010) A review of the potential climate change impact on insect populations-general and agricultural aspects. Appl Ecol Environ Res 8(2):143–152
Montserrat M, Ma Sahún R, Guzmán C (2013) Can climate change jeopardize predator control of invasive herbivore species? A case study in avocado agroecosystems in Spain. Exp Appl Acarol 59(1/2):27–42
Moore D, Bridge PD, Higgins PM et al (1993) Ultra-violet radiation damage to *Metarhizium flavo-viride* conidia and the protection given by vegetable and mineral oils and chemical sunscreens. Ann Appl Biol 122(3):605–616
Morecroft MD, Keith SA (2009) Plant ecology as an indicator of climate and global change. In: Letcher T (ed) Climate change: observed impacts on planet earth. Elsevier, Amsterdam, pp 297–306
Pelini SL, Prior KM, Parker DJ et al (2009) Climate change and temporal and spatial mismatches in insect communities. In: Letcher T (ed) climate change: observed impacts on planet earth. Elsevier, Amsterdam, pp 215–232
Sabelis MW (1985) Development. In: Helle W, Sabelis MW (eds) Spider mites: their biology, natural enemies and control world crop pests. Elsevier, Amsterdam, pp 43–53
Seo SN (2011) An analysis of public adaptation to climate change using agricultural water schemes in South America. Ecol Econ 70(4):825–834
Smits PH (1996) Post-application persistence of entomopathogenic nematodos. Biocontrol Sci Tech 6(3):379–387
Stavrinides MC, Daane KM, Lampinen BD et al (2010) Plant water stress, leaf temperature, and spider mite (Acari: Tetranychidae) outbreaks in California vineyards. Environ Entomol 39(4):1232–1241
Stireman JO, Dyer LA, Janzen DH et al (2005) Climatic unpredictability and parasitism of caterpillars: implications of global warming. Proc Natl Acad Sci U S A 102(48):17384–17387
Tylianakis JM, Didham RK, Bascompte J et al (2008) Global change and species interactions in terrestrial ecosystems. Ecol Lett 11(12):1351–1363
Vázquez LL (2007) Contribución de la sequía a los desastres fitosanitarios causados por insectos y recomendaciones para mitigar sus impactos. Portal del Medioambiente. http://www.portaldelmedioambiente.com. Accessed: 30 Oct 2007

Vázquez LL (2011) Cambio climático, incidencia de plagas y prácticas agroecológicas resilientes. In: Labrada HR, Blandino DV, Funes-Monzote FR Innovación agroecológica, adaptación y mitigación del cambio climático. Instituto Nacional de Ciencias Agrícolas, San José de Las Lajas, pp 75–101

Voigt W, Perner J, Davis AJ et al (2003) Trophic levels are differentially sensitive to climate. Ecology 84(9):2444–2453

Wallner WE (1987) Factors affecting insect population dynamics: differences between outbreak and non-outbreak species. Annu Rev Entomol 32:317–340

Walstad JD, Anderson RF, Stambaugh WJ (1970) Effects of environmental conditions on two species of muscardine fungi (*Beauveria bassiana* and *Metarrhizium anisopliae*). J Inverebr Pathol 16(2):221–226

Part II
Bioecology of Natural Enemies Used in Biological Control in the Neotropical Region

Chapter 7
Predatory Insects

Brígida Souza, Terezinha Monteiro dos Santos-Cividanes, Francisco Jorge Cividanes, and Ana Luiza Viana de Sousa

7.1 Introduction

Ecosystems involve complex and dynamic interactions between living organisms and their environment, which include mutual actions of attack and defense between individuals of different trophic levels (Polis 1999; Ohgushi 2005). Interactions involved in predation and avoiding predation in each trophic level are reflected in the morphological, physiological, and behavioral characteristics of a predator (Moraes et al. 2000). Larvae of some genus of Chrysopidae, for example, make a habit of covering themselves with small plant fragments, body parts of its prey, and their own exuviae – known as trash-carriers or debris-carrying larvae – what serves as camouflage to protect themselves from their natural enemies. Instead, larvae of another genus of this family show a not trash-carrier habit, and they use other protecting strategies, such as the agility and production of repellent secretions.

Several predator species show attributes that confer them a potential to act as biological control agents of arthropods in agroecosystems. The understanding of these interactions is found to be the basis for the successful implementation of biological control methods (Costanza et al. 1997; Daily 1997; Groot et al. 2002; Andrade and Romeiro 2009).

B. Souza (✉)
Department of Entomology, Federal University of Lavras (UFLA), Lavras, MG, Brazil
e-mail: brgsouza@ufla.br

T. M. dos Santos-Cividanes
Laboratory of Plant Parasitology, Biological Institute, Ribeirão Preto, SP, Brazil
e-mail: terezinha@biologico.sp.gov.br

F. J. Cividanes
Instituto Biológico, Laboratório de Parasitologia Vegetal, Campinas, SP, Brazil

A. L. V. de Sousa
Department of Entomology, Rutgers University, New Brunswick, NJ, USA

B. Souza et al. (eds.), *Natural Enemies of Insect Pests in Neotropical Agroecosystems*, https://doi.org/10.1007/978-3-030-24733-1_7

This chapter approaches aspects related to predator-prey interactions in the context of biological control of pests and the role of plant diversity used as a strategy to the increase of this ecosystemic service, showing especially results obtained from researches developed in Latin American countries. In this chapter, the term predation is used to infer on interactions in which individuals of given species obtain their food by killing and consuming various organisms, referred to as prey, along their life cycle.

7.2 Search for Prey

Generalist predators commonly show complex behavioral mechanisms related to the search, choice, and eating of prey, which affect the control of pests. For the majority of predatory insect species, the capture of a potential prey occurs in an active way, involving speed and agility in the process of searching for food. These insects usually use their specialized mouthparts to capture the victim, as occurs, for example, with ants (Formicidae), carabids (Carabidae), and some heteropterous species (e.g., Anthocoridae), or by using their front legs modified to raptor appendages (e.g., Belostomatidae). Predatory insects that catch their prey during flight, such as Odonata and Asilidae, for example, have a modified thoracic structure for this purpose. On the other hand, predatory insects that usually employ a passive hunting strategy, behavior that requires them less energy expenditure, tend to wait for the approaching of the prey. Mantis (Mantodea) are examples of this hunting strategy since they use raptorial prothoracic legs to catch prey and antlions larvae (Myrmeleontidae), which build underground chambers where they wait for prey to pass by.

7.3 Encounter of the Prey Within the Habitat

For natural enemies to find and establish in a cultivated area, they should be attracted to the habitat colonized by pests. The attraction of natural enemies is usually directly or indirectly associated with the food requirements for adults or their progeny. They can use colors, format, and movement of flowers or chemical signals emitted by plants, which indicate herbivory, presence of flower resources, or any other type of attractive characteristic (Warren and James 2008; Maffei 2010; Hogg et al. 2011; Salamanca et al. 2015; Silva et al. 2017; Sousa et al. 2019). For example, when specimens of *Podisus nigrispinus* (Dallas) (Pentatomidae) were released to control the leaf worm on eucalyptus [*Thyrinteina arnobia* (Stoll) (Geometridae)], they preferred staying in *Eucalyptus pellita* F. Muell. infested by the pest than in other eucalyptus species, and the effect of kairomones was marked as the possible attractive factor for these natural enemies (Cavalcanti et al. 2000). Females of *Chrysoperla externa* (Hagen) (Chrysopidae) released in roses growing in protected environment were more attracted to rose plants infested by *Macrosiphum euphorbiae* (Thomas) (Aphididae), where they performed the oviposition, than for those not infested by the aphid (Salamanca et al. 2015). The predatory mirids, *Macrolophus basicornis* (Stål), *Engytatus varians* (Distant), and *Campyloneuropsis infumatus* (Carvalho), are guided by volatile compounds produced by tomato plants to find the prey *Tuta absoluta* (Meyrick) (Gelechiidae) (Silva et al. 2017). Another example of guidance for search or hunt for prey refers to females of *Orius insidiosus* (Say) (Anthocoridae), which are attracted by rose plants under single herbivory, infested by *Tetranychus urticae* Koch (Tetranychidae) or *Frankliniella insularis* (Franklin) (Thripidae), and under multiple herbivory (*T. urticae* + *F. insularis*) (Sousa et al. 2019).

7.4 Characteristics of the Habitat

Various environmental factors influence the diversity, abundance, and activity of predatory insects within agroecosystems, among those are highlighted the following: climate conditions; resources from the habitat, such as availability of food, reproduction, and refuge sites; inter- and intraspecific competition; and presence of other organisms. Climate factors, such as temperature, relative humidity, and photoperiod, affect the population dynamic of green lacewing species within agroecosystems. Population peaks of *Ceraeochrysa* species and *Chrysoperla externa* (Hagen) (Chrysopidae) occurred respectively in months with the lowest temperature and rainfall recorded in citrus grown in Lavras, Minas Gerais state, Brazil (Gitirana Neto et al. 2001; Souza and Carvalho 2002). The survey on the carabid populations conducted for two crop seasons of soybean and maize in Jaboticabal, São Paulo State, Brazil, showed that these beetle were more numerous in the crop season with the level precipitation 2.2 times higher, showing clearly that the relative humidity favored the increase of the population density of these insects (Cividanes 2002). Carabids are found to be different from other groups of insects by their high ability to rapidly disperse in the soil. This behavior makes them survive in adverse climatic conditions and to the negative effects resulting from agricultural activities, such as plowing and application of pesticides (Wallin 2002).

The physical structure of the habitat may be important in the selection process of microhabitats for predator and prey and is found to be a determining factor in the regulation of density of local populations (Lima 1998). The architecture and morphological characteristics of plant organs, for example, may influence predator-prey interactions since they show structures that provide shelter for both predators and prey. In addition, these characteristics may modify the microclimate created in the plant, also affecting the lives of other associated organisms.

Honek (2012) highlighted the importance of studies on the habitat of lady beetles (Coccinellidae) to provide basic information for ecology, ecophysiology, and biogeography, such as the determination of ecological niches, preference for food and microclimate, and factors limiting the distribution of these beetles.

7.5 Predator-Prey Interactions

Predator-prey interactions affect the behavior of individuals and may interfere in the predation rate. The simple presence of the predator, for example, may cause changes in the prey's behavior, making it lesser accessible to attack by the predator. The adaptive flexibility in the prey's behavior, such as agility for escaping, thanatosis and camouflage, as well as structural and physiological modifications, may impact the interaction in the effective predatory action (Lima 1998).

Search Ability and Handling Time The search ability and handling time of a prey by a predator are characteristics that affect the dynamics of populations. Knowing these attributes contributes to the estimation of the predator's participation in the species population dynamic and allows determining the density, from which the predator would not be effective in the pest control. The handling time and food consumption vary according to many factors, such as the size of the predator and prey, and the predator starvation state.

The functional response, described as the relationship between the predator consumption rate and the prey density, is found to be an important parameter on the determination of the efficiency of predators in the control of populations of pest arthropods (Solomon 1949). Changes in the behavior of predators in function of the development of its prey may be reflected in the functional response. The coccinellid *Tenuisvalvae notata* (Mulsant) showed the functional response type III when preying first instar nymphs of *Ferrisia virgata* Cockerell (Pseudococcidae) and functional response type II when preying third instar nymphs and adult females of this mealybug (Barbosa et al. 2014). A similar pattern was reported by Costa et al. (2017) when determining the functional response of *Stethorus tridens* Gordon (Coccinellidae) to various densities of the mite-pest *Tetranychus bastosi* Tuttle, Baker and Sales. The predator preferred eggs and larvae of the prey and then showed functional response type III. However, when preying nymphs and adults of the mite, the coccinellid showed functional response type II.

Food Preference Some predator generalist species own a narrow range of acceptable prey, while others cover a wide range of prey species. Generalist predators are supposed to benefit from the greater diversity of prey available in their natural environment. Chrysopidae larvae, for example, usually have food habits associated with the ecological niche where they occur and may feed on any kind of prey available in the environment. The locomotion speed and size of the prey are important criteria used by larvae of these predators, which commonly prefer smaller-sized, slow-motion, or sessile prey (Carvalho and Souza 2009).

Most of the coccinellids prefer preying hemipteran Sternorrhyncha insects, such as aphids, scales, whiteflies, and psyllids (Hodek and Honek 2009; Martins et al. 2009a, b; Obrycki et al. 2009; Hodek et al. 2012). Other species prefer mites, thrips, eggs, and neonate larvae of Lepidoptera and Coleoptera (Triltsch 1997; Biddinger et al. 2009; Evans 2009). The Australian ladybird beetle, *Cryptolaemus montrouzieri* (Mulsant), prefers preying scale insects of the families Pseudococcidae, Coccidae, Diaspididae, Ericoccidae, Ortheziidae, Margarodidae, and Aleyrodidae (Kairo et al. 2013).

Aphidophagous coccinellids are highly dependent on aphids as a food source for larval development (Hodek and Honek 1996) and oviposition (Evans 2000), although other types of food may supplement their requirements. Studies have shown that generalist predators may reach better performance or lay a great number of eggs when provided with different kinds of foods, which provide particular food requirements.

Predator food preference affects directly its efficiency in searching and apprehension of the prey (Xu and Enkegaard 2009), which is essential in the determination of the predatory potential when there is more than one species of pests in the crop (Enkegaard et al. 2001). Thus, the food preference gains importance in biological control programs once the target prey should be preferred in relation to other species eventually found in the system.

Suitability and Palatability of Prey Species Although generalist predators may prey on different types of prey, they may have their range of prey restricted due to physical, physiological, and behavioral characteristics. Chrysopid larvae, for example, search for prey with less sclerotized integument in order to facilitate their piercing.

Prey species used as food by predators may also not supply their nutritional and energetic requirements for growth, survival, and reproduction. The quality of an herbivore ingested affects the population growth rates of the natural enemy and may be assessed by means of the effect on the fitness parameters, such as survival, development time, and fecundity of adults. Laboratory studies showed that the diet of *Eriopis connexa* (Germar) (Coccinellidae) based on the aphid *Macrosiphum euphorbiae* (Thomas) (Aphididae) results in better development rate than when the diet is based on *Tetranychus evansi* (Baker and Pritchard) (Tetranychidae), because this mite species does not provide sufficient required nutrients for development and reproduction of this coccinellid species (Sarmento et al. 2004, 2007).

The host plant may affect indirectly the natural enemy, besides the direct effect of the prey on the bioecology of the predator. Nymphs of *Bemisia tabaci* (Gennadius) biotype B (Aleyrodidae) from the milky weed [*Euphorbia heterophylla* (L.)] used as food for larvae of *C. externa* caused reduction of the reproductive period, production (for about 50%), and viability (for about 60%) of eggs of the predator, when compared to nymphs fed on cucumber [*Cucumis sativus* (L.)] and cabbage [*Brassica oleracea* (L.)] (Silva et al. 2004).

Density of Prey Researches on feeding behavior of predatory arthropods show that, besides prey species and their development stage, predators may feed themselves from different quantities of them, depending on their density. Therefore, the relationship between density and the number of preys consumed is essential in the predator-prey dynamic. Higher prey availability may lead to increased consumption due to increased encounter opportunities. The number of specimens of *Schizaphis* *graminum* (Rondani) (Aphididae) and *Myzus persicae* (Sulzer) (Aphididae) preyed by *C. externa* larvae increased as the aphid availability increased up to reach a threshold, from which reduced gradually down to a stable limit (Fonseca et al. 2000; Barbosa et al. 2006, 2008). The positive relationship between the density of *M. euphorbiae* nymphs in rose plants and the consumption by third instar *C. externa* larvae confirms these findings. The average consumption was 21 specimens at a density of 40 aphids per plant and 100 specimens at a density of 160 aphids per plant (Gamboa et al. 2016). The second, third, and fourth instar larvae of *Scymnus* (Pullus) *argentinicus* (Weise) (Coccinellidae) showed a similar behavior when fed on *S. graminum* (Vieira et al. 1997). The same pattern was also reported by Santa-Cecília et al. (2001) for larvae of *Cycloneda sanguinea* (L.) (Coccinellidae) fed on *S. graminum*.

7.6 Intraguild Interaction

The use of more than one species of a natural enemy to control one or various herbivores is a strategy that aims at increasing the biological control; however, the association of them may result in negative or positive effects on the prey density (Cakmak et al. 2009). When two predators share the same prey, a negative effect may occur due to competition, resulting in the death of one of the competitors. Larvae of *C. externa* and *Ceraeochrysa cubana* (Hagen) (Chrysopidae) kept together under different densities guaranteed greater survival of the first species, regardless of the instar and proportion among them (Souza et al. 2008). According to Messelink et al. (2012, 2013), the intraguild predation may affect the biological control of the target pest, leading to an increase of the herbivore population.

Various specialist or generalist natural enemies may be associated with other predatory arthropods for the control of herbivores. Species of *Orius* (Anthocoridae) from Nearctic and Palearctic regions have been commercialized to be used in biological control programs as single or associated with other predatory arthropods to control populations of thrips in protected crops (Cloutier and Johnson 1993; Chow et al. 2008, 2010; van Houten et al. 2016). Some associations of *Orius* spp. and the predatory mite are recommended since anthocorids prefer thrips instead of a predatory mite. However, some associations still need to be well studied, because anthocorids may prefer another predator than the target species of the control. The association of *Orius laevigatus* (Fieber) and *Phytoseiulus persimilis* Athias-Henriot (Phytoseiidae) to control *T. urticae* in Brazil did not prevent the intraguild predation of *P. persimilis* by *O. laevigatus* (Venzon et al. 2001). On the other hand, there was an additive response to the consumption of *T. urticae* when the predators *O. insidiosus* and *Neoseiulus californicus* (McGregor) (Phytoseiidae) were kept together, resulting in a higher consumption of the spotted spider mite specimens than that observed for each one predator alone (personal communication, Ana Luiza Viana de Sousa). According to Muniz et al. (2014), the simultaneous presence of *C. externa* and *C. sanguinea* in monoculture of tomato (*Solanum lycopersicum* L.) (Solanaceae) resulted in greater consumption of *M. persicae* even the intraguild predation of *C. sanguinea* being observed. However, the cultivation of tomato intercropped with coriander (*Coriandrum sativum* L.) (Apiaceae) avoided the intraguild predation of *C. sanguinea* due to the increase of foraging area and reduction of encounter rates among predators.

Some generalist predators may feed on many prey species. The variability in the availability and quality of the food, especially proteins, the population density of predators, as well as the size of cospecific individuals in natural populations may frequently interfere in the cannibalism rate. This behavior, which is considered a nutritional or competitive strategy, is found to be a characteristic affecting the coexistence of species and the intraguild predation (Costa et al. 2003; Crumrine 2010). Rosenheim (1998) and Rosenheim et al. (1999) reported that, besides the intraguild predation caused by heteropterous, the density of larvae of *Chrysoperla carnea* (Stephens) (Chrysopidae) in cotton plants infested by *Aphis gossypii* Glöver (Aphididae) was negatively affected due to the escape behavior of the green lacewing.

7.7 Habitat Manipulation on Predator Ecology

Food resources, such as pollen, nectar, and prey present in the crop and in surrounding areas, should be suitable to provide survival, development, and reproduction of natural enemies and therefore promote their population growth and establishment (Janssen et al. 2007). Coccinellids, for example, can also consume honeydew and parts of plants such as pollen and nectar that provide them to survive during periods of prey shortage and reduce mortality during diapause periods or increase the energy gain for migration and reproduction (Figueira et al. 2003; Oliveira et al. 2004; Michaud and Grant 2005; Lundgren 2009; Weber and Lundgren 2009).

The increase in abundance of the coccinellids *C. sanguinea*, *H. convergens*, and *E. connexa*, in Seropédica, Rio de Janeiro State, Brazil, was correlated with the presence of coriander, dill (*Anethum graveolens* L.), and sweet fennel (*Foeniculum vulgare* Mill.) (Apiaceae), which are used as survival and reproduction sites for predators, and provides food resources, such as pollen, and prey, a shelter for larvae, pupae, and adults, as well as substrate for mating and oviposition (Medeiros et al. 2009; Lixa et al. 2010; Resende et al. 2010). Resende et al. (2015) reported that *E. connexa* larvae feeding on nectar and/or pollen of coriander reached the adult stage. In addition, these authors emphasized that supplying *M. persicae* and floral resources of coriander provided greater longevity for *E. connexa*, which also survived to feed on flower structures of coriander when the prey availability was scarce.

Natural enemies that can be benefited by plant diversification also include the minute pirate bugs *Orius* spp. Bugs of the genus *Orius* belong to natural enemies group that are benefited by plant diversity. The increased of population density of *O. insidiosus*, which is an omnivore insect, was related to the greater availability of refuge sites provided by coriander flowers and food resources, such as pollen and nectar, enabling them to reduce the populations of *Frankliniella* sp. (Thripidae) and *Thrips tabaci* Lindeman (Thripidae) associated with this Apiaceae (Resende et al. 2012).

Chrysopidae larvae have great ecological plasticity, and they are associated with a wide range of plant types, although there is some specialization regarding plant species (Freitas 2002). In different cropping systems, Resende et al. (2014) observed a trend to Chrysopidae community to be more abundant in areas with Poaceae crops. Larvae of *C. externa* were able to feed on elephant grass pollen (*Pennisetum purpureum* Schum) and to develop into adulthood, although larval stage extended and the pupal viability decreased (Oliveira et al. 2010).

Refuge sites that favor the occurrence of carabids are essential to the survival of these beetles within agroecosystems. For example, higher abundance and diversity of carabids may occur in soybean and maize crops grown under the no-tillage system and orange orchards with soil surface covered by spontaneous vegetation. This is due to no-tillage practice and mulching with vegetation that generally changes soil properties, making it more favorable for sheltering carabid beetles (Cividanes et al. 2009).

The introduction and maintenance of habitats that favor the occurrence and survival of natural enemies to increase the efficiency of these organisms in pest control is called conservative biological control (Barbosa 1998; Lee and Landis 2002). Generally, the increase in carabid abundance and diversity in agroecosystems has been achieved by establishing refuge areas by introducing hedges, grasses, or wildflower species into crop margins and by building beetle banks (Thomas et al. 2002; van Alebeek et al. 2007). Thomas et al. (1991) and Collins et al. (2002) reported that beetle banks introduced in cereal fields provided higher occurrence of carabids, staphylinids, and spiders in crops, reducing pest population density. In northern temperate zone, carabids can cross the winter season in stable habitats and colonize crops when conditions become favorable (Lövei and Sunderland 1996; Wissinger 1997). Due to this behavior, forests, hedges, windbreaks, grasslands, and prairies serve as refuge sites, enabling these predators to colonize crops and act as biological control agents (Pfiffner and Luka 2000; Lazzerini et al. 2007; Picault 2011).

In Latin America there is lack information on conservative biological control. In Brazil, Martins et al. (2009b) introduced herbaceous plants into soybean crops to verify their effectiveness as a refuge for carabids, reporting the highest density of *Scarites* specimens that occurred in an area with *Stylosanthes* spp. Correia (2017) reported that herbaceous plants increased the occurrence of soil-associated predatory insects in soybean, contributing to conservative biological control.

7.8 Other Factors Affecting the Bioecology of Predators

Exotic Predator While exotic natural enemies may compensate for the loss of native species richness, populations of these exotic organisms usually reach higher densities and have a greater impact on prey than native species. These natural enemies displace many native predators and can cause negative interference with autochthonous species (Evans and Toler 2007; Finke and Snyder 2010). Native South American coccinellid species have been displaced by the Asian ladybird *Harmonia axyridis* (Pallas) (Coccinellidae) due to intraguild predation and competition for resources (Koch 2003; Mirande et al. 2015). In southern Brazil, from 2004 to 2006, *H. axyridis* represented 38% of coccinellids caught in fruit trees. After a year, this species corresponded to 91% of the total specimen collected (Milléo et al. 2008). Laboratory studies carried out in Brazil and Argentina by Santos et al. (2009) and Mirande et al. (2015) reported the performance of *H. axyridis* as an intraguild predator, being the dominant species in competition with *E. connexa*.

Zoophagy Although antagonistic interactions between predators, such as predation, cannibalism, and escape, may reduce the impact on shared prey, there are divergent situations about that. In Miridae (Heteroptera), for example, the first instar is often phytophagous, acquiring predatory habits as the size and searching ability increase. Heteropteran species *Macrolophus pygmaeus* (Rambur) and *Nesidiocoris tenuis* (Reuter), which are commonly found simultaneously in protected tomato crops in the Mediterranean region, are the most efficient natural enemies of *B. tabaci*, *Trialeurodes vaporariorum* (Westwood) (Aleyrodidae), and lepidopteran larvae. However, when both predators occur simultaneously, adult females of the zoophytophagous species, *N. tenuis*, cause major damage to tomato plants (Moreno-Ripoll et al. 2012). The variability of the salivary gland enzymes of these predatory bugs has been suggested as an adaptation to omnivory (Torres and Boyd 2009).

Interactions with Other Organisms The mutualistic interaction between ant species (Formicidae) and aphids can reduce the performance of natural enemies. In addition to works such as those performed by Powell and Silverman (2010) and Rice and Eubanks (2013) in the United States of America, a work developed in Brazil by Canedo-Júnior et al. (2018) can be mentioned. This work reported a negative effect caused by aphid-tending ants on *C. externa* larvae by excluding them from the host plant. Details on these interactions are particularly approached in Chap. 42 of this book.

Phytosanitary Products Regarding the effect of phytosanitary products in tests of toxicity, selectivity, and ecological effects, predators are generally less susceptible and have more variable responses than parasitoids. However, researches carried out in Sweden over 24 years, seeking to assess the impact of phytosanitary products on populations in Carabidae family, showed changes in several characteristics of these predators, which are important in the control of *Rhopalosiphum* spp. (Aphididae). The widespread application of insecticides led to a reduction in activity and change in the composition of the community of carabids in the field (Rusch et al. 2013). The assessment of the impact of systemic insecticides used in sunflower seeds treatment showed a significant reduction in survival and fecundity of *C. carnea* adults due to the consumption of extra-floral nectar (Gontijo et al. 2014). Details on this specific matter are particularly approached in Chap. 37 of this book.

7.9 Final Considerations

Biological control is based on the mechanism of reciprocal density, in which natural enemies are agents of mortality dependent on the population density of their prey/hosts. This control method plays a key role in sustainable agricultural ecosystems as natural agents acting to balance the arthropod-pest populations.

Predatory insects of major importance in the agricultural pests' control are included in different taxonomic groups, which have bioecological characteristics that highlight them in the performance of this ecosystem service. Knowledge of these particularities is fundamental to the promotion and strengthening of the biological control as controlling agents are vulnerable to the effects of environmental factors, whether biotic or abiotic. However, even though they are the basis for crop pest management, it is clear that these natural factors of biotic regulation are relatively underexplored in Latin America countries and should be the target of research seeking to increase the effectiveness of this control method.

Acknowledgments The authors are grateful to the CNPq for granting scholarships to B. Souza and CAPES and FAPEMIG for financial support.

References

Andrade DC, Romeiro AR (2009) Serviços ecossistêmicos e sua importância para o sistema econômico e o bem-estar humano. IE/UNICAMP, Campinas. (Texto para Discussão, 155)

Barbosa P (1998) Conservation biological control. Academic Press, San Diego

Barbosa LR, Carvalho CF, Souza B et al (2006) Influência da densidade de *Myzus persicae* (Sulzer) sobre alguns aspectos biológicos e capacidade predatória de *Chrysoperla externa* (Hagen). Acta Sci Agron 28(2):221–225

Barbosa LR, Carvalho CF, Souza B et al (2008) Eficiência de *Chrysoperla externa* (Hagen, 1861) (Neuroptera: Chrysopidae) no controle de *Myzus persicae* (Sulzer, 1776) (Hemiptera: Aphididae) em pimentão (*Capsicum annum* L.). Ciênc Agrotec 32(4):1113–1119

Barbosa PRR, Oliveira MD, Giorgi JA et al (2014) Predatory behavior and life history of *Tenuisvalvae notate* (Coleoptera: Coccinellidae) under variable prey availability conditions. Fla Entomol 97:1026–1034

Biddinger DJ, Weber DC, Hull LA (2009) Coccinellidae as predators of mites: Stethorini in biological control. Biol Control 51(2):268–283

Cakmak I, Janssen A, Sabelis MW et al (2009) Biological control of an acarine pest by single and multiple natural enemies. Biol Control 50(1):60–65

Canedo-Júnior EO, Santiago GS, Ribas CR et al (2018) The effect size of aphid-tending ants in an agricultural tri-trophic system. J Appl Entomol 142(3):349–358

Carvalho CF, Souza B (2009) Métodos de criação e produção de crisopídeos. In: Bueno VHP (ed) Controle biológico: produção massal e controle de qualidade, 2nd edn. UFLA, Lavras, pp 77–115

Cavalcanti MG, Vilela EF, Eiras AE et al (2000) Interação tritrófica entre *Podisus nigrispinus* (Dallas) (Heteroptera: Pentatomidae), *Eucalyptus* e lagartas de *Thyrinteina arnobia* (Stoll) (Lepidoptera: Geometridae): I Visitação. An Soc Entomol Bras 29(4):697–703

Chow A, Chau A, Heinz KM (2008) Compatibility of *Orius insidiosus* (Hemiptera: Anthocoridae) with *Amblyseius degenerans* (Acari: Phytoseiidae) for control of *Frankliniella occidentalis* (Thysanoptera: Thripidae) on cut roses. Biol Control 44(2):259–270

Chow A, Chau A, Heinz KM (2010) Compatibility of *Amblyseius* (*Typhlodromips*) *swirskii* (Athias-Henriot) (Acari: Phytoseiidae) and *Orius insidiosus* (Hemiptera: Anthocoridae) for biological control of *Frankliniella occidentalis* (Thysanoptera: Thripidae) on roses. Biol Control 53(2):188–196

Cividanes FJ (2002) Efeitos do sistema de plantio e da consorciação soja-milho sobre artrópodes capturados no solo. Pesqui Agropecu Bras 37(1):15–23

Cividanes FJ, Barbosa JC, Ide S et al (2009) Faunistic analysis of Carabidae and Staphylinidae (Coleoptera) in five agroecosystems in northeastern São Paulo state, Brazil. Pesqui Agropecu Bras 44(8):954–958

Cloutier C, Johnson SG (1993) Predation by *Orius tristicolor* (Hemiptera: Anthocoridae) on *Phytoseiulus persimilis* (Acarina: Phytoseiidae): testing for compatibility between biocontrol agents. Environ Entomol 22(2):477–482

Collins KL, Boatman ND, Wilcox A et al (2002) Influence of beetle banks on cereal aphid predation in winter wheat. Agric Ecosyst Environ 93(1/3):337–350

Correia ET (2017) Diversidade, padrão de distribuição e sazonalidade de besouros associados ao solo em cultivo de soja e plantas herbáceas. Universidade Estadual Paulista, Tese

Costa RIF, Carvalho CF, Souza B (2003) Influência da densidade de indivíduos na criação de *Chrysoperla externa* (Hagen, 1861) (Neuroptera: Chrysopidae). Ciênc Agrotec:1539–1545. (edição especial)

Costa JF, Matos CHC, Oliveira CRF et al (2017) Functional and numerical responses of *Stethorus tridens* Gordon (Coleoptera: Coccinellidae) preying on *Tetranychus bastosi* Tuttle, Baker & Sales (Acari: Tetranychidae) on physic nut (*Jatropha curcas*). Biol Control 111:1–5

Costanza R, D'Arge R, Groot R et al (1997) The value of the world's ecosystem services and natural capital. Nature 387:253–260

Crumrine PW (2010) Size-structured cannibalism between top predators promotes the survival of intermediate predators in an intraguild predation system. J North Am Benthol Soc 29(2):636–646

Daily GC (1997) Nature's services. Societal dependence on natural ecosystems. Island Press, Washington, DC

Enkegaard A, Brodsgaard HF, Hansen DL (2001) *Macrolophus caliginosus*: functional response to whiteflies and preference and switching capacity between whiteflies and spider mites. Entomol Exp Appl 101(1):81–88

Evans EW (2000) Egg production in response to combined alternative foods by the predator *Coccinella transversalis*. Entomol Exp Appl 94(2):141–147

Evans EW (2009) Lady beetles as predators of insects other than Hemiptera. Biol Control 51(2):255–267

Evans EW, Toler TR (2007) Aggregation of polyphagous predators in response to multiple prey: ladybirds (Coleoptera: Coccinellidae) foraging in alfalfa. Popul Ecol 49(1):29–36

Figueira LK, Toscano LC, Lara FM et al (2003) Aspectos biológicos de *Hippodamia convergens* e *Cycloneda sanguinea* (Coleoptera: Coccinellidae) sobre *Bemisia tabaci* biótipo B (Hemiptera: Aleyrodidae). Bol San Veg Plagas 29:3–7

Finke D, Snyder WE (2010) Conserving the benefits of predator biodiversity. Biol Conserv 143(10):2260–2269

Fonseca AR, Carvalho CF, Souza B (2000) Resposta funcional de *Chrysoperla externa* (Hagen) (Neuroptera: Chrysopidae) alimentada com *Schizaphis graminum* (Rondani) (Hemiptera: Aphididae). An Soc Entomol Bras 292:309–317

Freitas S (2002) O uso de crisopídeos no controle biológico de pragas. In: Parra JRP, Botelho PSM, Corrêa-Ferreira BS et al (eds) Controle biológico no Brasil: parasitóides e predadores. Manole, São Paulo, pp 209–224

Gamboa S, Souza B, Morales R (2016) Actividad depredadora de *Chrysoperla externa* (Neuroptera: Chrysopidae) sobre *Macrosiphum euphorbiae* (Hemiptera: Aphididae) en cultivo de Rosa sp. Rev Colomb Entomol 42(1):54–58

Gitirana Neto J, Carvalho CF, Souza B et al (2001) Flutuação populacional de espécies de *Ceraeochrysa* Adams, 1982 (Neuroptera: Chrysopidae) em citros, na região de Lavras-MG. Ciênc Agrotec 25(3):550–559

Gontijo PC, Moscardini VF, Michaud JP et al (2014) Non-target effects of chlorantraniliprole and thiamethoxam on *Chrysoperla carnea* when employed as sunflower seed treatments. J Pest Sci 87(4):711–719

Groot R, Wilson M, Boumans R (2002) A typology for the classification description and valuation of ecosystem functions, goods and services. Ecol Econ 41(3):393–408

Hodek I, Honek A (1996) Ecology of coccinellidae. Kluwer Academic, London

Hodek I, Honek A (2009) Scale insects, mealybugs, whiteflies and psyllids (Hemiptera, Sternorrhyncha) as prey of ladybirds. Biol Control 51(2):232–243

Hodek I, Van Emden HF, Honek A (2012) Ecology and behaviour of the ladybird beetles (Coccinellidae). Wiley-Blackwell, Chichester

Hogg BN, Bugg RL, Daane KM (2011) Attractiveness of common insectary and harvestable floral resources to beneficial insects. Biol Control 56(1):76–84

Honek A (2012) Distribution and habitats. In: Hodek I, Van Emden HF, Honek A (eds) Ecology and behaviour of the ladybird beetles (Coccinellidae). Wiley-Blackwell, Chichester, pp 110–140

Janssen A, Sabelis MW, Magalhães S et al (2007) Habitat structure affects intraguild predation. Ecology 88(11):2713–2719

Kairo MTK, Paraiso O, Gautam R et al (2013) *Cryptolaemus montrouzieri* (Mulsant) (Coccinellidae: Scymninae): a review of biology, ecology, and use in biological control with particular reference to potential impact on non-target organisms. CAB Rev 8:1–20

Koch RL (2003) The multicolored Asian lady beetle, *Harmonia axyridis*: a review of its biology, uses in biological control, and non-target impacts. J Insect Sci 3:1–16

Lazzerini G, Camera A, Benedettelli S et al (2007) The role of field margins in agro-biodiversity management at the farm level. Ital J Agron 2(2):127–134

Lee JC, Landis DA (2002) Non-crop habitat management for carabid beetle. In: Holland JM (ed) The agroecology of carabid beetles. Intercept, Andover, pp 279–303

Lima SL (1998) Nonlethal effects in the ecology of predator-prey interactions. Bioscience 48(1):25–34

Lixa AT, Campos JM, Resende ALS et al (2010) Diversidade de Coccinellidae (Coleoptera) em plantas aromáticas (Apiaceae) como sítios de sobrevivência e reprodução em sistema agroecológico. Neotrop Entomol 39(3):354–359

Lövei GL, Sunderland KD (1996) Ecology and behavior of ground beetles (Coleoptera: Carabidae). Annu Rev Entomol 41(1):231–256

Lundgren JG (2009) Nutritional aspects of non-prey foods in the life histories of predaceous Coccinellidae. Biol Control 51(2):294–305

Maffei ME (2010) Sites of synthesis, biochemistry and functional role of plant volatiles. S Afr J Bot 76(4):612–631

Martins CBC, Almeida LM, Zonta-de-Carvalho R et al (2009a) *Harmonia axyridis*: a threat to Brazilian Coccinellidae? Rev Bras Entomol 53(4):663–671

Martins ICF, Cividanes FJ, Augusto T et al (2009b) Avaliação de área de refúgio para manipular populações de Carabidae (Coleoptera). Rev Bras Agroecol 4(2):1479–1483

Medeiros MA, Sujii ER, Morais HC (2009) Efeito da diversificação de plantas na abundância da traça-do-tomateiro e predadores em dois sistemas de cultivo. Hortic Bras 27(3):300–306
Messelink GJ, Sabelis MW, Janssen A (2012) Generalist predators, food web complexities and biological pest control in greenhouse crops. In: Larramendy ML, Soloneski S (eds) Integrated pest management and pest control: current and future tactics. InTech, Rijeka, pp 191–214
Messelink GJ, Bloemhard CMJ, Vellekoop R (2013) Biological control of aphids in the presence of thrips and their enemies. Biol Control 58(1):45–55
Michaud JP, Grant AK (2005) Suitability of pollen resources for the development and reproduction of *Coleomegilla maculata* (Coleoptera: Coccinellidae) under simulated drought conditions. Biol Control 32:363–370
Milléo J, Souza JMT, Barbola IF et al (2008) *Harmonia axyridis* em árvores frutíferas e impacto sobre outros coccinelídeos predadores. Pesqui Agropecu Bras 43(4):537–540
Mirande L, Desneux N, Haramboure M et al (2015) Intraguild predation between an exotic and a native coccinellid in Argentina: the role of prey density. J Pest Sci 88(1):155–162
Moraes CM, Lewis WJ, Tumlinson JH (2000) Examining plant-parasitoid interactions in tritrophic systems. An Soc Entomol Bras 29(2):189–203
Moreno-Ripoll R, Agustí N, Berruezo R et al (2012) Conspecific and heterospecific interactions between two omnivorous predators on tomato. Biol Control 62(3):189–196
Muniz CA, Togni PHB, Souza LM et al (2014) Coexistências de predadores generalistas mediada pela diversificação vegetal planejada. In: Anais do 25° congresso brasileiro de entomologia, Centro de Convenções, Goiânia, 4–18 set 2014
Obrycki JJ, Harwood JD, Kring TJ et al (2009) Aphidophagy by Coccinellidae: application of biological control in agroecosystems. Biol Control 51(2):244–254
Ohgushi T (2005) Indirect interaction webs: herbivore-induced effects through trait change in plants. Ann Rev Ecol Evol Syst 36:81–105
Oliveira NC, Wilcken CF, Matos CAO (2004) Ciclo biológico e predação de três espécies de coccinelídeos (Coleoptera, Coccinellidae) sobre o pulgão-gigante-do-pinus *Cinara atlantica* (Wilson) (Hemiptera, Aphididae). Rev Bras Entomol 48(4):529–533
Oliveira AS, Souza B, Auad AM et al (2010) Can larval lacewings *Chrysoperla externa* (Hagen) (Neuroptera, Chrysopidae) be reared on pollen? Rev Bras Entomol 54(4):697–700
Pfiffner L, Luka H (2000) Overwintering of arthropods in soils of arable fields and adjacent semi-natural habitats. Agric Ecosyst Environ 78(3):215–222
Picault S (2011) Functional biodiversity: natural regulation of aphid populations in lettuce crops. Infos Ctifl 275:27–35
Polis GA (1999) Why are parts of the world green? Multiple factors control productivity and the distribution of biomass. Oikos 86(1):3–15
Powell BE, Silverman J (2010) Impact of *Linepithema humile* and *Tapinoma sessile* (Hymenoptera: Formicidae) on the natural enemies of *Aphis gossypii* (Hemiptera: Aphididae). Biol Control 54(3):285–291
Resende ALS, Viana AJS, Oliveira RJ et al (2010) Consórcio couve-coentro em cultivo orgânico e sua influência nas populações de joaninhas. Hortic Bras 28(1):41–46
Resende ALS, Haro MM, Silva VF et al (2012) Diversidade de predadores em coentro, endro e funcho sob manejo orgânico. Arq Inst Biol 79(2):193–199
Resende ALS, Souza B, Aguiar-Menezes EL et al (2014) Influência de diferentes cultivos e fatores climáticos na ocorrência de crisopídeos em sistema agroecológico. Arq Inst Biol 81(3):257–263
Resende ALS, Ferreira RB, Silveira LCP et al (2015) Desenvolvimento e reprodução de *Eriopis connexa* (Germar 1824) (Coleoptera: Coccinellidae) alimentada com recursos florais de coentro (*Coriandrum sativum* L.). Entomotropica 30(2):12–19
Rice KB, Eubanks MD (2013) No enemies needed: cotton aphids (Hemiptera: Aphididae) directly benefit from red imported fire ant (Hymenoptera: Formicidae) tending. Fla Entomol 96(3):929–932
Rosenheim JA (1998) Higher-order predators and the regulation of insect herbivore populations. Annu Rev Entomol 43:421–447

Rosenheim JA, Limburg DD, Colfer RG (1999) Impact of generalist predators on a biological control agent, *Chrysoperla carnea*: direct observations. Ecol Appl 9(2):409–417

Rusch A, Bommarco R, Chiverton P et al (2013) Response of ground beetle (Coleoptera: Carabidae) communities to changes in agricultural policies in Sweden over two decades. Agric Ecosyst Environ 176:63–69

Salamanca J, Pareja M, Rodriguez-Saona C et al (2015) Behavioral responses of adult lacewings, *Chrysoperla externa*, to a rose-aphid-coriander complex. Biol Control 80:103–112

Santa-Cecília LVC, Gonçalves-Gervasio RCR, Torres RMS et al (2001) Aspectos bilógicos e consumo alimentar de larvas de *Cycloneda sanguinea* (Linnaeus, 1763) (Coleoptera: Coccinelidae) alimentadas com *Schizaphis graminum* (Rondani, 1852) (Hemiptera: Aphididae). Ciênc Agrotec 25(6):1273–1278

Santos NRP, Santos-Cividanes TM, Cividanes FJ et al (2009) Aspectos biológicos de *Harmonia axyridis* alimentada com duas espécies de presas e predação intraguilda com *Eriopis connexa*. Pesqui Agropecu Bras 44(6):554–560

Sarmento RA, Oliveira HG, Holtz AM et al (2004) Fat body morphology of *Eriopis connexa* (Coleoptera, Coccinellidae) in function of two alimentary sources. Braz Arch Biol Technol 47(3):407–411

Sarmento RA, Pallini A, Venzon M et al (2007) Functional response of the predator *Eriopis connexa* (Coleoptera: Coccinellidae) to different prey types. Braz Arch Biol Technol 50(1):121–126

Silva CG, Auad AM, Souza B et al (2004) Aspectos biológicos de *Chrysoperla externa* (Hagen, 1861) (Neuroptera: Chrysopidae) alimentada com *Bemisia tabaci* (Gennadius, 1889) Biótipo B (Hemiptera: Aleyrodidae) criada em três hospedeiros. Ciênc Agrotec 28(2):243–250

Silva DB, Weldegergis BT, Van Loon JJA et al (2017) Qualitative and quantitative differences in herbivore-induced plant volatile blends from tomato plants infested by either *Tuta absoluta* or *Bemisia tabaci*. J Chem Ecol 43(1):55–65

Solomon ME (1949) The natural control of animal populations. J Anim Ecol 18(1):1–35

Sousa ALV, Silva DB, Silva GG et al (2019) Behavioral response of the generalist predator *Orius insidiosus* to single and multiple herbivory by two cell content-feeding herbivores in rose plants. Arthropod Plant Interact. https://doi.org/10.1007/s11829-019-09729-5

Souza B, Carvalho CF (2002) Population dynamics and seasonal occurrence of adults of *Chrysoperla externa* (Hagen, 1861) (Neuroptera: Chrysopidae) in a citrus orchard in Southern Brazil. Acta Zool Acad Sci Hung 48:301–310

Souza B, Costa RIF, Tanque RL et al (2008) Aspectos da predação entre larvas de *Chrysoperla externa* (Hagen, 1861) e *Ceraeochrysa cubana* (Hagen, 1861) (Neuroptera: Chrysopidae) em laboratório. Cienc Agrotec 32(3):712–716

Thomas MB, Wratten SD, Sotherton NW (1991) Creation of "island" habitats in farmland to manipulate populations of beneficial arthropods: predator densities and emigration. J Appl Ecol 28(3):906–917

Thomas CFG, Holland JM, Brown NJ (2002) The spatial distribution of carabid beetles in agricultural landscapes. In: Holland JM (ed) The agroecology of carabid beetles. Intercept, Andover, pp 305–344

Torres JB, Boyd DW (2009) Zoophytophagy in predatory Hemiptera. Braz Arch Biol Technol 52(5):1199–1208

Triltsch H (1997) Gut contents in field sampled adults of *Coccinella septempunctata* (col: Coccinellidae). Entomophaga 42(1/2):125–131

Van Alebeek F, Visser A, Van den Broek R (2007) Field margins as (winter) refuge for natural enemies. Entomolog Ber 67:223–225

Van Houten YM, Hoogerbrugge H, Lenferink KO et al (2016) Evaluation of *Euseius gallicus* as a biological control agent of western flower thrips and greenhouse whitefly in rose. Nihon Dani Gakkai Shi 259(suppl):147–159

Venzon M, Pallini A, Janssen A (2001) Interactions mediated by predators in arthropod food webs. Neotrop Entomol 30(1):1–9

Vieira GF, Bueno VHP, Auad AM (1997) Resposta funcional de *Scymnus (Pullus) argentinicus* (Weise) (Coleoptera: Coccinellidae) a diferentes densidades do pulgão verde *Schizaphis graminum* (Rondani) (Homoptera: Aphididae). An Soc Entomol Bras 26(3):495–502
Wallin H (2002) Foreword. In: Holland JM (ed) The agroecology of carabid beetles. Intercept, Andover, pp xiii–xxiv
Warren J, James P (2008) Do flowers wave to attract pollinators? A case study with *Silene maritime*. J Evol Biol 21(4):1024–1029
Weber DC, Lundgren JG (2009) Assessing the trophic ecology of the Coccinellidae: their roles as predators and as prey. Biol Control 51(2):199–214
Wissinger SA (1997) Cyclic colonization in predictably ephemeral habitats: a template for biological control in annual crop systems. Biol Control 10(1):4–15
Xu X, Enkegaard A (2009) Prey preference of *Orius sauteri* between Western Flower Thrips and spider mites. Entomol Exp Appl 132(1):93–98

Chapter 8
Predator Mites

Maurício Sergio Zacarias, Erika Carla da Silveira, and Leopoldo Ferreira de Oliveira Bernardi

8.1 Introduction

The great behavioral diversity of mites gives them an important role in the environment, such as cycling of nutrients, transmission of phytopathogens, and regulation of populations of other living beings. Faced with this, man eventually uses these organisms to his advantage, especially in agriculture, where many species may be important auxiliary tools in the control of certain pests (Carrillo et al. 2015; Gerson et al. 2003; Krantz 2009; Walter and Proctor 2013). The use of mites in agriculture was applied from the mid-nineteenth century, and since then pest control through these organisms has undergone modifications aimed at improving the used methods (Carrillo et al. 2015; Gerson et al. 2003; Shimer 1868).

With regard to Latin America, several specific successful and key examples of mite pest control are shown in different sections on this book. Coffee, beans, cotton, fruit, and ornamental plants as in the sections about mass production of predatory mites were object of detailed description. In this section, we will deal with the main bioecological factors that strategically determine the success in Applied Biological Control (ABC) of arthropod pests with predatory mites.

M. S. Zacarias (✉)
Embrapa Coffee, Brazilian Agricultural Research Corporation,
PqEB, W/3 norte (at the end), Brasília, Federal District, Brazil
e-mail: mauricio.zacarias@embrapa.br

E. C. da Silveira · L. F. de Oliveira Bernardi
Federal University of Lavras (UFLA), Department of Entomology, Lavras, MG, Brazil

B. Souza et al. (eds.), *Natural Enemies of Insect Pests in Neotropical Agroecosystems*, https://doi.org/10.1007/978-3-030-24733-1_8

8.2 Some Historical Remarks

Bioecology is a contraction of bio + ecology (or biological + ecology). According to the Webster's New World College Dictionary (2017), it is the science that deals with the interrelations of communities of animals and plants with their environment. The Oxford English Dictionary cites its earliest use in the 1920s by Frederic Edward Clements (1874–1945).

Although historically mites play a relevant role as biological control agents, they are still poorly understood and underutilized for this purpose. Studies aimed at the use of these arthropods in pest control began in the mid-nineteenth century, having intensified in the last decades (Shimer 1868; Gerson et al. 2003; Carrillo et al. 2015). The pioneering work of Shimer in 1868 revealed the potential of *Hemisarcoptes malus* (Shimer) (Hemisarcoptidae) as a control agent of *Lepidosaphes ulmi* (L.) (Diaspididae) as it was able to significantly reduce the population of the pest (Gerson et al. 2003).

> […] This acarian is the most formidable enemy of the apple bark louse extant, and is far more effectual than all other combined restraints in nature, the most zealous efforts of man included … Many a patent nostrum, as well as the more public remedies— washes, substances deposited in holes in the trunk of the tree, etc., etc., have stolen all their reputation from this little, microscopic enemy. In these nostrums, etc., thousands of dollars in money and labor have been cheerfully paid, while this the true working remedy has up to within the last year, passed unobserved (Shimer 1868, p. 231). [https://babel.hathitrust.org/cgi/pt?id=umn.31951d00031547m;view=1up;seq=1].

The recognition of the benefit of *H. malus* for *L. ulmi* control led to the export of this natural enemy (NE) to Canada in 1917 in order to reduce populations of this same pest (Moraes 2002). Another mite that was believed to be an important predator and also underwent an export process was *Tyrolyphus phylloxerae* Riley (= *Rhizoglyphus echinopus*) (Acaridae), which was transferred from the United States to France, in order to control the phylloxera of the vine, *Daktulosphaira vitifoliae* (Fitch) (Phylloxeridae), a root pest of this Vitaceae. However, after its release on European soil, no effect on control of the target species was observed (Gerson et al. 2003).

8.3 Mites

It is important to consider every aspect of bioecology in clarifying how to establish a successful strategy using most effective NEs to achieve the objective to maintain an undesirable population in acceptable levels. Habitus, habits, and habitat must be the focus in those studies.

Habitus is the morphology related to the lifestyle and behavior. Krantz (2009) stated that the remarkable morphological diversity in Acari is reflected in the variety of ecological and behavioral patterns the mites have adopted. Because morphological specialization often is paralleled by specialization in choice of niche, an understanding of the ecology and behavior of mites is essential. This can lead us to consider that the lifestyle and the general appearance of a species habitus are correlated and useful in the search for an effective biological control.

Habit is the behavioral pattern. In zoology, habit is usually referred to the more or less predictable behavioral aspects. There is no concern in the fact that we need a predator behavior in a mite NE.

Habitat is the ecological area or environment where the organism communities live and prosper. The term usually refers to the zone where an organism can live and find food, shelter, protection, and mates. In the case of this section, plants especially the phylloplane Acari adopted a large variety of feeding habits, and due to its variety of habitats, they have adopted (predation, phytophagy, mycophagy, saprophagy, coprophagy, and necrophagy) (Evans 1992; Krantz 2009).

8.4 Mites Important to Agriculture

Mites can be divided into seven orders: Opilioacarida, Holothyrida, Ixodida, Mesostigmata, Trombidiformes, Sarcoptiformes, and Endeostigmata. However, three of these groups stand out with most of the species that are important to agriculture.

Trombidiformes One of the most important plant mites to agriculture ecosystems belongs to the order Trombidiformes (containing most of the phytophagous species). In the Trombidiformes order, the families Tetranychidae, Tarsonemidae, and Tenuipalpidae contain the most important phytophagous species. It should also be noted the families Stigmaeidae, Bdellidae, Cheyletidae, and Cunaxidae, consisting of group predatory species.

Sarcoptiformes The order Sarcoptiformes includes the important superfamily Eriophyoidea. Especially, plant feeder group also has some species that vector viruses.

Mesostigmata In this order, there are many families of mites with a predatory habit that can assist in pest management, acting in the control of soil pests such as Ascidae, Melicharidae, Blasttisociidae, Laelapidae, Macrochelidae, and Rodacaridae (Wallace et al. 1979; Walter and Ikonen 1989; Evans and Campbell 2003; Castilho et al. 2009; Carrillo et al. 2015; Moreira et al. 2015), or the phylloplane, such as the Ascidae family, for example, which contains species active in the control of pest mites as well as edaphic pests. The latter is a poorly explored family, requiring further studies involving the role of these predators in the agricultural system (Moraes et al. 2015).

Predatory mites belonging to the order Mesostigmata are the most important agents of biological control due to the wide diversity of habitus, habits, and habitats since they can feed on a large number of species of small arthropods and nematodes as well as using alternative sources such as pollen, fungi, and plant exudates, among others. Its prey capture strategies are diversified, which provides ample possibility of use of its species (Koehler 1999; Yaghoub and Bahador 2016). However, among all the families of Mesostigmata, the Phytoseiidae stands out.

Phytoseid Mites The mites most studied as biological control agents of pests belong to the family Phytoseiidae, included in the order Mesostigmata (Koehler 1999, Yaghoub and Bahador 2016).

Phytoseiid mites are primary plant-inhabiting mites (McMurtry et al. 2015), and they can be separated into four categories based on their lifestyles habits (McMurtry et al. 2013). Two of these categories (I and III) were subdivided into subcategories. As the lifestyles are related to the preying habitus, habits, and habitats, they achieve the best performance. These categories are a relevant tool and can give us insights on the performance of a species in preying. These categories are listed below (McMurtry et al. 2013):

- Type I phytoseiids are considered specialized predators of *Tetranychus* spp., and they are primarily from the *Phytoseiulus* genus. All life function of the predators takes place in prey colonies (McMurtry et al. 2015). This type is divided into tree subcategories, named: Ia (*Tetranychus* spp. profuse webbing producers) restrict to *Phytoseiulus* spp., Ib (*Tetranychus* spp. that produce a web-nest), and Ic (Tydeoid mite predators).
- Type II phytoseiids are selective predators of tetranychid mites, not only *Tetranychus* spp. but also a wide range of acceptable prey mites. The genus *Galendromus*, few *Neoseiulus*, and some *Typhlodromus* species account for this type. Notably, the species *Neoseiulus californicus* (McGregor) is important in the biological control of mites worldwide and has its origins in the new world.
- Type III phytoseiids are generalist predators with broad feeding habits. For example, many *Neoseiulus*, most *Amblyseius* and *Thyphlodromus* species, and many other genera are included in Type III, which has five subdivisions based on their habitats: IIIa (lives in hairy leaves), IIIb (glabrous leaves), IIIc (confined spaces on dicotyledonous plants), IIId (confined spaces on monocotyledonous plants, and IIIe (soil and humus habitats).
- Type IV includes pollen feeders and generalist phytoseiid predators.

Although extremely useful, these categories do not have an absolute boundary line between them. Further studies in bioecology of a particular species should be carried out to corroborate the type and help in achieving success.

In the taxonomic catalog of the phytoseiids, a database available in the Internet (Demite et al. 2018, http://www.lea.esalq.usp.br/phytoseiidae), from the 2479 valid species, we found 572 species that were accounted for the Latin-American and Caribbean region (almost 24%). If we have in mind that this number represents only a fraction of the real number of species, and that we know few bioecological aspects of only few in this plethora of species and there we still have much hard work to do, we can assume that categorizing them may help to reduce the risk in a biological control project and play a relevant role in the decision for the biological control as it.

Among these numerous species of Phytoseiidae mites, many have potential use as ABC agents but lack studies aimed at the regularization and commercialization of these. However, few companies commercialize predatory mites for biological control in Latin America and the Caribbean, some of them of European origin (e.g., Koppert and Biobest), and other Brazilian ones (e.g., Promip). These companies have almost the same portfolio of predatory mites, such as *N. californicus* (Koppert, Biobest, and Promip), *Phytoseiulus persimilis* Athias-Henriot (Koppert and Biobest), *N. cucumeris* (Oudemans), *Iphiseius degenerans* (Berlese) (Biobest), *Amblydromalus limonicus* (Garman & McGregor) (Koppert), *Amblyseius swirskii* Athias-Henriot (Koppert and Biobest), *Phytoseiulusm acropilis* (Banks), and *Stratiolaelaps scimitus* (Womersley) (Promip). In Peru, *Euseius stipulatus* (Athias-Henriot) is the only species commercialized and focuses on the control of *Panonychus citri* (Koch).

Studies demonstrate the potential of other Phytoseiidae species in the population reduction of pest mites and small arthropods found in agricultural crops. Among these species are *N. idaeus* Denmark & Muma, *E. concordis* (Chant), *A. herbicolus* (Chant), *E. alatus* DeLeon, *Typhodromus* spp. and *Iphiseiodes zuluagai* Denmark & Muma. Despite the proven efficacy by using laboratory or field experiments, most species have not yet aroused the commercial interest of agricultural input companies.

Although Phytoseiidae presents a relatively large number of species for use in biological control, there is a need to consider other information such as previous efficacy tests. According to Bellows et al. (1992), the impact of NEs on a given prey population can be determined through the use of life tables. This helps in the understanding of the population dynamics of a NE, as well as in estimating its efficacy in ABC.

Another feature to be taken into consideration when choosing a predator mite is its habitus, such as number, positioning, and size of dorsal setae. The setae may facilitate or hinder the movement of a species in a particular environment (habitat), especially when one of the main pests targeted for biological control are mites of the genus *Tetranychus* (Tetranychidae), which is composed of phytophagous species capable of producing web (Sabelis and Bakker 1992). In addition, the host plant's own morphology (habitat) may favor or offer resistance due to the presence and shape of trichomes, domatia, exudates, or other structures (Rezende et al. 2014).

The use of NEs, in order to meet the premises of integrated pest management (IPM), depends on its association with other control tactics, such as the chemical. Therefore, the use and the search for the improvement of less toxic and selective molecules should be prioritized since in many cultures a great number of predators can be found that, when kept in these places, help in the control of natural form. Poletti et al. (2008) reported that *N. californicus* is well tolerant to most products used in strawberry cultivation; on the contrary, *P. macropilis* is very sensitive to these same products. This information suggests that when using the chemical control associated with the biological for the strawberry crop, *N. californicus* should be used. Despite the difficulties that exist during the implementation of a strategy based on the use of mites as control agents in the Latin-American and Caribbean region, reducing the population density of pests with the use of mites still needs to be further advanced. However, there are countless researches that indicate the success of this type of action. Some of them use Phytoseiidae mites for controlling thrips [*Frankliniella occidentalis* (Pergande)], two spotted spider mite [*Tetranychus urticae* (Koch)] and whitefly [*Trialeurodes vaporariorum* (Westwood)] (Rodríguez-Reina et al. 1992; Muñoz et al. 2009), besides the use of species of other families, such as Laelapidae (*Cosmolaelaps jaboticabalensis* Moreira, Klompen, & Moraes) and Rhodacaridae (*Protogamasellopsis posnaniensis* Wiśniewski & Hirschmann), which can be used as soil control agents of species of insects of the families Sciaridae (*Bradysia matogrossensis* (Lane) and Thripidae (*F. occidentalis*), mites of the families Acaridae (*Tyrophagus putrescentiae* (Schrank) and *Rhizoglyphus echinopus* (Fumouze & Robin), and nematodes of the family Rhabditidae (*Protorhabditis* sp.) (Castilho et al. 2009; Moreira et al. 2015).

Another strategy considered is the introduction of non-native NEs, which should be studied and analyzed with great caution. The African mite, *A. swirskii*, used in Europe for the control of whiteflies and thrips was introduced in Brazil to control the whitefly *Bemisia tabaci* (Gennadius) (Aleyrodidae) (Cavalcante et al. 2015a). However, actions such as this are not necessary, since, among native species, many possible biological control agents have been found, as or better than those known in other countries. Examples include the phytoseiid mites *A. herbicolus*, *A. tamatavensis* Blommers, and *N. tunus* (De Leon), as well as mites from other families (Cavalcante et al. 2015b).

8.5 Final Remarks

Regarding ABC, the main challenge is the economic value added to the commercial production cycle and distribution of predatory mites. Latin America and the Caribbean include countries that are predominantly poor, underdeveloped, and economically unstable. Thus, logistics and investments required for this activity are unpredictable and very vulnerable to the risks arising from the regional economic and social condition due to the impossibility of long-term storage of mites. When demand is low, losses can occur. This may explain the predominance of a few large companies in the business, since they have a greater capacity to absorb the impacts of the oscillations of demand.

Predatory mites are currently a viable alternative in the substitution of pesticides for the ABC of some agricultural pests. Its use has been consolidated in recent years, especially with species of the family Phytoseiidae, but its potential is still well below estimated, especially in the Neotropical region where the diversity of predators is very large, and few species have been studied.

References

Bellows TSJ, Morse JG, Gaston LK (1992) Residual toxicity of pesticides used for control of lepidopteran insects in citrus to the predaceous mite *Euseius stipulatus* Athias-Henriot (Acarina, Phytoseiidae). J App Entomol 113(3):493–501

Carrillo D, Moraes GJ, Peña JE (2015) Prospects for biological control of plant feeding mites and other harmful organisms. Springer International Publishing, New York

Castilho RC, Moraes GJ, Silva ES et al (2009) Predation potential and biology of *Protogamasellopsis posnaniensis* Wisniewski & Hirschmann (Acari: Rhodacaridae). Biol Control 48(2):164–167

Cavalcante AC, Borges LR, Lourenção AL et al (2015a) Potential of two populations of *Amblyseius swirskii* (Acari: Phytoseiidae) for the control of *Bemisia tabaci* biotype B (Hemiptera: Aleyrodidae) in Brazil. Exp Appl Acarol 67(4):523–533

Cavalcante AC, Santos VL, Rossi LC et al (2015b) Potential of five brazilian populations of phytoseiidae (Acari) for the biological control of *Bemisia tabaci* (Insecta: Hemiptera). J Econom Entomol 108(1):29–33

Demite PR, Moraes GJ, McMurtry JÁ et al (2018) Phytoseiidae database. www.lea.esalq.usp.br/phytoseiidae. Accessed 2 Nov 2018

Evans GO (1992) Principles of acarology. CAB International, Wallingford

Evans D, Campbell NJH (2003) Exotic vs endemic biocontrol agents: would the real *Stratiolaelaps miles* (Berlese) (Acari: Mesostigmata: Laelapidae), please stand up? Biol Control 26(2):253–269

Gerson U, Smiley RL, Ochoa R (2003) Mites (Acari) for pest control. Blackwell Science, Oxford

Koehler HH (1999) Predatory mites (Gamasina, Mesostigmata). Agric Ecosyst Environ 74(1/3):395–410

Krantz GW (2009) Habits and habitat. In: Krantz GW, Walter DE (eds) A manual of acarology. Texas Tech University Press, Lubbock, pp 64–82

McMurtry JA, Moraes GJ, Sourassou NF (2013) Revision of the lifestyles of phytoseiid mites (Acari: Phytoseiidae) and implications for biological control strategies. Syst Appl Acarol 18(4):297–320

McMurtry JA, Sourassou NF, Demite PR (2015) Thephytoseiidae (Acari: Mesostigmata) as biological control agents. In Carrillo D, Moraes GJ de, Peña EJ (eds) Prospects for biological control of plant feeding mites and other harmful organisms. Springer International, Cham, pp 133–149

Moraes GJ (2002) Controle biológico de ácaros fitófagos com ácaros predadores. In: Parra JRP, Botelho PSM, Corrêa-Ferreira BS et al (orgs) Controle biológico no Brasil. Manole, Barueri, pp 225–237

Moraes GJ, Venancio R, Santos VLV et al (2015) Potential of ascidae, blattisociidae and melicharidae (Acari: Mesostigmata) as biological control agents of pest organisms. In: Carrillo D, Moraes GJ, Peña JE (eds) Prospects for biological control of plant feeding mites and other harmful organisms. Springer International, New York, pp 36–76

Moreira GF, Morais MR, Busoli AC et al (2015) Life cycle of *Cosmolaelaps jaboticabalensis* (Acari: Mesostigmata: Laelapidae) on *Frankliniella occidentalis* (Thysanoptera: Thripidae) and two factitious food sources. Exp Appl Acarol 65(2):219–226

Muñoz K, Fuentes L, Cantor F et al (2009) Preferencia alimenticia del ácaro depredador *Balaustium* sp. en condiciones controladas. Agron Colomb 27(1):95–103

Poletti M, Collette P, Omoto C (2008) Compatibilidade de agrotóxicoscom os ácaros predadores *Neoseiulus californicus* (McGregor) e *Phytoseiulus macropilis* (Banks) (Acari: Phytoseiidae). BioAssay 3(3):1–14

Rezende JM, Lofego AC, Nuvoloni FM et al (2014) Mites from Cerrado fragments and adjacent soybean crops: does the native vegetation help or harm the plantation. Exp Appl Acarol 64(4):501–518

Rodríguez-Reina JM, García-Marí F, Ferragut F (1992) Actividad depredadora de varios ácaros fitoseidos sobre distintos estados de desarrollo del trips de las flores *Frankliniella occidentalis* (Pergande). Bol San Veg Plagas 18:253–263

Sabelis MW, Bakker FM (1992) How predatory mites cope with the web of their tetranychid prey: a functional view on dorsal chaetotaxy in the Phytoseiidae. Exp Appl Acarol 16(3):203–225

Shimer H (1868) The apple bark-louse in 1866. Birds vindicated from the charge preferred against them by the State Entomologist. Trans Illinois St Hort Soc 2:227–233

Wallace MMH, Tyndale-Bidcoe M, Holm E (1979) The influence of *Macrocheles glaber* on the breeding of the Australian bush fly, *Musca vetustissima*, in cow dung. In: Rodriguez JG (ed) Recent advances in acarology. Academic Press, London, pp 217–222

Walter DE, Ikonen EK (1989) Species, guilds, and functional groups: taxonomy and behavior in nematophagous arthropods. J Nematol 21(3):315–327

Walter DE, Proctor HC (2013) Mites: ecology, evolution & behaviour. Springer Netherlands, Dordrecht

Webster's New World College Dictionary (2017). http://www.yourdictionary.com/bioecology. Accessed 2 Nov 2017

Yaghoub F, Bahador M (2016) Mite predators. In: Omkar (ed) Ecofriendly pest management for food security. Academic Press, London, pp. 329–366

Chapter 9
Parasitoid Insects

Luis Cláudio Paterno Silveira, Ivana Lemos Souza, Vitor Barrile Tomazella, and Heisler Alexsander Gomez Mendez

9.1 Introduction

The term parasitoid can be defined as an organism that develops at the expense of a single host, nurturing from it and leading to its death, directly or indirectly, as the cause of its development. In practical terms we will be restricted to parasitoid insects (to the detriment of other groups, such as some Nematoda and fungi that fit the same definition), and we will focus on insects as hosts, since it is known that they also use several other classes of the phylum Arthropoda as hosts, such as Chilopoda, Diplopoda and Arachnida, as well as other phyla, such as Platyhelminthes, Mollusca (mainly Gastropoda) and Annelida (Eggleton and Gaston 1990). Also we will not consider here the cases where the host suffers only castration by the "parasitoid", nor the species that attack oothecae and feed on several embryos (considered predators), nor those species where a single individual is able to dominate the nest of a social insect, called "nest parasitoids". Finally, in terms of terminology, we prefer to use the term parasitism rather than parasitoidism to refer to the action of a parasitoid, since theoretically the correct would be parasite/parasitism and parasitoid/parasitoidism. However, this term is rarely used in entomology publications.

It is very difficult to determine the exact number of insects that exhibit parasitoid behaviour at some point in their life, or during all of it, but it is accepted that today about 10% of the described species of insects show parasitoid behaviour (Eggleton and Belshaw 1992) which would result in more than 100 thousand species. Therefore, this behaviour or way of life is very important in terms of species irradiation, which results in the great adaptation of this group to the most diverse terrestrial and, in some cases, aquatic habitats (Hanson and Gauld 2006).

L. C. P. Silveira (✉) · I. L. Souza · V. B. Tomazella
Federal University of Lavras (UFLA), Department of Entomology, Lavras, MG, Brazil
e-mail: lcpsilveira@ufla.br

H. A. G. Mendez
Protecciőn de Plantas Department, Universidad de San Carlos de Guatemala, Guatemala, Guatemala

B. Souza et al. (eds.), *Natural Enemies of Insect Pests in Neotropical Agroecosystems*, https://doi.org/10.1007/978-3-030-24733-1_9

There are families with parasitoid representatives in five orders of Insecta: 1 family in Neuroptera, 2 in Lepidoptera, 11 in Coleoptera, 21 in Diptera and 65 in Hymenoptera. Diptera and Hymenoptera are the most important, with 16,000 and 95,000 species already identified, respectively. For some reasons the order Hymenoptera clearly stands out from the rest: first, by the number of species described, since Hymenoptera correspond to 78% of all species mentioned as parasitoids, the remaining 22% being concentrated in Diptera and Coleoptera. Second, they present a perforating ovipositor, allowing them to deposit the eggs inside hosts, even in hidden places, as galls, and to explore small resources such as eggs of other insects (Gauld and Bolton 1996). Third, they have venom accessory glands, which allow them to dominate and subdue very active hosts and control their physiology in favour of their progeny. Finally, hymenopteran parasitoids are haplodiploid, allowing females to control the sex ratio of their progeny, which leads to less competition and adjustment according to the size of the exploited resource (Godfray and Shimada 1999). Due to these characteristics, the importance of Hymenoptera parasitoid as regulators of agricultural pests in many biological control programs applied throughout the world is undeniable.

Despite this, Diptera parasitoids are important in the regulation of many relevant species of insect pests, and it is estimated that they represent about 20% of all species with this behaviour. This indicates that the group is possibly underutilized as a pest regulator when compared to the use of Hymenoptera (Feener Jr and Brown 1997). On the other hand, the order Diptera presents a much wider range of adaptations and behaviours to exert its parasitism, allowing much more advanced studies in terms of evolution of parasitic behaviour (Gullan and Cranston 2017). For example, in Diptera there are structural adaptations, as for oviposition, ranging from a false ovipositor, where the abdominal segments just extend, to a perforating ovipositor, there are also structures like the respiratory funnels made by the larvae inside the host to avoid encapsulation and death by asphyxia. There are also changes in behaviour, such as the existence of planid larvae in many species, which are responsible for actively penetrating the body of the host, and host location detection by sounds emitted by it. Therefore, it is a very rich group for several important studies to understand parasitoid behaviour.

As for the other orders with parasitoid representatives, their applied importance is lower, but its study has revealed interesting aspects regarding the evolution of the parasitoid behaviour that, unlike what occurred in Hymenoptera (this behaviour appeared only once during the evolution of the group), arose independently for each family where this lifestyle is found, that is, 21 times in Diptera and 11 in Coleoptera.

9.2 General Characteristics

The parasitoids developed different strategies of parasitism along the coevolution with their hosts. According to the preference for hosts and the way in which they explore them during their life, a possible classification is as follows:

(a) *Host diversity* (Marshall 1981; Jameson 1985; Hofstede et al. 2004)

Monoxenes – group of parasitoids whose development cycle is restricted to only one species of host
Oligoxenes – when they are restricted to more than one species, but within the same genus
Pleioxenes – restricted to hosts within a single host family
Polyxenes – use hosts belonging to two or more families

(b) *Manipulation of the host*

Coinobionts – group of parasitoids that partially paralyze the hosts for oviposition, in such a way that the target is considered sessile, but soon recover its movements. The host then continues its development and is killed when the parasitoid reaches maturity, which usually happens in a few days (Godfray and Shimada 1999). Parasitoids with this behaviour differ very little, in fact, from predators, who devour their prey usually immediately after the attack, leading to their instantaneous death (Gullan and Cranston 2017). They can be divided into: ectoparasitoids, developing externally to the host, and endoparasitoids (Fig. 9.1), which develop internally in the host (Gordh and Headrick 2001).
Idiobionts – a group of parasitoids that prevent the future development of the host after initial paralysis. They usually attack hosts in their naturally immobile stage (e.g. eggs, pupae, immobile larvae and nymphs) (Gordh and Headrick 2001).

(c) *Preference for hosts parasitized or not*

Primary parasitoid – individuals that seek out non-parasitized hosts and develops on them (Gullan and Cranston 2017). The species with this behaviour are almost always selected to compose biological pest control programs, since they are able to locate healthy hosts (usually species known as agricultural pests), discriminating those that are already hosting other parasitoids, evaluating their size, sanity and nourish condition for their progeny, to finally parasite them and regulate their population.

Fig. 9.1 Endoparasitoid *Palmistichus elaeisis* (Eulophidae) inserting its ovipositor inside the pupae of *Tenebrio molitor* (Tenebrionidae). (Photo: Silveira LCP 2018)

Primary parasites, however, often confront the immune defence system of the host and may undergo the process called encapsulation, when haemolymph cells (haemocytes), as well as the pigment melanin, clump together on the egg or larva of the parasitoid, suffocating them (Salt 1963; Nappi 1975; Blumberg 1997). However, by the process of coevolution, many parasitoids developed mechanisms to avoid these defences (Salt 1968), such as:

- *Evasion*, a typical case of ectoparasitoids, which are not involved in the haemolymph of the host, as well as egg parasitoids, since eggs have no immune response
- *Molecular mimicry*, when the larva of the parasitoid produces and is surrounded in substances similar to the proteins of the host, passing unnoticed
- *Viral suppression*, when the parasitoid introduces into the host one or more viruses that deactivate its immune system
- *Destruction*, which can occur through vigorous feeding of host tissues by the larva of the parasitoid, or by the engorgement of haemolymph and defence cells, leading to the rapid weakening of the host, making it impossible to defend itself
- *Subversion*, divided into two possibilities: first, when the parasitoid maintains the host cell coverage used in the encapsulation, and the larva only opens a food channel, nourishing itself without the host noticing, and second, when giant cells (teratocytes) arise from fragments of the parasitoid egg and disrupts the host's immune system

As we can see, there are several ways for parasitoids to avoid being encapsulated and, as a rule, the older the host (e.g. caterpillars in more advanced instars), the greater the encapsulation. Therefore, for the regulation of a particular pest, it is important that the female parasitoid correctly discriminate the species, size, nutritional condition and age of the host, so that its progeny are successful, and biological control is effective.

Secondary parasitoid – develop on a primary parasitoid. This type of preference is also called hyperparasitism, found in many species, which have adapted to explore not the primary host (the organism being biologically regulated), but rather its primary parasitoid, leading to its death and impairing biological control programs. The host will fatally die, but there will be no progeny of the primary parasitoid, allowing future generations of the insect pest to escape control. Hyperparasitoids may also place their eggs inside or outside the larva of the primary parasitoid, receiving the following denominations:

- *Endophagous*, when the hyperparasitoid places its eggs inside the primary parasitoid
- *Ectophagous*, when it deposits the eggs on the surface of the larva of the primary parasitoid within the host being regulated (Fig. 9.2)

Facultative parasitoids – may act as a primary parasitoid, attacking healthy hosts or, optionally, attack the primary parasitoid within the already parasitized host (Salt 1968)

Fig. 9.2 Example of an ectophagous hyperparasitoid, *Asaphes* sp. (Pteromalidae) ovipositing on the larva of a primary Braconidae parasitoid within a mummified aphid *Lipaphis erysimi*. (Photo: Silveira LCP 2012)

9.3 Families of Hymenoptera and Diptera Parasitoids

In Hymenoptera currently 63 families belonging to 15 superfamilies present insect or spider parasitoids. The total number is bigger because some families also have predatory or phytophagous behaviour. One is considered a predator, not a parasitoid, and one is considered completely phytophagous, included in Table 9.1. The order Diptera has 16 families with insect or spider parasitoid species, mentioned in Table 9.2 with their respective primary hosts. It is quite common to find Diptera as parasitoids of other animal groups, not included here.

9.4 Foraging and Oviposition Behaviour in Parasitoids

To parasitize the host, a parasitoid first locates the host's potential habitat, locates the appropriate host and finally oviposits (Doutt 1964; Vinson 1975). Females of many species of parasitoids can use resources for feeding or oviposition throughout adult life. Their choices are mediated by external stimuli such as odours, tastes, colours and shapes and the individuals can be continuously influenced as they learn these stimuli during foraging. Successful foraging can increase longevity and search ability and lead to increased fecundity (Baggen and Gurr 1998; Winkler et al. 2006). Female parasitoids need to detect and respond to sensory signals indicating the occurrence of both host and food resources to achieve reproductive success (Schroeder and Hilker 2008).

Plants provide a variety of defences, which help to reduce the damage of insect pests in agroecosystems. They produce different important substances in the foraging process, such as volatile organic compounds and plant volatiles induced by herbivory (De Moraes et al. 1998). Such substances act through the detection of plant odours by parasitoids and consequently increase their foraging efficiency. The response to external stimuli in host locations for oviposition or food depends on the physiological state in which each female is (Lewis and Takasu 1990). In this way, when well fed they expand their foraging and reproduction capacity (Wäckers 1994).

Table 9.1 Superfamilies and families of hymenopteran parasitoids and their primary hosts

Superfamily/family	Primary hosts
1. *Apoidea*	
1.Sphecidae	Araneae, Orthoptera, larvae of Lepidoptera and Hymenoptera
2. *Ceraphronoidea*	
2. Ceraphronidae	Diptera, Hemiptera, Neuroptera, Thysanoptera
3. Megaspilidae	Coccoidea, Neuroptera, pupae of many Diptera
3. *Chalcidoidea*	
4. Aphelinidae	Hemiptera Sternorrhyncha; eggs of Lepidoptera, Hemiptera and Orthoptera
5. Chalcididae	Pupae of Lepidoptera; larvae of Diptera
6. Elasmidae	Larvae of Lepidoptera
7. Encyrtidae	Hemiptera, Coleoptera, Diptera, Lepidoptera
8. Eucharitidae	Larvae of Formicidae (Heraty 2002)
9. Eulophidae	Lepidoptera, Coleoptera, Hymenoptera, Diptera
10. Eupelmidae	Larvae and pupae of Coleoptera, Diptera and Lepidoptera
11. Eurytomidae (PH/PA)	Larvae and pupae of Coleoptera, Diptera and Lepidoptera
12. Leucospidae	Larvae and pupae in nests of Apoidea, Vespidae and Sphecidae
13. Mymaridae	Eggs of Auchenorrhyncha, Heteroptera, Coleoptera, Orthoptera and Diptera
14. Ormyridae	In galls of Diptera and Hymenoptera (Hanson 1992)
15. Perilampidae	Larvae of Tachinidae, Ichneumonoidea and Coleoptera
16. Pteromalidae (PH/PA/PR)	Coleoptera, Diptera, Lepidoptera, Araneae, Blattodea, Dermaptera, Hemiptera, Neuroptera and Hymenoptera
17. Rotoitidae	Unknown, possibly insect eggs (Bouček 1987)
18. Signiphoridae	Coccidae, Aleyrodidae, Aphididae and Psyllidae
19. Tanaostigmatidae (PH/PA)	Larvae of Cynipidae
20. Tetracampidae (PH/PA)	Larvae of Diptera, eggs of Diprionidae and Coleoptera
21. Torymidae (PH/PA)	Larvae and pupae of Diptera, nests of Hymenoptera, larvae of Lepidoptera
22. Trichogrammatidae	Eggs of Holometabola and of Hemiptera, Orthoptera and Thysanoptera
4. *Chrysidoidea*	
23. Bethylidae	Cryptic larvae of Coleoptera and microlepidoptera
24. Chrysididae	Larvae and prepupae of Symphyta, eggs of Phasmatodea
25. Scolebythidae	Larvae of woodborer Coleoptera
26. Sclerogibbidae	Nymphs and adults of Embioptera
27. Embolemidae	Nymphs of some of a few Auchenorrhyncha
28. Dryinidae	Nymphs of Auchenorrhyncha
5. *Cynipoidea*	
29. Austrocynipidae	Larvae of Lepidoptera in *Araucaria* sp.

Table 9.1 (continued)

Superfamily/family	Primary hosts
30. Cynipidae (PH)	Gall formers or invaders
31. Figitidae	In galls of Cynipidae and Chalcidoidea, larvae of Chrysopidae, Hemerobiidae and Diptera
32. Ibaliidae	Larvae of Siricidae
33. Liopteridae	Larvae of Buprestidae and Cerambycidae
6. *Evanioidea*	
34. Aulacidae	Larvae of Buprestidae, Cerambycidae and Xiphydriidae
35. Evaniidae (PRE)	Ootheca of Blattodea
36. Gasteruptiidae (PAR/PRE)	Nests of Apoidea and solitary wasps
7. *Ichneumonoidea*	
37. Braconidae (FI/PAR)	Larvae of Lepidoptera, Coleoptera, Diptera and Hymenoptera, nymphs and adults of Aphididae
38. Ichneumonidae	Eggs, larvae and pupae of Lepidoptera, Coleoptera, Diptera, Hymenoptera, Symphyta, Raphidioptera and Trichoptera, eggs and adults of Araneae
39. Apozygidae	Host unknown
8. *Megalyroidea*	
40. Megalyridae	Larvae of Bostrichidae, Buprestidae and Cerambycidae
9. *Mymarommatoidea*	
41. Mymarommatidae	Polyporales fungi (shelf fungi), possibly insect eggs (unconfirmed)
10. *Platygastroidea*	
42. Platygastridae	Eggs of Orthoptera, Mantodea, Coleoptera, Hemiptera and Arachnida, Cecidomyidae larvae, Psodococcidae nymphs and Aleyrodidae
11. *Proctotrupoidea*	
43. Austroniidae	Host unknown
44. Diapriidae	Larvae of Formicidae, terrestrial and aquatic Diptera
45. Heloridae	Larvae of Neuroptera
46. Maamingidae	Host unknown
47. Mesoserphidae	From Mesozoic fossils only
48. Monomachidae	Larvae of Stratiomyidae (Chiromyzinae) and Muscidae
49. Pelecinidae	Larvae of Scarabaeidae
50. Proctorenyxidae	Host unknown
51. Peradeniidae	Host unknown
52. Proctotrupidae	Larvae of Coleoptera and Mycetophilidae
53. Roproniidae	Symphyta pupae
54. Vanhorniidae	Larvae of Eucnemidae
12. *Stephanoidea*	
55. Stephanidae	Several woodborers Coleoptera and Siricidae

(continued)

Table 9.1 (continued)

Superfamily/family	Primary hosts
13. *Trigonaloidea*	
56. Trigonalidae	Larvae of Vespidae, Ichneumonidae and Tachinidae
14. *Vespoidea*	
57. Bradynobaenidae	Solifugae adults (Arachnida)
58. Mutillidae	Aculeate Hymenoptera (Apidae, Halictidae, Crabronidae, Megachilidae, Sphecidae, Pompilidae)
59. Pompilidae	Adults of Araneae
60. Rhopalosomatidae	Adults of Gryllidae
61. Sapygidae	Larvae of Vespidae, Apoidea
62. Scoliidae	Larvae of Scarabaeidae
63. Sierolomorphidae	Host unknown
64. Tiphiidae	Larvae of Cerambycidae, Carabidae, Curculionidae and Scarabaeidae in the soil
15. *Orussoidea*	
65. Orussidae	Larvae and pupae of woodborers (Buprestidae, Cerambycidae, Xiphydriidae) and Siricidae

Some families are parasitoids and predators, indicated by the acronym PA/PR, others are phytophagous and parasitoids (PH/PA), and one is considered predator (PR) and other phytophagous (FI). Host information without reference in the right column refers to Goulet and Huber (1993) and Hanson and Gauld (2006)

Table 9.2 Diptera families with insect or spider parasitoid representatives and their primary hosts

Family	Primary host
1. Acroceridae	Araneae (Schlinger 1987)
2. Anthomyiidae	Nymphs and adult Orthoptera
3. Asilidae	Scarabaeidae and Xylocopidae (Knutson 1972)
4. Bombyliidae	Hymenoptera, Coleoptera, Diptera, Neuroptera, Orthoptera and Lepidoptera (Davis 1919, Hull 1973)
5. Cecidomiidae	Hemiptera: adult of Aphididae and Psyllidae, nymphs of Tingidae (Eggleton and Belshaw 1992)
6. Chloropidae	Chrysomelidae eggs, larvae of Coleoptera and Tortricidae
7. Conopidae	Hymenoptera (Askew 1971)
8. Cryptochetidae	Coccoidea nymphs (Eggleton and Belshaw 1992)
9. Empididae	Trichoptera (Eggleton and Belshaw 1992)
10. Nemestrinidae	Acrididae and Scarabaeidae (Ricther 1997)
11. Pipunculidae	Auchenorryncha (Waloff 1975) and Tipulidae (Skevington 2005)
12. Phoridae	Hymenoptera, Diptera, Coleoptera, Isoptera (Disney 1994)
13. Pyrgotidae	Scarabaeidae (Davis 1919)
14. Rhinophoridae	Isoptera (Sutton 1980)
15. Sarcophagidae	Lepidoptera, Auchenorryncha, Sternorrhyncha, Coleoptera, Orthoptera, Diptera and Hymenoptera (Pape 1990, Eggleton and Belshaw 1992)
16. Tachinidae	Hymenoptera, Coleoptera, Hemiptera (Wood 1987)

Although some parasitoids feed on host larvae for protein, most of them feed also on carbohydrate and protein sources as floral pollen, nectar and honeydew harvested from Hemiptera (Jervis et al. 1996). In the case of nectar and pollen, diversified agricultural landscapes could provide these food resources to parasitoids, through an ecological engineering plan in order to have plant diversification within and around the crops. The ingestion of carbohydrates derived from flowers allows the parasitoids to deposit more eggs in the hosts while feeding freely in the fields (Lee and Heimpel 2008).

In contrast, in the absence of carbohydrate sources, females reabsorb the eggs and redirect energy for survival, thus reducing fertility (Rivero and Casas 1999). In a study carried out in Brazil, the females of the parasitoid *Aphidius platensis* Bréthes (Braconidae) parasitized a larger number of aphids *Myzus persicae* Sulzer or *Schizaphis graminum* Rondani (Aphididae) when the flowers of the yellow marigold *Tagetes erecta* L. (Asteraceae) were present. The lack of these flowers as food source led to a lower rate of parasitism, and females spend most of their time walking or remaining immobile (Souza et al. 2018a).

9.5 Influence of Plant Diversification on Parasitoids

Conservation biological control through plant diversification has been a subject of exploration in agricultural production. Several studies evaluate different spatial arrangements, such as vegetation corridors, plant strips and consortium between plants in general. In addition, it is important to assess the impact of plant diversification on the increase of biological control. The maintenance of vegetation adjacent to crops is important as a strategy of natural enemies' conservation, since it promotes the flow of energy, genes, plants and animals among the elements of the landscape (Altieri et al. 2003). For example, many plants have morphological structures such as hairs, domatia and floral and extrafloral nectaries that provide shelter and food sources for many entomophagous arthropods, many of which are effective in controlling various pests (Marquis and Whelan 1996; Agrawal et al. 2000).

Márquez et al. (2017) registered the beneficial entomofauna in Guatemala on the edges of sugarcane fields, in vegetation corridors among the fields and in the interior of the crop. They found that some families were more abundant in the vegetation corridor while others were more abundant inside the sugarcane fields, indicating that the presence of some landscape attributes was important in terms of abundance of parasitoid families.

In Colombia the major sugarcane pest *Diatraea saccharalis* Fabr. (Crambidae) is controlled using the tachinid fly *Jaynesleskia jaynesi* Aldrich (Tachinidae) which locate and find the borer by the hole left in the stems (Williams et al. 1969). Vargas et al. (2006) found that the weed *Bidens pilosa* L. (Compositae) were suitable as a nectar source for these parasitoids, recommending its maintenance within sugarcane fields as a tactic of conservation biological control.

According to Haro et al. (2015) important parasitoids of various vegetable pests, for example, of several aphid pests, were found in plants of the family Apiaceae as coriander (*Coriandrum sativum* L.) and dill (*Anethum graveolens* L.). In each of these plants, eight different species of parasitoids were found while the sweet fennel (*Foeniculum vulgare* Mill.) were suitable for four species. The main genus of parasitoids found were *Aphidius* and *Lysiphlebus* (Braconidae), *Copidosoma* (Encyrtidae), *Pediobius* and *Sympiesis* (Eulophidae) and *Trichogramma* sp. (Trichogrammatidae). Thus, these surveys are important to elucidate the role of different plants to attract and conserve natural enemies of horticultural pests and can be used in different spatial arrangements as a component to increase biological control and pest regulation in the tropics.

One of the most promising plants to promote vegetable diversification in horticultural systems in Latin America is the yellow marigold *Tagetes erecta*. In Brazil the maintenance of lines of *T. erecta* near the onion cultivation promoted a greater amount of parasitoids, specially from the families Braconidae, Mymaridae, Figitidae, Trichogrammatidae, Eulophidae and Scelionidae, resulting in a lower presence of phytophagous insects in the plants, helping to regulate the natural pests of the crop (Silveira et al. 2009).

In Guatemala studies by Gomez (2017) on corn (*Zea mays*) associated with *T. erecta* at different distances from the field showed that this attractive plant influences the composition of the parasitoid species found in the crop. The abundance and richness of parasitoids associated with the fall armyworm *Spodoptera frugiperda* (J.E.Smith) was higher near the *T. erecta* strips, resulting in a better biological control of the pest. The presence of representatives of the genus *Apanteles*, *Chelonus*, *Cotesia*, *Trichospilus*, *Anomalon*, *Telenomus* and *Trichogramma*, all directly associated with all stages of the pest, was found in the marigold strips and on corn near these strips, proving that the diversification with this plant was efficient.

In Brazil Silva et al. (2016) observed that plants of kale *Brassica oleracea* L. (Brassicaceae) associated with coriander (*Coriandrum sativum*), dill (*Anethum graveolens*), yellow marigold (*T. erecta*) and calendula (*Calendula officinalis*) influenced the abundance, species richness and diversity of parasitoids of the aphid pest *Lipaphis erysimi* (Kaltenbach) (Aphididae). The aphid parasitoid species *Diaeretiella rapae* (McIntosh), *Aphidius colemani* (Viereck) and *Praon volucre* (Haliday) were attracted to the entomophagous plants and disperse to kale beds, helping to control the aphids.

A survey of Souza et al. (2018b) in sweet pepper (*Capsicum annuum* L.) fields associated with *T. erecta* and basil (*Ocimum basilicum* L.) revealed a greater abundance of parasitoids when the crop was next to these attractive plants (abundance of 98 versus 130 parasitoids for marigold and basil, respectively). In sweet pepper alone (monoculture), the number of parasitoids was reduced to 40 individuals. Among the parasitoids that contributed to increase abundance, the genera *Didyctium* sp. (Figitidae), *Polynema* sp. (Mymaridae) and *Apanteles* sp. (Braconidae) stood out. The diversification of sweet pepper fields with basil and marigold increases the number of parasitoids who benefits the culture and it is, therefore, recommended.

Haro et al. (2018) observed that the presence of marigold flowers within lettuce *Lactuca sativa* L. fields mediate shifts in arthropod food webs. The presence of marigold flowers in the field successfully increased richness, body size and the numerical and biomass abundance of natural enemies in the lettuce arthropod community, which affected the number of links, vulnerability, generality, omnivory rate and food chain length in the community. These are key factors for the stability of relationships between species in food webs. In conclusion, this reinforces the need of having flowers distributed, i.e. within horticultural fields, as a tool for regulation of pests via conservation biological control.

Wyckhuys et al. (2013) made a survey about conservation biological control at developing countries and have found 390 papers related to the topic about a variety of things regarding the method. Many of them were from Brazil, Cuba, Mexico and Philippines. They found that despite being a method of control not highly supported by the government nor the industries, it is growing every year with more and more researches. By the time of the survey there were more than 50 plants already studied (cultivated or not), and they attested that this method has a great importance and impact at developing countries' agriculture and even to help minimize the greenhouse effect.

9.6 Final Considerations

Parasitoids are individuals of great importance as part of the ecosystem itself and as part of agroecosystems, where they stand a role as insect regulators. Most of those insects regulated by parasitoids are serious pests, so the ecological service made by them contribute to enhanced production. Since they offer this service, the need of chemical sprays on crops tends to minimize, bringing benefits to the environment. There is still a lot to be known about parasitoids in diversified agroecosystems, especially at the tropics, where the utilization of biological control as a tool is still a small portion of pest control. Recent research, however, has shown that habitat diversification with flowering plants increases the abundance, richness and survival of parasitoids in tropical agroecosystems and, therefore, must be increased over the next few years.

References

Agrawal AA, Karban R, Colfer RG (2000) How leaf domatia and induced plant resistance affect herbivores, natural enemies and plant performance. Oikos 89:70–80

Altieri MA, Ne S, Nicholls CI (2003) O papel da biodiversidade no manejo de pragas. Holos, Ribeirão Preto. 226p

Askew RR (1971) Parasitic insects. Elsevier, New York. 316 p

Baggen LR, Gurr GM (1998) The influence of food on *Copidoso makoehleri* (Hymenoptera: Encyrtidae), and the use of flowering plants as a habitat management tool to enhance biological control of potato moth, *Phthorimaea operculella* (Lepidoptera: Gelechiidae). Biol Control 11:9–17

Blumberg D (1997) Encapsulation of parasitoids. In World Crop Pests, Elsevier 7:375–387
Bouček NJ (1987) Rotoitidae, a curious new family of Chalcidoidea (Hymenoptera) from New Zealand. Syst Entomol 12(4):407–412
Davis JJ (1919) Contributions to knowledge of the natural enemies of Phyllophaga. Ill Nat Hist Surv Bull 13:53–133
Disney RHL (1994) Scuttle flies: the Phoridae. Chapman & Hall, London. 467p
De Moraes CM, Lewis WJ, Paré PW, Alborn HT, Tumlinson JH (1998) Herbivore-infested plants selectively attract parasitoids. Nature 393:570–573
Doutt RL (1964) Biological characteristics of entomophagous adults. In: De Bach P (ed) Biological control of insects pests and weeds. Reinhold, New York, pp 145, 844 p–167
Eggleton P, Belshaw R (1992) Insect parasitoids: an evolutionary overview. Philos Trans Biol Sci 337:1–20
Eggleton P, Gaston KJ (1990) "Parasitoid" species and assemblages: convenient definitions or misleading compromises? Oikos 59:417–421
Feener DH Jr, Brown BV (1997) Diptera as parasitoids. Annu Rev Entomol 42(1):73–97
Gauld ID, Bolton B (1996) The Hymenoptera, 2nd edn. Oxford University Press, Oxford
Godfray HCJ, Shimada M (1999) Parasitoids as modelorganisms for ecologists. Popul Ecol 41(1):3–10
Gomez H (2017) Vegetação de entorno e sua influência sobre insetos praga e parasitoides em cultivos de cana e milho na Guatemala. (Tesis de doctorado). Universidad Federal de Lavras, Brasil
Gordh G, Headrick D (Eds.) (2001). A dictionary of entomology. Cabi
Goulet H, Huber JT (1993). Hymenoptera of the world: an identification guide to families. Research Branch, Agriculture Canada. 668p
Gullan PJ, Cranston OS (2017) Insetos - Fundamentos da Entomologia – 5a Ed. Guanabara Koogan, 460p
Hanson P (1992) The Nearctic species of *Ormyrus* Westwood (Hymenoptera: Chalcidoidea: Ormyridae). J Nat Hist 26(6):1333–1365
Hanson PE, Gauld ID (2006). Hymenoptera de la Regiốn Neotropical. Gainesville: American Entomological Institute. (Memoirs of the American Entomological Institute, 77). 994 p
Haro MM, Resende ALS, Silva VF, Souza B, Silveira LCP (2015) Parasitoids of horticultural pests associated to commercial development stages of Apiaceae plants. Entomotropica 30:174–180
Haro MM, Silveira LCP, Wilby A (2018). Stability lies in flowers: Plant diversification mediating shifts in arthropod food webs. PLOS ONE | https://doi.org/10.1371/journal.pone.0193045 February 16, 2018.
Heraty JM (2002) Revision of the genera of Eucharitidae (Hymenoptera: Chalcidoidea) of the world. Mem Entomol Soc Can 68:1–359
Hofstede HM, Fenton MB, Whitaker JO Jr (2004) Host and host-site specificity of bat flies (Diptera: Streblidae and Nycteribiidae) on Neotropical bats (Chiroptera). Can J Zool 82:616–626
Hull FM (1973) Bee flies of the world: the genera of the family Bombyliidae. Smithsonian Institution Press, Washington, DC
Jameson EW (1985) Pleioxenous host-restriction in fleas. J Nat Hist 19:861–876
Jervis M, Hawkins B, Kidd NAC (1996) The usefulness of destructive host feeding parasitoids in classical biological control: theory and observation conflict. Ecol Entomol 21(1):41–46
Knutson LV (1972) Pupa of *Neomochtheru sangustipennis* (Hine), with notes on feeding habits of robber flies and a review of publications on immature stages (Diptera: Asilidae). Proc Biol Soc Wash 85:163–178
Lee JC, Heimpel GE (2008) Floral resources impact longevity and oviposition rate of a parasitoid in the field. J Anim Ecol 77:565–572
Lewis WJ, Takasu K (1990) Use of learned odours by a parasitic wasp in accordance with host and food needs. Nature 348:635–636
Márquez JM, Reyes E, Lemus JM, López E, Torres L, Duarte R, Leal S, Arroyo LC, Pec M, López E (2017) Entomofauna benéfica (Hymenoptera: Parasitica) en caña de azúcar, asociada a las áreas con infestación del barrenador del tallo. Cengicaña 16:223–241

Marquis RJ, Whelan C (1996) Plant morphology and recruitment of third trophic level: subtle and little recognized defenses? Oikos 75:330–334
Marshall AG (1981) The sex ratio in ectoparasitic insects. Ecol Entomol 6:155–174
Nappi AJ (1975) Parasite encapsulation in insects. In: Maramorosch K, Shope RE (eds) Invertebrate immunity. Academic Press, New York, NY, pp 293–326
Pape T (1990) Revisionary notes on American Sarcophaginae (Diptera: Sarcophagidae). Tijdschr Voor Entomol 133:43–74
Ricther VA (1997) Family Nemestrinidae. In: Papp L, Darvas B (eds) Contributions to a manual of Palearctic Diptera, vol 2. Science Herald, Budapest, pp 459–468
Rivero A, Casas J (1999) Incorporating physiology into parasitoid behavioral ecology: the allocation of nutritional resource. Res Popul Ecol 41:39–45
Salt G (1963) The defence reactions of insects to metazoan parasites. Parasitology 53:527–642
Salt G (1968) The resistance of insect parasitoids to the defence reactions of their hosts. Biol Rev 43:200–232
Schlinger EI (1987) The biology of Acroceridae (Diptera): true endoparasitoids of spiders. In: Nentwig W (ed) Ecophysiology of spiders. Springer, Berlin, Heidelberg
Schroeder R, Hilker M (2008) The relevance of background odor in resource location by insects: a behavioral approach. Bioscience 58:308–316
Silva V, Silveira L, Santos A, Santos AJN, Tomazella VB (2016) Companion plants associated with kale increase the abundance and species richness of the natural-enemies of *Lipaphis erysimi* (Kaltenbach) (Hemiptera: Aphididae). Afr J Agric Res 11(29):2630–2639
Silveira LCP, Berti-Filho E, Pierre LSR, Peres SC, Louzada JL (2009) Marigold (*Tagetes erecta* L.) as an attractive crop to natural enemies in onion fields. Sci Agric 66:780–787
Skevington JH (2005) Revision of Nearctic *Nephrocerus* Zetterstedt (Diptera: Pipunculidae). Zootaxa 977:1–36
Souza IL, Marucci RC, LCP S, de Paulo NCP, Lee JC (2018a) Effects of marigold on the behavior, survival and nutrient reserves of *Aphidius platensis*. BioControl 63(4):543–553
Souza IL, Tomazella VB, AJN S, Moraes T, LCP S (2018b) Parasitoids diversity in organic Sweet Pepper (*Capsicum annuum*) associated with Basil (*Ocimum basilicum*) and Marigold (*Tagetes erecta*). Braz J Biol (AHEAD)
Sutton SL (1980) Woodlice, 2nd edn. Pergamon Press, Oxford. 144 p
Vargas GA, Obando V, Gómez LA (2006) *Jaynesleskia jaynesi*: otra alternativa para el manejo de Diatraea spp. Carta Trimestral. Cenicaña 28(2):3–5
Vinson SB (1975) Biochemical coevolution between parasitoids and their host. In: Price PW (ed) Evolutionary strategies of parasitic insects and mites. Plenum Press, New York, pp 14–48. 224p
Wäckers FL (1994) The effect of food deprivation on the innate visual and olfactory preferences in the parasitoid *Cotesia rubecula*. J Insect Physiol 40:641–649
Waloff N (1975) The parasitoids of the nymphal and adult stages of leafhoppers (Auchenorrhyncha: Homoptera) of acidic grassland. Trans R Entomol Soc Lond 126(4):637–686
Wood, D.M. 1987. Tachinidae. In: Manual of Neartic Diptera. Vol. 1. Research Branch, Agriculture Canada, Monograph No. 27, pp 549–573
Winkler K, Wackers F, Bukovinszkine-Kiss G, van Lenteren J (2006) Sugar resources are vital for *Diadegma semiclausum* fecundity under field conditions. Basic Appl Ecol 7:133–140
Wyckhuys KAG, Lu Y, Morales H, Vazquez LL, Legaspi JC, Eliopoulos PA, Hernandez LM (2013) Current status and potential of conservation biological control for agriculture in the developing world. Biol Control 65(1):152–167
Williams JR, Metcalfe JR, Mungomery RW, Mathes RW (1969) Pests of sugarcane. Elsevier publishing company, Amsterdam-London-New York, p 367

Chapter 10
Entomopathogenic Nematodes

Vanessa Andaló, Juan Pablo Molina Acevedo, Aldomário Santo Negrisoli Júnior, and Viviane Araujo Dalbon

10.1 Introduction

Insect-associated nematodes are filiform non-segmented organisms that belong to the Nematoda phylum, which includes more than 30 families. These organisms may have forethic associations or parasitism with insects, which occurs with entomopathogenic nematodes (Rhabditida: Heterorhabditidae, Steinernematidae), considered pathogenic to different orders of insects (Kaya and Stok 1997). The family Heterorhabditidae comprises the genus *Heterorhabditis*, with 19 species (Nguyen 2017a). The family Steinernematidae includes the genus *Neosteinernema*, with a described species, and the genus *Steinernema*, with 84 described species (Nguyen 2017b).

In Latin America, studies on entomopathogenic nematodes have advanced in terms of species description, identification of isolates, isolation of native species, in vivo and in vitro multiplication, virulence evaluation, and pest management of economic importance, especially in Argentina, Bolivia, Brazil, Colombia, Cuba, Mexico, Peru, and Venezuela. However, the use of these organisms is still incipient.

V. Andaló (✉)
Federal University of Uberlândia, Monte Carmelo Campus, Institute of Agrarian Sciences, Monte Carmelo, MG, Brazil
e-mail: vanessaandalo@ufu.br

J. P. M. Acevedo
Colombian Agricultural Research Corporation (Agrosavia)/Embrapa CPATC, Rio Largo, AL, Brazil

A. S. Negrisoli Júnior
Embrapa Coastal Tablelands – CPATC, Research and Development Execution Unit, Rio Largo, AL, Brazil

V. A. Dalbon
Federal University of Alagoas, Laboratory of Research in Chemistry of Natural Products – LPQPN, Maceió, AL, Brazil

B. Souza et al. (eds.), *Natural Enemies of Insect Pests in Neotropical Agroecosystems*, https://doi.org/10.1007/978-3-030-24733-1_10

Despite the already developed research and climatic conditions favorable to the implementation of the control of agricultural pests by entomopathogenic nematodes, this practice has been rarely commercially carried out in Latin America (Leite 2011).

In the case of Brazil, in the last two decades, studies have been conducted with emphasis on the isolation of native nematodes, showing excellent characteristics regarding virulence patterns and adaptation to environmental conditions. This has enabled the development of studies related to taxonomic identification, biology, application technology applied with other microorganisms (such as entomopathogenic fungi), and compatibility with phytosanitary products, besides the control of insects of economic importance (Dolinski et al. 2017).

Therefore, the objective of this chapter is to show the main results of studies associated with the bioecology of entomopathogenic nematodes in Latin America in order to compile the information obtained so far and to disseminate these results. Moreover, researchers from different parts of the world might adapt it considering their respective conditions.

10.2 Diversity of Entomopathogenic Nematodes

Entomopathogenic nematodes are found in all continents with a great diversity of species adapted to different environmental conditions, which improves pest control in specific niches. Many of the studies developed in Brazil have been related to the survey of the native fauna of these organisms and include morphological, taxonomic, biological, and ecological studies, tests on target pests, and research involving in vivo and in vitro preservation and production.

The number of native isolates in Latin America does not reflect the reality of the amount of species, since it is assumed that several other species are yet to be discovered, especially considering the recognized fauna and flora diversity in the different ecosystems and areas of native vegetation. The use of native populations in biological control programs may lead to higher mortality rates of target insects than when using exotic nematodes due to the greater adaptation of species to the local conditions.

The first occurrence of a nematode of the genus *Heterorhabditis* was recorded in Brazil more than 80 years ago by Pereira (1937), and it was described as *Rhabditis hambletoni*, a parasite of "broca-do-algodoeiro" *Gasterocercodes brasiliensis* Hambleton (Curculionidae) (currently named *Eutinobothrus brasiliensis*). Pizano et al. (1985) recorded the occurrence of infective juveniles (JIs) of *Neoaplectana glaseri* Steiner (currently called *Steinernema glaseri*) in the egg of *Migdolus fryanus* (Westwood) (Cerambycidae) at the Amália Plant, São Paulo, Brazil. In 1990, Poinar (1990) recorded the occurrence of *Heterorhabditis bacte-*

riophora Poinar in the state of Pernambuco. Del Valle et al. (2005) obtained an isolate of *Heterorhabditis baujardi* Phan, Subbotin, Nguyen, & Moens in the state of Rondônia. Machado et al. (2005) isolated *S. glaseri* in soil samples in the Araras region, São Paulo, and *Heterorhabditis indica* Poinar, Karunakar, & David in Itapetininga, São Paulo.

In Amazonas state, Brazil, Andaló et al. (2006) isolated a new species of the genus *Heterorhabditis* in soil samples from Benjamin Constant using larvae of *Galleria mellonella* L. (Pyralidae), which was named *Heterorhabditis amazonensis*. Molina et al. (2005) isolated three nematode populations (JPM3, JPM3.1, and JPM4) in Lavras, Minas Gerais state, which was later identified as *H. amazonensis* by Andaló et al. (2013). In a survey carried out in Lavras, MG, Andaló et al. (2009) obtained other populations of the same species from sorghum (*Sorghum* sp.) and garlic (*Allium sativum*). Nguyen et al. (2010) conducted a study using soil samples from native vegetation areas in the state of Mato Grosso do Sul, in Brazil, and recorded a new species called *Steinernema brazilense*, which has not yet been found in other regions of the country. Barbosa-Negrisoli et al. (2009) isolated, in the state of Rio Grande do Sul, in Brazil, *Steinernema rarum* (Doucet) Mamiya, until then found only in Argentina; *H. bacteriophora*, *Steinernema feltiae* (Filipjev) Wouts, Mracek, Gerdin, & Bedding; and *Steinernema riobrave* Cabanillas, Poinar, & Raulston.

In other Latin American countries, efforts have also been made to isolate entomopathogenic nematodes. In the Southern Cone region, the Argentine Pampas revealed a great diversity of nematodes such as *Steinernema carpocapsae* (Weiser) Wouts, Mracek, Gerdin, & Bedding, *S. feltiae*, *Steinernema scapterisci* Nguyen & Smart, *H. bacteriophora*, and *Heterorhabditis argentinensis* Stock (Leite 2011). In Colombia, *H. bacteriophora* strains were commonly found associated with the cassava stink bug *Cyrtomenus bergi* Froeschner (Cydnidae) (Caicedo and Bellotti 1996) and, in the coffee region of that country, a new species *Steinernema colombiense* was also described by López-Núñez et al. (2008). *Steinernema cubanum* Mracek was isolated from the soil of citrus orchards in Cuba. Similarly, two *H. bacteriophora* (Tetuan and P2M) isolates were identified parasitizing *Cylas formicarius* (Fabricius) (Brentidae) and *Pachnaeus litus* (Germar) (Curculionidae) (Rodríguez et al. 1996). In the states of Jalisco, Colima, Michoacán, and Tamaulipas, in Mexico, different *Steinernema* and *Heterorhabditis* isolates have been found (Lezama-Gutierrez et al. 2001).

Thus, new *Heterorhabditis* and *Steinernema* species have been recorded for Latin America and native isolates have been obtained, providing greater potential of these organisms as pest control agents in this region.

10.3 Bioecology

10.3.1 Determination of Life Cycle in Native Nematodes

The life cycle and pathogenicity of entomopathogenic nematodes can vary between genera, species, and populations of the same species. This variability must be detected in order for nematodes to be successfully used in biological control programs. Molina et al. (2005) and Andaló et al. (2009) studied the population characteristics of *H. amazonensis* isolated from different locations in Brazil and detected a great variability among them, such as life cycle and insect-host pathogenicity, which underscores the importance of the knowledge of the biological and ecological characteristics of each population. The description of new native species requires the determination of biology to understand its life cycle, which may be short (one generation) or long (2–3 defined generations). An example of a complete life cycle of an entomopathogenic nematode is presented in the study of Molina et al. (2004) for *H. amazonensis* populations JPM3, JPM3.1, and JPM4, isolated in Lavras, Minas Gerais state, Brazil.

One of the factors influencing the life cycle of native isolates is the amount of inoculum used for the infection. *Heterorhabditis amazonensis* JPM4 infecting *G. mellonella* larvae may develop a short life cycle with nematode generation, which includes three development stages: egg, juvenile (J), and adult, with the juvenile phase consisting of four morphologically distinct stages (J1, J2, J3–IJ, and J4). In the adult phase, the first generation consists of hermaphrodite females and the second generation of females and amphimictic males. These first-generation of infective juveniles (IJs) are highly virulent, especially those that abandon the larvae in the first 3 days. The J3 that continue the development within the *G. mellonella* larvae determine a long life cycle, up to 456 hours (19 days), from the beginning of the initial infection to the total exhaustion of the host, determining three generations of nematode production. However, these nematodes of second and third generation are less virulent (Molina et al. 2005).

10.3.2 Agricultural Pests Susceptible to the Effect of Entomopathogenic Nematodes

An efficient evaluation of native and exotic *Steinernema* spp. and *Heterorhabditis* spp. nematodes for the control of insect pests in Latin America is applied to several orders, especially Coleoptera, aiming at the larval stages in the soil. However, studies on the action of nematodes on other orders, such as Diptera, Lepidoptera, and Hemiptera, and even other classes of arthropods and phyla have also been developed (Table 10.1).

Table 10.1 Pest/crop and entomopathogenic nematodes evaluated in laboratory and/or field experiments in Latin America

Pest/crop	Biocontrol agent (country)[a]	Reference
Lepidoptera		
Grapholita molesta – Peach	*H. bacteriophora* RS33 (BR)	Barbosa-Negrisoli et al. (2013)
Bonagota salubricola – Apple	*H. bacteriophora* RS107 (BR)	Barbosa-Negrisoli et al. (2010)
Spodoptera frugiperda – Maize	*H. amazonensis* RSC02 (BR)	Andaló et al. (2010)
Diatraea saccharalis – Sugarcane	*H. amazonensis* JPM4 + *Metarhizium anisopliae* LPP39 (BR)	Molina et al. (2008), Bellini and Dolinski (2012)
Coleoptera		
Conotrachelus psidii – Guava	*H. bacteriophora* LPP30 (BR) *H. indica* IBCB05 (BR)	Silva et al. (2010), Dolinski et al. (2012)
Sphenophorus levis – Sugarcane	*S. brazilense* (BR)	Leite et al. (2012), Gionetti et al. (2011), Tavares et al. (2007, 2009)
Alphitobius diaperinus – Aviary	*Steinernema arenarium* (BR)	Alves et al. (2005, 2012), Rodrigueiro et al. (2009)
Diabrotica speciosa	*H. amazonensis* RSC01 (BR)	Santos et al. (2011)
Rhynchophorus palmarum	*Heterorhabditis* spp. and *Steinernema* spp. (BR)	Sabino (2014)
Cosmopolites sordidus – Banana	*S. carpocapsae*, *S. glaseri*, and *S. feltiae* (PR)	Figueroa (1990)
Premnotrypes vorax – Potato	*Steinernema* sp. (CO) and *Steinernema carpocapsae*	Garzon et al. (1996)
Hypothenemus hampei – Coffee	*Heterorhabditis* sp. and *Steinernema* (CO)	Molina and Lopez (2009)
Diptera		
Anastrepha fraterculus – Peach	*H. bacteriophora* RS88 (BR)	Barbosa-Negrisoli et al. (2009)
Ceratitis capitata	*S. carpocapsae* all, *Heterorhabditis* sp. JPM4, and *S. carpocapsae* + extracts (BR)	Rohde et al. (2013), Rohde et al. (2012a, b), Silva et al. (2010)
Bradysia spp. (Sciaridae)	*Heterorhabditis* spp. and *Steinernema* spp. (BR)	Leite et al. (2007), Tavares et al. (2012)
Hemiptera		
Dysmicoccus texensis	*H. amazonensis* JPM3 (BR)	Alves et al. (2009a, b)
Mahanarva spp. – sugarcane and pastures	*S. carpocapsae* Mexican and *Heterorhabditis* spp.	Georgis and Hom (1992), Batista and Auad (2010), Batista et al. (2011a, b, 2014)
Cyrtomenus bergi – Cassava	*H. bacteriophora* LFR 92 and SQC 92 (CO)	Barberena and Bellotti (1998)

(continued)

Table 10.1 (continued)

Pest/crop	Biocontrol agent (country)[a]	Reference
Isoptera		
Cornitermes cumulans	*Steinernema carpocapsae* (BR)	Rosa et al. (2007, 2008)
Arachnida		
Rhipicephalus microplus	*Heterorhabditis bacteriophora* HP88 *H. indica* LPP1 *H. amazonensis* RSC5 + associations (extracts, miticides, and fungi) (BR)	Monteiro et al. (2012, 2013, 2014a, b), Silva et al. (2012)
Plant-parasitic nematodes		
Meloidogyne mayaguensis	*H. baujardi* LPP7, *S. feltiae* Sn, and *H. amazonensis* JPM4 (BR)	Molina et al. (2008, 2010a, 2010b)

[a]*BR* Brazil, *CO* Colombia, *PR* Puerto Rico

10.3.3 Factors Influencing the Efficacy of Entomopathogenic Nematode Control

Entomopathogenic nematodes are vulnerable to a number of environmental conditions, as they suffer from the action of biotic and abiotic factors. Unfavorable conditions, such as low moisture, solar radiation, soil texture, and natural enemies, can affect the success in pest control due to reduced activity or death of IJs (Cutler and Webster 2003).

The behavioral characteristics of entomopathogenic nematodes are closely related to the success of the use of these organisms as control agents. Therefore, studies on ecology and biology must first be conducted in order to ensure that JI will survive according to the biotic and abiotic factors of the environment in which they develop and thus be able to locate the host and cause its death (Lewis 2002).

In the soil, nematodes actively search for their host, and their low mobility may favor the exposure to unfavorable biotic and abiotic factors, reducing the chance of finding and reproducing the target host (Portillo-Aguilar et al. 1999).

10.3.3.1 Biotic Factors

Natural enemies, such as predatory fungi capable of producing mycelial structures to capture nematodes, are included among the unfavorable biotic factors. Andaló et al. (2008) studied the susceptibility of *H. amazonensis* to the fungi *Arthrobotrys oligospora* Fresenius, *A. conoides* Drechsler, and *Duddingtonia flagrans* (Duddington) Cooke by evaluating their predatory capacity. The authors observed the occurrence of predation after 8 days of contact with IJs, concluding that these fungi are capable of predating *H. amazonensis*. However, an initial contact period was required to stimulate not only the vegetative growth of the fungus, but also the formation of traps. Thus, it is important to consider that possible antagonistic fungi in the soil should be analyzed in pest control programs with the use of nematodes since many JIs can be predated.

Molina and Lopez (2009) attributed the nondevelopment of *S. cubanum* IJs in late-instar larvae of *Bombyx mori* L. (Bombycidae) to the low efficiency of the multiplication of their symbiotic bacterium, *Xenorhabdus* spp., which does not produce sufficient quantities of antibiotics, such as xenocoumacine, bacteriocin, and xenorababine, in the infection process.

Mertz et al. (2014) studied the forethic dispersion of entomopathogenic nematodes in larvae and adults of *Calosoma granulatum* Perty (Carabidae). Since entomopathogenic nematodes have limited ability to travel long distances, they can use strategies to target hosts in different locations, such as attaching to an organism that will act as a dispersing agent from one location to another. Thus, the authors verified that *C. granulatum* is a good dispersing agent of *H. amazonensis*, and the transport of IJs at distances greater than 40 cm was observed. The ability of *C. granulatum* to transport the nematode to even longer distances is important, since this carabid is a predator of caterpillars inhabiting the soils of agroecosystems in Brazil.

Valle et al. (2008) evaluated the dispersion of *H. baujardi* LPP7 in field guava culture and found nematodes up to 90 cm from the point of release of the entomopathogenic nematode-containing cadavers and up to 10 cm deep. Andaló et al. (2012) evaluated the horizontal and vertical displacement of *H. amazonensis* RSC2 in the search of *Spodoptera frugiperda* (Smith) (Noctuidae) and verified that insect mortality was inversely proportional to the distance between it and the IJs; the nematodes were able to locate and kill the host at distances up to 60 cm (horizontal) and 20 cm (vertical). The authors emphasized the importance of understanding the behavior of entomopathogenic nematodes in the host search, since several factors are involved in this process and behavioral studies are essential to ensure the success of control programs using entomopathogenic nematodes.

10.3.3.2 Abiotic Factors

There are several critical abiotic factors for soil nematodes, such as texture, moisture, temperature, and agrochemical residues (McCoy et al. 2002). Dolinski et al. (2010) tested the horizontal displacement of *H. baujardi* LPP7 in soil with different textural classes and found that the highest rates of host mortality and infectivity occurred at a distance of 30 cm for the sandy and sandy clay loam classes, in which mortality was also observed at a distance of 60 and 90 cm. No mortality was observed in the clay loam at 60 and 90 cm. For the clay class, no infection was found for any of the distances evaluated. The authors concluded that *H. baujardi* LPP7 showed better host-seeking behavior on sandy texture substrates, confirming that soil class may be an important indicator of the presence of nematodes and that with increasing clay content, the dispersion, persistence, and efficacy of entomopathogenic nematodes are generally reduced.

Soil temperature is another factor that influences the behavior of nematodes. Kung et al. (1991) stated that entomopathogenic nematodes tolerate temperatures between 3 °C and 35 °C. Low temperatures (5 °C to 10 °C) may affect the occurrence of some entomopathogenic species in the soil, such as *H. bacteriophora*, *S. glaseri*, and *S. carpocapsae*, but it does not affect *S. feltiae* (Brown and Gaugler 1997). Kaya (1990) found that *S. feltiae* could infect its host in the soil at temperatures between 2 °C and 30 °C. Finally, saline soils and soils with high aluminum contents affect the cellular structure and viability of entomopathogenic nematodes (Glazer 1996). Another important factor in the infection and survival of entomopathogenic nematodes is soil pH. Extreme pH values (3 or 11) limit infection capacity, but not its infectivity (Glazer 1996). Andaló et al. (2018) observed that moisture, organic matter, and the levels of phosphorus in the soil are factors that influence the presence of entomopathogenic nematodes in the soil and that knowing these factors is fundamental to enable greater persistence and efficacy of their populations in the field, providing greater chances of pest control.

10.4 Final Considerations

The diversity of entomopathogenic nematodes is a subject not fully explored in the ecological studies conducted in Latin American countries, which may be due to the few specialists who are dedicated to the subject, since there is a vast potential research in Latin America, in view of the biodiversity existing in its various biomes.

Many species of entomopathogenic nematodes marketed worldwide are massively produced in vitro in laboratories in the United States, Germany, and the Netherlands. However, since they are species from other regions, in addition to the high costs for production and importation and the regulatory measures of some countries that prevent the entry of exotic agents, their massive use in biological control programs in Latin America becomes unfeasible. Thus, despite the efforts made in the last three decades, which have contributed to the progress in the isolation and perspectives of the use of these organisms against pests of economic importance with potential results, it is necessary to invest in studies related to in vitro production and formulation of commercial native species, since many have high virulence against pest insects.

We believe that it is only through a greater involvement of partnership experts through Latin American research networks, coupled with both government and private initiatives, that we can advance our knowledge of the bioecology of these control agents. Therefore, it is of great importance that more research ought to be carried out in laboratory and field conditions worldwide.

References

Alves VS, Alves LFA, Rohde C (2005) Patogenicidade de *Steinernema glaseri* e *S. carpocapsae* (Nematoda: Rhabdita) contra o cascudinho, *Alphitobius diaperinus* (Panzer) (Coleoptera: Tenebrionidae). Neotrop Entomol 34(1):139–141

Alves VS, Oino AJ, Santa-Cecilia LVC et al (2009a) Patogenicidade de nematoides entomopatogênicos a cochonilha-da-raiz-do-cafeeiro *Dysmicoccus texensis* (Tinsley) (Hemiptera: Pseudococcidae) em laboratório. Arq Inst Biol 76(1):67–73

Alves VS, Oino AJ, Santa-Cecilia LVC et al (2009b) Testes para o controle de *Dysmicoccus texensis* (Tinsley) (Hemiptera, Pseudococcidae) em cafeeiro com nematoides entomopatogênicos do gênero *Heterorhabditis* (Rhabditida, Heterorhabditidae). Rev Bras Entomol 53(1):139–143

Alves VS, Neves PMOJ, Alves LFA et al (2012) Entomopathogenic nematodes (Rhabditida: Heterorhabditidae and Steinernematidae) screening for lesser mealworm *Alphitobius diaperinus* (Coleoptera: Tenebrionidae) control. Rev Colomb Entomol 38(1):73–80

Andaló V, Nguyen KB, Oino AJ (2006) *Heterorhabditis amazonensis* n. sp. (Rhabditida: Heterorhabditidae) from Amazonas, Brazil. Nematology 8(6):853–867

Andaló V, Moreira GF, Maximiniano C et al (2008) Suscetibilidade de *Heterorhabditis amazonensis* (Rhabditida: Heterorhabditidae) a fungos predadores de nematoides. Nematol Bras 32(3):177–184

Andaló V, Moreira GF, Cavalcanti RS et al (2009) Observations on the life cycle and pathogenicity of *Heterorhabditis amazonensis* (Rhabditida: Heterorhabditidae). Nematol Bras 33(2):194–197

Andaló V, Santos V, Moreira GF et al (2010) Evaluation of entomopathogenic nematodes under laboratory and greenhouses conditions for the control of *Spodoptera frugiperda*. Ciênc Rural 40(9):1860–1866

Andaló V, Santos V, Moreira GF et al (2012) Movement of *Heterorhabditis amazonensis* and *Steinernema arenarium* in search of corn fall armyworm larvae in artificial conditions. Sci Agric 69(3):226–230

Andaló V, Acevedo JPM, Oino AJ et al (2013) Populações de *Heterorhabditis amazonensis* (Rhabditida: Heterorhabditidae) da região de Lavras – MG. In: Anais do 13° Simpósio de Controle Biológico, Embrapa Agropecuária Oeste, Bonito, 15–18 set 2013

Andaló V, Mieko J, Carvalho FJ et al (2018) Entomopathogenic nematode distribution and edaphoclimatic conditions in the Cerrado of Minas Gerais, Brazil. Appl Entomol Zool 53(1):129–136

Barberena MF, Bellotti AC (1998) Parasitismo de dos razas del nematodo *Heterorhabditis bacteriophora* sobre la chinche *Cyrtonemus bergi* en laboratorio. Rev Colomb Entomol 24:7–11

Barbosa-Negrisoli CRC, Garcia MS, Dolinski C et al (2009) Efficacy of indigenous entomopathogenic nematodes (Rhabditida: Heterorhabditidae, Steinernematidae), from Rio Grande do Sul Brazil, against *Anastrepha fraterculus* (Wied.) (Diptera: Tephritidae) in peach orchards. J Invertebr Pathol 102(1):6–13

Barbosa-Negrisoli CRC, Negrisoli Júnior AS et al (2010) Efficacy of entomopathogenic nematodes (Nemata: Rhabditida) to control Brazilian apple leafroller *Bonagota salubricola* (Meyrick, 1937) (Lepidoptera: Tortricidae). Crop Prot 29(11):1274–1279

Barbosa-Negrisoli CRC, Negrisoli Júnior AS, Garcia MS et al (2013) Control of *Grapholita molesta* (Busck, 1916) (Lepidoptera: Tortricidae) with entomopathogenic nematodes (Rhabditida: Heterorhabditidae, Steinernematidae) in peach orchards. Exp Parasitol 135(2):466–470

Batista ESP, Auad AM (2010) Application methods of entomopathogenic nematodes for control of *Mahanarva spectabilis* (Hemiptera: Cercopidae). Biocontrol Sci Tech 20(10):1079–1085

Batista ESP, Auad AM, Frias MP et al (2011a) Nematoides entomopatogênicos infectam ovos e adultos de cigarrinha-das-pastagens? Arq Inst Biol 78(3):475–478

Batista ESP, Auad AM, Rezende TT et al (2011b) Screening of entomopathogenic nematodes to control *Mahanarva fimbriolata* (Hemiptera: Cercopidae). Rev Colomb Entomol 37(2):198–202

Batista ESP, Auad AM, Andaló V et al (2014) Virulence of entomopathogenic nematodes (Rhabditida: Steinernematidae, Heterorhabditidae) to spittlebug *Mahanarva spectabilis* (Hemiptera: Cercopidae). Arq Inst Biol 81(2):145–149

Bellini LL, Dolinski C (2012) Foliar application of entomopathogenic nematodes (Rhabditida: Steinernematidae and Heterorhabditidae) for the control of *Diatraea saccharalis* in greenhouse. Sem, Ciênc Agrár 33(3):997–1004

Brown IM, Gaugler R (1997) Temperature and humidity influence emergence and survival of entomopathogenic nematodes. Nematologica 43(5):363–375

Caicedo AM, Bellotti AC (1996) Survey of native nematodes associated with *Cyrtonemus bergi* in eight colombian sites. Rev Colomb Entomol 22:19–24

Cutler GC, Webster JM (2003) Host-finding ability of three entomopathogenic nematode isolates in the presence of plant roots. Nematology 5(4):601–608

Del Valle EE, Dolinski C, Souza RM et al (2005) Performance de *Heterorhabditis baujardi* LPP7(28) (Nematoda: Rhabditida), selecionada para tolerância a elevadas temperaturas, no controle de *Conotrachelus psidii* (Coleoptera: Curculionidae). Nematol Bras 29:199–205

Dolinski C, Pinto CCS, Robaina RR et al (2010) Efeito de substratos com diferentes classes texturais na mobilidade do nematoide entomopatogênico *Heterorhabditis baujardi* LPP7 (Rhabditida: Heterorhabditidae). Nematol Bras 34:123–128

Dolinski C, Choo HY, Duncan LW (2012) Grower acceptance of entomopathogenic nematodes: case studies on three continents. J Nematol 44(2):226–235

Dolinski C, Monteiro C, Andaló V et al (2017) Studies on entomopathogenic nematodes in Brazil: past and future. Nematoda 4:1–14

Figueroa W (1990) Biocontrol of the banana root borer weevil, *Cosmopolites sordidus* (Germar), with steinernematid nematodes. J Agric Univ P R 74:15–19

Garzon MY, Aza BO, Jimenez GJ et al (1996) Potencial del nematodo *Steinernema* sp. para el control biologico del gusano blanco de la papa. Rev Colomb Entomol 22(1):25–30

Georgis R, Hom A (1992) Introduction of entomopathogenic nematode products into Latin America and the Caribbean. Nematropica 22(1):81–98

Gionetti FHC, Leite LG, Tavares FM et al (2011) Virulência de nematoides entomopatogênicos (Nematoda: Rhabditida) à *Sphenophorus levis* (Coleoptera: Curculionidae). Bragantia 70(1):81–86

Glazer I (1996) Survival mechanisms of entomopathogenic nematodes. Biomed Sci Technol 6(3):373–378

Kaya HK (1990) Soil ecology. In: Gaugler R, Kaya HK (eds) Entomopathogenic nematodes in biological control. CRC Press, Boca Raton, pp 93–115

Kaya HK, Stok P (1997) Techniques in insect nematology. In: Lacey L (ed) Manual of techniques in insect pathology. Academic Press, New York, pp 281–324

Kung S, Gaugler R, Kaya HK (1991) Effects of soil temperature, moisture, and relative humidity on entomopathogenic nematode persistence. J Invertebr Pathol 57(2):242–249

Leite LG (2011) Experiencias com nematodos entomopatógenos: retos e oportunidades de seu uso em Latinoamérica. Cyted 1:19–37

Leite LG, Tavares FM, Bussola RA et al (2007) Virulência de nematoides entomopatogênicos (Nemata: Rhabditida) contra larvas da mosca dos fungos, *Bradysia mabiusi* (Lane, 1959) e persistência de *Heterorhabditis indica* Poinar *et al* 1992, em substratos orgânicos. Arq Inst Biol 74(4):337–342

Leite LG, Tavares FM, Botelho PSM et al (2012) Eficiência de nematoides entomopatogênicos e inseticidas químicos contra *Sphenophorus levis* e *Leucothyreus* sp. em cana-de-açúcar. Pesqui Agropecu Trop 42(1):40–48

Lewis EE (2002) Behavioral ecology. In: Gaugler R (ed) Entomopathogenic nematology. CABI, Wallingford, pp 205–224

Lezama-Gutierrez R, Hamm JJ, Ochoa JM et al (2001) Occurrence of entomopathogens of *Spodoptera frugiperda* in the Mexican states of Michoacán, Colima, Jalisco and Tamaulipas. Fla Entomol 84(1):23–30

López-Núñez JC, Plichta K, Góngora-Botero CE et al (2008) A new entomopathogenic nematode, *Steinernema colombiense* n. sp. (Nematoda: Steinernematidae), from Colombia. Nematology 10(4):561–574

Machado LA, Habib M, Leite LG et al (2005) Pathogenicity of entomopathogenic nematodes to eggs and larvae of *Migdolus fryanus* (Westwood, 1863) (Coleoptera: Vesperidae). Arq Inst Biol 72:221–226

McCoy CW, Stuart RJ, Duncan LW et al (2002) Field efficacy of two commercial preparations of entomopathogenic nematodes against larvae of *Diaprepes abbreviatus* (Coleoptera: Curculionidae) in alfisol type soil. Fla Entomol 85(4):537–544

Mertz NR, Agudelo EJG, Sales FS (2014) Phoretic dispersion of the entomopathogenic nematode *Heterorhabditis amazonensis* (Rhabditida: Heterorhabditidae) by the beetle *Calosoma granulatum* (Coleoptera: Carabidae). Phytoparasitica 42:179–187

Molina AJP, Lopez NJC (2009) Efeito da aplicação de nematoides entomopatogênicos sobre frutos infestados com broca-do-café, *Hypothenemus hampei* (Coleoptera: Scolytidae). Nematol Bras 33(2):115–122

Molina JP, Oino AJ, Cavalcanti RS (2004) Produção *in vivo* de nematoides entomopatogênicos em diferentes insetos hospedeiros. Arq Inst Biol 71(3):347–354

Molina JP, Oino AJ, Cavalcanti RS et al (2005) Patogenicidade, multiplicação e biologia de isolados nativos de nematoides entomopatogênicos (Rhabditida: Heterorhabditidae) provenientes de Lavras, MG. Nematol Bras 29(1):25–30

Molina JP, Dolinski C, Souza RM et al (2008) Negative effect of entomopathogenic nematodes (Rhabditida: Steinernematidae and Heterorhabditidae) on *Meloidogyne mayaguensis* (Nematoda: Tylenchida) infection on tomato. J Nematol 4:338–342

Molina JP, Dolinski C, Souza RM et al (2010a) Efecto de nematodos entomopatógenos (Rhabditida: Steinernematidae y Heterorhabditidae) en la multiplicación de *Meloidogyne mayaguensis* (Tylenchida) en plantas de tomate. Fitopatol Colomb 1:23

Molina JP, Dolinski C, Souza RM et al (2010b) Efecto de nematodos entomopatógenos (Rhabditida: Steinernematidae y Heterorhabditidae) en la mortalidad de juveniles y en la inhibición de la eclosión de juveniles de *Meloidogyne mayaguensis* (Tylenchida). Fitopatol Colomb 34(2):53–56

Monteiro CMO, Prata MCA, Furlong J et al (2012) *Heterorhabditis bacteriophora* (Rhabditida: Heterorhabditidae) isolate HP88 for biological control of *Rhipicephalus microplus* (Acari: Ixodidae): the effect of different exposure times of engorged females to the nematodes. Vet Parasitol 185(2/4):364–367

Monteiro CMO, Araujo LX, Matos RS et al (2013) Association between entomopathogenic nematodes and fungi for control of *Rhipicephalus microplus* (Acari: Ixodidae). Parasitol Res 112(10):3645–3651

Monteiro CMO, Gomes GA, Araujo LX et al (2014a) Entomopathogenic nematodes associated with essential oil of *Lippia sidoides* for control of *Rhipicephalus microplus* (Acari: Ixodidae). Parasitol Res 113(1):189–195

Monteiro CMO, Matos RS, Araujo LX et al (2014b) Entomopathogenic nematodes in insect cadaver formulations for the control of *Rhipicephalus microplus* (Acari: Ixodidae). Vet Parasitol 203(3/4):310–317

Nguyen KB (2017a) Species of *Heterorhabditis*. Entomology & Nematology Department, Flórida. http://nematology.ifas.ufl.edu/nguyen/morph/HETEROSP.htm. Accessed 18 Nov. 2017

Nguyen KB (2017b) Species of *Steinernema*. Entomology & Nematology Department, Flórida. http://nematology.ifas.ufl.edu/nguyen/morph/steinsp1.htm. Accessed 18 Nov. 2017

Nguyen KB, Ginarte CM, Leite LG et al (2010) *Steinernema brazilense* n. sp. (Rhabditida: Steinernematidae), a new entomopathogenic nematode from Mato Grosso, Brazil. J Invertebr Pathol 103(1):8–20

Pereira C (1937) *Rhabditis hambletoni* n.sp. nema apparentemente semiparasito da "broca do algodoeiro" (*Gasterocercodes brasiliensis*). Arq Inst Biol 8:215–230

Pizano MA, Aguillera MM, Monteiro AR et al (1985) Incidencia de *Neoaplectana glaseri* parasitando ovos de *Migdolus fryanus* (Coleoptera: Cerambycidae). Nematol Bras 9:9–10

Poinar GO (1990) Biology and taxonomy of steinernematidae and heterorhabditidae. In: Gaugler R, Kaya HK (eds) Entomopathogenic nematodes in biological control. CRC Press, Boca Raton, pp 23–62

Portillo-Aguilar C, Villani MG, Tauber MJ et al (1999) Entomopathogenic nematode (Rhabditida: Heterorhabitidae and Steinernematidae) response to soil texture and bulk density. Environ Entomol 28:1021–1035

Rodrigueiro TSC, Ginarte CMA, Leite LG et al (2009) Eficiência de *Heterorhabditis indica* IBCB n5 (Rhabditida: Heterorhabditidae) no controle de *Alphitobius diaperinus* (Coleoptera: Tenebrionidae) sob comedouro de granja avícola. Arq Inst Biol 75(3):279–284

Rodríguez MG, Rodríguez I, Sánchez L et al (1996) Identificación y caracterización morfológica de tres cepas de nematodos entomopatógenos de Cuba. Rev Prot Veg 11:159–163

Rohde C, Oino AJ, Carvalho FD et al (2012a) Selection of entomopathogenic nematodes for the control of the fruit fly *Ceratitis capitata* (Diptera: Tephritidae). Rev Bras Ciênc Agrar 7(supl):797–802

Rohde C, Oino AJ, Silva MAT et al (2012b) Effect of *Heterorhabditis* sp. and *Steinernema carpocapsae* applied in different periods of soil infestation with larvae of *Ceratitis capitata* (Wiedemann) (Diptera: Tephritidae). Pesqui Apl Agrotec 5(3):79–84

Rohde C, Oino AJ, Mertz NR et al (2013) Compatibilidade de nematoides entomopatogênicos e extratos vegetais aquosos visando o controle da mosca-das-frutas *Ceratitis capitata* (Wiedemann) (Diptera: Tephritidae). Semin, Ciênc Agrár 34(3):1033–1042

Rosa JMO, Wilcken SRS, Aguillera MM et al (2007) Suscetibilidade de *Cornitermes cumulans* (Kollar, 1832) a isolados de nematoides entomopatogênicos. Nematol Bras 31:210–221

Rosa JMO, Wilcken SRS, Wilcken CF et al (2008) Patogenicidade de *Steinernema carpocapsae* (Rhabditida: Steinernematidae) ao cupim de montículo *Cornitermes cumulans* (Isoptera: Termitidae). Nematol Bras 32(4):260–269

Sabino AR (2014) Utilização de nematoides entomopatogênicos (Nematoda: Rhabditida) e inseticidas botânicos visando ao controle de adultos de *Rhynchophorus palmarum* L., 1764 (Coleoptera: Curculionidae). Dissertação, Universidade Federal de Alagoas

Santos V, Oino AJ, Andaló V et al (2011) Virulence of entomopathogenic nematodes (Rhabditida: Steinernematidae and Heterorhabditidae) for the control of *Diabrotica speciosa* Germar (Coleoptera: Chrysomelidae). Ciênc Agrotec 35(6):1149–1156

Silva AC, Batista Filho A, Leite LG et al (2010) Efeito de nematoides entomopatogênicos na mortalidade da mosca-do-mediterrâneo, *Ceratitis capitata*, e do gorgulho-da-goiaba, *Conotrachelus psidii*. Nematol Bras 34(1):31–40

Silva ER, Monteiro CMO, Reis-Menini C et al (2012) Action of *Heterorhabditis indica* (Rhabditida: Heterorhabditidae) strain LPP1 on the reproductive biology of engorged females of *Rhipicephalus microplus* (Acari: Ixodidae). Biol Control 62(3):140–143

Tavares FM, Batista Filho A, Leite LG et al (2007) Efeito de *Heterorhabditis indica* e *Steinernema* sp. (Nemata: Rhabditida) sobre larvas do bicudo da cana de açúcar, *Sphenophorus levis* (Coleoptera: Curculionidae), em laboratório e casa de vegetação. Nematol Bras 31:12–19

Tavares FM, Batista Filho A, Leite LG et al (2009) Efeito sinérgico de combinações entre nematoides entomopatogênicos (Nemata: Rhabditida) e inseticidas químicos na mortalidade de *Sphenophorus levis* (Vaurie) (Coleoptera: Curculionidae). BioAssay 4(7):1–10

Tavares FM, Batista Filho A, Leite LG et al (2012) Avaliação de isolados de nematoides entomopatogênicos sobre a mosca-dos-fungos, *Bradysia mabiusi* (Diptera: Sciaridae), praga em estufas. BioAssay 7:1–6

Valle EE, Dolinski C, Souza RM (2008) Dispersal of *Heterorhabditis baujardi* LPP7 (Nematoda: Rhabditida) applied to the soil as infected host cadavers. Int J Pest Manag 54(2):115–122

Chapter 11
Entomopathogenic Fungi

Lorena Barra-Bucarei, Andrés France Iglesias, and Carlos Pino Torres

11.1 Introduction

Biological control, defined as the reduction of pest populations by natural enemies, is an alternative to synthetic chemical pesticides and usually involves human intervention (Hoffmann and Frodsham 1993). Biological control agents are classified as predators, parasitoids, and pathogens (the latter are produced by microorganisms and cause diseases). Many microorganisms, such as biopesticides, have been used because they offer additional benefits beyond their objective function (Glare et al. 2012). Furthermore, they are easy to handle, multiply, and formulate, exhibiting high pest control effectiveness levels. Entomopathogenic fungi (EPFs) are highlighted within this group and have been widely studied.

More than 1000 species included in approximately 100 genera are currently known as EPFs; they affect insects of different orders and their use as biopesticides has increased during the last decades (Shah et al. 2009; Vega et al. 2012). Most EPF species are found in the Hypocreales (Ascomycota) group, which have a wide range of hosts, and Entomophthoromycota (Zygomycota), which are more specific. Although different in some aspects, both groups produce conidia or other asexual spores that are infective units (Furlong and Pell 2005; Roy et al. 2006), and they constitute a fundamental element when developing biocontrollers. An important aspect to consider in the biology of these fungi is that they can act as obligate or facultative arthropod pathogens (Goettel et al. 2005). The latter condition allows them to live as saprophytes, in which they are able to survive at the expense of

L. Barra-Bucarei (✉) · A. France Iglesias
Instituto de Investigaciones Agropecuarias, INIA, Banco de Recursos Genéticos Microbianos Department, Quilamapu, Chillán, Chile
e-mail: lorena.barra@inia.cl

C. Pino Torres
Centro I+D Agroecología, Curicó, Chile

B. Souza et al. (eds.), *Natural Enemies of Insect Pests in Neotropical Agroecosystems*, https://doi.org/10.1007/978-3-030-24733-1_11

organic matter from the soil or other substrate and as endophytes in plants (Vega et al. 2008).

One of the most relevant EPF groups for practical pest control purposes consists of the *Metarhizium* and *Beauveria* genera. They have been most studied worldwide because they are frequently found in nature and have a large number of hosts, their control is very effective, and they are simple to use in the development of commercial products. They are characterized for producing free conidia from reproductive structures (conidiophores) that are easily transportable by the wind. EPFs are widely distributed in different ecosystems and can be found in different geographic and climatic locations in both cultivated and natural soils (Vega et al. 2008). They can infect at different stages of insect development, such as eggs, larvae, pupae, and adults. Although these fungi are known as pest control agents, they have also demonstrated their ability to protect their host from diseases and limit damage caused by pathogenic microorganisms (Arnold et al. 2003; Ownley et al. 2010).

The aim of this chapter is to review the importance of entomopathogenic fungi in the context of sustainable agriculture and crop production for biological control of pests.

11.2 How Do Entomopathogenic Fungi (EPFs) Behave?

EPFs are related to insects because they use chitin as a source of carbon, which is the main component of their exoskeleton; their conidia adhere to the bodies of the insects and the cycle begins (Fig. 11.1). The life cycle is divided into a parasitic phase, which starts with infection and lasts until the host dies, and a saprophytic phase, which takes place after the death of the insect. In the first phase, conidia make contact with the insect and the germination tube is formed; the appressorium or haustorium, through mechanical (hyphal pressure) and enzymatic (lipases, proteases, and chitinases) action, penetrates the cuticle and a micropore is produced by which it advances toward the interior of the insect and reaches the hemocele (Vega et al. 2012). When it reaches the hemocele, a transition from germinative growth to vegetative growth occurs, in which metabolic changes allow the fungus to utilize the nutrients for growth and reproduction. After producing substantial biomass, the fungus generates toxins and degrading enzymes that saturate the immune system of the insect and alter its metabolism by primarily affecting its nervous system, Malpighian tubules (excretory system), and other organs because it is unable to curtail fungal development (Roberts and Humber 1981). The secondary metabolites produced by EPF have antifungal and antibacterial properties (Wagner and Lewis 2000; Parine et al. 2010). Hundreds of small molecules with insecticidal activity have been identified from EPF, such as destruxins, oosporein, beauvericin, bassianolide, bassianin, beauveriolides, bassiacridin, cordycepin, and ciclosporin among others (Hamil et al. 1969; Susuki et al. 1977; Quesada-Moraga et al. 2006; Wang and Wang 2017). The action of these metabolites induces to the host-mediated behavioral changes of the insect, such as behavioral fever, elevation seeking, reduced or increased activity,

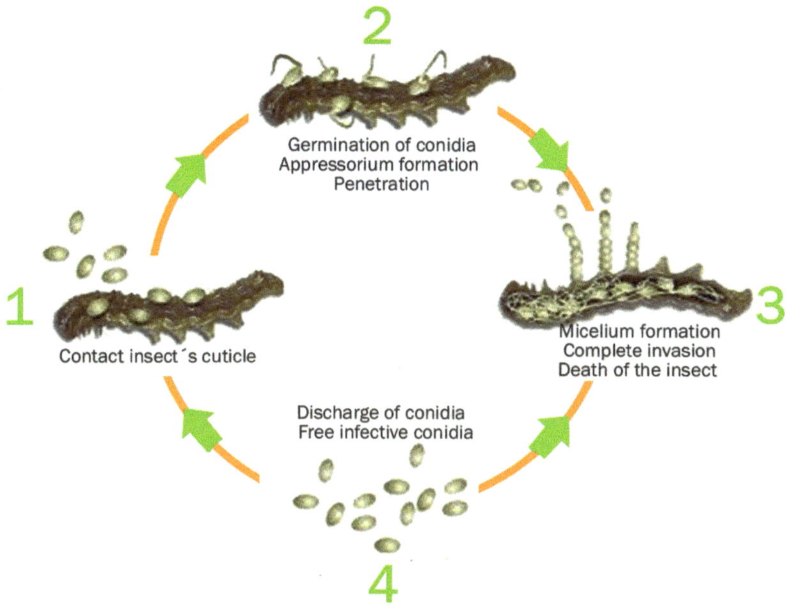

Fig. 11.1 Life cycle of entomopathogenic fungi. (Figure from INIA Quilamapu)

reduced response to semiochemicals, changes in reproductive behavior, and death (Roy et al. 2006).

The death of the insect can occur from the second day after infection, thus initiating the second phase of the cycle. The fungus has the ability to go through the cuticle of the insect and go outside where it continues its saprophytic development on the cadavers, which rapidly produces conidia or other asexual propagules that are converted into new foci of fungal dissemination (Meyling and Eilenberg 2007). EPFs have a life cycle that is synchronized with the insect stages and environmental conditions (Shah and Pell 2003).

11.3 Characteristics of Entomopathogenic Fungi (EPFs) for Biological Pest Control

During the last decades, attention has been focused on EPF development, mainly *Beauveria bassiana* (Balsamo-Crivelli) Vuillemin and *Metarhizium anisopliae* (Metchnikoff) Sorokin as inundative biopesticides (Faria and Wraight 2007). Licensed commercial products exist that have formulations containing conidia and/or mycelium for field application (Faria and Wraight 2007; Lacey et al. 2015). They are very important for the development of sustainable agriculture, significantly contribute to integrated pest management, and respond to an increasing world demand for ecologically compatible and environmentally friendly products. Some characteristics of this ecological implication include:

- Safety: EPFs are usually safe for humans and it is still unknown if they have a negative impact on the environment (Vega et al. 2008). However, there are some reports of problems caused by *Metarhizium* spp. in immunodepressed individuals (Nourrisson et al. 2017).
- Resistance: No reports have yet been found of insects that have developed resistance to EPF. However, there are studies that demonstrate insect response to infection as a result of their ability to produce detoxification enzymes, antibiotic secretions, and immune responses (Serebrov et al. 2006; Vega et al. 2012).
- Selectivity: This point is debatable because strains have been detected that are pathogens of one species; however, with the commercial development of EPF strains in the last few years, some have been found that can be parasites for an important group of insects.
- Persistence: Given the condition of facultative parasite organisms, the possibility of these fungi to persist in the soil as saprophytes awaiting new insects to appear is quite probable.
- Residues: They do not produce residues that can contaminate the environment.
- Mass production: They can be produced in large volumes and easily applied with machinery or irrigation water. These microorganisms can be produced and formulated as mycoinsecticides to be used as bait or in liquid and solid applications. They can be cultured in the laboratory and taken to the field in different substrates and conditions to cause damage to pest insects (Jaronski and Mascarin 2017).
- Handling: Handling products developed based on EPF must be similar to the handling of chemical products; however, few reports exist of human pathologies associated with EPF, which would allow reducing protection requirements during applications.
- Compatibility: They can be applied with chemical insecticides, herbicides, and some fungicides (Yáñez and France 2010; Schumacher and Poehling 2012; Machado 2016). However, it is important to consider that some products, primarily fungicides, could cause problems in its effectiveness. It is therefore necessary to know the chemical groups with which they are compatible.

11.4 Recommendations for Use of Entomopathogenic Fungi (EPFs)

Similarly to chemical products, biological products also require certain use recommendations in order to obtain expected control results. One important recommendation is to try using products developed with native or local strains that are adapted to the conditions where they will be used; this is considered as an essential factor for the success of this type of control under field conditions (Gutiérrez and Maldonado 2010).

EPFs are live organisms and are therefore susceptible to a set of factors that can affect their effectiveness, such as:

Temperature Fungi have a temperature range for optimal growth; most EPFs can germinate and develop adequately between 22 and 28 °C. Outside this range, development problems can occur, which would not allow them to reach the expected results.

Humidity One of the most important conditions for conidial germination is humidity. The majority of EPF species are hydrophilic; they reach high mortality levels in insects when humidity is high. However, some species can also achieve low mycosis levels. This is very relevant when selecting the strains to be used. Humidity on the surface of the integument of the insect, in the foliage, or on the soil can limit germination and penetration of the fungus in the host. Water combined with other environmental factors may or may not limit the persistence and effectiveness of EPF at field level.

Ultraviolet Radiation Solar radiation has wavelengths, such as UVA and UVB, which affect EPF effectiveness. Short wavelengths are able to delay or, in some extreme cases, suppress conidial germination mainly because of the damage that these wavelengths cause in fungal DNA. There is direct damage associated with the formation of photoproducts, such as pyrimidine dimmers, pyrimidine hydrates, and crosslinks between DNA and proteins. There is also indirect damage produced by the appearance of reactive oxygen molecules (e.g., hydrogen peroxide and free radicals), which damage DNA (Diffey 1991; Cerdá-Olmedo et al. 1996). The geographic zone where strains are obtained therefore becomes relevant. Studies conducted with several *Beauveria* sp. isolates showed high variability in the resistance to UVB rays among isolates obtained at different latitudes. Resistance can vary from 0% to 80% and is higher in isolates obtained at lower latitudes (Fernandes et al. 2007).

Synthetic Chemical Pesticides The use of some pesticides can affect EPF effectiveness. Studies conducted by Machado (2016) demonstrated that the tebuconazole molecule inhibits sporulation and germination of *B. bassiana*. Other studies probed that azoxystrobin, benomyl, and chlorothalonil are incompatible with different strains of *Metarhizium anisopliae* (Yáñez and France 2010). Mancozeb applications can also reduce sporulation of this fungus as much as 75%. This same study revealed that some herbicides such as glufosinate ammonium could affect EPF spore production and their performance as biocontrollers. On the other hand, studies that tested the effect of chemical pesticides (fipronil, permethrin, imidacloprid, and amitraz) on EPF concluded that they have no negative effects if applied at the recommended doses (Schumacher and Poehling 2012).

In order to improve control effectiveness using EPF, it is recommended to consider the following aspects:

1. Identify the target pest: It is necessary to have knowledge about the pest to be controlled because many EPFs are specific. Once the pest is identified, the most susceptible stage of the life cycle of the insect must be identified for control.
2. Monitoring: Before and after applying EPF, monitoring must be done with the aim of knowing pest behavior; this will help to determine the best time to make applications and increase the probabilities of success in pest control.
3. Time of application: As previously mentioned, different conditions exist in the medium that affect EPF effectiveness. For this reason, it is recommended to make applications of many EPF-based products first thing in the morning or before sunset. Most fungi require high humidity to be able to infect the insect (80% to 100%), so that natural epizootics are more common during high humidity conditions.
4. How to apply: There are different ways to apply EPF-based products, and the most effective one will depend on the nature of the pest to be controlled, the crop, and the environmental conditions.
5. Application equipment: It is recommended that the equipment be specially designed for the use of EPF because there is the risk of affecting fungal effectiveness when using equipment containing chemical pesticide residues, especially if fungicides have been used.
6. Phytosanitary management of nearby fields: It is also recommended to have knowledge about the phytosanitary management of fields bordering those managed by EPF. Fungicide applications in neighboring fields under wind conditions toward the field threatened the EPF and could affect its effectiveness.
7. Storing the product: Just as for all pesticides, EPF-based products have an expiry date; it is recommended to strictly follow manufacturer conditions because many of these products rapidly lose their viability. On the other hand, storage conditions affect their effectiveness and it is usually recommended to store them in cool, dry, and dark places.
8. Transportation and handling: It is necessary to always take into account that fungi are live organisms affected by the environmental conditions surrounding them. It is therefore recommended to transport and handle these EPF-based products under conditions that allow maintaining their effectiveness, preventing exposure to solar radiation and high temperatures.

11.5 Successful Experiences in the Development and Use of Entomopathogenic Fungi (EPFs)

11.5.1 Experience in Argentina

One of the most important cotton pests of the Americas is the cotton boll weevil (*Anthonomus grandis* Boheman) (Curculionidae), which is found in Argentina since 1993. The National Institute of Agricultural Technology (INTA) has performed evaluations to control it; the objective was to select native isolates of *B. bassiana* and *M. anisopliae* that are virulent for this pest. Isolates were obtained from soil and insect samples as well as from the mycology collection of the Entomopathogenic Fungus Laboratory (IMYZA, INTA). The pathogenicity and virulence of 28 strains of *M. anisopliae* and 66 strains of *B. bassiana* were evaluated on adult boll weevils, and the sublethal effects on feeding and oviposition produced by these fungi were studied. The Ma 50 and Ma 20 strains were the most virulent; there was decreased feeding of females infected with Ma 20 and Bb 23 and decreased oviposition with Ma 20. The evaluated *M. anisopliae* strains were compatible with pyrethroid insecticides and more tolerant to high temperatures; these results will allow the incorporation of EPF strains to the integrated management programs of this pest (Nussenbaum 2014).

Another important pest in Argentina is the fruit fly *Anastrepha fraterculus* (Wiedemann) (Tephritidae) (native) and *Ceratitis capitata* (Wiedemann) (Tephritidae) (exotic), which is a quarantine pest that limits fresh fruit and vegetable exports. Studies conducted by Albornoz (2014) allowed the selection of EPF strains to control these species in walnut, peach, and guava trees. A search of different EPFs resulted in four strains of *B. bassiana* whose pathogenicity was later evaluated on different fly stages. Larva and pupa mortality was greater than 55% for *C. capitata* and less than 27% for *A. fraterculus* in sprinkling applications. Adult mortality was 22% for both species. Results were promising to control this species, but field evaluations are necessary. Other studies conducted under laboratory conditions in the Institute of Agricultural Microbiology and Zoology INTA-Castelar evaluated the lethal concentration (90), mean survival, and conidial production of *B. bassiana* strains to control *C. capitata*. Furthermore, the combativeness of the fungus was studied with the Mercaptothion 100% CE and Dimetoato 40% CE insecticides. The study results concluded that the Bb 238, Bb 259, and Bb 132 strains were promising for the development of a commercial product to control the insect. These strains were also compatible with Mercaptothion, but exhibited compatibility limitations with Dimetoato (Porras and Lecuona 2008).

Argentina has a pilot plant within INTA called MICOPLAR that mass-produces biopesticides. This plant was initially intended to manufacture mycoinsecticide bait against leaf-cutter ants and to develop more efficient methodologies to produce EPF. It is hoped that the research developed in the plant will result in technology transfer to different productive sectors of the country in such a way as to promote local (public and/or private) micro-ventures and/or generate a network of mycoinsecticide pilot plants in strategic locations to benefit small- and medium-sized producers.

11.5.2 Experience in Brazil

Among the reasons to develop biological control in Brazil, the ample use of agrochemicals is highlighted; this has caused serious biological imbalances, such as water and soil contamination, as well as the appearance of secondary pests. Brazil is one of the Latin American countries that has achieved great progress in the use of EPF to control pests in pasturelands, sugarcane, coffee, citrus fruits, and vegetables. Just as in other countries of the region, *M. anisopliae* and *B. bassiana* are the most used fungi. The Brazilian Agricultural Research Corporation (Embrapa) has made significant contributions to the development of biological control technologies using EPF. It has an important fungus collection currently conserved in the National Genetic Resources and Biotechnology Research Center (Cenargen). It has also implemented technology transfer programs that include consultancy, training, and publications.

Various biological control strategies have been developed for the pest *Bemisia tabaci* (Gennadius) (Aleyrodidae); Brazil has been seriously affected by this insect since 1995 and accumulated losses exceed 5 billion dollars (Maranhão and Maranhão 2009). Among the most affected crops are tomato, cucumber, melon, watermelon, cotton, and other ornamental plants. The insect was initially detected in the south-east region (São Paulo) and was rapidly disseminated to almost all the regions of the country (Villas Bôas et al. 1997; Lima et al. 2000). Vicentini et al. (2001) evaluated 50 strains of *B. bassiana* to control nymphs of *B. tabaci* biotype B in melon. Results of strains CG 136 and CG 149 for nymph mortality were approximately 90%. These strains were isolated from insects of the same taxon as *B. tabaci* and performed better than the strains contained in commercial products recommended to control this insect.

11.5.3 Experience in Chile

There has been increasing interest in Chile since the 1990s to develop this technology, and several institutions have conducted research in this field. Important collections were established, especially those provided by the Institute of Agricultural Research (INIA) through its Technology Center for Biological Control, which consists of more than 1000 fungal strains that were collected throughout the country and currently conserved in the Microbial Genetic Resources Bank. Currently, INIA is the institution that has conducted the most studies with EPF and is developing many initiatives to control economically important pests.

Fruit growing for exportation faces a significant group of pests that are economically important that consist of Curculionidae (weevils), especially the species *Aegorhinus superciliosus* (Guérin-Méneville), *Aegorhinus nodipennis* Hope, and *Naupactus xanthographus* (Germar). One of the main limiting factors in grapevine production and other fruit trees in Chile is related to the control of the vine weevil (*N. xanthographus*) (Artigas 1994). It is a native quarantine species conventionally controlled by repeated applications of organophosphate pesticides in the aerial part and in the past by the establishment of sticky bands with azinphos-methyl on the trunk of each plant (González 1983). It is critical to attack this pest in vineyards under organic management where the use of synthetic chemical insecticides is prohibited. In 2002, the vine weevil attack was massive in organic vineyards and provoked serious economic damage, exhibiting a mean of 12 adults per plant. Consequently, biological control alternatives in the soil began to be explored.

After large laboratory screening tests using different native strains of *B. bassiana* and *M. anisopliae* against fruit tree weevil larvae, the strains 1 CET and B323 of *B. bassiana* and M82 and M430 of *M. anisopliae* (Fig. 11.2) were selected. Field evaluations were then performed by applying lyophilized conidia, which controlled larvae by more than 90% as compared to the control; strains M430 and B323 were the most effective (Mejías 2004). The Agroecology R&D Center made large-scale applications of these strains through a drip irrigation system, which directly arrived at the area where the larvae were attacking the vine roots. Applications took place

Fig. 11.2 Biological control experience using entomopathogenic fungi (EPFs) in organic grapevines: (**a**) vine weevil adult female, (**b**) vine weevil larvae in the soil, (**c**) larvae parasitized with EPF. (Photo: Carlos Pino, Agroecological R&D Center)

in autumn and spring 2003 and control levels were approximately 90% of larvae; in the aerial part, the level decreased to less than 1 adult per plant. These evaluations became common management practice for this Chilean native pest and stopped being one of the main limiting factors of ecological management. This is one of the milestones of successful large-scale biological control, which facilitated the conversion of more than 3000 ha to organic management and proved that ecological management requires science, proactivity, and collective ingenuity for its development (Pino 2013).

Another pest that is currently provoking serious damage in Chilean fruit growing is *Lobesia botrana* (Denis & Schiffermüller) (Tortricidae), which affects *Vitis vinifera* and other fruit trees in urban and productive sectors of the country. Formulations are being evaluated to control this quarantine pest, which are based on different strains of the genera *Beauveria* and *Metarhizium* management at field level.

For horticulture, evaluations with native strains of *B. bassiana* and *M. anisopliae* are in progress to control the tomato moth [*Tuta absoluta* (Meyrick) (Gelechiidae)], which is an important pest in the tomato crop (*Solanun lycopersicon*) because it can cause yield losses of approximately 90% of the fruit (Apablaza 1988). Strains of *Metarhizium* exhibit high control levels, 80% mortality in eggs and 90% in larvae, higher than *B. bassiana* strains which had 60% mortality in eggs and 50% in larvae (Rodríguez et al. 2006a, b).

11.5.4 Experience in Colombia

The history of insect biological control in Colombia using EPF arise from the need of producers to have access to more sustainable and economical products to control important agricultural pests. Together with research on the use of EPF on different pests of economic interest, several small artisanal companies were established to mass-produce these agents. Colombia currently has both large national and international companies that produce, commercialize, and export several EPF strains. The development of EPFs has been associated with the implementation of integrated pest management programs for different crops, such as cotton, maize, soy, sorghum, yucca, tomato, bean, banana, and fruit trees.

As demonstrated in different research studies, Colombia is working hard to evaluate the potential of EPF to control extensive crop pests, such as the coffee berry borer (*Hypothenemus hampei* (Ferrari) (Curculionidae) and the white potato worm (*Premnotrypes vorax* Hustache) (Curculionidae). For example, efforts have been made in recent years to reduce the losses caused by the white worm in potato crops, which can reach 100% depending on the infestation level and crop management. Studies conducted by Villamil et al. (2016) demonstrated that commercial strains combined with Metaril® W.P + *B. brongniartii*® W.P (T6) and the native isolate of *Beauveria* sp. Bv01 (T1) reduced damage caused by the insect with significant control levels (77.0% and 77.6%), which represents a promising alternative to control *P. vorax* when it is incorporated in an integrated pest management scheme (Delgado 2015). In the case of coffee, the most important pest is the berry borer, which was introduced from Africa and arrived without any natural enemies. The National Coffee Research, Innovation, and Technological Development Center (Cenicafé) has established an EPF collection (native and exotic strains) with important biocontrol activity on the insect. Numerous studies have been conducted over time using EPF with excellent control results. The EPF introduction program sponsored by the National Coffee Growers Federation is highlighted; the total berry borer population in 1995 decreased by 49% on the average because of *Beauveria* (Ruíz 1996).

In the last 20 years, the Colombian Agricultural Institute (ICA) has concentrated on research to find biological alternatives for the integrated pest management programs. This is coupled with the effort by the Colombian Agricultural Research Corporation (Corpoica, now Agrosavia) to conduct studies of biocontrol applications in which EPFs are emphasized in different economically important crops for the country.

11.6 Conclusion

Entomopathogenic fungi have evolved from the "calcinaccio" described from the *Beauveria bassiana*-covered spore caterpillars, by the "Father of Insect Pathology" Agostino Bassi (Davidson 2012), to a complex science with a tremendous potential for insect control. Thus, EPFs can serve as effective biological control agents for pest control and also as plant-beneficial microbes by protecting plants against insects or microbial plant pathogens. EPFs can affect insects by changing its behavior, ability to feed, reproduce, and survive. Moreover, some EPFs can colonize various tissues and persist for months, even pass to the next generation such as endophytic microorganisms, protecting plants through direct interactions with insect herbivores by producing insecticidal metabolites or by inducing systemic resistance.

It is clear that EPFs play an important role as effective biocontrol agents and pest management worldwide as well as in Latin American countries, but our understanding of their effectiveness, mass production, effective deliveries, and impacts on plant and insect communities is still at the beginning stages. Research is needed in order to understand their additional roles in nature, as endophytes, antagonists to plant pathogens, in association with the rhizosphere, and even plant growth-promoting agents. Thus, these multiple roles deserve to the EPF an auspicious future in the development of agriculture.

References

Albornoz P (2014) Hongos patógenos asociados a moscas de la fruta (Diptera: Tephritidae) en el sector austral de Las Yungas del noroeste argentino. Dissertation, Universidad Nacional de La Plata

Apablaza J (1988) Avances en el control de la polilla del tomate. Evaluación del insecticida cartap. Cienc Tecnol 35(9):27–30

Arnold A, Mejía L, Kyllo D et al (2003) Fungal endophytes limit pathogen damage in a tropical tree. Proc Natl Acad Sci U S A 100(26):15649–15654

Artigas J (1994) Entomología económica: insectos de interés agrícola, forestal, médico y veterinario, vol 2. Ediciones Universidad de Concepción, Concepción

Cerdá-Olmedo E, Martín-Rojas V, Cubero B (1996) Causes of cell death following ultraviolet B and C exposures and the role of carotenes. Photochem Photobiol 64(3):547–551

Davidson EW (2012) History of insect pathology. In: Vega F, Kaya H (eds) Insect pathology, 4th edn. Academic Press, New York, pp 13–28

Delgado XLM (2015) Control biológico con hongos entomopatógenos: una estrategia de aula para promover el cuidado del medio ambiente. Dissertation, Universidad Nacional de Colombia

Diffey BL (1991) Solar ultraviolet radiation effects on biological systems. Phys Med Biol 36(3):299–328

Faria MR, Wraight SP (2007) Mycoinsecticides and mycoacaricides: a comprehensive list with worldwide coverage and international classification of formulation types. Biol Control 43(3):237–256

Fernandes EK, Rangel DE, Moraes AM et al (2007) Variability in tolerance to UV-B radiation among *Beauveria* spp. isolates. J Invertebr Pathol 96(3):237–243

Furlong MJ, Pell JK (2005) Interactions between entomopathogenic fungi and arthropod natural enemies. In: Vega FE, Blackwell M (eds) Insect-fungal associations: ecology and evolution. Oxford University Press, New York, pp 51–73

Glare T, Caradus J, Gelernter W et al (2012) Have biopesticides come of age? Trends Biotechnol 30(5):250–258

Goettel MS, Eilenberg J, Glare T (2005) Entomopathogenic fungi and their role in regulation of insect populations. In: Gilbert L, Iatrou K (eds) Comprehensive molecular insect science. Elsevier, Amsterdam, pp 361–406

González R (1983) Manejo de plagas de la vid. Universidad de Chile, Santiago

Gutiérrez CG, Maldonado MBG (2010) Uso de bioinsecticidas para el control de plagas de hortalizas en comunidades rurales. Ra Ximhai 6(1):17–22

Hamil R, Higgens C, Boaz H et al (1969) The structure of beauvericin, a new depsipeptide antibiotic toxic to *Artemia salina*. Tetrahedron Lett 49(49):4255–4258

Hoffmann M, Frodsham A (1993) In: Cooperativa de la Universidad de Cornell (ed) Los enemigos naturales de las plagas de insectos vegetales, Ithaca

Jaronski ST, Mascarin GM (2017) Mass production of fungal entomopathogens. In: Lacey L (ed) Microbial control of insect and mite pests: from theory to practice. Academic Press, New York, pp 141–155

Lacey LA, Grzywacz D, Shapiro-Ilan DI et al (2015) Insect pathogens as biological control agents: back to the future. J Invertebr Pathol 132:1–41

Lima LHC, Návia D, Inglis PW et al (2000) Survey of *Bemisia tabaci* (Gennadius) (Hemiptera: Aleyrodidae) biotypes in Brazil using RAPD markers. Genet Mol Biol 23(4):781–785

Machado L (2016) Efectos de plaguicidas químicos empleados en *Phaseolus vulgaris* L. sobre el crecimiento, colonización y patogenicidad de *Beauveria bassiana* (Bals.) Vuill y *Metarhizium anisopliae* (Mest.) Sorok. Dissertation, Universidad Central Marta Abreu de Las Villas

Maranhão EAA, Maranhão EHA (2009) Hongos entomopatógenos: importante herramienta para el control de "moscas blancas" (Homoptera: Aleyrodidae). An Acad Pernamb Ciênc Agron 5(6):209–242

Mejías P (2004) Efectividad de *Beauveria bassiana*, *Metarhizium anisoliae* y *Steinernema* sp. en el control de *Naupactus xanthographus*. Tesis, Universidad Católica del Maule

Meyling NV, Eilenberg J (2007) Ecology of the entomopathogenic fungi *Beauveria bassiana* and *Metarhizium anisopliae* in temperate agroecosystems: potential for conservation biological control. Biol Control 43(2):145–155

Nourrisson C, Dupont D, Lavergne RA et al (2017) Species of *Metarhizium anisopliae* complex implicated in human infections: retrospective sequencing study. Clin Microbiol Infect 23(12):994–999

Nussenbaum AL (2014) Aislamientos de *Beauveria bassiana* y *Metarhizium anisopliae* virulentos para el control del picudo del algodonero, *Anthonomus grandis* (Coleoptera: Curculionidae). Tesis, Universidad de Buenos Aires

Ownley B, Gwinn K, Vega F (2010) Endophytic fungal entomopathogens with activity against plant pathogens: ecology and evolution. BioControl 55(1):113–128

Parine NR, Devinder K, Khan PAA et al (2010) Antifungal efficacy of secondary metabolites from entomopathogenic fungi *Beauveria bassiana*. J Pharm Res 3(4):855–856

Pino C (2013) Manual de vitivinicultura orgánica. Cespedes, Curicó

Porras L, Lecuona R (2008) Estudios de laboratorio para el control de *Ceratitis capitata* (Wiedmann) (Diptera: Tephritidae) (Mosca del Mediterráneo) con *Beauveria bassiana*. Agron Costarricense 32(2):119–128

Quesada-Moraga E, Maranhao E, Valverde-García P et al (2006) Selection of *Beauveria bassiana* isolates for control of the whiteflies *Bemisia tabaci* and *Trialeurodes vaporariorum* on the basis of their virulence, thermal requirements and toxicogenic activity. Biol Control 36(3):274–287

Roberts DW, Humber RA (1981) Entomogenous fungi. In: Cole GT, Kendrick B (eds) Biology of conidial fungi. Academic Press, New York, pp 201–236

Rodríguez M, Gerding M, France A (2006a) Efectividad de aislamientos de hongos entomopatógenos sobre larvas de polilla del tomate *Tuta absoluta* Meyrick (Lepidoptera: Gelechiidae). Agric Téc 66(2):159–165

Rodríguez M, Gerding M, France A (2006b) Entomopathogenic fungi isolates selection for egg control of tomato moth Tuta absoluta Meyrick (Lepidoptera: Gelechiidae) eggs. Agric Téc 66(2):151–158

Roy HE, Steinkraus DC, Eilenberg J et al (2006) Bizarre interactions and endgames: entomopathogenic fungi and their arthropod hosts. Annu Rev Entomol 51:331–357

Ruíz R (1996) Efecto de la fenología del fruto del café sobre los parámetros de la tabla de vida de la broca del café, *Hypothenemus hampei* (Ferrari). Tesis, Universidad de Calda

Schumacher V, Poehling HM (2012) In vitro effect of pesticides on the germination, vegetative growth, and conidial production of two strains of *Metarhizium anisopliae*. Fungal Biol 116(1):121–132

Serebrov VV, Gerber ON, Malyarchuk AA et al (2006) Effect of entomopathogenic fungi on detoxification enzyme activity in greater wax moth *Galleria mellonella* L. (Lepidoptera, Pyralidae) and role of detoxification enzymes in development of insect resistance to entomopathogenic fungi. Biol Bull 33:581

Shah PA, Pell JK (2003) Entomopathogenic fungi as biological control agents. Appl Microbiol Biotechnol 61(5/6):413–423

Shah FA, Ansari MA, Watkins J et al (2009) Influence of commercial fungicides on the germination, growth and virulence of four species of entomopathogenic fungi. Biocontrol Sci Tech 19(7):743–753

Susuki A, Kanaoka M, Isogai A et al (1977) Bassianolide, a new insecticidal cyclodepsipeptide from *Beauveria bassiana* and *Verticillium lecanni*. Tetrahedrom Lett 18(25):2167–2170

Vega F, Posada F, Aime M et al (2008) Entomopathogenic fungal endophytes. Biol Control 46(1):72–82

Vega FE, Meyling NV, Luangsa-Ard JJ et al (2012) Fungal entomopathogens. In: Vega FE, Kaya HK (eds) Insect pathology, 2th edn. Academic Press, New York, pp 171–122

Vicentini S, Faria M, Oliveira MR (2001) Screening of *Beauveria bassiana* (Deuteromycotina: Hyphomycetes) isolates against nymphs of *Bemisia tabaci* (Genn.) biotype B (Hemiptera: Aleyrodidae) with description of a new bioassay method. Neotrop Entomol 30(1):97–103

Villamil JE, Martínez JW, Pinzón EH (2016) Actividad biológica de hongos entomopatógenos sobre *Premnotrypes vorax* Hustache (Coleoptera: Curculionidae). Rev Ciênc Agr 33(1):34–42

Villas Bôas GL, França FH, Ávila AC et al (1997) Manejo integrado da mosca-branca *Bemisia argentifolii*. Embrapa, Brasília. (Circular Técnica da Embrapa Hortaliças, 9)

Wagner BL, Lewis LC (2000) Colonization of corn, *Zea mays*, by the entomopathogenic fungus *Beauveria bassiana*. Appl Environ Microbiol 66(8):3468–3473

Wang C, Wang S (2017) Insect pathogenic fungi: genomics, molecular interactions, and genetic improvements. Annu Rev Entomol 62:73–90

Yáñez M, France A (2010) Effects of fungicides on the development of the entomopathogenic fungus *Metarhizium anisopliae* var. anisopliae. Chil J Agric Res 70(3):390–398

Chapter 12
Entomopathogenic Viruses

Fernando Hercos Valicente

12.1 Introduction

Brazil is essentially a tropical country with continental dimensions and several agricultural frontiers. The most recent MATOPIBA border includes part of the states of **Ma**ranhão, **To**cantins, **Pi**auí, and **Ba**hia. The average annual temperature in these regions is always favorable to the appearance of insect pests in many different crops. Faced with the current Brazilian landscape, these regions reach up to three annual crops with a supply of "green" food for insect pests from planting to harvest due to the overlapping of crops throughout the year. This factor is called "green bridge." As large areas of agricultural frontiers mainly plant soybeans, cotton, maize, and beans, insect pests such as caterpillars migrate easily between crops, for example, the soybean looper, *Chrysodeixis* sp., which is also a major cotton pest. Fall armyworm, *Spodoptera frugiperda* (J. E. Smith) (Noctuidae), is one of the most important corn pests, but it is also a pest of soy, cotton, and beans. The same happens with the cotton bollworm *Helicoverpa armigera* (Hübner) (Noctuidae) that attacks corn, soybean, and cotton. The control of these agricultural pests is essentially done with the indiscriminate use of chemical insecticides, which has generated environmental pollution throughout the planet, besides causing the intoxication of applicators, rivers, and springs and the contamination of the final product to be sold in the market, both *in natura* and in processed ones. Among the biological control agents, viruses, especially those from the baculovirus group, are a viable alternative for pest control of agricultural importance and are a fundamental tool within the context of integrated pest management (IPM). Baculoviruses are host-specific restricted to arthropods (Organisation for Economic Co-operation and Development 2002). So far, no cases of pathogenicity of a baculovirus to a vertebrate have been reported (Krieg et al. 1980; Entwistle 1983).

F. H. Valicente (✉)
Embrapa Maize and Sorghum, Sete Lagoas, MG, Brazil
e-mail: fernando.valicente@embrapa.br

B. Souza et al. (eds.), *Natural Enemies of Insect Pests in Neotropical Agroecosystems*, https://doi.org/10.1007/978-3-030-24733-1_12

More than 17 families or groups of invertebrate viruses have been recognized as pathogenic to insects, of which baculoviruses have been the most documented (Harrison et al. 2016). Baculoviruses are genetically and morphologically distinct from other invertebrate virus families and are the largest group of insect-attacking viruses. According to Miller (1997) and Eberle et al. (2012a, b), the number of baculoviruses described to date, over 600, is from the family Baculoviridae infecting Lepidoptera (butterflies and moths), Hymenoptera (sawflies), and Diptera (mosquitos) (Herniou et al. 2012), although Federici (1997) suggested that many more virus can be identified from Lepidoptera. The diseases associated with this type of viral structure are named polyhedrosis and their transmission can occur via occlusion bodies (OBs) present in foods – usually sprayed on the leaves like a biopesticide, or horizontal transmission, when present on the surface of eggs, and vertical transmission, inside the ovum of infected adults (females or males). The baculovirus group is a viable alternative for pest control of agricultural importance and is an important tool within the context of IPM. Many biological control programs use baculovirus as the main biological pesticide. Baculoviruses are host-specific restricted to arthropods. So far, no cases of pathogenicity of a baculovirus to a vertebrate have been reported. The objective of this chapter is to demonstrate the diversity of the Baculoviridae family, to which one of the most important virus used in biological control belongs. We indicate the advantages and disadvantages of using the baculovirus for the biological control of agricultural pests and the safety towards mammals, vertebrates, and other animals of using biological products based on baculovirus. We also list biological control programs worldwide that use baculovirus as a biological pesticide and considerations on the use and application of baculovirus products in the field.

12.2 Diversity of the Family Baculoviridae

A great diversity of invertebrate viruses has been documented. However, baculoviruses have been highlighted as one of the most important entomopathogenic viruses with great potential to be used in IPM.

According to the Organisation for Economic Co-operation and Development (2002) and Ikeda et al. (2015), baculoviruses are arthropod-specific viruses and are the only ones lacking homology with viruses found in other organisms such as plants, animals, fungi, and bacteria. Baculoviruses belong to the family Baculoviridae and are one of the largest groups of viruses showing high complexity of forms and functions, besides diversity related to their size, organization, and the gene content of their genomes. The high genomic variation may reflect the phenotypic diversity observed among the four genera of the Baculoviridae family (Herniou et al. 2003; Jehle et al. 2006; Van Oers and Vlak 2007; Miele et al. 2011). In addition to the morphological differences of the granulovirus (GV) and nucleopolyhedrovirus (NPVs), OBs are found in lepidopteran-specific baculoviruses. Figure 12.1 shows NPV and GVs identified from fall armyworm (Valicente et al. 1989).

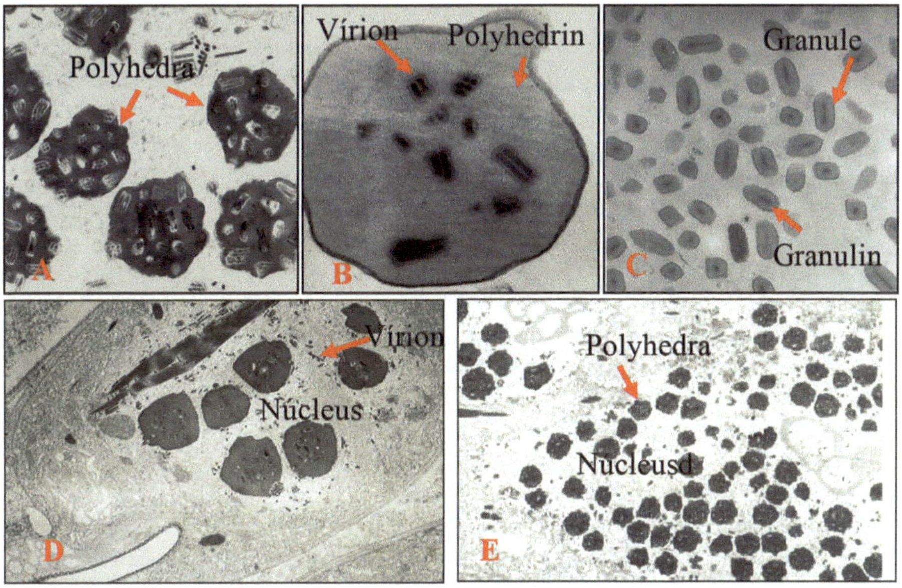

Fig. 12.1 Electron microscope photographs of infected tissues of fall armyworm, *Spodoptera frugiperda*, with baculovirus. (**a**, **b**) Infection of fall armyworm tissue with nucleopolyhedrovirus (NPV) where polyhedra, virion, and polyhedrin can be observed. (**c**) Granulovirus (GV), where only one capsid per envelope and the granulin it is observed; (**d**, **e**) infected tissues with nucleopolyhedrovirus, virions, and polyhedra in the cell nucleus. (Photos: Dr. E.W. Kitajima)

Alphabaculovirus presents cell infection in virtually all tissues of the host insect (Katsuma et al. 2012). The heterogeneity of phenotypes is commonly maintained in field populations, and it is believed that genetic diversity has advantages for the adaptation, evolution, and survival time of baculovirus in the field.

According to Miller and Ball (1998) and Harrison and Hoover (2012), the family Baculoviridae has double-stranded circular DNA with genomes ranging from 80 to 180 kbp, which are packaged in rod-shaped infective particles or nucleocapsids. These are found within proteinaceous bodies called occlusion bodies. For the NPV, polyhedrin is the main constituent of their crystalline protein structure, presenting polyhedral forms called polyhedrins, or granulin with ovoid shape (granules) for the granulovirus.

The VIII Report of the International Committee of Taxonomy of Viruses (Fauquet et al. 2005) classified the family Baculoviridae in two genera until 2015: Nucleopolyhedrovirus and Granulovirus. A new taxonomic subdivision was proposed, which has resulted in the current classification of the Baculoviridae family into four genera: Alpha-, Beta-, Gamma-, and Deltabaculovirus. Alphabaculovirus includes all lepidopteran NPVs that form the viral phenotypes budded virus (BV) and occlusion-derived virus (ODV), Betabaculovirus comprises lepidopteran GVs that also form viral particles during infection, Gammabaculovirus are

Hymenoptera-specific and encompasses NPVs that do not have genes corresponding to the specific proteins of the BV particle, and Deltabaculovirus includes dipteran baculoviruses that do not have in its genome a gene homologous to that encoding polyhedrin expression, characteristic of the other NPVs. According to Eberle et al. (2012a, b), most species of baculovirus are found within the Alpha- and Betabaculoviruses. The largest number of the over 600 known species is from the Lepidoptera.

Figure 12.1 shows the basic structure of a nucleopolyhedrovirus (NPV) and a granulovirus (GV) in an electron microscope of fall armyworm, *Spodoptera frugiperda*, infected tissues.

The life cycle of a baculovirus is pH dependent and the caterpillar ingests the contaminated food with polyhedral inclusion bodies (PIBs). The alkaline pH of the midgut triggers the polyhedra to dissolve and release of one or more nucleocapsids into the midgut lumen (Adams and McClintock 1991; Hasse et al. 2015). According to Hasse et al. (2015) the life cycle starts when a baculovirus occlusion body (OB) is ingested with contaminated food, and when these OBs reach the alkaline midgut, the proteinaceous matrix is dissolved releasing occlusion-derived virus (ODV). The peritrophic membrane is degraded by virus- and host-encoded enzymes present in the OB allowing the ODV to enter the cell, and the ODV enters the cell by fusion with epithelial cell microvilli releasing nucleocapsids (NCs) into the cytoplasm. Nucleocapsids may enter the nucleus, disassemble, and release the genome. At this point, early genes are transcribed and translated, and some of the proteins translocate into the nucleus and take part in genome transcription/replication, NC, and virion assembly. In the first stages of viral infection, NC is transported to the cytoplasm, approaches the basolateral cell membrane (CM), and emerges as budded virus (BV) in the spots where the viral envelope fusion protein (EFP) accumulates, using the secretory pathway. In the very late stages of infection, NCs are enveloped in the nucleus and occluded in the polyhedral-shaped protein matrix (OB) (Hasse et al. 2015).

12.3 Advantages and Disadvantages of Using Baculovirus for the Biological Control of Pests

The great advantage of baculoviruses, which are natural control agents, is that they do not cause any harm to the health of applicators; do not kill natural enemies of insect pests; do not destabilize the environment; do not pollute forests, springs; and rivers; and do not contaminate products *in natura* to be sold in the shelves of supermarkets, leaving no residues in flowers, fruits, and vegetables. All these factors combined with the specificity and ease of handling of baculovirus in relation to chemical pesticides and the target insect make it one of the best biological control agents. The specificity of baculoviruses is a great advantage; however, it can be considered a disadvantage, sometimes infecting only one insect species at a time.

Another advantage of baculovirus-based biopesticide is that they can be sprayed using the same equipment for application of chemicals, but respecting the final volume per hectare. This is a factor that contributes to low-cost application of the baculovirus that does not require special equipment. Several baculovirus-based products have already been tested with chemicals with some good results (Valicente and Costa 1995). Most chemicals have compatibility to be mixed up with baculoviruses, so the use of the same equipment is not a limiting factor, which facilitates the handling within the rural properties.

The action of the baculoviruses is slower than chemical insecticides, taking more time to kill the target insect. Although baculovirus-contaminated larvae reduced their feed by 93% (Valicente 1989), this insect still feeds on the sprayed plant for a couple of days, until it drastically reduces feeding. This factor may contribute to the fact that some farmers do not see the immediate effect of baculovirus in controlling caterpillars, sometimes delaying the incorporation of this control agent into the IPM. In general, baculoviruses infect caterpillars, e.g., fall armyworm, and cause their death as long as they are up to 1 cm in length or at most in the third instar. Thus, soon after planting, crop areas should be monitored to prevent caterpillars from growing and lodging in the corn whorl. In this case, it is very difficult for a biological product and even chemical to reach the caterpillar inside the whorl.

The tendency of a disease caused by baculovirus to become an epizootic depends on the scale of dispersion of this pathogen, the persistence of the virus inside and outside the host, which requires several generations of the host to develop and spread the disease, and it also depends on the number of the target insects present in the field and its stage of development. The instar of the insect is directly related with the rate of infection in the field. Dead larvae represent a relevant source of inoculum for the occurrence and maintenance of epizootics in epidemic populations. Other insects and birds can feed on these dead insects, promoting the dispersion of viral particles into the environment, as well as the rain and wind. The OBs, which present themselves in an environmentally stable way, persist dormant and viable in soil, surviving for many years (Jaques 1967; Thompson et al. 1981). Thus, insect migration and host population fluctuations are events that strongly influence the persistence of the virus in the field, hence assuming an important role in the ecology of baculoviruses.

The population dynamics of insects, considering more broadly the mechanisms of interaction between pathogen and host or the impacts of these pathogens on insect populations in the field, have been little explored. However, the action and influences of parasitoids, predators, and pathogens on maintaining pest density below the threshold of economic damage or reducing their outbreaks have been known and documented.

In most cases, baculoviruses are very efficient because they are characterized as highly virulent and specific to their hosts, as well as being safe for human health and the environment, since they do not cause negative impacts to plants, mammals, birds, fish, or even nontarget insects. In the past 20 years, safety tests have been done and no health or environmental problems have been documented (Hauschild et al. 2011). Baculoviruses have been included in lists of low-risk biocontrol agents, as described in the Organisation for Economic Co-operation and Development (OECD) document entitled Consensus Documentation Information Used in the Assessment of Environmental Applications Involving Baculovirus (http://www.rebecanet.de).

12.4 Safety of the Use of Biological Products Based on Baculovirus

Burges et al. (1980) and the Organisation for Economic Co-operation and Development (2002) published about the safety of baculoviruses stating that NPVs are harmless or are unable to replicate in microorganisms, noninsect cell cultures of invertebrates, vertebrate cell culture, non-arthropod plants and invertebrates. At the end of the 1960s, a NPV isolate for the control of *Helicoverpa zea* (Boddie) (Noctuidae) was subjected to a series of tests as stringent as the chemicals that are submitted by the World Health Organization (WHO) and the U.S. Environmental Protection Agency (USEPA). Tests were performed on primates and on human for carcinogenic and teratogenic effects. The tests were extremely rigorous in order to be sure of noninfection of vertebrate and human animals with insect viruses (Burges et al. 1980). Organisms tested include sparrows, rats, and nontarget organisms (predatory insects, etc.). No adverse effects were detected in any of the organisms tested. Summers and Kawanishi (1978) and the Organisation for Economic Co-operation Development in 2002 published extensive studies confirming biosafety in the use and application of baculovirus-based products in agriculture.

12.5 Biological Control with Baculovirus

Worldwide, there are several biological control programs. There are many examples of Betabaculoviruses, Alphabaculovirus, and Gammabaculovirus that have been developed as microbial control agents and used in a wide range of crops such as corn, potato, citrus, cotton, pasture grass, and tomato (Grzywacz 2017).

12.6 Some Examples of Biological Control Programs Using Baculovirus

It is worth mentioning that all the production process of a baculovirus-based biopesticide is done in the laboratory using healthy larvae from artificial rearing. It is not always easy to complete the life cycle of the caterpillar in the laboratory, having an artificial diet as a limiting factor, and the incubation temperature of the insect after infection with baculovirus is a challenge. Herein, we describe some new biological control programs.

12.7 *Anticarsia gemmatalis* Nucleopolyhedrovirus (AgMNPV) in Soybean Crops

According to Sosa-Goméz (2017), Brazil treats the largest area of soybean with microbial agents, and it includes the baculovirus. This was the largest biological control program using entomopathogenic viruses in Brazil (Moscardi 1999; Moscardi et al. 2011; Sosa-Gómez et al. 2014; Sosa-Goméz 2017). The soybean caterpillar, *Anticarsia gemmatalis* Hübner (Noctuidae), is one of the main pests of soybean cultivation in Brazil, occurring from Argentina to southeastern of the United States. The application of the product reached almost two million hectares of soybeans in the country (Moscardi 1999, 2007; Moscardi and Sosa-Gomez 2007; Moscardi et al. 2011) and counted with the help of researchers from Embrapa and extensionists. However, due to some technical problems, the production by the company Coodetec was closed (Dr. Flávio Moscardi – personal information). Nowadays, *Baculovirus anticarsia* is produced for an area of approximately 300,000 ha/year (Dr. Bráulio Santos – personal information). This virus (AgMNPV) is considered the prototype of baculovirus in relation to biological control because of its importance in both basic and applied research. Extensive studies have been carried out regarding the morphological identification, genetic and molecular characterization, pathology, and biological activity of different AgMNPV isolates. The complete genome of the purified clone AgMNPV-2D has been published since 2006 (Oliveira et al. 2006).

12.8 *Spodoptera frugiperda* Nucleopolyhedrovirus (SfMNPV) in Corn Crops

A very important project started in the 1980s and has developed some baculovirus-based biopesticides. The work with the baculovirus to control fall armyworm began in 1984 at Embrapa Maize and Sorghum in Brazil. A survey of the natural enemies of this pest was carried out in several maize-producing regions of the state of Minas Gerais. During the survey, between 1984 and 1989, more than 14,000 caterpillars were collected, where several parasitoids of the Diptera and Hymenoptera orders were found, including several dead larvae killed by baculoviruses (Valicente 1989). Currently, the baculovirus collection for fall armyworm has 22 isolates sampled in several regions of Brazil. These isolates were studied and characterized and their efficiency evaluated in relation to the cartridge caterpillar (Barreto et al. 2005).

Among the most studied and efficient isolates in the control of this pest, isolate 19 had its genome already fully sequenced (Wolff et al. 2008). The baculovirus that infects fall armyworm larvae causes the disruption of the integument of the insect immediately after their death due to the action of two genes: cathepsin (v-cath) and chitinase (chiA) (Hawtin et al. 1997). Isolate 6 from the Baculovirus Collection of Embrapa Maize and Sorghum presents a unique characteristic of not causing liquefaction of the integument immediately after larva death. Sequencing of the chitinase A (v-chiA) gene from this isolate (SfMNPV-6) showed a mutation in the gene that generates a premature stop codon, considerably reducing the size of the putative enzyme (Valicente et al. 2007a, b, 2008; Vieira 2012). When the insect tegument ruptures, dead larvae infected with baculovirus need to be frozen so that they are harvested with forceps and frozen again until processing and/or formulation. This factor implies greater expenditure of labor, electricity, freezers, and physical space, which results in a final product with a higher price. The advantages of an isolate that does not cause the disruption of the integument is that they are easy to harvest and there is no loss of PIBs.

Many bioassays were performed to characterize this baculovirus isolate such as lethal concentration 50 (LC_{50}), lethal time 50 (TL_{50}), and several other factors such as polyhedra produced per caterpillar, larval equivalent (which is the number of larvae needed to spray 1 ha – LE/ha), and weight larval equivalent (which can be defined as the weight of larvae required to be sprayed in 1 ha). Thus, the main objective is to produce more polyhedra per larvae and, consequently, to reduce the number of equivalent larvae (LE) (Valicente et al. 2013). Valicente et al. (2013) also found a strong correlation between weight and the number of dead larvae that needed to be sprayed in 1 ha. Thus, between 12 and 14 g of dead larvae equivalent is needed to be sprayed in 1 ha. Nowadays, there are contracts with 3 large companies for this product to be on the market. The first Brazilian company to register the *Baculovirus spodoptera* for fall armyworm was VR Biotech®, and the name of the commercial product is Cartuchovit®. This isolate was also registered by Simbiose®, also a Brazilian company, named VirControl® (Fig. 12.2).

12.9 Baculovirus for the Control of the Soybean Looper, *Chrysodeixis* sp.

Embrapa Maize and Sorghum in Brazil has a collection of more than 50 baculovirus isolates that cause typical symptoms (Fig. 12.3). Each isolate has already been tested in the laboratory and there are projects with partner companies in which there is a development for large-scale production of commercial products (Simbiose® company). The characterization of the best isolate has been done and this biopesticide should be on the market in 2018/2019. Another research unit of Embrapa (CENARGEN), since 2008, has also been developing studies of biological and molecular characterization and pathogenicity evaluation (in collaboration with Embrapa Soybean).

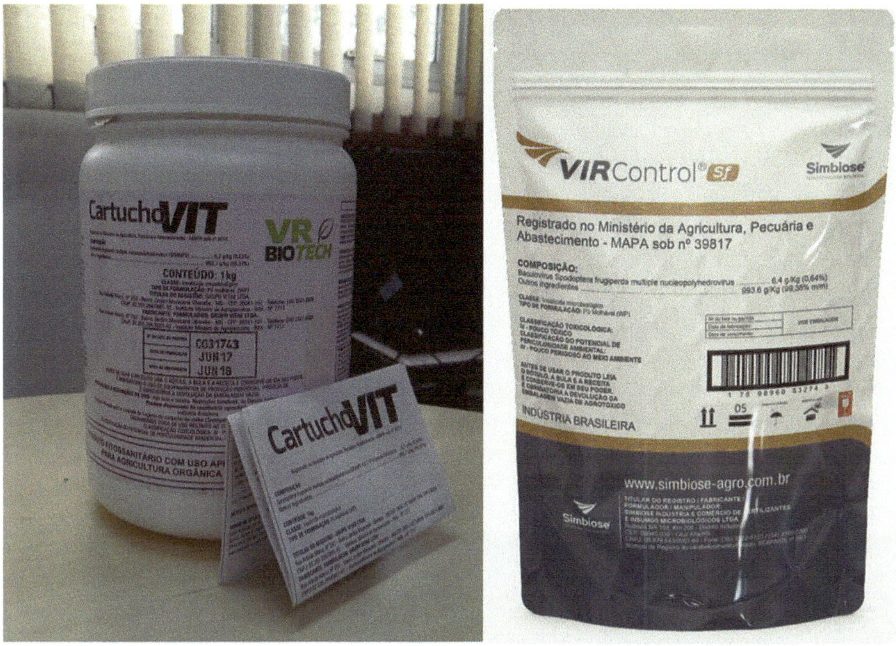

Fig. 12.2 Baculovirus-based biopesticides registered in Brazil for fall armyworm, *Spodoptera frugiperda*. Both formulations are wettable powder. (Photo: Fernando H. Valicente)

Fig. 12.3 Dead larvae of soybean looper, *Chrysodeixis* sp., with a typical symptom caused by baculovirus. (Photo: Priscila Marques de Paiva)

12.10 Nucleopolyhedrovirus for the Control of *Helicoverpa armigera*

Helicoverpa armigera larvae were collected in some regions of Brazil as soon as the outbreak of this pest occurred between 2012 and 2013. Larvae were brought to the laboratory and several baculovirus isolates were discovered. These isolates were tested on healthy caterpillars, obtaining the same initial symptoms, and some isolates were identified through primers and were sequenced. The comparative analysis of the sequencing for the genes LEF-8 and LEF-9 showed that the isolates found in Brazil are closely related to the isolates from Australia and India. All baculovirus isolates tested caused a good mortality rate in larvae of *H. armigera* third instar. However, lethal concentration 50 and lethal time 50 varied among these isolates. All of our isolates showed to be HearNPV and not HzNPV (Gemstar) according to the DNA sequencing. HearNPV-BR2 showed the best results for LC_{50} and LT_{50}. Thus, this was the first report of baculovirus isolates infecting *H. armigera* larvae in Brazil, and also the first report of *H. armigera* baculovirus isolates to be identified as HearNPV in Brazil. Isolate BR2 showed the best results and will be used in *H. armigera* biological control programs due to its characteristics. An agreement with Simbiose® company was signed for large-scale production of this biological pesticide. Figure 12.4 shows a caterpillar killed by the baculovirus, with typical symptoms.

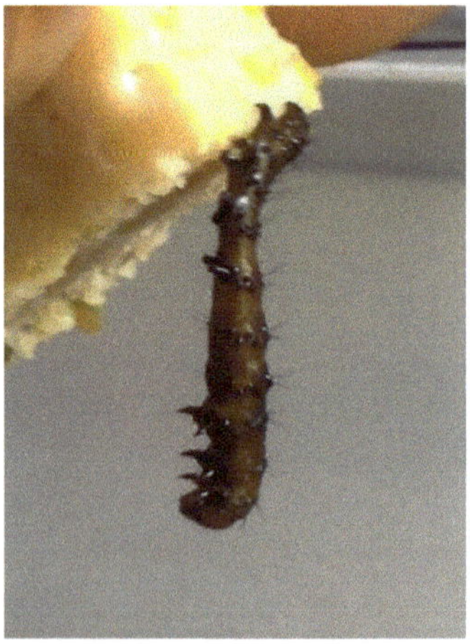

Fig. 12.4 Dead larva of *Helicoverpa armigera* with typical symptoms of baculovirus. (Photo: Arthur Torres)

12.11 Considerations in the Use and Application of Field Baculovirus Products

Some considerations should be made for the best use and storage of baculovirus in the form of biopesticide. The baculovirus-based bioinsecticide should be stored in a cool dry place without direct sunlight for better preservation, maintaining the product quality. Wettable powder formulations do not need to be stored in the freezer. Timing of the application of the baculovirus should be done according to the pest and with the respective crop. For example, the timing of application of *H. armigera* baculovirus is different in maize, soybean, and cotton. *H. armigera* attacks at different times, different parts of the plant, and different stages in relation to the development of each crop. In the specific case of the fall armyworm, observe the first injured leaves or the location presence of this pest through the years. In very warm regions with a history of pest attack, the first application of the baculovirus against the fall armyworm should be made always between 10 and 15 days after germination. It is important to monitor the presence of the insect pest, since the timing of the application of the baculovirus for the first application is very important; the smaller the caterpillar, the greater the chances of control. Overlapping of larval stages should be avoided. Spraying should be performed after 4 pm due to the lower incidence of ultraviolet (UV) rays. In large areas where there is a need for continuous spraying during the day, it is best to use a baculovirus formulation, if available, with protection against UV rays. Viral particles exposed to sunlight or high temperatures can be inactivated quickly. The amount of the water to be sprayed in the field should be adequate according to the spraying technology and for each crop. However, it must be ensured that there has been adequate deposition of the product on the leaf of the crop in question, specifically for maize, especially in the region of the whorl. It is not efficient to use a very low volume of water to spray in order to have a high field efficacy, and the product is not deposited on the leaves. In the case of organic products, the biological pesticide to be sprayed and its mode of action must be respected. The insect must ingest part of the sprayed leaf to become contaminated, so an inadequate spraying technology cannot be used. To use a smaller amount of water, one should have the appropriate equipment for low volume, but check the relative humidity of the air because if it is very low, you can lose much with the evaporation of the water before the product reaches the leaves (Fig. 12.5). All applications of the baculovirus-based biopesticide should be used with an adhesive spreader that is compatible with the biological product as it improves the distribution and adhesion of the product on the leaves.

The application of baculovirus in corn, soybean, and cotton crops should follow some important factors. The first one is that the crop should be monitored weekly to detect the level of attack of the target pest in the crop. The attack of larvae begins at different times of the growing stage of each crop. Therefore, the timing of the baculovirus application is important for each pest and is specific for each crop.

Another important factor is the architecture of the plant and the leaves to be sprayed by the baculovirus or any other biological product. In the case of maize that grows vertically, if necessary, weekly applications of the biopesticide should be carried

Fig. 12.5 Application of baculovirus with an appropriate equipment and volume of water. (Photo: Fernando H. Valicente)

out in the same way, as needed, as it is done with the use of chemicals. Because every week there will be new leaves without any biological product coverage.

12.12 Final Remarks

Baculoviruses are arthropod-specific viruses and are the only ones lacking homology with viruses found in other organisms such as plants, animals, fungi, and bacteria. Although the action of the baculoviruses is slower than chemical insecticides, taking more time to kill the target insect, it is considered an important and powerful tool in the integrated pest management. Some important aspects should be considered for the use and field application of baculovirus products, as well as storage. There are many biological control programs in the world that use baculovirus as biopesticide with positive results. These programs have been expanding as the insect pests are moving throughout the world such as fall armyworm that is present in African countries and now in India. Fall armyworm is present in Latin American countries.

References

Adams JR, McClintock JT (1991) Baculoviridae nuclear polyhedrosis viruses Part 1: nuclear polyhedrosis viruses of insects. In: Adams JR, Bonami JR (eds) Atlas of invertebrate viruses. CRC Press, Boca Raton,. Chap 6, pp 87–180

Barreto MR, Guimarães CT, Teixeira FF et al (2005) Effect of *Baculovirus spodoptera* isolates in *Spodoptera frugiperda* (J. E. Smith) (Lepidoptera: Noctuidae) larvae and their characterization by RAPD. Neotrop Entomol 34(1):67–75

Burges HD, Croizier G, Huber J (1980) A review of safety tests on baculoviruses. Entomophaga 25(4):329–340

Eberle KE, Wennmann JT, Kleespies RG et al (2012a) Basic techniques in insect virology. In: Lacey LA (ed) Manual of techniques in invertebrate pathology, 2th edn. Academic Press, San Diego, pp 15–74

Eberle KE, Jehle JA, Hüber J (2012b) Microbial control of crop pests using insect viruses. In: Abrol DP, Shankar U (eds) Integrated pest management: principles and practice. CABI Publishing, Wallingford, pp 281–298

Entwistle PF (1983) Viruses for insect pest control. e-Spania 26:59–62

Fauquet CM, Mayo MA, Maniloff J et al (2005) Virus taxonomy: VIIIth report of the international committee on taxonomy of viruses. Academic Press, Amsterdam

Federici BA (1997) Baculovirus pathogenesis. In: Miller LK (ed) The baculoviruses. Springer, Califórnia, pp 33–59

Grzywacz D (2017) Basic and applied research: baculovirus. In: Lacey LA (ed) Microbial control of insect and mite pests: from theory to practice. Elsevier, Amsterdam, pp 27–46

Harrison R, Hoover K (2012) Baculoviruses and other occluded insect viruses. In: Vega F, Kaya H (eds) Insect pathology. Elsevier, Amsterdam, pp 73–131

Harrison RL, Rowley DL, Funk CJ (2016) The complete genome sequence of *Plodia interpunctella* granulovirus: evidence for horizontal gene transfer and discovery of an unusual inhibitor-of-apoptosis gene. PLoS One 11(7):e0160389

Hasse S, Sciocco-Cap A, Romanowski V (2015) Baculovirus insecticides in Latin America: historical overview, current status and future perspectives. Viruses 7(5):2230–2267

Hauschild R, Speiser B, Tamm L (2011) Regulation according to EU directive 91/414: data requirements and procedure compared with regulation practice in other OECD countries. In: Ehlers R-U (ed) Regulation of biological control agents. Springer, Dordrecht, pp 25–77

Hawtin RE, Zarkowska T, Arnold K et al (1997) Liquefaction of Autogra- pha californica nucleopolyhedrovirus-infected insects is dependent on the integrity of virus-encoded chitinase and cathepsin genes. Virology 238(2):243–253

Herniou EA, Olszewski JA, Cory JS et al (2003) The genome sequence and evolution of baculoviruses. Annu Rev Entomol 48:211–234

Herniou EA, Arif BM, Becnel JJ et al (2012) Baculoviridae. In: King AMQ, Adams EB, Carstens EJ et al (eds) Virus taxonomy: classification and nomenclature of viruses. Elsevier Academic Press, San Diego, pp 163–173

Ikeda M, Hamajima R, Kobayashi M (2015) Baculoviruses: diversity, evolution and manipulation of insects. J Entomol Sci 18(1):1–20

Jaques RP (1967) The persistence of a nuclear polyhedrosis virus in habitat of host insect *Trichoplusiani* I. Polyhedra deposited on foliage. Can Entomol 99(8):820–829

Jehle JA, Blissard GW, Bonning BC et al (2006) On the classification and nomenclature of baculoviruses: a proposal for revision. Arch Virol 151(7):1257–1266

Katsuma S, Kobayashi J, Koyano Y et al (2012) Baculovirus-encoded protein BV/ODV-E26 determines tissue tropism and virulence in lepidopteran insects. J Virol 86(5):2545–2555

Krieg A, Franz JM, Groner A et al (1980) Safety of entomopathogenic viruses for control of insect pests. Environ Conserve 7:158–160

Miele SA, Garavaglia MJ, Belaich MN et al (2011) Baculovirus: molecular insights on their diversity and conservation. Int J EvolBiol 2011:379–424

Miller LK (1997) The baculoviruses. Plenum Press, New York

Miller L, Ball LA (eds) (1998) The insect viruses. Plenum Press, New York

Moscardi F (1999) Assessment of the application of baculoviruses for the control of Lepidoptera. Annu Rev Entomol 44:257–289
Moscardi F (2007) Development and use of the nucleopolyhedrovirus of the velvetbean caterpillar in soybeans. In: Vincent C, Goettel MS, Lazarovits G (eds) Biological control: a global perspective. CAB International, Wallingford, pp 344–353
Moscardi F, Sosa-Gomez D (2007) Microbial control of insect pests of soybean. In: Lacey LA, Kaya HK (eds) Field manual of techniques in invertebrate pathology: application and evaluation of pathogens for control of insects and other invertebrate pests, 2rd edn. Springer, Dordrecht, pp 411–426
Moscardi F, Souza ML, Castro MEB et al (2011) Baculovirus pesticides: present state and future perspectives. In: Ahmad I, Ahmad F, Pichtel J (eds) Microbes and microbial technology. Springer, Dordrecht, pp 415–445
Oliveira JVC, Wolff JLC, Garica-Maruniak A et al (2006) Genome of the most widely used viral biopesticide: *Anticarsia gemmatalis* multiple nucleopolyhedrovirus. J GenVirol 87(11):3233–3250
Organization for Economic Co-operation and Development (2002) Consensus document on information used in the assessment of environmental applications involving baculovirus. Organization for Economic Co-operation and Development, Paris. (Series on Harmonization of Regulatory Oversight in Biotechnology, n° 20)
Sosa-Goméz DR (2017) Microbial control of soybean pest insects and mites. In: Lacey LA (ed) Microbial control of insect and mite pests: from theory to practice. Elsevier, Amsterdam, pp 199–208
Sosa-Gómez DR, Côrrea-Ferreira BS, Hoffmann-Campo CB et al (2014) Manual de identificação de insetos e outros invertebrados da cultura da soja. 3ª ed. Embrapa Soja, Londrina
Summers MD, Kawanishi CY (eds) (1978) Viral pesticides: present knowledge and potential effects on public and environmental health. U. S. EPA, Washington
Thompson CG, Scott DW, Wickman BE (1981) Long-term persistence of the nuclear polyhedrosis virus of the Douglas-fir tussock moth, *Orgyia pseudotsugata* (Lepidoptera: Lymantriidae), in Forest soil. Environ Entomol 10(2):254–255
Valicente FH (1989) Levantamento dos inimigos naturais de *Spodoptera frugiperda* (J. E. Smith) (Lepidoptera: Noctuidae) em diferentes regiões do estado de Minas Gerais. An Soc Entomol Bras 18(1):119–130
Valicente FH, Costa EF (1995) Controle da lagarta do cartucho *Spodoptera frugiperda* (J.E. Smith) com baculovirus spodoptera, aplicado via água de irrigação. AnSocEntomol Bras 24(1):61–67
Valicente FH, Peixoto MJVVD, Paiva E, Kitajima EW (1989) Identificação e purificação de um vírus da poliedrose nuclear da lagarta *Spodoptera frugiperda* (J.E. Smith 1797) (Lepidoptera: Noctuidae). An Soc Entomol Bras 18:71–81
Valicente FH, Tuelher ES, Pena RC et al (2007a) The use of Baculovirus to control fall armyworm, *Spodoptera frugiperda*, in Brazil. In: Proceedings of the 16th international plant protection congress, British Crop Protection Council, Alton, 15–18 Oct 2007
Valicente FH, Tuelher ES, Pena RC et al (2007b) The use of Baculovirus to control fall armyworm, *Spodoptera frugiperda*, in Brazil. In: Proceedings of the 40th annual meeting of the society for invertebrate pathology, University of Warwick, Quebec, 12–16 Aug 2007
Valicente FH, Tuelher ES, Paiva CEC et al (2008) New baculovirus isolate that does not cause the liquefaction of the integument in *Spodoptera frugiperda* dead larvae. Rev Bras Milho Sorgo 7(1):77–82
Valicente FH, Tuelher ES, Pena RC et al (2013) Cannibalism and virus production in *Spodoptera frugiperda* (J.E. Smith) (Lepidoptera: Noctuidae) larvae fed with two leaf substrates inoculated with Baculovirus spodoptera. Neotrop Entomol 42(2):191–199
Van Oers MM, Vlak JM (2007) Baculovirus genomics. Curr Drug Targets 8(10):1051–1068
Vieira CM, Tuelher ES, Valicente FH et al (2012) Characterization of a *Spodoptera frugiperda* multiple nucleopolyhedrovirus isolate that does not liquefy the integument of infected larvae. J Invertebr Pathol 111(2):189–192
Wolff JLC, Valicente FH, Martins R et al (2008) Analysis of the genome of *Spodoptera frugiperda* nucleopolyhedrovirus (SfMNPV-19) and of the high genomic heterogeneity in group II nucleopolyhedroviruses. J Gen Virol 89(5):1202–1211

Chapter 13
Bacillus thuringiensis

Fernando Hercos Valicente

13.1 Introduction

Bacillus thuringiensis (*Bt*) is an aerobic Gram-positive bacterium from the Bacillaceae family that produces protein crystalline inclusions named Cry proteins during the stationary phase encoded by different *cry* genes (Angus 1954; Bechtel and Bulla 1976). *Bacillus thuringiensis* is a ubiquitous bacterium that can be found in different substrates such as soil, water, plant surfaces, dead insects, grain dust, spiderwebs, and stored grain (Federici 1999; Glare and O'Callaghan 2000; Valicente and Barreto 2003). Crystal proteins are composed of one or more proteins, Cry or Cyt (cytolytic protein), and they are named delta (δ) endotoxins. These are the factors that determine Bt pathogenicity (Schnepf et al. 1998) and may show different forms as shown in Fig. 13.1 (Valicente and Souza 2004). Many *Bt* strains also produce other types of insecticidal proteins, such as the Vip proteins (vegetative insecticidal proteins) that are synthesized during the vegetative phase growth not forming any crystals, which were identified by Estruch et al. 1996. Some other important proteins are also produced, such as Cyt, β-exotoxins, and Sip proteins. However, the most studied are the *cry* genes/Cry proteins. The identification of a *Bt* strain to subspecies is done using the flagellar antigen H, e.g., *Bacillus thuringiensis* sv *kurstaki*. However, this type of characterization does not consider the genes present in these strains, e.g., strain HD-1 (*Bt* sv *kurstaki*) harbors the genes *cry1Aa*, *cry1Ab*, *cry1Ac*, *cry2A*, and *cry2B*, and strain HD-73 *(Bt sv kurstaki)* harbors only *cry1Ac* gene. These toxic proteins from Bt are used in pest control as a biological pesticide or as transgenic plant expressing these proteins. This is a very useful and powerful tool in integrated pest management (IPM). In this work, we describe the genetic variability

F. H. Valicente (✉)
Embrapa Maize and Sorghum, Sete Lagoas, MG, Brazil
e-mail: fernando.valicente@embrapa.br

B. Souza et al. (eds.), *Natural Enemies of Insect Pests in Neotropical Agroecosystems*, https://doi.org/10.1007/978-3-030-24733-1_13

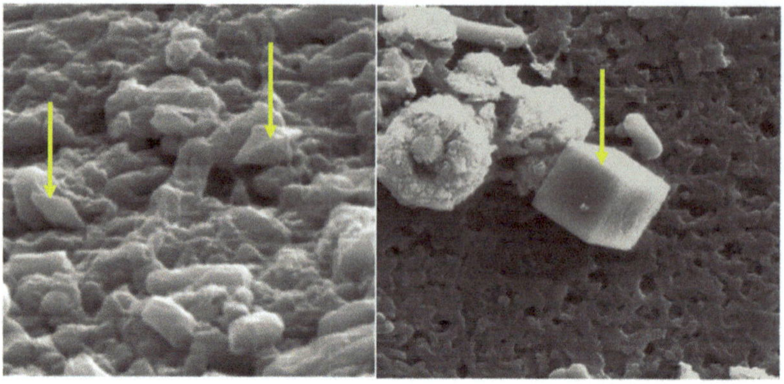

Fig. 13.1 Crystal of *Bacillus thuringiensis* showing a bipyramidal shape (20.000 ×) (**a**) and cuboid shape (30.000 ×) (**b**). (Valicente and Souza 2004)

and molecular characterization of *Bacillus thuringiensis*, and the importance of finding the appropriate strain for a specific insect pest. We also provide the description of the Cry proteins and the nomenclature of Cry proteins, as well as their importance for insect specificity, their mode of action, and how *Bacillus thuringiensis* is used as biological pesticides.

13.2 Genetic Variability and Molecular Characterization of *Bacillus thuringiensis*

Molecular characterization of *Bt* strains may be used to characterize DNA, protein, and genetic variability among *Bt* isolates. Different techniques have been used to discriminate different isolates with different purposes. The most common techniques used are polymerase chain reaction (PCR) (Cerón et al. 1994, 1995; Bravo et al. 1998; Shangkuan et al. 2001; Lima et al. 2002; Valicente et al. 2010), repetitive element polymorphism (REP-PCR) (Higgins et al. 1982; Stern et al. 1984; Sharples and Lloyd 1990; Versalovic et al. 1991, 1994; Rademaker and de Bruijin 1997; Louws et al. 1999; Shuhaimi et al. 2001; da Silva and Valicente 2013), amplified fragment length polymorphism (AFLP) (Vos et al. 1995; Arnold et al. 1999; Ridout and Donini 1999; Mueller and Wolfenbarger 1999; Grady et al. 2001; Ticknor et al. 2001; Burke et al. 2004; Hill et al. 2004; Abreu et al. 2007; Valicente and da Silva 2014), and plasmid pattern characterization (Gilson et al. 1984; Birge 1994; Berry et al. 2002; Gitahy et al. 2005; Loeza-Lara et al. 2005; Roh et al. 2007; Ramírez and Ibarra 2008; Fagundes et al. 2011). *Bt* proteins can be characterized according to their size in SDS PAGE.

13.3 Cry Proteins

Cry proteins are the most studied toxins and they are toxic to insects from the orders Lepidoptera, Coleoptera, Hemiptera, Neuroptera, Orthoptera, Siphonaptera, Thysanoptera, and Blattodea (Isoptera) (Glare and O'Callaghan 2000) and also to nematodes. *Bacillus thuringiens*is may be used as biological pesticides in transgenic plants (*Bt* plants). In 1998, Crickmore et al. proposed a new nomenclature to Cry and Cyt proteins. Cry and Cyt protein nomenclature is based on the identity of the primary sequences among proteins. The nomenclature and sequences are available at the website http://www.lifesci.susx.ac.uk/home/Neil_Crickmore/Bt/. Most of *cry* genes are present in the plasmids and not in the chromosomes. *Bt* strains may be used to produce biological pesticides in fermentation systems, and *Bt* genes can be used in plant transformation against insect pests.

13.4 Nomenclature of Cry Proteins and Their Importance Towards Insect Specificity

Crystal proteins receive a classification based on four hierarchical categorizations consisting of numbers, uppercase letters, and lowercase letters (e.g., Cry1Ab1). The classification is given depending on the sequence homology that is shared with other Cry proteins. A different number (first category) is given to a protein if it has less than 45% identity or homology with all other Cry proteins (e.g., Cry1, Cry2, Cry3 (...) Cry65, Cry74) (Crickmore et al. 2016). The second category is the capital letter, which is given if the protein has less than 78% but more than 45% homology with a specific group of Cry proteins (e.g., Cry1A, Cry1B, Cry1C). The third category, which is a lowercase letter, is given to distinguish proteins sharing more than 78% and less than 95% identity with the other Cry proteins (e.g., Cry1Aa, Cry1Ab, Cry1Ac). The number added at the end is to distinguish proteins that share more than 95% identity but are not identical and should be considered variants of the same protein (Cry1Ab1, Cry1Ab2, Cry1Ab3, etc.) (Crickmore et al. 1998, 2016). This classification is also used for Vip proteins.

Cry proteins show specific toxicity to certain insect orders. Cry1 proteins are active against Lepidoptera, Cry2 against Lepidoptera and Diptera, Cry3 against Coleoptera, etc. With the evolution of molecular techniques, PCR (polymerase chain reaction) can differentiate the *Bt* genes that are present in a certain strain or *Bt* isolate. Cry2Aa protein is toxic to lepidopterans and dipterans, and the Cry2Ab2 and Cry2Ac3 proteins are toxic only against lepidopterans. Another example is the Cry5Ac1 protein that acts against Hymenoptera (ants) and the Cry5Ba3 and Cry5Ca1 proteins have a toxic action against nematodes. There is no *Bt* protein that has insecticidal action against bees, although bees are insects of the order Hymenoptera (Dr. Neil Crickmore – personal information, 2016). This factor is important in determining the toxicity of a particular pest, insect, or group of insects.

Cry proteins that have three domains (domains I, II, and III) are called 3d-Cry. Cry toxins belonging to the 3-domain Cry toxin family show clear differences in their amino acid sequences, but all have in common a conserved 3-domain structure (de Maagd et al. 2001, 2003; Bravo et al. 2007; Pardo-López et al. 2013). Toxins are considered viable alternatives for the control of insect pests in agriculture and vectors of importance in public health (Crickmore et al. 2016). These proteins are highly specific to target insects, killing a limited number of species. The toxins are innocuous to humans, vertebrates, and plants and are completely biodegradable (Pardo-López et al. 2013). Bt 3d-Cry toxins show toxic activity against insects of orders Lepidoptera, Diptera, Coleoptera, and Hemiptera (low to moderate toxicity in aphids) and nematodes (Van Frankenhuyzen 2009, 2013). In general, domain I is the domain of the perforation and is subject to proteolytic cleavage, domain II or central domain is involved in toxin-receptor interactions, and domain III is involved in the junction of galactose, receptor junction, and pore formation (Bravo et al. 2007; de Maagd et al. 2001, 2003; Pardo-López et al. 2013).

13.5 Mode of Action of Cry Proteins

The mode of action of *Bt* toxins described in the literature refers to Cry1 proteins, which have a toxic action against Lepidoptera. The mode of action of *Bt* toxins has been well studied by some research groups; this way one can better understand the steps for the death of the insect as well as the resistance of this insect to a certain protein. The mode of action of 3d-Cry toxins against lepidopterans can be understood at the molecular level as follows: the interaction of Cry1 toxins with different proteins present in the midgut of Lepidoptera in a complex process involving several proteins such as cadherin, aminopeptidase N, and alkaline phosphatase (Pigott and Ellar 2007; Soberón et al. 2009). According to Pardo-López et al. 2013, the main steps of this mode of action are as follows:

1. Caterpillars ingest 3d-Cry protoxins, which are solubilized in the midgut due to the high pH and reducing conditions and, activated by proteases of the intestine, generating the toxic fragment.
2. The 3d-Cry monomeric toxin binds to alkaline phosphatase and aminopeptidase receptors, and with a low interaction and affinity, the toxin is then located near the membrane.
3. The 3d-Cry monomeric toxins bind to the cadherin receptor in a high affinity binding, and this interaction induces a proteolytic cleavage at the N-terminus at the end of the toxin, including the α-helix domain.
4. The cleaved 3d-Cry protein is then capable of oligomerizing in a pre-pore oligomer toxin.
5. The oligomeric 3d-Cry structure binds to the alkaline phosphatase and aminopeptidase receptors with high affinity.
6. The pre-pore inserts into the membrane causing the pore to form.

The mode of action of the Cry proteins mentioned above is always related to Cry1 proteins – which have activity against Lepidoptera. These steps should not be used as a mode of action for insects from other insect orders. There are orders of insects that do not have intestinal receptors and not all insects have a medium intestine with alkaline pH.

These *Bt* genes may be used in plant transformation, generating transgenic crops protected against specific insect pests.

13.6 *Bacillus thuringiensis* Used as Biological Pesticides

Bacillus thuringiensis needs sources of nitrogen, carbon, mineral salts, and oxygen to grow. Various agricultural and industrial by-products, such as maize glucose, soybean flour soy extract, peanuts, sugarcane molasses, and liquid swine manure, are carbon- and nitrogen-rich and may be used as sterilized raw materials in biopesticide production. Tirado-Montiel et al. (2001) first tested the use of wastewater sludge for biopesticide production, although the entomotoxicity level reported was low. Many authors reported the use of by-products and raw materials to develop *Bt*-based biopesticides (Salama et al. 1983; Obeta and Okafor 1984; Morris et al. 1997). Moreover, the low cost of by-products as nutrient source in fermentation media for *Bt* biopesticide production has received little attention. However, a higher level of entomotoxicity is desired to reduce the production costs of biopesticides. Valicente and Mourão (2008) also reported that a carbon/nitrogen proportion and mineral salts (all expressed in 0.1–0.002 gl^{-1} of $FeSO_4$, 0.02 gl^{-1} of $ZnSO_4$, 0.02 gl^{-1} of $MnSO_4$, 0.3 gl^{-1} of $MgSO_4$) are essential to promote adequate *Bt* growth.

Bacillus thuringiensis production in Cuba is done in 185 "Centros de Reproducción de Entomófagos y Entomopatógenos (Cree)" distributed all over the country (Pérez and Vásquez 2001), and the *Bt* produced is stored for 3 months in temperatures up to 25 °C. According to Aranda et al. (2000), *Bt* is the most studied entomopathogen in Cinvestav (Centro de Investigaciones y de Estudios Avanzados de IPN), in México, and UNAM (Universidad Nacional Autónoma de México). Also according to Aranda et al. (2000), Peru produces *Bt* with activity against *Anopheles albimanus* (Wiedemann) (Culicidae).

Bravo and Ceron (2004) reported the production of *Bt*-based biopesticide in different formulations (wettable powder, granules, etc.) to control mosquitos.

Arcas et al. (1984, 1987) studied the first aspects of *Bt* fermentation in Argentina at the University of La Plata. In the 1990s, Dr. Graciela Benintende and Jorge Cozzi developed technology to produce a *Bt*-based biopesticide against lepidopteran pest.

In Brazil, one *Bt*-based biopesticide was registered for fall armyworm, *Spodoptera frugiperda* (J. E. Smith) (Noctuidae). The biopesticide "Crystal®" (Fig. 13.2) produced by Farroupilha/Lallemand is the result of an agreement with Embrapa Maize and Sorghum Research Center.

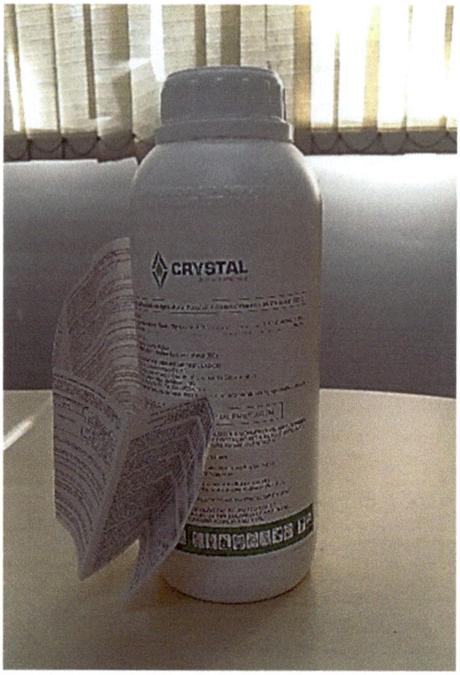

Fig. 13.2 *Bacillus thuringiensis*-based biopesticide "Crystal®" produced by Farroupilha/Lallemand in Agreement with Embrapa Maize and Sorghum. (Photo: Fernando H. Valicente)

13.7 Final Considerations

Bacillus thuringiensis (*Bt*) is a Gram-positive bacterium that produces protein crystalline inclusions named Cry proteins during the stationary phase encoded by different *cry* genes. *Bacillus thuringiensis* can be found in different substrates such as soil, water, plant surfaces, dead insects, grain dust, spiderwebs, and stored grain. Crystal proteins are composed of one or more proteins, Cry or Cyt (cytolytic protein), and they are named delta (δ) endotoxins. These are the factors determining Bt pathogenicity to insects and nematodes. Many *Bt* strains also produce other types of insecticidal proteins, such as the Vip proteins (vegetative insecticidal proteins), and do not form any crystals. Some other important proteins are also produced, such as β-exotoxins and Sip proteins. These toxic proteins from Bt are used in pest control as a biological pesticide or as transgenic plant expressing these proteins. This is a very useful tool in integrated pest management (IPM).

References

Abreu IL, Guidelli AM, Lemos MVF (2007) Análise da diversidade genética de isolados *Bacillus thuringiensis* por fAFLP (fluorescent amplified fragment length polymorphism). Científica 35:25–27

Angus TA (1954) A bacterial toxin paralysing silkworm larvae. Nature 173:545–546

Aranda E, Lorence A, Del Refugio TM (2000) Rural production of *Bacillus thuringiensis* by solid state fermentation. In: Charles JF, Delécluse A, Roux CNL (eds) Entomopathogenic bacteria: from laboratory to field application. Springer, Dordrecht, pp 317–332
Arcas J, Yantorno O, Arrarás E et al (1984) A new medium for growth and delta-endotoxin production by *Bacillus thuringiensis* var. kurstaki. Biotechnol Lett 6(8):495–500
Arcas J, Yantorno O, Ertola R (1987) Effect of high concentration of nutrients on *Bacillus thuringiensis* cultures. Biotechnol Lett 9(2):105–110
Arnold C, Metherell L, Willshaw G et al (1999) Predictive fluorescent amplified-fragment length polymorphism analysis of *Escherichia coli*: high-resolution typing method with phylogenetic significance. J Clin Microbiol 37(5):1274–1279
Bechtel DB, Bulla LA (1976) Electron microscope study of sporulation and parasporal Crystal formation in *Bacillus thuringiensis*. J Bacteriol 127(3):1472–1481
Berry C, O'Neil S, Ben-Dov E et al (2002) Complete sequence and organization of Bt toxins, the toxin-coding plasmid of *Bacillus thuringiensis* subsp. *israelensis*. Appl Environ Microbiol 68(10):5082–5095
Birge EA (1994) Bacterial and bacteriophage genetics. Springer, New York
Bravo A, Ceron J (2004) *Bacillus thuringiensis* en el control biológico. Buena Semilla, Bogotá
Bravo A, Gill SS, Soberón M (2007) Mode of action of *Bacillus thuringiensis* Cry and Cyt toxins and their potential for insect control. Toxicon 49(4):423–435
Bravo A, Sarabia S, Lopez L et al (1998) Characterization of cry genes in a Mexican *Bacillus thuringiensis* strain collection. Appl Environ Microbiol 64(12):4965–4972
Burke SA, Wright MK, Robinson MK et al (2004) Detection of molecular diversity in *Bacillus atrophaeus* by amplified fragment length polymorphism analysis. Appl Environ Microbiol 70(5):2786–2790
Cerón J, Covarrubias L, Quintero R et al (1994) PCR analysis of the cryI insecticidal crystal family genes from *Bacillus thuringiensis*. Appl Environ Microbiol 60(1):353–356
Cerón J, Ortiz A, Quintero R et al (1995) Specific PCR primers to identify cryI and cryIII genes within a *Bacillus thuringiensis* strain collection. Appl Environ Microbiol 61(11):3826–3831
Crickmore N, Baum J, Bravo A et al (2016) *Bacillus thuringiensis* toxin nomenclature. http://www.lifesci.susx.ac.uk/home/Neil_Crickmore/Bt/. Accessed 5 Mar 2018
Crickmore N, Zeigler DR, Feitelson J et al (1998) Revision of the nomenclature for the *Bacillus thuringiensis* pesticidal crystal proteins. Microbiol Mol Biol Rev 62(3):807–813
Estruch JJ, Warren GW, Mullins MA et al (1996) Vip3A, a novel *Bacillus thuringiensis* vegetative insecticidal protein with a wide spectrum of activities against lepidopteran insects. Proc Natl Acad Sci U S A 93(11):5389–5394
Fagundes RBS, Picoli EAT, Lana UGP et al (2011) Plasmid patterns of efficient and inefficient strains of *Bacillus thuringiensis* against *Spodoptera frugiperda* (J.E. Smith) (Lepidoptera: Noctuidae). Neotrop Entomol 40(5):600–606
Federici BA (1999) *Bacillus thuringiensis*. In: Bellows TS, Fischer TW (eds) Handbook of biological control. Academic Press, New York, pp 517–548
Gilson E, Clément JM, Brutlag D et al (1984) A family of dispersed repetitive extragenic palindromic DNA sequences in *E. coli*. EMBO J 3(6):1417–1421
Gitahy PM, Lima GMS, Araújo SMT et al (2005) Purificação de DNA plasmidial de *Bacillus thuringiensis* por ultracentrifugação em gradiente de cloreto de césio. Embrapa Agrobiologia, Seropédica. (Boletim de Pesquisa e Desenvolvimento Agrobiologia, 8)
Glare TR, O'Callaghan M (2000) *Bacillus thuringiensis*: biology, ecology and safety. Wiley, New York
Grady R, Blanc D, Hauser P et al (2001) Genotyping of European isolates of methicillin-resistant Staphylococcus aureus by fluorescent amplified-fragment length polymorphism analysis (FAFLP) and pulsed-field gel electrophoresis (PFGE) typing. J Med Microbiol 50(7):588–593
Higgins CF, Ames GFL, Barnes WM et al (1982) A novel intercistronic regulatory element of prokaryotic operons. Nature 298:760–762
Hill KK, Ticknor LO, Okinaka RT et al (2004) Fluorescent amplified fragment length polymorphism analysis of *Bacillus anthracis*, *Bacillus cereus* and *Bacillus thuringiensis* isolates. Appl Environ Microbiol 70(2):1068–1080

Lima ASG, Guidelli AM, Abreu IL et al (2002) Identification of new isolates of *Bacillus thuringiensis* using rep-PCR products and δ-endotoxin electron microscopy. Genet Mol Biol 25(2):225–229
Loeza-Lara PD, Benintende G, Cozzo J et al (2005) The plasmid pBMBt1 from *Bacillus thuringiensis* subsp. *darmstadiensis* (INTA Mo14-4) replicates by the Rolling-circle mechanism and encodes a novel insecticidal crystal protein-like gene. Plasmid 54(3):229–240
Louws FJ, Rademaker JLW, de Bruijn FJ (1999) The three Ds of PCR-based genomic analysis of phytobacteria: diversity, detection and disease diagnosis. Annu Rev Phytopathol 37:81–125
de Maagd RA, Bravo A, Berry C et al (2003) Structure, diversity, and evolution of protein toxins from spore-forming entomopathogenic bacteria. Annu Rev Genet 37:409–433
de Maagd RA, Bravo A, Crickmore N (2001) How *Bacillus thuringiensis* has evolved specific toxins to colonize the insect world. Trends Genet 17(4):193–199
Morris ON, Kanagaratnam P, Converse V (1997) Suitability of 30 agricultural products and by-products as nutrient sources for laboratory production of *Bacillus thuringiensis* subsp. *Aizawai* (HD 133). J Invertebr Pathol 70(2):113–120
Mueller UG, Wolfenbarger LL (1999) AFLP genotyping and fingerprinting. Trends Ecol Evol 14(10):389–394
Obeta JAN, Okafor N (1984) Medium for the production of primary powder of *Bacillus thuringiensis* subsp. *israelensis*. Appl Environ Microbiol 47(4):863–867
Pardo-López L, Soberón M, Bravo A (2013) *Bacillus thuringiensis* insecticidal three-domain Cry toxins: mode of action, insect resistance and consequences for crop protection. FEMS Microbiol Rev 37(1):3–22
Pérez N, Vásquez LL (2001) Manejo ecológico de plagas. In: Funes F (ed) Transformando el campo cubano: avances de la agricultura sostenible. Universidad Agrária de La Habana, La Habana, pp 191–226
Pigott CR, Ellar DJ (2007) Role of receptors in *Bacillus thuringiensis* crystal toxin activity. Microbiol Mol Biol Rev 71(2):255–281
Rademaker JJW, de Bruijin FJ (1997) Characterization and classification of microbes by rep-PCR genomic fingerprinting and computer assisted patterns analysis. In: Anolles GG, Gresshoff PM (eds) DNA markers: protocols, applications & overviews. Willey, New York, pp 151–171
Ramírez R, Ibarra J (2008) Plasmid patterns of *Bacillus thuringiensis* type strains. Appl Environ Microbiol 74(1):125–129
Ridout CJ, Donini P (1999) Use of AFLP in cereal research. Trends Plant Sci 4(2):76–79
Roh JY, Choi JY, Li MS et al (2007) *Bacillus thuringiensis* as a specific, safe, and effective tool for insect pest control. J Microbiol Biotechnol 17(4):547–559
Salama HS, Foda MS, Dulmage HT et al (1983) Novel fermentation media for production of σ-endotoxins from *Bacillus thuringiensis*. J Invert Pathol 41:8–19
Schnepf E, Crickmore N, Van Rie J et al (1998) *Bacillus thuringiensis* and its pesticidal crystal proteins. Microbiol Mol Biol Rev 62(3):775–806
Shangkuan Y, Chang Y, Yang JF et al (2001) Molecular characterization of *Bacillus anthracis* using multiplex PCR, ERIC-PCR and RAPD. Lett Appl Microbiol 32(3):139–145
Sharples GJ, Lloyd RG (1990) A novel repeated DNA sequence located in the intergenic regions of bacterial chromosomes. Nucleic Acids Res 18(22):6503–6508
Shuhaimi M, Ali AM, Saleh NM et al (2001) Utilization of enterobacterial repetitive intergenic consensus (ERIC) sequence-based PCR to fingerprint the genomes of *Bifidobacterium* isolates and other probiotic bacteria. Biotechnol Lett 23(9):731–736
da Silva RB, Valicente FH (2013) Molecular characterization of *Bacillus thuringiensis* using rep-PCR. Springerplus 2:641
Soberón M, Gill SS, Bravo A (2009) Signaling versus punching hole: how do *Bacillus thuringiensis* toxins kill insect midgut cells? Cell Mol Life Sci 66(8):1337–1349
Stern MJ, Ames GF, Smith NH et al (1984) Repetitive extragenic palindromic sequences: a major component of the bacterial genome. Cell 37(3):1015–1026
Ticknor TO, Kolsto AB, Hill KK et al (2001) Fluorescent amplified fragment length polymorphism analysis of norwegian *Bacillus cereus* and *Bacillus thuringiensis* soil isolates. Appl Environ Microbiol 67(10):4863–4873

Tirado-Montiel MI, Tyagi RD, Valero JR (2001) Wastewater treatment sludge as a raw material for the production of *Bacillus thuringiensis* based biopesticides. Water Res 35(16):3807–3816

Valicente FH, Barreto MR (2003) *Bacillus thuringiensis* survey in Brazil: geographical distribution and insecticidal activity against *Spodoptera frugiperda* (J.E. Smith) (Lepidoptera: Noctuidae). Neotrop Entomol 32(4):639–644

Valicente FH, Souza IRP (2004) Cultivo e preparo de *Bacillus thuringiensis* para microscopia eletrônica de varredura. In: Resumos. XXV Congresso Nacional de Milho e Sorgo, Cuiabá, p 146

Valicente FH, Mourão AHC (2008) Use of by-products rich in carbon and nitrogen as a nutrient source to produce *Bacillus thuringiensis* (Berliner)-Based biopesticide. Neotrop Entomol 37(6):702–708

Valicente FH, Picoli EAT, Vasconcelos MJV et al (2010) Molecular characterization and distribution of *Bacillus thuringiensis cry1* genes from Brazilian strains effective against the fall armyworm, *Spodoptera frugiperda*. Biol Control 53(3):360–366

Valicente FH, da Silva RB (2014) AFLP analysis of Brazilian *Bacillus thuringiensis* isolates. Springerplus 3:256

Van Frankenhuyzen K (2013) Cross-order and cross phylum activity of *Bacillus thuringiensis* pesticidal proteins. J Invertebr Pathol 114:76–85

Van Frankenhuyzen K (2009) Insecticidal activity of *Bacillus thuringiensis* crystal proteins. J Invertebr Pathol 101:1–16

Versalovic J, Koeuth T, Lupski JR (1991) Distribution of repetitive DNA sequences in eubacteria and application to fingerprinting of bacterial genomes. Nucleic Acids Res 19:6823–6831

Versalovic J, Schneider M, de Bruijn FJ (1994) Genomic fingerprinting of bacteria using repetitive sequence-based polymerase chain reaction. Methods Mol Cell Biol 5:25–40

Vos P, Hogers R, Bleeker M et al (1995) AFLP a new technique for DNA fingerprinting. Nucleic Acids Res 23:4407–4414

Chapter 14
Interactions of Entomopathogenic Fungus with Entomophagous Insects in Agroecosystems

Luis L. Vázquez

14.1 Introduction

The use of entomopathogenic biopesticides, be them spore suspensions or formulated solid products, can have different collateral effects in the agroecosystems since they are generally applied with conventional aspersion equipments that cover the cultivated surface (soil, crops), with possible drifts toward adjacent fields and auxiliary vegetation.

The fungal entomopathogens are typically applied as inundative sprays with the expectation of short-term pest control (Inglis et al. 2001). However, they do not only affect particular pest, but they are also species that occupy the same habitat and related species, such as parasitoids, predators, and decomposers (nontargeted organisms) (Sosa-Gómez et al. 1998). This situation implies that the possible effects of entomopathogens should be valued by their interactions, positive, negative, or neutral, more than for their collateral effects on nontargeted species.

The behavioral response of an insect to a fungal pathogen will have a direct effect on the efficacy of the fungus as a biological control agent. Two processes have a significant effect on the interactions between insects and entomopathogenic fungi: (a) the ability of targeting insects to detect and avoid fungal pathogens and (b) the transmission of fungal pathogens between host insects (Baverstock et al. 2010).

Natural enemies of insects have evolved and function in a complex multitrophic environment (Vet and Dicke 1992; Poppy 1997). Organisms are influenced by a range of abiotic and biotic factors, which affect their physiology, ecology, and behavior. Interactions between organisms within and between trophic levels are important factors that affect the structure of populations and communities (Roy and Pell 2000). Multiple biological control species may act synergistically, additively, or antagonistically (Ferguson and Stiling 1996).

L. L. Vázquez (✉)
Instituto de Investigaciones de Sanidad Vegetal (INISAV), Miramar, La Habana, Cuba

B. Souza et al. (eds.), *Natural Enemies of Insect Pests in Neotropical Agroecosystems*, https://doi.org/10.1007/978-3-030-24733-1_14

The World Health Organization (WHO) decided, in 1967, that any material of biological origin that is sought to be introduced as pesticide will be subject to the same toxicity studies that are applied to the synthetic products (Repetto 1992), not only for its possible effect on human or animal health, but for its environmental impact.

The use of entomopathogenic microorganism for controlling agricultural pests shows a tendency to increasing. This chapter seeks to assess many important aspects to understand the interactions of these microorganisms, especially entomopathogenic fungus, using entomophagues in the agroecosystems.

14.2 Interactions of the Entomopathogens in Agroecosystems

Many factors influence the effectiveness and establishment of entomopathogens. However, abiotic factors are known to affect the ability and speed with which entomopathogenic fungi can infect and colonize their insect hosts, thus limiting the "window of opportunity" for insect infection and the total efficacy for pest control (Yeo et al. 2003).

For better understanding the agroecosystems, it is important to study the existent relationships between the species and their food chains that compose a complex of trophic interactions, with interest in tritrophic interactions when seeking the use of pathogens, parasitoids, and predators, along with the host plant/insecticide (Knaak et al. 2009).

The understanding of the tritrophic interactions should be used as a tool for developing integrated pest management programs, looking to conserve the relationships between the organisms belonging to the trophic system, verifying the dynamic best applied to the agroecosystems, and avoiding the negative effects reaching high levels and causing damages to the agricultural production (Freitas et al. 2007).

It is important to distinguish between the physiological and ecological susceptibility of predators and parasitoids to pathogens when considering these natural enemies as biological control agents (Roy and Pell 2000). A laboratory bioassay on one particular isolate (strain GHA) of *Beauveria bassiana* (Bals.-Criv.) Vuill. demonstrated that some predators and parasitoids were susceptible, but in the field, the impact on the natural enemy was minimal (Jaronski et al. 1998).

Many entomopathogenic fungi appear to have a limited range of hosts. In general, fungi from the Entomophthorales order commonly cause epizootics in insect populations, which can eradicate populations at a local scale (Glare and Milner 1989).

The intraguild interactions between *Dicyphus hesperus* Knight (Miridae) and *Isaria fumosorosea* Wise (=*Paecilomyces fumosoroseus*) Apopka-97 (PFR-97™) (Clavicipitaceae) on *Trialeurodes vaporariorum* (Westwood) (Aleyrodidae) populations in tomato greenhouse microenvironment revealed a nonsignificant interaction effect between *D. hesperus* and *I. fumosorosea*, indicating that their actions were independent. The reduction of 48% and 35% of whitefly population was achieved by PFR-97™ and *D. hesperus*, respectively (Colleen et al. 2007).

According to Zaki (2011), the effect of the biopesticide based on *B. bassiana* spores on predators was adverse to the predator *Coccinella undecimpunctata* L. (Coccinellidae) and its biological aspects. A high concentration of the pathogen (2, 4, and 8 g/L) eliminated the egg deposition in treated females, and spores at 4 g/L affected the longevity of adults and the duration of larvae and pupal period.

The importance of predators, parasitoids, and pathogens in the suppression of pest populations has been the motivation for many studies to quantify field populations of natural enemies (Roy and Pell 2000). These studies have generally concentrated on either predators (Booij and Noorlander 1992), parasitoids (Muller et al. 1999), or pathogens (Ekbom and Pickering 1990) in isolation. Assessments of the temporal and spatial occurrence of different natural enemies have largely been neglected (Roy and Pell 2000).

According to Vázquez et al. (2008), the occurrence of epizootic in the coffee berry borer [*Hypothenemus hampei* (Ferrari) (Curculionidae)] for *B. bassiana* is the result of a process favored by the edafoclimatic conditions of coffee plantations, the reduction or no use of chemical pesticides, and the presence of the pest populations. The authors also state that the soil and shade management of the coffee crop (intensity of solar radiations, relative humidity, and temperature) may favor the epizootic since it contributes to creating a microclimate where the microorganism develops its saprophytic and pathogenic cycle, as well as its dispersion in the field.

14.3 Interactions of Endophytic Entomopathogens with Entomophagues

A growing number of studies demonstrate the potential of entomopathogenic fungi to play additional ecological roles including the endophytic colonization of a wide variety of plants (Vega et al. 2009). Inoculation of entomopathogenic fungi into plants and their establishment as endophytes could help overcome such constraints and improve their deployment for a more effective pest control (Vega et al. 2009; Jaber and Ownley 2018).

Studies conducted by Jaber and Araj (2018) demonstrated for the first time the negative effects of two endophytic entomopathogenic fungi, *B. bassiana* (Naturalis) and *Metarhizium brunneum* Petch (Bipesco5), on the development and fecundity of the green peach aphid *Myzus persicae* Sulzer (Aphididae), throughout the two successive generations (first- and second-generation aphid). The study also provides one of very few reports of the compatible use of endophytic fungal entomopathogens with other biological control agents, particularly parasitoids.

Furthermore, the endophytic colonization of plants with the tested fungal entomopathogens did not have a significant effect on the percentage of female *Aphidius colemani* Viereck (Braconidae) progenies among the treatments. Furthermore, adult longevity of male and female parasitoid progenies parasitizing second-generation aphids reared on inoculated or control plants was not influenced by plant colonization with fungal entomopathogens. Akutse et al. (2014) found no significant differences in the parasitism rate and survival of two parasitoids of the pea leafminer *Liriomyza huidobrensis* Blanchard (Agromyzidae), *Phaedrotoma scabriventris* Nixon (Braconidae), and *Diglyphus isaea* Walker (Eulophidae), parasitizing second and third instar larvae reared on broad bean plants inoculated by several fungal isolates of *B. bassiana* and *Hypocrea lixii* Patouillard or the control.

Moreover, higher numbers of flies emerged from pupae in the control when compared to endophytically colonized plants (Akutse et al. 2014). Gathage et al. (2016) investigated the effects of seed inoculation with *B. bassiana* and *H. lixii* on the pea leafminer and six associated parasitoid species [*Phaedrotoma scabriventris* Nixon (Braconidae), *Diglyphus isaea* Walker (Eulophidae), *Opius dissitus* Muesebeck (Braconidae), *Neochrysocharis formosa* Westwood (Eulophidae), *Hemiptarsenus varicornis* Girault (Eulophidae), and *Halticoptera arduine* (Walker) (Pteromalidae)] under field conditions. The authors found no differences in the number of parasitoids that emerged from the pupae between endophyte and control treatments. Therefore, both of these studies indicated that endophytic fungal entomopathogens and parasitoids can be used in combination for controlling insect pests (Jaber and Araj 2018). On the other hand, the endophytic colonization of cauliflower plants with *Aspergillus flavus* Link and *Aspergillus niger* van Tieghem (Eurotiales) had a negative effect on parasitism and the development of the ectoparasitoid *Bracon hebetor* Say (Braconidae), parasitizing larvae of *Spodoptera litura* Fabricius (Noctuidae) fed endophyte-colonized plants (Kaur et al. 2015). It is thus essential to note that the results of studies investigating multitrophic interactions between endophytic fungi, parasitoids, and host plants may vary according to the fungal strain (Bultman et al. 2003; Akutse et al. 2014), parasitoid species (Bixby-Brosi and Potter 2012), or the tested organism system as a whole.

14.4 Interactions of Entomopathogens with Entomophagues

The entomopathogenic interactions with the entomophague groups have been demonstrated in laboratory studies. These studies show negative effects of different magnitudes (Table 14.1) against predators (Coccinellidae, Chrysopidae, Formicidae) and parasitoids (Aphelinidae, Braconidae, Bethylidae, Platygasteridae), results that could be influenced by the research method used and the characteristics of the entomopathogenic strain, among others.

Despite a higher number of compiled studies leading to results different from the previous ones due to the characteristics of the natural enemy species and entomopathogens, it is evident that negative and neutral, as well as positive, interactions exist for the dissemination of these microorganisms. The confirmation under field conditions would allow greater security in the integration of these bioproducts with the natural enemies that inhabit the agroecosystems.

Table 14.1 Mortality effects of entomopathogenic biopesticides with different groups of entomophagues in laboratory studies

Natural enemies	Pests and cultivation	Entomopathogens	Mortality (%)	References
Coccinellidae				
Cycloneda sanguinea sanguinea (L.) (Coccinellidae)	Homopterous in citrus	*Bacillus thuringiensis*	0	Almaguel et al. (1993)
Cycloneda sanguinea limbifer Casey (Coccinellidae)	Aphids in horticulture, root, and tuber	*B. thuringiensis* (Turingin) *B. thuringiensis* (Betoxibacilin) *Verticillium lecanii* (Verticilin)	90 25 15	Burkova and Krasavina (1997)
Coccidophilus citricola Brethes (Coccinellidae)	Coccoideos in citrus	*Hirsutella thompsonii* fisher	0	Sosa et al. (1985)
Chrysopidae				
Chrysoperla carnea Stephens (Chrysopidae)	Aphids in horticulture, root, and tuber	*B. thuringiensis* (Turingin) *B. thuringiensis* (Betoxibacilin) *V. lecanii* (Verticilin)	100% 25 0	Burkova and Krasavina (1997)
	Bemisia tabaci (Aleyrodidae) horticulture, grains	*Paecilomyces* spp.	12,5	Hernández and Berlanga (1995)
Nodita fermini Navás (Chrysopidae)	*Heliothis virescens* (Noctuidae) in tobacco	*Verticillium lecanii* *Bacillus thuringiensis*	0 0	Acosta and Rijo (1997), Acosta and Rijo (1998)
Formicidae				
Pheidole megacephala (F) (Formicidae)	*Cylas formicarius* (Fabricius) (Curculionidae) in sweet potato	*Beauveria bassiana* *B. thuringiensis* (Dipel) *B. thuringiensis* (Bitoxibacilin 202)	1,64 1,34 2,24	Castiñeiras and Calderón (1982)
Phytoseiidae				
Phytoseiulus macropilis (Phytoseiidae)	*Tetranychus tumidus* (Tetranychidae) in banana	*Bacillus thuringiensis*	0	Almaguel et al. (1993)
Hymenoptera				

(continued)

Table 14.1 (continued)

Natural enemies	Pests and cultivation	Entomopathogens	Mortality (%)	References
Encarsia Formosa (Gahan) (Aphelinidae)	Mosca blanca (Aleyrodiae) in horticulture, grains	*B. thuringiensis* (Betoxibacilin) *Bt* (Dendrobacilin) *Bt* (Entobacterin) *Bt* (Lepidocid) *V. lecanii* (Verticilin) *B. bassiana* (Boverin)	20 20 8 5 5 14	Burkova and Krasavina (1997)
	Trialeurodes vaporariorum Westwood (Aleyrodidae) (nymphs of 3° and 4° instar parasitized) in horticulture	*Aschersonia aleyrodis* Webber	0	Vargas et al. (1995)
Encarsia transvena (Timberlake) (Aphelinidae)	*Bemisia tabaci* (Gennadius) (Aleyrodidae)	*Verticillium lecanii* (Mycotal and vertisol) *Isaria fumosorosea* *B. bassiana* (naturalist) *Entomophthora virulenta* (Vektor)	0	Serra et al. (1997)
Telenomus sp. (Platygastridae)	*Spodoptera frugiperda* (Noctuidae) in corn	*Metarhizium rileyi* (=*Nomuraea*) *B. thuringiensis*	0 0	Blanco (1997)
Cotesia americanus (Lepelitier) (Braconidae)	*Erinnyis ello* (L.) (Sphingidae) in cassava	*Verticillium lecanii* *B. bassiana*	95 84	Vázquez (2002)
Diadegma sp. (Ichneumonidae)	*Heliothis virescens* (Noctuidae) in tobacco	*B. thuringiensis* (BTB) Dendrobacillin (Dipel)	0	Jiménez (1996)
Cephalonomia stephanoderis Betren (Bethylidae)	*Hypothenemus hampei* (Ferrari) (Curculionidae) in coffee	*B. bassiana* *Metarhizium anisopliae*	48 40	Reyes et al. (1995)

Additionally, entomopathogenic fungi may not be selective and may directly infect key nontargeted organisms. Thus, the parasitoids may be susceptible to diseases caused by entomopathogenic fungi. Although, in most cases, parasitoids are not susceptible to entomopathogenic fungal infections, knowledge on the sensitivity of different parasitoid life stages to fungal infections, as well as on the possibility of host competition between the two types of biological control agents, is important (Magalhães et al. 1988; Sosa-Gómez et al. 1998).

Studies focusing on verifying the interaction between entomopathogenic fungi with egg parasitoids have reported that *Metarhizium anisopliae* (Hypocreales) did not interfere in the reproductive and survival characteristics of *Trichogramma pretiosum* Riley (Trichogrammatidae) (Polanczyk et al. 2009; Amaro et al. 2015; Pazini et al. 2016). Nevertheless, Potrich et al. (2017) verified that *M. anisopliae* (Unioeste 22 strain) caused a reduction in the emergence of *T. pretiosum* (females and males) and led to the rapid mortality of parasitic females when compared to the nonparasitic females.

Broglio-Micheletti et al. (2006) verified that a strain of *M. anisopliae* (IPA 159E) caused a reduction in the number of *Diatraea saccharalis* (Fabricius) (Crambidae) eggs parasitized by *Trichogramma galloi* Zucchi (Trichogrammatidae), also reducing the emergence and longevity of the second generation. In contrast, two different strains (IPA 211 and IPA139E) of this same species did not interfere in the evaluated parameters. Negative effects of *M. anisopliae* (commercial strain E9) were not observed on the reproduction and survival of *Trichogramma atopovirilia* Oatman and Platner (Trichogrammatidae) (Polanczyk et al. 2010). The same was reported for *M. anisopliae* on *T. galloi* in *D. saccharalis eggs* treated with this entomopathogen (Oliveira et al. 2013).

Despite the *T. pretiosum* grooming itself (cleaning behavior) by continuously scraping its legs over its wings and body (Vinson 1998), possibly removing conidia and avoiding infection, many parts of the insect's body are difficult to reach and conidia that have adhered to the cuticle may begin to germinate. Susceptible regions include the intersegmental membranes, the wing folding area, the mouth, and the ventral regions, which are very difficult for *T. pretiosum* to reach and clean. The "self-grooming" behavior of this parasitoid has been previously reported (Basibuyuk and Quicke 1999). In this study, we observed that, when *T. pretiosum* detects microorganisms or toxic substances on its body, parasitism rates are reduced as the insect invests more time in self-grooming (Potrich et al. 2017).

An interesting case is that of the endosymbiont bacteria of the *Wolbachia* (Anaplasmataceae) genera, which, according to Rodriguero (2013), is the most widespread infection in animals. Its ability to manipulate insect reproduction places it at the center of organismal biology, affecting such important processes like sex determination, cell cycle, generation and extinction of species, and the behavior of arthropods, including several pest and disease vector insects. Attributes like vertical transmission, the speed at which it dissipates in the population, the ability to block the pathogenic activity of several microorganisms, and the ability to shorten the lifespan of its host point to this bacteria as a potential tool to control harmful insects and nematodes.

For example, in most cases, thelytokous *Trichogramma* wasps are infected with parthenogenesis-inducing (PI) *Wolbachia* (Stouthamer et al. 1993). Endosymbiotic *Wolbachia* bacteria are known to affect the fecundity and dispersion of infected strains (Stouthamer and Luck 1993). In some cases, infection with PI-*Wolbachia* also have severe negative effects on the competitive ability under conditions of superparasitism and on the survival of immature stages (Miura and Tagami 2004).

Farrokhi et al. (2010) showed that the *Wolbachia*-infected and uninfected *Trichogramma brassicae* Bezdenko (Trichogrammatidae) strains had only a slight difference in their functional type II response. In conclusion, infection with *Wolbachia* does not seem to have a significant impact on the type of functional response and attack rate of the studied parasitoids under controlled laboratory conditions. However, the infected strain of *T. brassicae* had a lower quality in comparison to the B-strain regarding the time of hosting and the magnitude of parasitism capacity.

To date, there is very little information on the compatible use of endophytic fungal entomopathogens and other groups of natural enemies, particularly parasitoids, for integrated pest management (IPM) (Akutse et al. 2014).

14.5 Final Remarks

In the programs of augmentative biological control, the interactions between entomophagues and entomopathogens are generally considered. However, little attention is given to the interactions with the natural enemies that inhabit the agroecosystems. When the microbiological agents are applied, be them inundative or inoculative, two effects are generally achieved: (a) the direct control of the pest population and (b) their establishment in the agroecosystem to act as a regulator.

On the other hand, the use of biological control agents as if they were a product (focus of substitution of inputs) is not sustainable; instead, they must be used as part of a pest management system to increase the capacity of pest self-regulation.

References

Acosta N, Rijo E (1997) Compatibilidad de *Nodita fermini* (Neuroptera: Chrysopidae) con el hongo *Verticillium lecanii* (Zimm) Viegas y la bacteria *Bacillus thuringiensis* (Berliner). In: Anais do 3° seminario científico internacional de sanidad vegetal, ALF, Ciudad de La Habana, 23–27 June 1997

Acosta N, Rijo E (1998) Compatibilidad de *Nodita firmini* Navás (Neuroptera: Chrysopidae) con la bacteria *Bacillus thuringiensis* (Berliner). Fitosanidad 2(3/4):49–52

Akutse KS, Fiaboe KKM, van den Berg J et al (2014) Effects of endophyte colonization of *Vicia faba* (Fabaceae) plants on the life-history of leafminer parasitoids *Phaedrotoma scabriventris* (Hymenoptera: Braconidae) and *Diglyphus isaea* (Hymenoptera: Eulophidae). PLoS One 9(10):e109965

Almaguel L, González N, Fernández-Larrea O et al (1993) Utilización de *Bacillus thuringiensis* (BT-13) en programas de lucha contra ácaros en cítricos, plátano y papa, informe final de proyecto. Inisav, La Habana

Amaro JT, Bueno AF, Pomari-Fernandes A et al (2015) Selectivity of organic products to *Trichogramma pretiosum* Riley (Hymenoptera: Trichogrammatidae). Neotrop Entomol 44:489–497

Basibuyuk HH, Quicke DLJ (1999) Grooming behaviours in the Hymenoptera (Insecta): potential phylogenetic significance. Zool J Linnean Soc 125(3):349–382

Baverstock J, Roy HE, Pell JK (2010) Entomopathogenic fungi and insect behaviour: from unsuspecting hosts to targeted vectors. BioControl 55(1):89–102

Bixby-Brosi AJ, Potter DA (2012) Endophyte-mediated tritrophic interactions between a grass-feeding caterpillar and two parasitoid species with different life histories. Arthropod Plant Interact 6(1):27–34

Blanco E (1997) Pesticide compatibility with *Telenomus* sp. eggs parasitoid of *Spodoptera frugiperda* (J.E. Smith). In: Anais do 3° seminario científico internacional de sanidad vegetal, ALF, Ciudad de La Habana, 23–27 June 1997

Booij CJH, Noorlander J (1992) Farming systems and insect predators. Agric Ecosyst Environ 40(1/4):125–135

Broglio-Micheletti SMF, Santos AJN, Pereira-Barros JL (2006) Ação de alguns produtos fitossanitários para adultos de *Trichogramma galloi* Zucchi, 1988 (Hymenoptera: Trichogrammatidae). Cienc Agrotec 30(6):1051–1055

Bultman TL, McNeill MR, Goldson SL (2003) Isolate-dependent impacts of fungal endophytes in a multitrophic interaction. Oikos 102(3):491–496

Burkova LA, Krasavina LP (1997) Toxicidad de los biopreparados para los insectos útiles. Phytoma 85:9–10

Castiñeiras A, Calderón A (1982) Suceptibilidad de *Pheidole megacephala* a tres insecticidas microbianos: Dipel, Bitoxibacillin 202 y *Beauveria bassiana* en condiciones de laboratorio. Cienc Téc Agric Prot Plantas Suppl:61–66

Colleen RA, Goettel MS, Roitberg BD et al (2007) Combined effects of the entomopathogenic fungus, *Paecilomyces fumosoroseus* Apopka-97, and the generalist predator, *Dicyphus hesperus*, on whitefly populations. BioControl 52(5):669–681

Ekbom B, Pickering J (1990) Pathogenic fungal dynamics in a fall population of the blackmargined aphid (*Monellia caryella*). Entomol Exp Appl 57(1):29–37

Farrokhi S, Ashouri A, Shirazi J et al (2010) A comparative study on the functional response of *Wolbachia*-infected and uninfected forms of the parasitoid wasp *Trichogramma brassicae*. J Insect Sci 10:167

Ferguson KI, Stiling P (1996) Non-additive effects of multiple natural enemies on aphid populations. Oecologia 108(2):375–379

Freitas JMS, Clery-Santos MP, Pérez-Maluf R (2007) Abundância de himenópteros parasitoides em diferentes perfis de paisagens. In: Anais do 8° congresso de ecologia do Brasil, Sociedade de Ecologia do Brasil, Caxambu, 23–28 Sept 2007

Gathage JW, Lagat ZO, Fiaboe KKM et al (2016) Prospects of fungal endophytes in the control of *Liriomyza* leafminer flies in common bean *Phaseolus vulgaris* under field conditions. BioControl 61(6):741–753

Glare TR, Milner RJ (1989) Ecology of entomopathogenic fungi. In: Arora DK, Ajello L, Mukerji KG (eds) Handbook of applied mycology, humans, animals and insects. Marcel Dekker, New York, pp 547–612

Hernández VM, Berlanga AM (1995) Selección de aislamientos de *Paecilomyces* spp. y su interacción con otros agentes de control de *Bemisia tabaci* En: Memorias do 16° congreso nacional control biologico, SMCB, Tapachula, 9–10 Nov 1995

Inglis GD, Goettel MS, Butt TM et al (2001) Use of hyphomycetous fungi for managing insect pests. In: Butt TM, Jackson J, Magan N (eds) Fungi as biocontrol agents. CAB International, UK, pp 23–69

Jaber LR, Araj SE (2018) Interactions among endophytic fungal entomopathogens (Ascomycota: Hypocreales), the green peach aphid *Myzus persicae* Sulzer (Homoptera: Aphididae), and the aphid endoparasitoid *Aphidius colemani* Viereck (Hymenoptera: Braconidae). Biol Control 116:53–61

Jaber LR, Ownley BH (2018) Can we use entomopathogenic fungi as endophytes for dual biological control of insect pests and plant pathogens? Biol Control 116:36–45

Jaronski ST, Lord J, Rosinska J et al (1998) Vect of *Beauveria bassiana*-based mycoinsecticide on beneficial insects under field conditions. In: the 1998 Brighton conference: pest & diseases:

proceedings of an international conference held at the Brighton Centre and the Stakis Brighton metropole hotel, British Crop Protection Council, Brighton, 16–19 Nov 1998

Jiménez J (1996) Lucha biológica contra el cogollero del tabaco *Heliothis virescens* (F.) (Lepidoptera: Noctuidae) con biopreparados nacionales y comerciales a base de *Bacillus thuringiensis* Berliner en Cuba. Tesis, Instituto de Investigaciones en Sanidad Vegetal

Kaur T, Singh B, Kaur A et al (2015) Endophyte-mediated interactions between cauliflower, the herbivore *Spodoptera litura*, and the ectoparasitoid *Bracon hebetor*. Oecologia 179(2):487–494

Knaak N, Azambuja AO, Lucho APR et al (2009) Interações de *Bacillus thuringiensis* e o controle de fitopatógenos. Biotecnologia Ciênc Desenvol 38:48–53

Magalhães BP, Lord JC, Wraight SP et al (1988) Pathogenicity of *Beauveria bassiana* and *Zoophthora radicans* to the coccinellid predators *Coleomegilla maculata* and *Eriopsis connexa*. J Invertebr Pathol 52(3):471–473

Miura K, Tagami Y (2004) Comparison of life history characters of arrhenotokous and *Wolbachia*-associated thelytokous *Trichogramma kaykai* Pinto and Stouthamer (Hymenoptera: Trichogrammatidae). Ann Entomol Soc Am 97(4):765–769

Muller CB, Adriaanse ICT, Belshaw R et al (1999) The structure of an aphidparasitoid community. J Anim Ecol 68(2):346–370

Oliveira HN, Antigo MR, Carvalho GA et al (2013) Seletividade de inseticidas utilizados na cana-de-açúcar a adultos de *Trichogramma galloi* Zucchi (Hymenoptera: Trichogrammatidae). Biosci J 29(5):1267–1274

Pazini JB, Grützmacher AD, Martins JFS et al (2016) Selectivity of pesticides used in rice crop on *Telenomus podisi* and *Trichogramma pretiosum*. Pesqui Agropecu Trop 46(3):327–335

Polanczyk RA, Grecco ED, Andrade GS et al (2009) Desempenho de *Trichogramma* spp. (Hymenoptera: Trichogrammatidae) em ovos de *Diaphania hyalinata* (L.) (Lepidoptera: Pyralidae) tratados com *Metarhizium anisopliae* e *Beauveria bassiana*. Arq Inst Biol 76(3):495–499

Polanczyk RA, Pratissoli D, Dalvi LP et al (2010) Efeito de *Beauveria bassiana* (Bals.) Vuillemin e *Metarhizium anisopliae* (Metsch.) Sorokin nos parâmetros biológicos de *Trichogramma atopovirilia* Oatman & Platner, 1983 (Hymenoptera: Trichogrammatidae). Cienc Agrotec 34(6):1412–1416

Poppy GM (1997) Tritrophic interactions: improving ecological understanding and biological control? Endeavour 21(2):61–64

Potrich M, Alves LFA, Lozano ER et al (2017) Potential side effects of the entomopathogenic fungus *Metarhizium anisopliae* on the egg parasitoid *Trichogramma pretiosum* (Hymenoptera: Trichogrammatidae) under controlled conditions. J Econ Entomol 110(6):2318–2324

Repetto M (1992) Evaluación de la toxicidad de los plaguicidas biológicos. Rev Toxicol 9(1):3–9

Reyes IC, Enrique A, Chávez B (1995) Efecto de *Beauveria bassiana* y *Metarrhizium anisopliae* sobre el parasitoide de la broca del café *Cephalonomia stephanoderis*. Rev Colomb Entomol 21(4):199–204

Rodriguero MS (2013) *Wolbachia*, una pandemia con posibilidades. Rev Soc Entomol Argent 72(3/4):117–137

Roy HE, Pell JK (2000) Interactions between entomopathogenic fungi and other natural enemies: implications for biological control. Biocontrol Sci Tech 10(6):737–752

Serra CA, Ortiz M, Ferreras R (1997) Evaluación de los efectos secundarios de insecticidas químicos y biológicos sobre *Encarsia transvena* (Timberlake). In: 33th proceedings of the Caribbean food crops society, Caribbean Food Crops Society, Puerto Rico, 6–12 July 1997

Sosa DR, Ricci JG, Nasca AJ (1985) Efecto de *Hirsutela thompsonii* Fisher var. *thompsonii* sobre larvas y adultos de *Coccidorhilus citricola* Brethes y *Lindorus lophanthae* (Blaisdell) (Coleoptera: Coccinellidae). CIRPON 3:73–77

Sosa-Gómez DR, Pereira RM, Alves SB (1998) Impacto ambiental de entomopatógenos. In: Alves SB (ed) Controle microbiano de insetos. FEALQ, Piracicaba, pp 1075–1096

Stouthamer R, Luck RF (1993) Influence of microbe-associated parthenogenesis on the fecundity of *Trichogramma deion* and *T. pretiosum*. Entomol Exp Appl 67(2):183–192

Stouthamer R, Breeuwer JAJ, Luck RF et al (1993) Molecular identification of microorganisms associated with parthenogenesis. Nature 361(6407):66–68

Vargas MM, Rodríguez DA, Sanabria J et al (1995) Ensayo de diferentes dosis de *Aschersonia aleyrodis* Webber y parasitismo de *Encarsia formosa* Gahan en ninfas de tercer y cuarto instar de la mosca blanca de los invernaderos. Rev Colomb Entomol 21(3):159–170

Vázquez LL (2002) Efecto de *Verticillium lecanii* y *Beauveria bassiana* sobre *Cotesia americanus* Lepeletier (Hymenoptera: Braconidae) parasitoide de larvas de la primavera de la yuca (*Erinnys ello* L.). Fitosanidad 6(1):25–27

Vázquez LL, Elósegui O, Alfonso J et al (2008) Regulación natural de la broca del café. Agric Org 14:32–34

Vega FE, Goettel MS, Blackwell M et al (2009) Fungal entomopathogens: new insights on their ecology. Fungal Ecol 2(4):149–159

Vet LEM, Dicke M (1992) Ecology of infochemical use by natural enemies in a tritrophic context. Annu Rev Entomol 37:141–172

Vinson SB (1998) The general host selection behavior of parasitoid Hymenoptera and a comparison of initial strategies utilized by larvaphagous and oophagous species. Biol Control 11(2):79–96

Yeo H, Pell JK, Alderson PG et al (2003) Laboratory evaluation of temperature effects on the germination and growth of entomopathogenic fungi and on their pathogenicity to two aphid species. Pest Manag Sci 59(2):156–165

Zaki FN (2011) Side effects of the entomopathogenic fungus *Beauveria bassiana* on the predator *Coccinella undecimpunctata*. Arch Phytopathol Plant Protec 44(19):1887–1893

Part III
Mass Production of Biocontrol Agents in Latin America: Rearing Techniques and Releasing Strategies

Chapter 15
Predatory Insects

Brígida Souza and Carlos Eduardo Souza Bezerra

15.1 Introduction

The reduction of arthropod pest populations by naturally occurring predators in agroecosystems is recognized and accepted as a strategy to control these undesirable organisms. However, acceptable population levels might not be achieved by these natural enemies. In addition to the increase in plant diversity discussed in several chapters of this book, the release of laboratory mass-reared organisms can increase the number of predators in crops.

Mass production of predatory insects, as well as other natural enemies, requires appropriate laboratories, efficient rearing techniques, adequate diet, and skilled labor in each activity of the productive process. Production cost is an important aspect of augmentative biological control, and all efforts must be directed towards achieving economically viable production. In addition to the steps that involve the entire productive process *per se*, the rearing of these natural enemies in laboratory requires prior knowledge of well-adapted species, as well as appropriate procedures for transport and release in the field.

In this chapter, rearing techniques and strategies for releasing predatory insects will be addressed aiming to emphasize the species involved in research in Brazil and other Latin American countries.

B. Souza (✉)
Department of Entomology, Federal University of Lavras (UFLA), Lavras, MG, Brazil
e-mail: brgsouza@ufla.br

C. E. S. Bezerra
Department of Agricultural, Biological and Engineering Sciences,
Várzea Grande University Center, Várzea Grande, MT, Brazil
e-mail: carlos.bezerra@univag.edu.br

B. Souza et al. (eds.), *Natural Enemies of Insect Pests in Neotropical Agroecosystems*, https://doi.org/10.1007/978-3-030-24733-1_15

15.2 Mass Rearing

The mass rearing of predatory insects faces many challenges and potential obstacles, for example, the development of rearing techniques and adequate diets, as well as appropriate release methods, combined with quality control and adequate procedures to assess the performance in the field (Sørensen et al. 2012). However, the entire process begins with the selection of the biocontrol species.

15.3 Selection of the Control Agent

The selection of the species must be based on its biology, ecology, and ethology, including its relationship with the target prey and its host plant. The selection of resistant insects ought to be prioritized when releasing the predator as part of an integrated control program, where the use of pesticides is included (Grafton-Cardwell and Hoy 1986). Inundative releases of predators do not primarily seek the establishment of these organisms in the area for more than a generation (DeBach and Hagen 1964). The establishment of predatory agents is three times smaller than the releases of parasitoids (Greathead 1986). Thus, the most important characteristics when choosing the agent in order to enable their multiplication on a large scale include the ability to control the population of the arthropod pests after release and the ease of rearing – along with a low cost – of these insects. The ability to remain in the environment should also be considered if there is more than one candidate to be massively produced.

The use of native predators may represent a promising alternative to introduced insects, since they naturally have a higher chance of survival after release and of establishing in the environment for longer periods. For example, in Brazil, the native predator *Supputius cincticeps* Stål (Pentatomidae) has been reported feeding on *Thaumastocoris peregrinus* Carpintero & Dellape (Thaumastocoridae), an exotic common pest in eucalyptus crops (Souza et al. 2012). Moreover, this predator can be easily reared in laboratory (Zanuncio et al. 2005, 2014), presenting great potential for use in biological control programs.

Induction of phenotypic plasticity or change of genetic composition of a population via selection is necessary in laboratory productions. However, no genotype or rearing protocol will be able to maintain the quality of the insects released in all possible situations to be found in the field, which thus requires the use of different strains (Sørensen et al. 2012). Therefore, the use of the same lineage of a particular predator species, especially in countries with great territorial extension and varied climatic conditions, presents strong risks related to low performance in particular environmental situations.

15.4 Establishment of Initial Population

After knowing the potential of predatory species to be reared, it is important to invest in research on methods for laboratory multiplication. This stage involves fundamental preliminary knowledge such as the number of specimens that will compose the initial population. This factor is important in species exposed to long rearing periods, which begin to manifest genetic degenerations and is attributed to inbreeding due to the low gene flow, among other factors. One way to avoid genetic degeneration or restoring genotypic variability consists in initiating the laboratory population with as many individuals as possible. The increase in the probability of the founding colony containing the different genotypes present in the species in the natural environment increases the frequency of heterozygotes and genetic variability in the population (Pires et al. 2016b). This particularity refers to the need to periodically include individuals of natural populations, which will be responsible for the maintenance of the attributes of the species across generations.

15.5 Predatory Insects Feeding and Cost of Production in Mass Rearing

For decades, the cost of mass production of predatory insects has been the main obstacle to its implementation. The labor and time required in the production of food are the main components responsible for the onerousness of the productive process. For this reason, artificial diets or, at least, the use of alternative prey that have a lower cost of production are recommended (King et al. 1985). Rearing predators on their natural prey involve the continuous cultivation of the host plant of these herbivores, which is costly, impractical, and performed only under research level (De Clercq et al. 2014). In the Entomology Department (DEN) of the Federal University of Lavras (UFLA), state of Minas Gerais, Brazil, host plants of several species of aphids are cultivated. However, pumpkins (*Cucurbita maxima* L.) are used as alternative hosts for *Planococcus citri* (Risso) (Pseudococcidae), a natural prey of larvae and adults of *Cryptolaemus montrouzieri* Mulsant (Coccinellidae). A large number of specimens of *C. montrouzieri* were also reared on *Dysmicoccus brevipes* Cockerel (Pseudococcidae) or *P. citri* using *C. maxima* cv. Jacarezinho as host (Sanches and Carvalho 2010).

Species of generalist predators can feed on preys other than those that are part of their natural diet. For example, in DEN/UFLA, several investigations involving generalist natural enemies have been performed using eggs and/or larvae of *Spodoptera frugiperda* (J.E. Smith) (Noctuidae) as prey/host. The larvae of this noctuid can be satisfactorily reared in artificial diet, making the cultivation of host plants unnecessary. The development of *Tynacantha marginata* (Pentatomidae) is favorable when fed with pupae of *Tenebrio molitor* L. (Tenebrionidae), which guarantee more than 60% survival of the nymphal stage (Dalvi et al. 2009). In addition to *T. molitor*, Asopinae nymphs can feed on alternative prey such as *Musca domestica* L. (Muscidae) and *Bombyx mori* L. (Bombycidae) larvae supplied in rearing vessels (Neves et al. 2010) or on larvae of natural prey such as *Anticarsia gemmatalis* Hübner (Noctuidae), *Thyrinteina arnobia* Stoll, and *Thyrinteina leucocerae* Rindge (Geometridae) (Ferreira et al. 2008; Oliveira et al. 2011).

Another example is the use of Mediterranean flour moth eggs, *Ephestia kuehniella* Zeller (Pyralidae), as a prey for *Orius insidiosus* (Say) (Anthocoridae), increasing the fecundity and longevity of the predator when compared to natural prey. Unlike other species of the genus, the addition of pollen to the diet of *O. insidiosus* is not a requirement for its production, which facilitates its rearing and reduces the costs (Calixto et al. 2013). Although in many cases the use of alternative prey is considered a facilitating strategy in the large-scale production process, in some situations, it is still not enough to meet mass production.

Rearing of preys, whether natural or alternative, is in many cases more difficult than the natural enemy itself. Besides the difficulty in the rearing of the prey, there is a relevant problem related to the labor involved in the management of the rearing and, consequently, the cost of production.

The labor factor was considered the most costly component of natural enemy production costs in Brazil, accounting for between 60 and 80% of the total (Parra 2002; Mendes et al. 2005; Vieira et al. 2013). The rearing of *O. insidiosus* on *Plutella xylostella* (L.) (Plutellidae) costs US $ 0.0029 per specimen produced, which also includes labor costs, equipment and energy expenses, materials used, and depreciation of specific materials that were recorded, not considering the costs of implementing the bio-factory, since an existing structure was used (Vieira et al. 2013). Considering the recommendation of 2 predators/m^2 (Silveira et al. 2004), the cost of producing *O. insidiosus* to cover a 1000-m^2 greenhouse would be US $ 5.76, of which 62% of this value is related to labor.

In Brazil, nearly 43 mg of eggs per larvae are required for rearing *Chrysoperla externa* (Hagen) (Chrysopidae) on the alternative prey *E. kuehniella* (Bortoli et al. 2006). Since each gram of *E. kuehniella* eggs costs the equivalent of US $ 2.20, the cost of food for each produced individual is US $ 0.10. In order to reduce these costs, Bezerra et al. (2016) developed an artificial diet capable of satisfactorily promoting the development of the second and third instars of *C. externa*, requiring only 5 mg of *E. kuehniella* eggs for first instar feeding. The use of this diet provided a reduction of the total food expenditure of each individual to US $ 0.02, maintaining the survival, size, and predatory capacity of the insects produced during more than seven generations. Therefore, artificial diets have been preferred for use as food in the mass production of natural enemies, constituting a key component in the production process. Researchers in various parts of the world have invested in the discovery of artificial foods that can replace the use of prey in order to promote the economic viability of the productive process. However, the development of an adequate recipe demands continuous tests involving the chemical composition of the diet produced and its texture (Grenier 2012). In parallel, the forms of supply must be studied, which find their greatest challenge in administration to the larval stage.

Regarding the rearing of predators with artificial diets, the use of a diet for *Brontocoris tabidus* (Signoret) (Pentatomidae) based on liver and beef, which caused a 52% survival in the immature phase of this Asopinae, did not differ from that obtained with the supply of *T. molitor* pupae (Zanuncio et al. 1996). The artificial diet composed of broiler's food, wheat bran, brewer's yeast, powdered milk, and the preservative methylparaben (Nipagin) was used in the rearing of *Euborellia annulipes* (Lucas) (Anisolabididae) (Costa et al. 2007). However, the viability of the nymphal stage was greater than 70% for *E. annulipes* fed on cat food with 26% protein, brewer's yeast, wheat germ, powdered milk, and methylparaben (Costa et al. 2011).

In Brazil, artificial diets have also been investigated for large-scale production of predatory bugs (Pentatomidae: Asopinae) in an attempt to replace the preys used to rear these predators and to reduce costs. However, a completely satisfactory diet has not yet been obtained and research is still being developed in search of the ideal food, so that these predators continue to be produced with the use of natural/alternative prey and leaves of plants are supplied as a food supplement (Lemos et al. 2005; Pires et al. 2016a, b).

Many predators are able to alternatively feed on plant materials, which eliminates the need of rearing an alternative prey or developing an artificial diet for the rearing of either the prey or the natural enemy. However, the use of plant parts in the feeding of these natural enemies also requires the continuous maintenance of the used botanical species. As an example, we can mention the mite *Neoseiulus californicus* (McGregor) (Phytoseiidae), predator of *Tetranychus urticae* (Koch) (Tetranychidae), which develops and reproduces in jack bean plants, *Canavalia ensiformis* L. (Fabaceae) (Marafeli et al. 2014), and *O. insidiosus* grown on *Bidens pilosa* L. (Asteraceae) during the flowering phase (Bueno et al. 2006; Carvalho et al. 2010). In this aspect, the variation of the type of food offered to predatory insects is highly important, even when using natural feeding, which can be a fundamental point for the success of production. In laboratory tests, larvae of *Coleomegilla maculata* DeGeer (Coccinellidae) fed only on natural prey consumed up to three times more plant tissues than those supplied with a mixed diet, suggesting that the larvae compensate for the nutritional deficiency of their prey through the consumption of additional plant tissue (Lundgren et al. 2011). The use of plant material together with an alternative prey in the diet of Asopinae had a positive effect on the development and reproduction of the predator. On the other hand, the loss of fitness in bugs of *Orius* genus reared with cysts of *Artemia franciscana* (Kellogg) for a long period was associated with the administration of only this diet (De Clercq et al. 2005).

15.6 Rearing Techniques

The use of currently available mass-rearing technologies has economically enabled a limited number of species of predatory insects (Leppla et al. 2014). Although many predators can be easily reared in the laboratory, the automation of some stage of the production process can favor the reduction of costs with labor, providing greater efficiency of the system. It is not often necessary to reproduce the natural environment of the predatory species, which can be reared by simply maintaining the basic conditions, such as constant photoperiod and temperature, and adequate rearing containers, avoiding to exceed the number of individuals per unit area. For example, in the rearing of *Geocoris punctipes* (Say) (Geocoridae), simulation tests of diurnal and nocturnal temperature variations did not indicate improvement in development or survival time of the predator (Calixto et al. 2014).

Species of Chrysopidae are the most researched predators to be used as pest control agents in Latin American countries. Several studies have been conducted with species of this family, particularly those of the genus *Chrysoperla*, some of which can be obtained commercially. *Chrysoperla externa* (Hagen) stands out among the species occurring in Latin American countries by concentrating the largest number of basic and applied studies, which include improvement in several stages of laboratory rearing and release tests in experimental and commercial production areas, besides post-release monitoring (e.g., Gamboa et al. 2016). Most of these studies have been conducted at DEN/UFLA, in which the methodologies for *C. externa* production are at an advanced stage.

15.6.1 *Rearing of* Chrysoperla externa

This species has been studied for a long time at DEN/UFLA, regarding its biology, behavior, interactions with different types of prey and crops, as well as interactions with other predators. Another objective of our studies is to develop and improve methodologies for the rearing of this predator on a large scale. It is important to carry out quality control tests, as well as to know the necessary precautions for storing and sending the material to the producer, monitoring the post-release populations of the pest and control agent.

The rearing of *C. externa* in laboratory requires the same care necessary for the implementation of the rearing of any other predator. However, some peculiarities must be observed. In situations where the predator presents cannibal behavior, such as Chrysopidae larvae, additional care should be taken, which may include the individualization of the insects in the development stage in which such behavior occurs. Eggs from *C. externa* are destalked (Sousa et al. 2016) and individualized in cardboard hives composed of approximately 400 cells, to which eggs of *E. kuehniella* are added in sufficient quantity to reach the second instar. The hives are closed at the bottom with paper (~ A4 size) and at the top with a "voile"-type fabric. When the larvae reach the second instar, an artificial diet sachet made with Parafilm® having the same dimensions of the hive is placed on the fabric side. The sachet is replaced every two days until the pupation of all individuals (Bezerra 2014).

The adults obtained by this methodology are conditioned in PVC cages of 100 mm diameter and 210 mm height, maintaining a density close to 20 couples per cage (Amaral 2015). Tests on the substrate color and texture preferred for oviposition by *C. externa* have identified the red color and the suede texture with the highest concentration of deposited eggs (Amaral 2011). However, due to the high cost of the suede paper and the incompatibility of the texture of this paper with the egg collection methodology defined by Sousa et al. (2016), sulfite paper has been used, which was the second most preferred among the materials used in the tests of oviposition by *C. externa*. For egg harvesting, the paper containing the eggs is removed, and after 48 hours from oviposition it is gently wiped by a piece of organza fabric folded several times until it forms a kind of sponge. This methodology allows the eggs to be removed without the stalks, facilitating their release or use in the distribution in new rearing hives (Sousa et al. 2016). In situations where there is no demand for immediate use, collected eggs can be stored for up to 11 days at 12 °C, which guarantees more than 90% of the embryonic viability (Amaral et al. 2013).

By the results presented for *C. externa*, it is possible to have an idea of the dimension of the studies necessary for the development of an efficient rearing methodology, from the rearing of larvae up to the obtention of adults for egg harvesting.

15.7 Quality Control

The rearing of predators on alternative prey or artificial diets is a facilitating strategy in the production process. However, concern arises about the reduction in the quality of the produced organisms. A relevant point is related to the production capacity of effective biocontrol individuals that are as competitive as those found in nature. For successive generations, there is a need to achieve a biological product that contains the desirable characteristics present in the original population, aiming to obtain products with the maximum effectiveness in the field (Bigler 1989; Bueno et al. 2006; Leppla 2014). According to Leppla (2014), quality control must be carried out in a similar way to the quality control of industrialized products.

In addition to the quality of the biological agent produced in the laboratory, the farmer generally does not receive the product with the quality he/she has paid for. It happens mostly because of improper conditions in the packing and shipping of the organisms. Tests on the suitability of substrates to transport *O. insidiosus* indicate vermiculite + rice husk (ratio 1:1) as the most efficient substrate, ensuring a survival of more than 93% after arrival at destination (72 h transport) (Bueno et al. 2014). For *C. externa*, the shipping of larvae at the end of the first instar inside containers having crushed corncob or wood sawdust allowed high predator survival after a simulation of the shipping process (Amaral 2015). In tests to verify the quality of this predator reared by seven generations in artificial diet, it was verified the maintenance of the high predatory capacity of the species and the capacity of the larvae to find and to feed easily on *Aphis gossypii* Glover (Aphididae) available in plants of cucumber (*Cucumis sativus* L.) (Bezerra et al. 2016).

15.8 Commercialization

In several Latin American countries, including Brazil, Chile, Colombia, Ecuador, Mexico, Peru, and Cuba, biological control agents are marketed by mass production companies or are reared by the farmers. In order to have the appropriate species at the desired moment, the commercialization of beneficial insects and their large-scale production requires prior planning. In Latin American countries, an increasing number of predatory arthropods have been produced. The most common groups include mites and Coccinellidae, Chrysopidae, Anthocoridae, and Miridae species (Table 15.1).

In Chile, several research institutions and universities have been working on biological pest control programs, such as the Instituto Nacional de Investigaciones Agrícolas (INIA-Chile), Universidad Católica de Santiago, Universidad de Chile, Universidad Católica de Valparaiso, and the Servicio Agrícola y Ganadero, which is responsible for the latest introductions of natural enemies for the control of forest pests in the country. The great majority of the companies producing biological agents in Chile are small scaled. Among the predatory species, Biobichos has commercialized *Chrysoperla defreitasi* (Brooks) (Chrysopidae) and, since 2017, *Tupiocoris cucurbitaceus* (Spinola) (Miridae) for the control of whiteflies (Aleyrodidae) in greenhouses (Marcos Gerding, personal communication).

In recent years, the biological control in forest crops has grown in Brazil. The release of predatory organisms has been carried out with satisfactory results for more than 10 years. Currently, several companies operating in the reforestation sector, mainly with *Eucalyptus*, have reared and released *Podisus* (Pentatomidae: Asopinae) species for the control of defoliators, *Atopozelus* (Reduviidae) for the control of red gum lerp psyllid *Glycaspis brimblecombei* Moore (Psyllidae), and bronze bug (*T. peregrinus*) (Ronald Zanetti, personal observation).

15.9 Release Strategies

The methodologies for release of predatory insects in Latin America are still manually performed, by mixing the predators in substrates for dispersion of eggs or larvae and later release in the crops. Results of research conducted with *C. externa* at DEN/UFLA are examples of the advances obtained in the use of this predator as a biological control agent. The release of eggs of this green lacewing glued on paper plates affixed to rose bushes for the control of *Macrosiphum euphorbiae* (Thomas) (Aphididae) caused a low rate of cannibalism when compared to the other tested

Table 15.1 Species of predatory insects (or species used in the production of these natural enemies), control pests, producer/trader, and country

Group/traded species	Target pests	Production company/ marketer	Country
Chrysoperla defreitasi (Chrysopidae)	Larvae used against aphids (Aphididae), mites (Tetranychidae), mealybugs (Pseudococcidae), and thrips (Thysanoptera)	Biobichos[a]	Chile
Chrysoperla carnea (Chrysopidae)	Not mentioned	Biocol[b]	Colombia
Green lacewings (Chrysopidae)	Not mentioned	SENASA[c]	Peru
Sympherobius barberi (Hemerobiidae)	Not mentioned	SENASA[c]	Peru
Cryptolaemus montrouzieri (Coccinellidae)	Not mentioned	SENASA[c]	Peru
Heteropterous (Pentatomidae: Asopinae)	Used against defoliator lepidopterans	Reforestation companies[d,e]	Brazil
Atopozelus (Reduviidae)	*Glycaspis brimblecombei* (Psyllidae) *Thaumastocoris peregrinus* (Thaumastocoridae)	Reforestation companies[e]	Brazil
Orius insidiosus (Anthocoridae)	Used against thrips, aphids, whiteflies, small caterpillars, and mites	Promip[f]	Brazil
Orius insidiosus (Anthocoridae)	Not mentioned	Scientia Colombia[b]	Colombia
Orius insidiosus (Anthocoridae)	Not mentioned	SENASA[c]	Peru
Tupiocoris cucurbitaceus (Miridae)	Used against whiteflies (Aleyrodidae) in all cultures infested by the pest	Biobichos[g]	Chile
Sitotroga cerealella (Gelechiidae)	Eggs and larvae used as alternative prey/host of natural enemies	Biobichos[h]	Chile
Sitotroga cerealella (Gelechiidae), *Spodoptera frugiperda* (Noctuidae) and *Galleria mellonella* (Pyralidae)	Eggs of *S. cerealella* and eggs and larvae of *S. frugiperda* and *G. mellonella* used as alternative prey/ host of natural enemies	SENASA[c]	Peru

Sources:
[a]http://biobichos.cl/es_chrysoperla_defreitasi.php
[b]Jordano Salamanca, personal communication
[c]https://www.senasa.gob.pe/senasa/precios-de-productos-y-servicios/
[d]Pires et al. (2016a)
[e]Ronald Zanetti, personal communication
[f]http://produtos.promip.agr.br/insidiomip/
[g]http://biobichos.cl/es_tupiocoris_cucurbitaceus.php
[h]http://biobichos.cl/es_insumos_biologicos.php

methods; the release of second instar larvae from this predator in cups containing rice husks + *E. kuehniella* eggs, as well as in cardboard hives, both attached to the plants, provided a reduction of more than 90% in the aphid population after 7 days (Sousa 2013). Eggs and larvae released in commercial watermelon cultivation for the control of whitefly *Bemisia tabaci* (Gennadius) (Aleyrodidae) promoted a significant reduction of the population density of the pest; the larvae were heaved directly from the cardboard hives (where they were raised until they reached the second instar) on the plants, and the eggs were applied with crushed popcorn (Bezerra 2014).

15.10 Final Remarks

Only few Latin American agricultural systems apply augmentative biological control. An exhaustive literature search resulting from scientific research on the subject and developed in Latin America evidences the scarcity of information on the species of predators produced and marketed by companies, especially regarding rearing techniques, quality control, and commercialization. Progress in research on large-scale predator production finds its biggest obstacles in rearing methodologies. The quality of the food to be supplied to adults and larvae, as well as the form of administration to the insects, still requires many studies. The lack of methodologies involving adequate production techniques at the various stages of the process induces a greater demand for labor and, therefore, higher costs of rearing. The automation of rearing aiming at cost reduction and production of quality control agents are the key to the intensive use of predators on pest control programs in agricultural crops in Latin America.

Acknowledgments The authors gratefully thank the National Council for Scientific and Technological Development (CNPq) for granting scholarships to B. Souza, the Coordination of Improvement of Higher Education Personnel (CAPES), and the Foundation for Research Support of the State of Minas Gerais (FAPEMIG) for financial support.

References

Amaral BB (2011) Otimização da criação de *Chrysoperla externa* (Hagen, 1861) (Neuroptera: Chrysopidae) visando sua produção em escala comercial. Dissertação, Universidade Federal de Lavras

Amaral BB (2015) Melhoria no processo de produção e transporte de *Chrysoperla externa* (Hagen, 1861) (Neuroptera: Chrysopidae) para uso no controle biológico aplicado. Tese, Universidade Federal de Lavras

Amaral BB, Souza B, Bezerra CES et al (2013) Storing eggs of *Chrysoperla externa* (Hagen, 1861) (Neuroptera: Chrysopidae) for management of large-scale rearing. Açoreana suppl 9:103–109

Bezerra CES (2014) *Chrysoperla externa* (Hagen): dietas artificiais na alimentação larval e controle de *Bemisia tabaci* (Gennadius) na cultura da melancia. Tese, Universidade Federal de Lavras

Bezerra CES, Amaral BB, Souza B (2016) Rearing *Chrysoperla externa* larvae on artificial diets. Neotrop Entomol 46(1):93–99

Bigler F (1989) Quality assessment and control in entomophagous insects used for biological control. J Appl Entomol 108(1/5):390–400

Bortoli SA, Caetano AC, Murata AT et al (2006) Desenvolvimento e capacidade predatória de *Chrysoperla externa* (Hagen) (Neuroptera: Chrysopidae) em diferentes presas. Rev Biol Ciênc Terra 6(1):145–152

Bueno VHP, Mendes SM, Carvalho LM (2006) Evaluation of a rearing-method for the predator *Orius insidiosus*. Bull Insectol 59(1):1–6

Bueno VHP, Carvalho LM, Van Lenteren JC (2014) Performance of *Orius insidiosus* after storage, exposure to dispersal material, handling and shipment processes. Bull Insectology 67(2):175–183

Calixto AM, Bueno VHP, Montes FC et al (2013) Effect of different diets on reproduction, longevity and predation capacity of *Orius insidiosus* (Say) (Hemiptera: Anthocoridae). Biocontrol Sci Tech 23(11):1245–1255

Calixto AM, Bueno VHP, Montes FC et al (2014) Development and thermal requirements of the Nearctic predator *Geocoris punctipes* (Hemiptera: Geocoridae) reared at constant and alternating temperatures and fed on *Anagasta kuehniella* (Lepidoptera: Pyralidae) eggs. Eur J Entomol 111(4):521–528

Carvalho LM, Bueno VHP, Castañé C (2010) Avaliação de substratos de oviposição para *Orius insidiosus* (Say) (Hemiptera, Anthocoridae). Rev Bras Entomol 54(1):115–119

Costa NP, Oliveira HD, Brito CH et al (2007) Influência do nim na biologia do predador *Euborellia annulipes* e estudo de parâmetros para sua criação massal. Rev Biol Ciênc Terra 7(2):1–10

Costa JVB, Cruz I, Silva RB et al (2011) Desenvolvimento de *Euborellia annulipes* (Lucas) (Dermaptera: Carcinophoridae) em dietas artificiais com diferentes teores de proteína. In: Resumos do 7° congresso brasileiro de agroecologia, Associação Brasileira de Agroecologia, Fortaleza, 12–16 dez 2011

Dalvi LP, Andrade GS, Pratissoli D et al (2009) Desenvolvimento do predador *Tynacantha marginata* (Heteroptera: Pentatomidae) alimentado com pupas de *Tenebrio molitor* (Coleoptera: Tenebrionidae). In: Anais do 9° encontro latino americano de pós-graduação e Anais 8° encontro latino americano de iniciação científica da Universidade do Vale do Paraíba, Univap, São José dos Campos, 3–9 ago 2009

De Clercq P, Arijs Y, Van Meir T et al (2005) Nutritional value of brine shrimp cysts as a factitious food for *Orius laevigatus* (Heteroptera: Anthocoridae). Biocontrol Sci Tech 15(5):467–479

De Clercq P, Coudron TA, Riddick EW (2014) Production of heteropteran predators. In: Morales-Ramos JA, Rojas MG, Shapiro-Iian DI (eds) Mass production of beneficial organisms invertebrates and entomopathogens. Academic Press, London, pp 57–100

DeBach P, Hagen KS (1964) Manipulation of entomophagous species. In: DeBach P (ed) Biological control of insect pests and weeds. Chapman and Hall, London, pp 429–455

Ferreira JAM, Zanuncio JC, Torres JB et al (2008) Predatory behaviour of *Podisus nigrispinus* (Heteroptera: Pentatomidae) on different densities of *Anticarsia gemmatalis* (Lepidoptera: Noctuidae) larvae. Biocontrol Sci Tech 18(7):711–719

Gamboa S, Souza B, Morales R (2016) Actividad depredadora de *Chrysoperla externa* (Neuroptera: Chrysopidae) sobre *Macrosiphum euphorbiae* (Hemiptera: Aphididae) en cultivo de *Rosa* sp. Rev Colomb Entomol 42(1):54–58

Grafton-Cardwell EE, Hoy MA (1986) Genetic improvement of common green lacewing *Chrysoperla carnea* (Neuroptera: Chrysopidae): selection for carbaryl resistance. Environ Entomol 15(6):1130–1136

Greathead DJ (1986) Parasitoids in classical biological control. In: Waage J, Greathead DJ (eds) Insect parasitoids. Academic Press, New York, pp 289–318

Grenier S (2012) Artificial rearing of entomophagous insects, with emphasis on nutrition and parasitoids – general outlines from personal experience. Karaelmas Sci Eng J 2(2):1–12

King EG, Hopper KR, Powell JE (1985) Analysis of systems for biological control of crop arthropod pests in the U.S. by augmentation of predators and parasites. In: Hoy MA, Herzog DC (eds) Biological control in agricultural IPM systems. Academic Press, Orlando, pp 201–227

Lemos WP, Ramalho FS, Serrão JE et al (2005) Morphology of female reproductive tract of the predator *Podisus nigrispinus* (Dallas) (Heteroptera: Pentatomidae) fed on different diets. Braz Arch Biol Technol 48(1):129–138

Leppla NC (2014) Concepts and methods of quality assurance for mas-rearing parasitoids and predators. In: Morales-Ramos JA, Rojas MG, Shapiro-Iian DI (eds) Mass production of beneficial organisms invertebrates and entomopathogens. Academic Press, London, pp 277–317

Leppla NC, Morales-Ramos JA, Shapiro-Iian DI et al (2014) Introduction. In: Morales-Ramos JA, Rojas MG, Shapiro-Iian DI (eds) Mass production of beneficial organisms invertebrates and entomopathogens. Academic Press, London, pp 3–16

Lundgren JG, Moser SE, Hellmich RL et al (2011) The effects of diet on herbivory by a predaceous lady beetle. Biocontrol Sci Tech 21(1):1–74

Marafeli PP, Reis PR, Silveira EC et al (2014) Life history of *Neoseiulus californicus* (McGregor, 1954) (Acari: Phytoseiidae) fed with castor bean (*Ricinus communis* L.) pollen in laboratory conditions. Braz J Biol 74(3):691–697

Mendes SM, Bueno VHP, Carvalho LM et al (2005) Custo da produção de *Orius insidiosus* como agente de controle biológico. Pesq Agropec Bras 40(5):441–446

Neves RCD, Torres JB, Zanuncio JC (2010) Production and storage of mealworm beetle as prey for predatory stinkbug. Biocontrol Sci Tech 20(10):1013–1025

Oliveira HN, Espindula MC, Duarte MM et al (2011) Development and reproduction of *Podisus nigrispinus* (Hemiptera: Pentatomidae) fed with *Thyrinteina arnobia* (Lepidoptera: Geometridae) reared on guava leaves. Braz Arch Biol Technol 54(3):429–434

Parra JRP (2002) Comercialização de inimigos naturais no Brasil, uma área emergente. In: Parra JRP, Botelho PSM, Correa-Ferreira BS et al (eds) Controle biológico no Brasil: parasitoides e predadores. Manole, São Paulo, pp 343–347

Pires EM, Nogueira RM, Fernandes BV et al (2016a) Predadores Asopinae: sua importância econômica e ambiental. In: Pires EM (ed) Controle biológico: estudos, aplicações e métodos de criação de predadores asopíneos no Brasil. Ed. UFV, Viçosa, pp 82–91

Pires EM, Soares MA, Zanuncio JC (2016b) Estrutura física, material e técnicas para criar percevejos predadores Asopinae. In: Pires EM (ed) Controle biológico: estudos, aplicações e métodos de criação de predadores asopíneos no Brasil. Ed. UFV, Viçosa, pp 111–138

Sanches NF, Carvalho RS (2010) Procedimentos para manejo da criação e multiplicação do predador exótico *Cryptolaemus montrouzieri*. Embrapa, Cruz das Almas. Circular Técnica, 99

Silveira LCP, Bueno VHP, Van Lenteren J (2004) *Orius insidiosus* as biological control agent of thrips in greenhouse chrysanthemums in the tropics. Bull Insectol 57(2):103–109

Sørensen JG, Addison MF, Terblanche JS (2012) Mass-rearing of insects for pest management: challenges, synergies and advances from evolutionary physiology. Crop Prot 38:87–94

Sousa ALV (2013) Métodos de liberação de *Chrysoperla externa* (Hagen, 1861) (Neuroptera: Chrysopidae) visando o controle de *Macrosiphum euphorbiae* (Thomas, 1878) (Hemiptera: Aphididae) em roseiras sob cultivo protegido. Dissertação, Universidade Federal de Lavras

Sousa ALV, Souza B, Bezerra CES et al (2016) Facilitated harvesting of eggs from laboratory-reared Chrysoperla externa (Neuroptera: Chrysopidae). Rev Colomb Entomol 42(2):133–136

Souza GK, Pikart TG, Pikart FC et al (2012) First record of a native heteropteran preying on the introduced *Eucalyptus* pest, *Thaumastocoris peregrinus* (Hemiptera: Thaumastocoridae) in Brazil. Fla Entomol 95(2):517–520

Vieira NF, Vacari AM, Bortoli AS et al (2013) Custo de produção de *Orius insidiosus* (Say) (Hemiptera: Anthocoridae) criado com ovos de *Plutella xylostella* (L.) (Lepidoptera: Plutellidae). In: Anais do 2° Congresso Brasileiro de Fitossanidade, Unesp, Jaboticabal, 31 Jul-2 ago 2013

Zanuncio TV, Zanuncio JC, Saavedra JLD et al (1996) Desenvolvimento de *Podisus nigrispinus* (Dallas) (Heteroptera, Pentatomidae) com *Zophobas confusa* Gebien (Coleoptera, Tenebrionidae) comparado à duas outras presas alternativas. Rev Bras Zool 13(1):159–164
Zanuncio JC, Beserra EB, Molina-Rugama AJ et al (2005) Reproduction and longevity of *Supputius cincticeps* (Het.: Pentatomidae) fed with larvae of *Zophobas confusa*, Tenebrio molitor (Col.: Tenebrionidae) or Musca domestica (Dip.: Muscidae). Braz Arch Biol Technol 48(5):771–777
Zanuncio JC, Tavares WS, Fernandes BV et al (2014) Production and use of heteroptera predators for the biological control of *Eucalyptus* pests in Brazil. Ekoloji 23(91):98–104

Chapter 16
Predatory Mites for Biological Control of Phytophagous Mites

Paulo Rebelles Reis

16.1 Introduction

Most predatory mites of phytophagous mites from the family Phytoseiidae are highly effective and mainly used to control the two-spotted spider mite, *Tetranychus urticae* Koch, 1836 (Tetranychidae), in many different crops and countries worldwide.

Several methods of mass rearing, in particular those for phytoseiid mites, originated from those initially developed for use in research. The rearing method to be chosen should be selected carefully in order to avoid the occurrence of factors that could negatively interfere in this process. In the early stages of multiplication, methods that are more rigorous, developed in conditions more controlled, even if they imply a greater care and, consequently, higher expenses for maintenance should be preferred.

Most methods of mass rearing are based on the multiplication of predatory mites on an inert substrate placed inside a tray, upon a constantly moistened sponge where the predaceous mites are mainly fed on eggs of phytophagous mites, and/or pollen derived from various species of plants.

Rearing in artificial medium is very rare, primarily because this technology is not sufficiently developed for mass production and because this way of production can lead to a poor performance of natural enemies in their target hosts.

A traditional technique to predatory mite rearing, especially Phytoseiidae, for laboratory studies, includes flooded arenas in Petri dishes, with detached leaves infested with spider mites or other species of mites as food. As an alternative, their mass rearing is performed in spider mite-infested live bean plants housed in greenhouses.

P. R. Reis (✉)
Empresa de Pesquisa Agropecuária de Minas Gerais – EPAMIG Sul/EcoCentro, Lavras, Minas Gerais, Brazil
e-mail: paulo.rebelles@epamig.ufla.br

B. Souza et al. (eds.), *Natural Enemies of Insect Pests in Neotropical Agroecosystems*, https://doi.org/10.1007/978-3-030-24733-1_16

The production of predatory mites directly on plants, where their prey is also reared, is an efficient and less expensive method among others that can be used. In this method, common bean plants (*Phaseolus vulgaris* L., Fabaceae) or jack bean plants [*Canavalia ensiformis* (L.) DC., Fabaceae] can be used; jack bean plants are very conducive to this finality because they are less affected by root diseases when compared to the common bean. A population of mite remains on these plants (e.g., two-spotted spider mite), and when it reaches high levels, the initial population of the predatory mites are released on the plants, multiplying it until virtually eliminating the population of its prey. At that point, the predaceous mites can be collected or taken together with the bean leaves and spread on the plants where you want to control the mite pests.

The environmental conditions at rearing should be as close as possible to the conditions in which the predaceous mites will be subject during the release, including climatic conditions, soil characteristics, the food offered, etc., in order to avoid the risk of occurrence of undesirable changes in the genetic and biological characteristics of these predaceous mites.

Taking into consideration the above stated, the objective of this chapter is to provide basic notions of mass production of predatory mites for biological control of phytophagous mites, especially for scientific research and small agricultural producers.

16.2 Phytoseiidae Predatory Mites

The most important phytoseiid predatory mites commercially produced in the world include *Phytoseiulus persimilis* Athias-Henriot, *Amblyseius swirskii* Athias-Henriot, *Neoseiulus californicus* (MacGregor), *Neoseiulus cucumeris* (Oudemans), *Neoseiulus fallacis* (Garman), *Iphiseius degenerans* (Berlese), *Galendromus helveolus* (Chant), *Galendromus occidentalis* (Nesbitt), and *Mesoseiulus longipes* (Evans), among others. Belonging to the family Laelapidae, the most common commercially available species are *Hypoaspis aculeifer* (Canestrini) and *Stratiolaelaps scimitus* (Womersley) (Laelapidae) formerly *Hypoaspis miles* (Berlese) (Parra 2002; Zhang 2003; Lepla 2014; Morales-Ramos and Rojas 2015).

In Brazil, the most produced and marketed predatory mites are *N. californicus*, *Phytoseiulus macropilis* (Banks) (Phytoseiidae), and *S. scimitus* (Laelapidae) (Promip 2016).

Production methods of predatory mites change from open systems in greenhouses to fully enclosed systems using acclimatized chambers (Bolckmans 2007).

Some predatory mites can be reared on alternative food, eliminating the need of cultivation of plants for spider mite or other prey mite production, such as *N. californicus*, which, if necessary, can be reared exclusively with pollen. However, some of the most important phytoseiid predatory mites, such as *P. persimilis* or *P. macropilis*, must be produced on spider mites (Tetranychidae), their natural prey, because they are specialized. Thus, this requires the use of at least three isolated greenhouses, one for bean (common bean or jack bean plants) free of mites, another for bean infested only with spider mites, and another one for bean infested with spider mites and predatory mites (Gilkeson 1992; Monteiro 2002a; Kovaleski et al. 2006).

Phytoseiids are best known as predators of small arthropods (including mites) and nematodes, but many species are also known to feed on fungi, plant exudates, and pollen.

Scientists have proposed a classification of the Phytoseiidae based on their food sources into four types (McMurtry and Croft 1997):

- *Type I*: specialized predatory mites of *Tetranychus* species – includes species that are specialized predatory mite. This type is further subdivided into three subgroups determined by the type of prey (McMurtry et al. 2013), e.g., *Subtype I-a* contains Phytoseiidae, like *Phytoseiulus* species, that have adapted to attacking spider mites producing the so-called complicated web (CW).
- *Type II*: selective predatory mites of tetranychid (most frequently associated with dense-web-producing species) – includes species that feed on tetranychid mites, meaning mites that are capable of spinning webs, e.g., *Galendromus* and one group of *Neoseiulus* and also (tentatively) a few species in *Typhlodromus*, subgenus *Anthoseius*.
- *Type III*: generalist predatory mites – they can feed on mites of many families, as well as thrips, whiteflies, nematodes, and even pollen. Type III is further subdivided into five subgroups based on the habitat where the Phytoseiids can be found (McMurtry et al. 2013), genus *Amblyseius* and *Typhlodromalus*.
- *Type IV*: specialized pollen feeders or generalist predatory mite phytoseiids depend on pollen as their primary food source. These species can also act as generalist predatory mites, but they are most successful when feeding on pollen, e.g., *Euseius* species.

16.3 Lab-Rearing Methods of Predatory and Phytophagous Mites

Basically, the methods of rearing predaceous mites in laboratory may also be classified as the medium used to support the mites, in: (1) natural (leaves and fruit) and (2) artificial (glass, plastic, paper, metal, etc.). Only the natural medium is used for the phytophagous mites.

The systems initially proposed for rearing predatory mites used small plastic cages and transparent or closed cells (Munger 1942) and afterwards were modified (Huffaker 1948; Ballard 1953). Later, breeding techniques were developed in detached bean leaves (*P. vulgaris*) placed on moistened filter paper, where the mite was confined by a viscous substance fillet (e.g., Tanglefoot™) (Ristich 1956). Leaf disc on blotting paper and sponge soaked in water in Petri dishes (Hoyt and Harries 1961), rearing mites sheets floating on water (Chant 1961), and sheets of wire mesh in water (Herne and Chant 1965) have also been reported.

In Brazil, rearing of predatory mites in natural arenas with barrier of cotton soaked in water was reported by several authors (Moraes and Lima 1983; Komatsu and Nakano 1988; Moreira 1993; Yamamoto and Gravena 1996). Artificial arenas floating in water was reported by (Reis and Alves 1997).

The rearing of predatory or phytophagous mites in the laboratory is advantageous in relation to the rearing in plants grown under the greenhouse or in any other type of protected cultivation, because the control that species are being reared is better, which is important, in addition to the biological control, for tests of selectivity of agrochemicals, predation tests, etc. Furthermore, it takes up little space and there is no need to grow plants, which overloads the rearing process.

The system of predatory mite rearing for maintenance or stock of species or mass production for biological control use is dependent on the habits of the species to be reared, and the job is facilitated if other food sources, other than preys, can be used.

The various methods described to rearing predatory mites in the laboratory generally use three forms of containment to the mites: (1) water barrier, open rearing units, usually with the use of cotton soaked in water as barrier; (2) viscous or adhesive substances as barriers, also in units of open rearing; and (3) cages or jail cell. The first method has been used more in stock or maintenance rearing, and the last two methods, beyond rearing, can be used for biology studies, tests with plant protection products, and agrochemical selectivity, among others. The first two types of barriers, water and sticky substances, can be also used in combination.

16.3.1 Rearing of Pest Mites in Natural Arenas

For rearing phytophagous or pest mites (prey mites), uncapped Petri dishes (15 cm in diameter) and 1-cm-thick foam can be used, which have to be maintained moist with distilled water and entirely covered. A detached plant leaf, or leaflet, is placed on the top of the foam and surrounded by strips of cotton in contact with the damp foam in order to inhibit mites from escaping and to better conserve the leaf (Fig. 16.1). For the rearing, prey mites (e.g., *T. urticae*) are placed on the upper surface of the jack bean detached leaflet (*C. ensiformis*); the leaflet must be weekly replaced (Reis et al. 1997; Souza-Pimentel et al. 2014).

16.3.2 Rearing of Specialized Phytoseiid (Type I) in Artificial Arenas

For this purpose, arena of black flexible plastic sheets (26 × 22 cm) is put over Styrofoam of equal size and placed over water on plastic trays (32 × 26.5 × 5.5 cm). Cotton is placed around the Styrofoam and the arena, which is in contact with water from the tray. In addition to preserving the jack bean leaflets, the cotton is used to

Fig. 16.1 Natural arena made with detached leaflet of jack bean plant with cotton soaked in water as barrier. (Photo: Paulo Rebelles Reis)

Fig. 16.2 Arena of black flexible plastic sheets (26 × 22 cm) put on Styrofoam of equal size and placed over water on plastic trays (32 × 26.5 × 5.5 cm). (Photo: Paulo Rebelles Reis)

inhibit mites from escaping. The leaflets of jack bean infested by *T. urticae,* which will serve as a food for predatory mites, are placed on arenas with the petiole under the damp cotton (Fig. 16.2); as the leaves become withered, other new leaves infested by *T. urticae* are placed over the old leaves (McMurtry and Scriven 1964; Mesa and Bellotti 1986; Souza-Pimentel et al. 2014).

16.3.3 Rearing of Generalist Phytoseiids (Types II, III, and IV) in Artificial Arenas

Some predatory mites can be reared on alternative food, eliminating the need of plant cultivation for spider mite or other prey as food, such as *N. californicus,* which can be reared only with pollen, e.g., castor bean pollen (*Ricinus communis* L., Euphorbiaceae), cattail (*Typha domingensis* (Pers.), Typhaceae), or any other plant pollen.

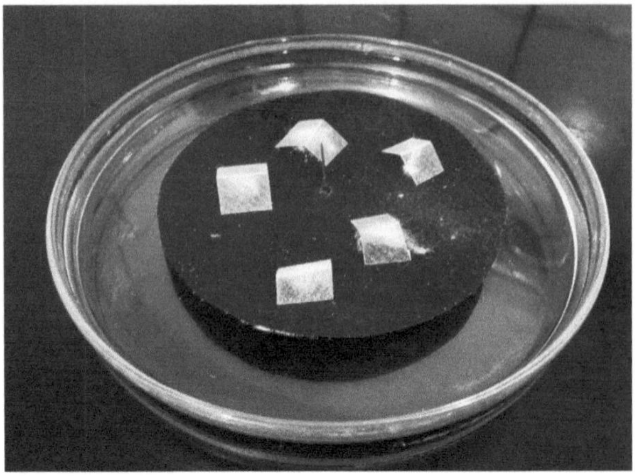

Fig. 16.3 Artificial arena made of thin black plastic sheet, with transparent shelter, cotton threads under the shelter, and castor bean pollen, floating over water in an uncapped Petri dish of 15 cm diameter. (Photo: Paulo Rebelles Reis)

A thin flexible black plastic sheet (or another color), with transparent shelter (made with transparency sheet), cotton threads (for the oviposition), and pollen as food, floating in water in a Petri dish of 15 cm diameter, can be used for rearing generalist phytoseiids. A hole is made in the center of the arena for the passage of a sewing needle (N 3), or a sewing pin, fixed in the center bottom of the Petri dish by a portion of silicone adhesive with the tip facing up. Thus, the arena remains in the center of the dish without touching the plate wall, moving only up and down with the water level variation. In addition to the water barrier, the water is meant to be ingested by predatory mites, because they are only fed with pollen (Fig. 16.3).

16.4 Mass Rearing of Phytoseiid Predatory Mites in Greenhouse

A mass production system of mites belonging to the family Phytoseiidae always aims to obtain a maximum amount of specimens with minimal investment (cost, space, and work) and, in general, involves four steps: (1) maintenance of phytoseiid mites in pure populations; (2) host plant growing in greenhouses for the prey mite, of which the common bean plants, jack bean plants, and soybean plants [*Glycine max* (L.) Merrill, Fabaceae] are the most commonly used species; (3) rearing of the prey mite, in the host plant, as food for the phytoseiids (two-spotted spider mite as prey is the most widely used); and (4) rearing of the phytoseiid predatory mites over the infested plants with the prey mites in greenhouses.

The predatory mites rearing in protected crop usually requires the use of three isolated greenhouses or other ways of protecting crops: one for the bean crop (common bean or jack bean plants) free of predatory mites, another one with bean infested only with spider mites (pest mites), and another greenhouse for bean infested with spider mites and mites (Gilkeson 1992; Monteiro 2002a, b; Monteiro et al. 2008; Kovaleski et al. 2006).

Fig. 16.4 Potted jack bean plants in a greenhouse and not yet infested with two-spotted spider mite. (Photo: Paulo Rebelles Reis)

In general, the predatory mites that occur naturally in all environments only need food to multiply. Therefore, it is desirable that before the establishment of the main crop, where it will be done the biological control of pest mites, a culture that is attacked by two-spotted spider (e.g., common bean, jack bean plant, etc.) is sown in the greenhouse (specially dedicated to it) for the multiplication of predatory mites, if possible near of the greenhouse with the main crop. The predatory mite will only be transferred to the main crop after the beginning of the spider mite infestation. Thus, it is likely that the first attack causes damage to the main culture, but later there will be equilibrium between the pest mite and the predatory mite.

A simple procedure to be followed to multiply the predatory mites is through the rearing of spider mite on bean, as follows: (1) sow beans in plastic pots or plastic bags using high density (Fig. 16.4); (2) approximately 15 days after sowing, infest plants with spider mite; (3) when you see that all the leaves are already attacked by the two-spotted spider mite, release the predatory mite; (4) after the infestation, periodically analyze the bean leaves, and when you have more predatory mites than spider mites, it is time to take the leaves for the main crop; and (5) the ideal is to always keep an isolated "colony" of predatory mites (Monteiro 2002a, b; Monteiro et al. 2008; Kovaleski et al. 2006).

16.5 Alternative Prey Mites as Food for Phytoseiids

Due to the difficulty to rear prey mites on plants that will be served as prey for the predatory mites, some studies show the possibility of the use of other mites that can be reared in large amounts in lab-culture media, or stored grains, and that are good foods to predatory mites.

The predatory mite phytoseiid *Iphiseiodes zuluagai* Denmark & Muma is an important predator of *Brevipalpus phoenicis* (Geijskes) (Tenuipalpidae) in citrus (*Citrus* spp.) and coffee (*Coffea* spp.) in Brazil. The suitability of the storage mite

Tyrophagus putrescentiae (Schrank) (Acaridae), lab-reared in a semimoist dog food, as a food source of *I. zuluagai* in laboratory rearing was investigated, and the results show that *T. putrescentiae* is a favorable food source for the development of *I. zuluagai* (Albuquerque and Moraes 2008).

A study result indicated that the photoperiod of 12:12 is optimal for the development and reproduction of *Neoseiulus barkeri* Hughes (Phytoseiidae) fed on *T. putrescentiae* and that *N. barkeri* may serve most efficiently as a biological control agent under this regimen (Zou et al. 2016). The storage mite *Aleuroglyphus ovatus* (Troupeau) (Acaridae) is also considered great food for *N. barkeri* (Xia et al. 2012). This phytoseiid is an excellent natural enemy of mites and insect pests in various crops of economic importance, such as broad mite, *Polyphagotarsonemus latus* (Banks) (Tarsonemidae), strawberry-cyclamen mite or strawberry stunting mite *Phytonemus pallidus* (Banks) (Tarsonemidae), and thrips *Thrips tabaci* Lindeman (Thripidae), among other agricultural pests (Xia et al. 2012; Zou et al. 2016).

16.6 Final Considerations

The domain of a methodology for predatory mite rearing is essential for the success of the biological control.

In the process of rearing mites, a fundamental aspect to be considered is the environmental condition, such as temperature, humidity, and the photoperiod, that interferes with the performance of the plants, and of the predatory and phytophagous mites, in function of its variation.

Greater effectiveness is achieved in a mass production system when all steps of production are rigorously controlled, especially those which may be directly affected by factors such as temperature and humidity. The production of these natural enemies throughout the year can only be achieved if the greenhouses where the predatory mites are being bred do not suffer the direct influence of the weather.

It is important to make an assessment of the development of mites at each step of rearing, which can be through weekly surveys. These surveys enable the quantification of the average values of mites produced in the greenhouse, monitor the development of the mites, and identify potential problems in the production stages. In addition, this information allows determining the amount of predatory mites required in the system of rearing in order to obtain maximum performance in less time.

In conducting, a mass rearing can face some problems as the death of phytophagous mite because of entomopathogenic agents, excess moisture in the air, etc.; it can still occur the death of the predatory mite due to improper application of chemicals or the lack of food, among others; a nonuniform production of predator and/or prey mite can also hinder the process.

The predatory mites should be released on the plants grown together with the leaves of bean plants and pest mites. Although, in most cases, the natural enemies in the greenhouse are released when the pest is already on the crop plant, phytoseiid mites of the genus *Amblyseius* (Phytoseiidae) should be introduced before the pest mite is detected.

The amount of material deposited in the growing area should be proportional to the mass of the leaf crop plants and the percentage of the mite present that was previously determined by monitoring. The leaves of bean plants remain longer turgid and opened during the afternoon, which facilitates the passage of the predatory mites to the infested crops with the phytophagous mites, being the most suitable time to perform releases.

References

Albuquerque FA, Moraes JG (2008) Perspectivas para a criação massal de *Iphiseiodes zuluagai* Denmark & Muma (Acari: Phytoseiidae). Neotrop Entomol 37(3):328–333

Ballard RC (1953) A modification of the Huffaker cage for confining mites or small insects. J Econ Entomol 46(6):1099

Bolckmans KJF (2007) Mass-rearing phytoseiid predatory mites. In: Van Lenteren JC, DeClercq P, Johnson MW (eds) Proceedings of the working group AMRQC. IOBC Global, Saint Paul, pp 12–15

Chant DA (1961) The effect of prey density on prey consumption and oviposition in adults of *Typhlodromus* (*T.*) *occidentalis* Nesbitt (Acarina: Phytoseiidae) in the laboratory. Can J Zool 39(3):311–315

Gilkeson LA (1992) Mass rearing of phytoseiid mites for testing and commercial application. In: Anderson TE, Leppla NC (eds) Advances in insect rearing for research and pest management. Westview Press, Boulder, pp 489–506

Herne DC, Chant DA (1965) Relative toxicity of parathion and Kelthane to the predaceous mite *Phytoseiulus persimilis* Athias-Henriot and its prey, *Tetranychus urticae* Koch, (Acarina: Phytoseiidae, Tetranychidae) in the laboratory. Can Entomol 97(2):172–176

Hoyt SC, Harries FH (1961) Laboratory and field studies on orchard-mite resistance to Kelthane. J Econ Entomol 54(1):12–16

Huffaker CB (1948) An improved cage for works with small insects. J Econ Entomol 41(4):648–649

Komatsu SS, Nakano O (1988) Estudos visando o manejo do ácaro da leprose em citros através do ácaro predador *Euseius concordis* (Acari: Phytoseiidae). Laranja 9:125–146

Kovaleski A, Ferla NJ, Botton M et al (2006) Produção de morangos no sistema semi-hidropônico: criação de ácaros predadores. Epagri, Florianópolis

Lepla NC (2014) Concepts and methods of quality assurance for mass-reared parasitoids and predators. In: Morales-Ramos JA, Rojas MG, Shapiro-Ilan DI (eds) Mass production of beneficial organisms, invertebrates and entomopathogens. Academic Press, Waltham, pp 277–317

McMurtry JA, Croft BA (1997) Life-styles of phytoseiid mites and their roles in biological control. Annu Rev Entomol 42:291–321

McMurtry JA, Scriven GT (1964) Studies on the feeding, reproduction, and development of *Amblyseius hibisci* (Acarina: Phytoseiidae) on various food substances. Ann Entomol Soc Am 57(5):649–655

McMurtry JA, Moraes JG, Sourassou NF (2013) Revision of the lifestyles of phytoseiid mites (Acari: Phytoseiidae) and implications for biological control strategies. System Appl Acarol 18(4):297–320

Mesa NC, Bellotti AC (1986) Ciclo de vida y hábitos alimenticios de *Neoseiulus anonymus*, predador de ácaros Tetranychidae en yuca. Rev Colomb Entomol 12(1):54–66

Monteiro LB (2002a) Criação de ácaros fitófagos e predadores: um caso de produção de *Neoseiulus californicus* por produtores de maçã. In: Parra JRP, Botelho PSM, Corrêa-Ferreira BS et al (eds) Controle biológico no Brasil: parasitoides e predadores. Manole, São Paulo, pp 351–365

Monteiro LB (2002b) Manejo integrado de pragas em macieira no Rio Grande do Sul II. Uso de *Neoseiulus californicus* para o controle de *Panonychus ulmi*. Rev Bras Frutic 24(2):395–405

Monteiro LB, Doll A, Boing LF (2008) Densidade de *Neoseiulus californicus* (McGregor, 1954) (Acari: Phytoseiidae) no controle do ácaro-vermelho da macieira, Fraiburgo-SC. Rev Bras Frutic 30(4):902–906

Moraes GJ, Lima HC (1983) Biology of *Euseius concordis* (Chant) (Acarina: Phytoseiidae) a predator of the tomato russet mite. Acarologia 24:251–255

Morales-Ramos JA, Rojas MG (2015) Production of predatory mites as biocontrol agents. USDA, Washington, DC

Moreira PHR (1993) Ocorrência, dinâmica populacional de ácaros predadores em citros e biologia de *Euseius citrifolius* (Acari: Phytoseiidae). Dissertação, Universidade Estadual Paulista

Munger F (1942) A method of rearing thrips in the laboratory. J Econ Entomol 35(3):373–375

Parra JRP (2002) Comercialização de inimigos naturais no Brasil: uma área emergente. In: Parra JRP, Botelho PSM, Corrêa-Ferreira BS et al (eds) Controle biológico no Brasil, parasitoides e predadores. Manole, São Paulo, pp 343–349

Promip (2016) Manejo integrado de pragas. http://promip.agr.br/produtos. Accessed 12 Aug 2016

Reis PR, Alves EB (1997) Laboratory rearing of *Iphiseiodes zuluagai* Denmark & Muma (Acari: Phytoseiidae). An Soc Entomol Bras 26(3):565–568

Reis PR, Alves EB, Sousa EO (1997) Biologia do ácaro-vermelho do cafeeiro *Oligonychus ilicis* (McGregor, 1917). Ciênc Agrotec 21:260–266

Ristich SS (1956) Mass rearing and testing techniques for *Typhlodromus fallacis* (Gar.). J Econ Entomol 49(4):476–479

Souza-Pimentel GC, Reis PR, Silveira EC et al (2014) Biological control of *Tetranychus urticae* (Tetranychidae) on rosebushes using *Neoseiulus californicus* (Phytoseiidae) and agrochemical selectivity. Rev Colomb Entomol 40(1):80–84

Xia B, Zou Z, Li P et al (2012) Effect of temperature on development and reproduction of *Neoseiulus barkeri* (Acari: Phytoseiidae) fed on *Aleuroglyphus ovatus*. Exp Appl Acarol 56(1):33–41

Yamamoto PT, Gravena S (1996) Influência da temperatura e fontes de alimento no desenvolvimento e oviposição de *Iphiseiodes zuluagai* Denmark & Muma (Acari: Phytoseiidae). An Soc Entomol Bras 25:109–115

Zhang Z-Q (2003) Mites of greenhouses identification, biology and control. CABI Publication, Oxon, UK

Zou Z, Min Q, Xiao S et al (2016) Effect of photoperiod on development and demographic parameters of *Neoseiulus barkeri* (Acari: Phytoseiidae) fed on *Tyrophagus putrescentiae* (Acari: Acaridae). Exp Appl Acarol 70(1):45–56

Chapter 17
Parasitoids Insects

Tiago Cardoso da Costa-Lima, Aloisio Coelho Jr., Alexandre José Ferreira Diniz, and Marcus Vinicius Sampaio

17.1 Introduction

The entomophagous rearing domain is one of the most important steps in augmentative biological control (BC) programs. In Latin America, BC programs occurred in 4 countries in the 1970s and was registered in 14 countries in 2003 (Van Lenteren and Bueno 2003). Nowadays, BC companies that started in Europe increased their distribution in many Latin American countries, such as Brazil, Mexico, Costa Rica, Chile, Colombia, and others. Furthermore, local companies aiming at these growing BC market have arisen.

The entomophagous rearing challenge is to produce in large scale and with low cost and maintaining its quality. Many commercialized parasitoid species are multiplied over hosts reared in plants (natural hosts) (van Lenteren 2000). This is the case of leafminers, whiteflies, psyllids, and aphids parasitoids. However, BC programs greatly improved after the development of artificial diets, especially for pest rearing in the 1960–1970 decade (Parra 2002). It is no coincidence that the most released parasitoids in Brazil are multiplied over pests species reared in artificial diet. That is the case of *Cotesia flavipes* (Fabricius) (Braconidae), *Trichogramma*

T. C. da Costa-Lima (✉)
Empresa Brasileira de Pesquisa Agropecuária (Embrapa Semiárido), Petrolina, PE, Brazil
e-mail: tiago.lima@embrapa.br

A. Coelho Jr.
Escola Superior de Agricultura "Luiz de Queiroz"/Universidade de São Paulo (USP/ESALQ), Piracicaba, SP, Brazil
e-mail: aloisiocoelho@usp.br

A. J. F. Diniz
Faculdade de Ensino Superior Santa Bárbara (FAESB), Tatuí, SP, Brazil

M. V. Sampaio
Federal University of Uberlândia, Agricultural Sciences Institute, Uberlândia, MG, Brazil
e-mail: mvsampaio@ufu.br

B. Souza et al. (eds.), *Natural Enemies of Insect Pests in Neotropical Agroecosystems*, https://doi.org/10.1007/978-3-030-24733-1_17

galloi Zucchi, and *Trichogramma pretiosum* Riley (Trichogrammatidae). Nowadays, Brazil has 3,000,000 hectares of sugarcane under control with *C. flavipes*, 500,000 hectares with *T. galloi*, and 250,000 ha of corn, tomato, fruits, and cotton with *T. pretiosum* (Parra et al. 2015). Rear parasitoids without their live hosts are still only in research level, as the examples with artificial eggs for *Trichogramma* species (Cônsoli and Grenier 2010; Dias et al. 2010a). However, it is still not a reality for the upcoming years.

The quality control of the parasitoids produced should be a continuous concern to avoid mistakes and to ensure that the biological control is effective. The first step is to obtain the correct parasitoid identification and to be certain that there is no contamination with other species. Danks (1988) and Smith (1996) reported several examples of costly biocontrol releases with limited effect on target pest populations. The authors attributed the lack of success due to misclassification and/or cross contamination of species during the mass-rearing process. Considering that morphological classification is time-consuming and requires a high level of technical experience, other methods can be applied. The DNA barcoding is already available for many species (Wilson et al. 2017), and even other techniques, such as hyperspectral image, can be an option (Nansen et al. 2014). The parasitoid quality control also includes a periodic evaluation of their flight ability and biological traits, such as parasitism capacity, survivorship, and others (van Lenteren 2003; Leppla 2014). Considering the insect mass rearing system, even the evaluation of gases concentration can be important regarding their influence over the organism (Coelho Jr and Parra 2013a; Coelho Jr et al. 2017).

Some other issues are important considering the insect rearing optimization:

(i) Process automatization – As already mentioned, the low cost to produce the parasitoid is extremely important to make it a commercial product. Considering the total cost for entomophagous production, the labor is responsible for 80% of the costs (Parra 2002). The rearing process automatization can boost the use of BC.
(ii) Resistance to insecticides – Selecting parasitoid strains resistant to certain chemical molecules is also a line of research, considering applying BC in crops that still demand high number of spraying (Whitten and Hoy 1999).
(iii) Parasitoid genetic background – Haplo-diploid wasps suffer much lower levels of inbreeding depression than diploid species (Legner 1979; Antolin 1999; Luna and Hawkins 2004; Prezotti et al. 2004). Even so, genetic adaptation to the laboratory will occur, and consequently the prediction is that with increasing effective population size during mass rearing, the performance of the parasitoid wasps should decrease under field conditions (Hufbauer 2001).

In this chapter, a better approach in four groups of parasitoids *Trichogramma* species, *Tamarixia radiata* (Kuwayama) (Eulophidae), leafminers parasitoids, and aphids parasitoids is presented. These entomophagous were selected considering their diverse rearing methodologies and current importance in different crops.

17.2 Parasitoids Rearing Systems

17.2.1 Trichogramma *Species*

Parasitoids of *Trichogramma* genus are one of the most studied BC agents worldwide. Many studies have evaluated the effects of biotic and abiotic factors on *Trichogramma*, and the results have helped to expand the use of these organisms as biocontrol agents (Wajnberg and Hassan 1994; Cônsoli et al. 2010). The trichogrammatids require well-defined microclimate conditions; thus, *Trichogramma* strains must be kept isolated according to the sampling locale and host on which they were collected, as a species collected in a cold region may not be efficient in a warmer region and vice versa (Bleicher and Parra 1990). The strains selection on laboratory are one of the first steps in BC programs. The laboratory performance (fecundity and offspring sex ratio) measures are good predictors of field success in *T. pretiosum* Riley (Coelho Jr et al. 2016a). Parasitoids collected from eggs of the target pest must be preferred when selecting a parasitoid for mass production. The species most often used in biological control in Brazil are *T. pretiosum* and *T. galloi* Zucchi, while *Trichogramma atopovirilia* Oatman and Platner has been used in a preliminary study (Parra et al. 2015).

The correct identification of *Trichogramma* species is really important for BC programs. Continuous taxonomic monitoring of *Trichogramma* populations is important both in academic and commercial rearing facilities (Zucchi et al. 2010). *Trichogramma* species have traditionally been identified based on morphological traits (Zucchi et al. 2010), but molecular tools (Ciociola Jr et al. 2001a, b; Stouthamer 2006) and, more recently, the use of reflectancy (hyperspectral image analysis) have shown promise in distinguishing among species (Nansen et al. 2014).

The *Trichogramma* use boosted after discovering the possibility of using factitious hosts for rearing, which substantially reduced costs and facilitated its production. Moth, *Sitotroga cerealella* (Olivier) (Gelechiidae), became the preferred factitious host after Flanders (1927) publication and is still in use in countries such as Colombia, Peru, and Germany (Parra et al. 2015). However, the choice of factitious host in the rearing of *Trichogramma* changed after Lewis et al. (1976) (Fig. 17.1) demonstrated that eggs of *Anagasta kuehniella* Zeller are more suitable than *S. cerealella*, allowing the development of wasps that were more fertile and longer-lived. In Brazil, *A. kuehniella* has been shown to be most suitable factitious host for mass rearing and use of *T. pretiosum, T. galloi*, and *T. atopovirilia*. Eggs of *A. kuehniella* have been proved to be the most suitable for rearing *Trichogramma bruni* Nagaraja and *Trichogrammatoidea annulata* De Santis (Dias et al. 2010b) compared to *S. cerealella*, most likely due to differences in their nutritional composition (Fig. 17.1).

The rearing system developed in Brazil for mass production of *A. kuehniella* followed the French system (Daumal et al. 1975), after a number of modifications to attend the conditions available in the country (Parra 2010; Parra et al. 2014). Mediterranean flour moth larvae are reared on artificial diets based on whole wheat

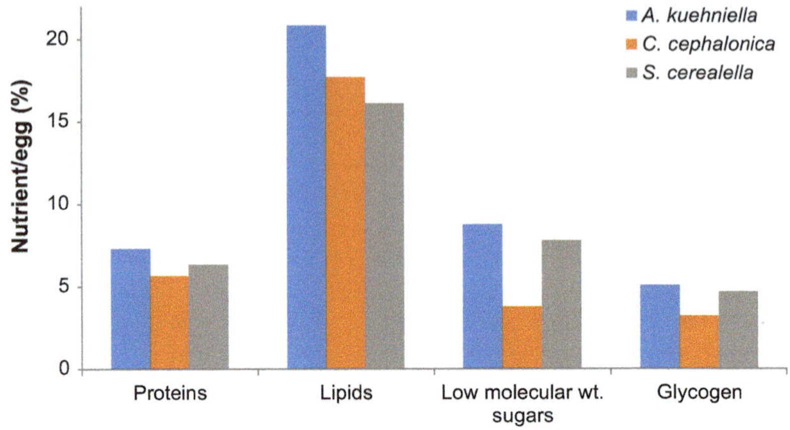

Fig. 17.1 Nutrient composition (%) in relation to the egg weight of three factitious hosts used in rearing systems of *Trichogramma* (Coelho Jr and Parra unpublished data)

flour (97%) and yeast (3%) or a mixture of wheat (40%) and corn flour (60%) (Parra et al. 2014). The sequence used in the production system of *A. kuehniella* was described by Parra (2010). Recent studies have confirmed the work done by Cerutti et al. (1992), who found that high larval densities are deleterious as they may increase the temperature in the rearing unit from 7 to 9 °C at the end of their larval development due to their metabolic activity (Coelho Jr et al. 2016b). This increase in temperature in the rearing unit not only extends the larval period development but also yields smaller adults that are short-lived and less fertile (Coelho Jr and Parra 2013b). The CO_2 concentration in the rearing rooms, which may reach 1200 PPM, might reduce egg production (Coelho Jr and Parra 2013a).

Regarding *Trichogramma*, biological parameters, such as development time, survivorship, sex ratio, parasitization capacity, flight capacity, and longevity, among others, are always used and subjected to group analysis for the selection of the most suitable *Trichogramma* strains against target pests. Flight capacity is a good quality indicator of trichogrammatids (Dias-Pini et al. 2014). The thermal requirements of most Brazilian studies on *Trichogramma* species have been determined in order to facilitate parasitoid rearing and to improve the predictability of the production system (Parra et al. 2015).

The quality control of mass-reared insects by assessing biological traits or by using molecular techniques is key to assure that laboratory-reared parasitoids will be competitive (van Lenteren 2003; Coelho Jr et al. 2016b). The International Organization for Biological Control (IOBC) recommendations for quality control assessment of mass-reared parasitoids were implemented in Brazil, including the flight assay, and have been in use to evaluate the quality of mass-reared *Trichogramma* (Prezotti et al. 2002). Concerning the atmospheric gases conditions, it is recommended ventilation in *T. pretiosum* rearing conditions to maintain CO_2 below 4.3% and O_2 above 18.5%, allowing maximal parasitoid quality (Coelho Jr et al. 2017).

17.2.2 Liriomyza *Parasitoids*

Leafminers flies from the genus *Liriomyza* (Agromyzidae) are major pests in many vegetables and ornamental crops in Latin America. Three key species are highly polyphagous and cause more damages to crops: *Liriomyza sativae* Blanchard, *Liriomyza trifolii* (Burgess), and *Liriomyza huidobrensis* (Blanchard) (Murphy and Lasalle 1999; Costa-Lima et al. 2015). Since the 1980s, *Liriomyza* parasitoids are commercialized mainly in Europe and North America directed to protected crops (van Lenteren 2012). The most common species used are *Diglyphus isaea* (Walker) (Eulophidae) and *Dacnusa sibirica* Telenga (Braconidae). In Latin America, nowadays, *D. isaea* is commercialized in Mexico and Costa Rica. In Colombia, there is a reported experience with *Diglyphus begini* (Ashmead) to control *L. huidobrensis* in ornamentals protected crops (Cure and Cantor 2003). In Brazil, researches directed to *L. sativae* biological control increased in consequence of the growing importance of this pest in melon crops (Costa-Lima et al. 2014, 2017). Nextly, we detail the rearing procedure for a very common *Liriomyza* larval-pupal parasitoid in Brazil that also occurs in Argentina and Peru, *Phaedrotoma scabriventris* Nixon (Braconidae). Originally, the following results were published as a technical report in Portuguese by Costa-Lima et al. (2017).

To multiply *P. scabriventris* or any other *Liriomyza* parasitoid, we need to rear the leafminer as host. Until now, no artificial diet was developed for rearing leafminers; thus, the first step is choosing the plant species. The cowpea plant [*Vigna unguiculata* (L.) Walp] is a natural host for *L. sativae* that it is easy to start a population in cages. Another option as plant host is jack bean [*Canavalia ensiformis* (L.) D.C], which is not a *L. sativae* major host; however, it has a large leaf area and it is more resistant to pests and diseases. The plants can be sown in pots or in cell trays at least two times per week. Maintaining the plants free from pests and diseases is a crucial step for the rearing system success. In this case, it is important to have greenhouses with antechambers and air curtain. It also recommended installing yellow sticky traps to detect any infestation in the greenhouse. When cowpea plants present two trifoliate leaves fully expanded, it is time to direct the pots for infestation. This process is made in cages with laterals and superior sides closed with a fine mesh, which allows air exchange and the microclimate inside not become saturated with high RH. If the cages are maintained indoor, two tubular LED lights have to be positioned connected to a timer with the region fotophase (e.g., 12 h). To reduce the insect escape during the plants exchange, a similar light can be installed in the cage posterior part. When the cage door is opened, the superior lights need to be turned off and the posterior turned on. The proportion of adults per plants needs to be adjusted to avoid high infestations (can cause leaf drop) and maximize the leaf consumption. As food source, 10% honey solution can be offered to the adults. The cages can also be maintained in greenhouses. However, in hot weather regions, systems to reduce temperature need to be adapted. During the plant exchange, the cage opening has to be darkened with the aid of a fabric to reduce insect loss. After 24 h in the leafminers adult cages, the plants are directed to a separate compartment

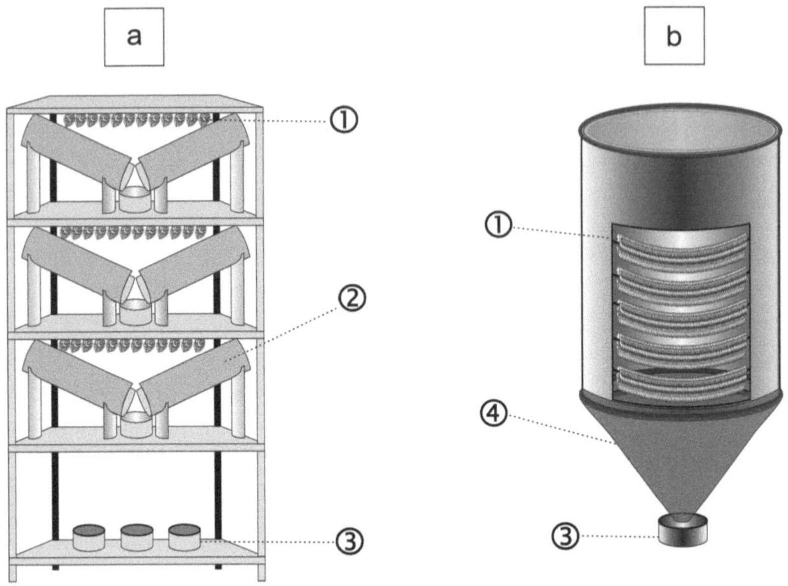

Fig. 17.2 Two types of *Liriomyza* prepupae collector. (a) Shelf collector: metallic shelf, wires to support leaves (1), plastic chute (2), and recipient (3); (b) cylindrical collector: metallic 200 liters cylinder, wires to support leaves (1), plastic funnel (4), and recipient. (3). (Adapted from Costa-Lima et al. (2017))

in a greenhouse for egg and larvae development. When larvae are midsize, corresponding to 4–5 days at 25 °C, leaves need to be cut off. In this step, 70% will be directed to parasitoids and 30% for leafminer maintenance production. Collecting leafminers prepupae is a critical step in this system. Collectors that reduce labor and pupae manipulation were developed, as shown in Fig. 17.2. Leaves are detached and a wire is passed through in a way they are maintained suspended over a funnel or chutes. The larvae will drop, being collected in a single recipient located at the bottom. For the cylindrical type, each wire can support 100 leaves and the entire cylinder supports a total 1.500 leaves. Considering a 30 larvae/leaf average, the cylinder can collect 45.000 pupae at every 3 days. The prepupae recipient can be exchanged daily. After the adult emergence start, just direct the recipient to the adult's cage. For the parasitoids, the cages must have similar structure as the leafminers; however, the superior face can be made of plastic or acrylic. This way, pure honey can be offered in this area as food source. Detached leaves can be exposed to larval-pupal parasitoids, such as *P. scabriventris*. For this method, floral sponge can be used to maintain the leaves turgid. The other option is just to transfer the plant from the pot or tray to the parasitoid cage. Midsize larvae are ideal to offer for this parasitoid. Leaves with *L. sativae* larvae are exposed for 24 h to the parasitoids. After that, leaves are directed to a similar prepupae collector, as previously described for the leafminers. The emerged parasitoids will be divided, one part to maintain the rearing system and another part for commercialization. The proportion will depend on the demand. It has to be considered that for most of *Liriomyza* parasitoids, the female also kills larvae by host feeding. For *P. scabriventris*, for each three larvae parasitized, one is host-fed. This is an aspect that can increase the rearing cost.

17.2.3 Aphid Parasitoids

Aphids (Aphididae) are worldwide pests in different crops of economic importance. For example, cosmopolitan species *Aphis gossypii* Glover, *Aphis spiraecola* Patch, *Aulacorthum solani* (Kaltenbach), *Macrosiphum euphorbiae* (Thomas), and *Myzus persicae* (Sulzer) are frequently attacked by parasitoids from the Aphelinidae and Braconidae family (Aphidiinae subfamily). These natural enemies insert their eggs inside the aphids without paralyzing the host. After several days, the parasitoid larvae consume the host and leave the exoskeleton that serves as a puparium (mummy). Inside the mummy, the parasitoid transforms into pupa that will originate a new adult.

The parasitoids *Aphidius colemani* Viereck, *Diaeretiella rapae* (McIntosh), and *Lysiphlebus testaceipes* (Cresson) are the dominant species in South America (Starý et al. 2007). These are generalists parasitoids that attack a high number of aphids species, both in Aphidini and Macrosiphini aphids tribes (Starý et al. 1993, 2007; Starý 1995). However, aiming the BC of many Macrosiphini aphid species, *Aphidius ervi* Haliday and *Praon volucre* (Haliday) are more indicated (Sidney et al. 2010; De Conti et al. 2011; Lins Junior et al. 2011). Over 10 aphids parasitoids species were commercialized or are still applied in Europe, North America, Asia, Africa, and Oceania (van Lenteren 2012; van Lenteren et al. 2017). In Latin America, some countries also have this entomophagous available, such as Mexico, Costa Rica, and Chile.

Parasitoid rearing can be conducted in greenhouses or laboratory; however, cages are always indicated to prevent contaminations. The aphid initial rearing must be monitored daily to avoid natural enemies, different aphid species and ants. For the parasitoid initial population, it is recommended to collect in the mummy stage and separate in small tubes. This caution will allow to remove different species and, mainly, hyperparasitoids. The hyperparasitoids are a serious problem for aphids parasitoids mass rearing. The contamination reduces the parasitoid population and the BC efficiency. To rear aphids and its parasitoids, the cages sizes must be planned according to the host plant used. The chosen plant species must be easy to obtain and maintain. During the aphids and parasitoids rearing, the host plants need to be exchanged periodically to new non-infested plants. High aphid populations can negatively affect the host plant and, consequently, reduce the parasitoid quality. The degree of aphid infestation will indicate the time to exchange the host plant. Depending on the demand for parasitoids use, the adult amount that will return to the rearing cages and the number that will be commercialized is decided.

17.2.4 Tamarixia radiata

The Asian citrus psyllid (ACP) *Diaphorina citri* Kuwayama (Liviidae) has become the most important pest on world citriculture since it was confirmed as vector of causal agents of huanglongbing (HLB), the most destructive bacterial disease of citrus plants. As there are not curative measures for HLB, its management is focused on vector control.

There are records of biological control of *D. citri* by 95 different species (arachnids and insects) (Kondo et al. 2015). Among these natural enemies, the parasitoids are the best candidates for augmentative biological control programs. There are two registered hymenopteran species parasitizing ACP, the endoparasitoid *Diaphorencyrtus aligarhensis* Shafee, Alam, and Agarwal and the ectoparasitoid *Tamarixia radiata* (Waterston) (Grafton-Cardwell et al. 2013). *T. radiata* was used successfully on different cases around the world like Taiwan (Chien and Chu 1996), the USA (Hoy and Nguyen 1998), Guadalupe and the Reunion islands (Etienne et al. 2001), Mexico (Miranda-Salcedo and López-Arroyo 2010), and Brazil (Parra et al. 2016).

In Brazil, primary inoculum of HLB occurs on abandoned and organic groves (with no psyllid control), urban areas with orange jasmine (*Murraya paniculata*, a psyllid host), and residential backyards with citrus trees (Bergamin Filho et al. 2016). On these areas a new approach on biological control has been used since 2014 by releasing *T. radiata* to control ACP before migration to commercial groves. The mass rearing of *T. radiata* was developed to supply this program with large number of parasitoids (Diniz 2013). Since there are no artificial diets, plant maintenance is required to rear *D. citri*, the host to multiply the parasitoid.

Orange jasmine was considered the most suitable host for ACP (Nava et al. 2007). *M. paniculata* seedlings should be fertilized frequently to promote vigor; these plants must be pruned to stimulate sprouting of axillary buds once the psyllid only lays eggs on flushes. The mean length of 1 cm with the leaflets still closed is considered to be preferred by the psyllids for oviposition (Diniz 2013). ACP rearing is conducted in greenhouses with a fan and pad evaporative cooling system, programmed to trigger when the internal temperature reaches 28 °C. However, on the hottest summer days, even with the system turned on, the internal temperature of the greenhouse can reach 35 °C for a few hours. This temperature increase does not seem to affect the rearing since the largest populations could be obtained during this period.

The rearing cages are cub of 45 cm of side constructed using fine nylon netting. Each cage contains orange jasmine with preferred flushes length for *D. citri* oviposition. Three hundred adult psyllids, about 12 days after emergence (i.e., at the beginning of the reproductive phase), are placed in each cage. These insects are kept in the cage for 7 days to allow oviposition, after which they are collected with the use of an electric aspirator. Not all insects are recovered; on average, 60% of the insects initially placed are collected. This reduction is attributed to natural mortality. The plants with *D. citri* eggs are transferred to another part of the greenhouse to allow the nymphs development. After 12 days (25 °C) the nymphs reach 4th and 5th instars, preferred phase for *T. radiata* parasitism. In approximately 18 days, it occurs the adult emergence. In this method, 70% of ACP production is directed to *T. radiata* multiplication and 30% to maintain *D. citri* rearing system.

The parasitoid is reared in rooms with controlled conditions (25 ± 2 °C and 14 h photophase). The rearing cages are 97 × 45 × 45 cm (length × width × height), with fine nylon netting on panel sides and polystyrene on front and top. Inside are placed orange jasmine plants with ACP nymphs. The cages are accommodated on shelves with four fluorescent lamps each. A mixture of honey and pollen (2:1) is offered as supplement food for *T. radiata* adults. Parasitoids (about 24 h old) are released at a ratio of 1:10 nymphs of *D. citri*. The parasitoid emergence begins with 12 days. Emergence boxes could be used to facilitate the parasitoid collection. This can be made using a completely sealed box (to exclude the light) with a single opening on the top, attached to a transparent bottle. Ten days after parasitism flushes with parasitized nymphs are cute and placed in the box; emerged parasitoids are attracted to the light on the top and get trapped in the bottle facilitating their collection. On each box should be placed a tray with humid vermiculite to keep high humidity preventing wasp mortality by desiccation. Around 70% of *T. radiata* emergence occurs in the first 3 days. The parasitoids are collected with electric aspirators to facilitate the process. In this system, 17% of the wasps return to the rearing and 83% used for releases or laboratory tests. A summary of all rearing process can be observed in Fig. 17.3. Insects for biological control programs could be stored at 18 °C on bottles with supplement food (honey + pollen 2:1); releases should be carried out in late afternoon or early morning for better environment conditions for the parasitoids.

This methodology was transferred to the orange-juice companies and citrus growers in Brazil that currently produce their own parasitoids on six biofactories. More than 10 million wasps were released on that primary inocula areas promoting parasitism rates of 70% on average.

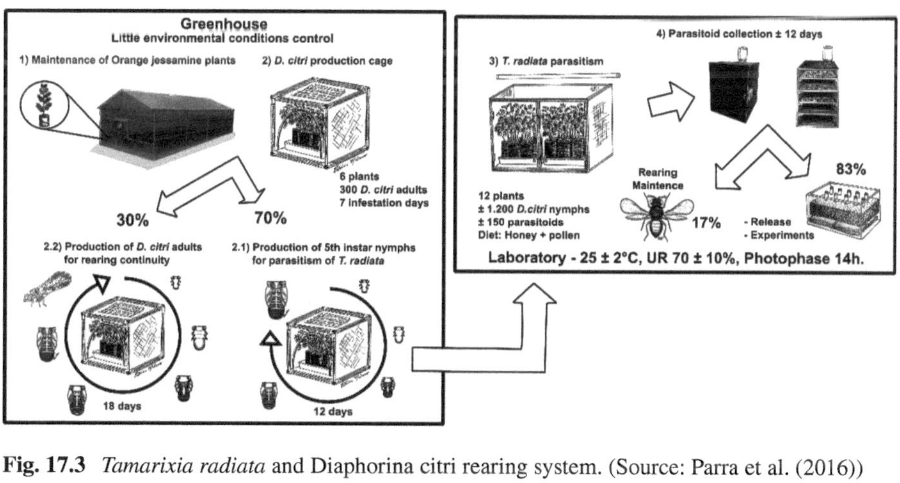

Fig. 17.3 *Tamarixia radiata* and Diaphorina citri rearing system. (Source: Parra et al. (2016))

17.3 Final Remarks

The BC market is increasing in all Latin America, with new local BC companies or by new branches from companies arriving mainly from Europe. The producers acceptance to adopt BC will only increase the demand for new entomophagous options. To achieve this, the research effort with parasitoids rearing for different pests needs to continue. The future BC programs success starts with new strategies for entomophagous mass rearing quality with low cost.

References

Antolin MF (1999) A genetic perspective on mating systems and sex ratios of parasitoid wasps. Res Popul Ecol 41(1):29–37

Bergamin Filho A, Inoue-Nagata AK, Bassanezi RB et al (2016) The importance of primary inoculum and area-wide disease management to crop health and food security. Food Sec 8(1):221–238

Bleicher E, Parra JRP (1990) Espécies de *Trichogramma* parasitoides de *Alabama argillacea*: II. tabela de vida de fertilidade e parasitismo de três populações. Pesq Agropec Bras 25(2):207–214

Cerutti F, Bigler F, Eden G et al (1992) Optimal larval density and quality control aspects in mass rearing of the Mediterranean flour moth, *Ephestia kuehniella* Zeller (Lepidoptera: Phycitidae). J Appl Entomol 114(1/5):353–361

Chien CC, Chu YI (1996) Biological control of citrus psyllid, *Diaphorina citri* in Taiwan: biological control pest control in systems of integrated pest management. Food Fertil Tech Cent 47:93–114

Coelho Jr A, Parra JRP (2013a) Effect of carbon dioxide (CO2) on mortality and reproduction of *Anagasta kuehniella* (Zeller, 1879), in mass rearing, aiming at the production of *Trichogramma* spp. An Acad Bras Ciênc 85(2):823–831

Coelho Jr A, Parra JRP (2013b) The effect of rearing in different temperature regimes on the reproduction of *Anagasta kuehniella* (Lepidoptera: Pyralidae). Environ Entomol 42(4):799–804

Ciociola Júnior AI, Querino RB, Zucchi RA et al (2001a) Molecular tool for identification of closely related species of *Trichogramma* (Hymenoptera: Trichogrammatidae): *T. rojasi* Nagaraja and Nagakarti and *T. lasallei* Pinto. Neotrop Entomol 30(4):575–578

Ciociola Júnior AI, Zucchi RA, Stouthamer R (2001b) Molecular key to seven Brazilian species of *Trichogramma* (Hymenoptera: Trichogrammatidae) using sequence of the IT S2 region of restriction analysis. Neotrop Entomol 30(2):259–261

Coelho Jr A, Reigada C, Haddad ML et al (2016a) Effects of temperature increase caused by larval metabolism on the reproduction of *Anagasta kuehniella* (Lepidoptera: Pyralidae), a factitious host for *Trichogramma*. Biocontrol Sci Tech 26(5):630–639

Coelho Jr A, Rugman-Jones PF, Reigada C et al (2016b) Laboratory performance predicts the success of field releases in inbred lines of the egg parasitoid *Trichogramma pretiosum* (Hymenoptera: Trichogrammatidae). PLoS One 11(1):e0146153

Coelho Jr A, Geremias LD, Alves GR et al (2017) The biology of *Trichogramma pretiosum* as atmospheric O_2 becomes depleted and CO_2 accumulates. Biol Control 105:1–5

Cônsoli FL, Grenier S (2010) The in vitro rearing of egg parasitoids. Prog Biol Control 9:293–313

Cônsoli FL, Parra JRP, Zucchi RA (2010) Egg parasitoides in agroecosystems with emphasis on *Trichogramma*. Springer, Dordrecht

Costa-Lima TC, Chagas MCM, Parra JRP (2014) Temperature-dependent development of two neotropical parasitoids of *Liriomyza sativae* (Diptera: Agromyzidae). J Insect Sci 14:1–4

Costa-Lima TC, Silva AC, Parra JRP (2015) Moscas-minadoras do gênero *Liriomyza* (Diptera: Agromyzidae): aspectos taxonômicos e biologia. Embrapa Semiárido, Petrolina
Costa-Lima TC, Geremias LD, Begiato AM et al (2017) Sistema de criação de parasitoide de mosca-minadora. Embrapa Semiárido, Petrolina
Cure JR, Cantor F (2003) Predacious and parasitic activity of *Diglyphus begini* (Ashm.) (Hymenoptera: Eulophidae) on *Liriomyza huidobrensis* (Blanch.) (Diptera: Agromyzidae), in *Gypsophila paniculata* L. (Monocotiledonea: Caryophyllaceae) plants. Neotrop Entomol 32(1):85–89
Danks HV (1988) Systematics in support of entomology. Annu Rev Entomol 33:271–294
Daumal J, Voegelé J, Brun P (1975) Les Trichogrammes: II. Unité de production massive etquotidienne d'un hôte de substitution *Ephestia kuehniella* Zell. (Lepidoptera, Pyralidae). Ann Zool Ecol Anim 7:45–49
De Conti BF, Bueno VHP, Sampaio MV et al (2011) Biological parameters and thermal requirements of the parasitoid *Praonvolucre* (Hymenoptera: Braconidae) with *Macrosiphum euphorbiae* (Hemiptera: Aphididae) as host. Biocontrol Sci Tech 21:497–507
Dias NS, Parra JRP, Cônsoli FL (2010a) Egg laying and development of Neotropical trichogrammatid species in artificial eggs. Entomol Exp Appl 137(2):126–131
Dias NS, Parra JRP, Dias CTS (2010b) Tabela de vida de fertilidade de três espécies neotropicais de Trichogrammatidae em ovos de hospedeiros alternativos como critério de seleção hospedeira. Rev Bras Entomol 54(1):120–124
Dias-Pini NS, Silva CSB, Peñaflor MF et al (2014) Does host determine short-range flight capacity of trichogrammatids? J Appl Entomol 138(9):1–6
Diniz AJF (2013) Otimização da criação de *Diaphorina citri* Kuwayama, 1908 (Hemiptera: Liviidae) e de *Tamarixia radiata* (Waterston, 1922) (Hymenoptera: Eulophidae), visando a produção em larga escala do parasitoide e avaliação do seu estabelecimento em campo. Tese, Escola Superior de Agricultura "Luiz de Queiroz", Universidade de São Paulo
Etienne J, Quilici S, Marvival D et al (2001) Biological control of *Diaphorina citri* (Hemiptera: Psyllidae) in Guadeloupe by imported *Tamarixia radiata* (Hymenoptera: Eulophidae). Fruits 56(5):307–315
Flanders SE (1927) Biological control of the codling moth (*Carpocapsa pomonella*). J Econ Entomol 20:644
Grafton-Cardwell EE, Stelinski LL, Stansly PA (2013) Biology and management of Asian citrus psyllid, vector of the huanglongbing pathogens. Annu Rev Entomol 58:413–432
Hoy MA, Nguyen R (1998) Citrus psylla: in Florida an action plan. Pest Alert. http://extlab7.entnem.ufl.edu/PestAlert/hoy-0615.htm. Accessed 10 March 2018
Hufbauer RA (2001) Pea aphid-parasitoid interactions: have parasitoids adapted to differential resistance? Ecology 82(3):717–725
Kondo T, Gonzalez G, Tauber C et al (2015) A checklist of natural enemies of *Diaphorina citri* Kuwayama (Hemiptera: Liviidae) in the department of Valle del Cauca, Colombia and the world. Insecta Mundi 457:1–14
Legner EF (1979) Prolonged culture and inbreeding effects on reproductive rates of 2 pteromalid parasites of muscoid flies. Ann Entomol Soc Am 72(1):114–118
Leppla NC (2014) Concepts and methods of quality assurance for mass-reared parasitoids and predators. In: Morales-Ramos JA, Roras MG, Shapiro-Ilan D (eds) Mass production of beneficial organisms: invertebrates and entomopathogens. Elsevier Science, New York, pp 277–305
Lewis WJ, Nordlund DA, Gross HR Jr et al (1976) Production and performance of *Trichogramma* reared in eggs of *Heliothis zea* and other hosts. Environ Entomol 5(3):449–452
Lins Júnior JC, Bueno VHP, Silva DB et al (2011) *Praon volucre* (Hymenoptera: Braconidae, Aphidiinae), a natural enemy of *Macrosiphum euphorbiae* (Hemiptera: Aphididae): life table and intrinsic rate of population increase. Eur J Entomol 108(4):575–580
Luna MG, Hawkins BA (2004) Effects of inbreeding versus outbreeding in *Nasonia vitripennis* (Hymenoptera: Pteromalidae). Environ Entomol 33(3):765–775

Miranda-Salcedo MA, López-Arroyo LI (2010) Avance de investigación para el manejo del psilido asiático de los cítricos *Diaphorina citri* Kuwayama (Hemiptera: Psyllidae) en Michoacán. In: Proceedings of the 3er congreso nacional mitigación del daño ambiental en el sector agropecuario de México, México

Murphy ST, Lasalle J (1999) Balancing biological control strategies in the IPM of New World invasive *Liriomyza* leafminers in field vegetable crops. Biocontrol News Inf 20(3):91–104

Nansen C, Coelho Júnior A, Vieira MV et al (2014) Reflectance-based identification of parasitized host eggs and adult *Trichogramma* specimens. J Exp Biol 217:1187–1192

Nava DE, Torres MLG, Rodrigues MDL et al (2007) Biology of *Diaphorina citri* (Hem., Psyllidae) on different hosts and at different temperatures. J Appl Entomol 131(9/10):9–10

Parra JRP (2002) Criação massal de inimigos naturais. In: Parra JRP, Botelho PSM, Corrêa-Ferreira BS et al (eds) Controle biológico no Brasil: parasitóides e predadores. Manole, Barueri, pp 143–164

Parra JRP (2010) Mass rearing of egg parasitoids for biological control programs. In: Cônsoli FL, Parra JRP, Zucchi RA (eds) Egg parasitoids in agroecosystems with emphasis on *Trichogramma*. Springer, New York, pp 267–292

Parra JRP, Coelho Júnior A, Geremias LD et al (2014) Criação de *Anagasta kuehniella*, em pequena escala, para produção de *Trichogramma*. Occasio, Piracicaba

Parra JRP, Zucchi RA, Coelho Júnior A et al (2015) *Trichogramma* as a tool for IPM in Brazil. In: Vinson B, Greenberg SM, Liu T et al (eds) Augmentative biological control using Trichogramma spp.: current status and perspectives. Northwest A&F University Press, Shaanxi, pp 472–496

Parra JRP, Alves GR, Diniz AJF et al (2016) *Tamarixia radiata* (Hymenoptera: Eulophidae) × *Diaphorina citri* (Hemiptera: Liviidae): mass rearing and potential use of the parasitoid in Brazil. J Integr Pest Manag 7(1):1–11

Prezotti L, Parra JRP, Vencovsky R et al (2002) Teste de vôo como critério de avaliação da qualidade de *Trichogramma pretiosum* Riley (Hymenoptera: Trichogrammatidae): adaptação de metodologia. Neotrop Entomol 31(3):411–417

Prezotti L, Parra JRP, Vencovsky R et al (2004) Effect of the size of the founder population on the quality of sexual populations of *Trichogramma pretiosum*, in laboratory. Biol Control 30(2):174–180

Sidney LA, Bueno VHP, Lins Junior JC et al (2010) Qualidade de diferentes espécies de pulgões como hospedeiros do parasitoide *Aphidius ervi* (Haliday) (Hymenoptera: Braconidae, Aphidiinae). Neotrop Entomol 39(5):709–713

Smith SM (1996) Biological control with *Trichogramma*: advances, successes, and potential of their use. Annu Rev Entomol 41:375–406

Starý P (1995) The aphidiidae of Chile (Hymenoptera, Ichneumonoidea, Aphidiidae). Dtsch Ent Z NF 42(1):113–138

Starý P, Gerding M, Norambuena H et al (1993) Environmental research on aphid parasitoid biocontrol agents in Chile (Hym., Aphidiidae; Hom., Aphidoidea). J Appl Entomol 115(1/5):292–306

Starý P, Sampaio MV, Bueno VHP (2007) Aphid parasitoids (Hymenoptera, Braconidae, Aphidiinae) and their associations related to biological control in Brazil. Rev Bras Entomol 51(1):107–118

Stouthamer R (2006) Molecular methods for the identification of biological control agents at the species and strain level. In: Bigler F, Babendreier D, Kuhlman U (eds) Environmental impact of invertebrates for biological control of arthropods. CABI Publishing, Wallingford, pp 187–201

Van Lenteren JC (2000) Success in biological control of arthropods by augmentation of natural enemies. In: Gurr G, Wratten S (eds) Biological control: measures of success. Kluwer Academic, Dordrecht, pp 77–104

Van Lenteren JC (2003) Need for quality control of mass produced biological control agents. In: Van Lenteren JC (ed) Quality control and production of biological control agents: theory and testing procedures. CABI, Wallingford, pp 1–18

Van Lenteren JC (2012) The state of commercial augmentative biological control: plenty of natural enemies, but a frustrating lack of uptake. Biol Control 57(1):1–20

Van Lenteren JC, Bueno VHP (2003) Augmentative biological control of arthropods in Latin America. Biol Control 48(2):123–139
Van Lenteren JC, Bolckmans K, Kohl J et al (2017) Biological control using invertebrates and microorganisms: plenty of new opportunities. Biol Control 62(1):1–25
Wajnberg E, Hassan SA (1994) Biological control with egg parasitoids. CABI, London
Whitten MJ, Hoy MA (1999) Genetic improvement and other genetic considerations for improving the efficacy and success rate of biological control. In: Bellows TS, Fisher TW (eds) Handbook of biological control: principles and applications of biological control. Academic Press, New York, pp 271–296
Wilson JJ, Sing KW, Floyd RM et al (2017) DNA barcodes and insect biodiversity. In: Foottit RG, Adler PH (eds) Insect biodiversity: science and society. Wiley, Chichester, pp 575–592
Zucchi RA, Querino RB, Monteiro RC (2010) Diversity and hosts of *Trichogramma* in the New World, with emphasis in South America. In: Cônsoli FL, Parra JRP, Zucchi RA (eds) Egg parasitoids in agroecosystems with emphasis on *Trichogramma*. Springer, Dordrecht, pp 219–236

Chapter 18
Entomopathogenic Nematodes

Luís Garrigós Leite, José Eduardo Marcondes de Almeida, Julie Giovanna Chacon-Orozco, and Clara Yalexy Delgado

18.1 Introduction

Sustainable farming is one of the main achievement of modern agriculture, seeking to meet the requirements for healthy farm living and environmental protection as well as to offer competitive and safe products that are free from chemical residues. In this context, integrated pest management has been gaining prominence, with biological control being the main tool by exploring predator and parasitoid insects as well as microbial control agents such as viruses, fungi, bacteria, and nematodes.

Unfortunately, microbial control agents are often unfavorably compared with broad spectrum pesticides in terms of cost and spectra of target insects. However, their benefits are far more numerous and include efficacy, safety to nontarget organisms (including natural enemies of pests), to farm workers, to food supply, and the environment (Lacey 2008). Currently, more than three million hectares in Brazil are treated with microbial agents, which have target mainly insect pests above ground or on the surface of the soil (Leite et al. 2017a).

Insect pests that live below ground or in cryptic habitats pose special problems in pest management (Jackson 1999). A group of entomopathogens that has been recognized as potential agents to target insects in the soil or in cryptic environments are entomopathogenic nematodes (EPNs) from the genera *Steinernema* and *Heterorhabditis* (Rhabditida: Steinernematidae and Heterorhabditidae). These nematodes are the only ones associated with insects that are widely used for pest

---

L. G. Leite (✉) · J. E. M. de Almeida
Instituto Biológico, Agência Paulista de Tecnologia dos Agronegócios (APTA), Campinas, SP, Brazil

J. G. Chacon-Orozco
Instituto Biológico, Campinas, SP, Brazil

C. Y. Delgado
TicoFrut, Alajuela, Costa Rica

B. Souza et al. (eds.), *Natural Enemies of Insect Pests in Neotropical Agroecosystems*, https://doi.org/10.1007/978-3-030-24733-1_18

control. They have an unusual mutualistic association that they establish with the bacterium *Xenohabdus* and *Photorhabdus*, respectively, resulting invariably in the rapid death of parasitized insects. This leads these agents to be known as entomopathogens and makes them more suitable for biological control compared to other groups of nematodes; they are used successfully in many countries to control various agricultural pests (Wouts 1981). In Brazil, new strains have been selected with potential to control several insect pests. However, the production process still requires improvement to meet the growing demand in order to provide products with good qualities and to suit the use to various crop systems (Leite et al. 2017a). Another nematode highlighted in Brazil is the facultative parasite *Deladenus siricidicola* Bedding (Tylenchida: Neotylenchidae), which is used for the management of the European wood wasp, *Sirex noctilio* Fabricius (Siricidae), an important pest in *Pinus* spp. planted stands (Ferraz 1998).

In general, the success and expansion of microbial pest control programs depends on the development of efficient mass production techniques and the formulation of these agents. The present review aims to report the different techniques of mass production of entomopathogenic nematodes of the families Steinernematidae and Heterorhabditidae.

18.2 In Vivo Production

The first study on the production of EPNs in Brazil focused on multiplication in vivo, using mainly *Galleria mellonella* L. (Pyralidae) larvae as the host insect. Other hosts were also used, such as *Diatraea saccharalis* (F.) (Crambidae) and *Tenebrio molitor* L. (Tenebrionidae) larvae (Folegatti et al. 1988; Leite et al. 1990).

In vivo production of EPNs was performed using *G. mellonella* larvae as the main host because it may be used for the multiplication of the majority of species from the genera *Steinernema* and *Heterorhabditis,* generating higher yields (ranging from 100,000 to 350,000 infective juveniles/larva) compared to other hosts (Dutky et al. 1964; Milstead and Poinar 1978). However, breeding of *G. mellonella* is considered costly, raising costs for the production and marketing of EPNs. When searching for alternative hosts for the in vivo production of EPNs, Molina et al. (2004) evaluated the multiplication of *Steinernema carpocapsae* (Weiser) Wouts, Mracek, Gerdin, and Bedding, *Steinernema glaseri* (Steiner), and *Steinernema arenarium* (Artyukhovsky) in the last larval instar of *G. mellonella, Tenebrio molitor* L. (Tenebrionidae), *Spodoptera frugiperda* (J.E. Smith) (Noctuidae), and *Bombyx mori* (L.) (Bombycidae). The highest production was obtained in the first 3 days after the emergence of the infective juveniles (IJs). *G. mellonella* larvae generated the highest yield, with 302,124 and 149,213 IJs/larva for the nematodes *S. carpocapsae* and *S. arenarium,* respectively. In comparison, *B. mori* and *S. frugiperda* generated low yields of the tested nematodes. To optimize the production of *Heterorhabditis baujardi* Phan, Subbotin, Nguyen, Moens, 2003, strain LPP7, in *G. mellonella* larvae, Dias et al. (2008) tested different concentrations of inoculum and different sizes of inoculated larvae, obtaining the highest yields at concentrations of 100 and 300 IJs/larvae inoculated on the larger larvae (200 mg).

Fig. 18.1 Production of *Heterorhabditis* sp. in Cuba: (**A–C**) *Galleria mellonella* rearing; (**D**) inoculation of larvae with nematode; (**E**) incubation; (**F–G**) IJs harvesting. (Photos: José Eduardo M. de Almeida)

Some countries of Latin America have commercially produced nematodes on *G. mellonella*, using a technique that was developed in Cuba (Fig. 18.1). The production in Cuba still relies on inoculation and harvest (via modified White trap). *G. mellonella* larva is reared in ~100 L metal buckets and fed with corn flower. Larvae are dipped in the nematode suspension and held inside Petri dishes to be incubated at room temperature. After the insect dies, cadavers are transferred to large trays containing the modified White traps (Almeida, J.E., personal information). In Cuba, EPNs have been used successfully for biological control due the development of several studies related to in vivo production (García et al. 2007), strain selection (Lucila et al. 2001; Arahis et al. 2006; Pozo et al. 2006, 2007; Evans et al. 2009), virulence and efficacy (Rodríguez et al. 1997; Bernal et al. 1998a, b; Valdés et al. 2005; Sanchez and Rodriguez 2008), persistence (Liens et al. 1998), and to other issues.

In Costa Rica, the nematode *Heterorhabdtis* sp. has been mass produced in *G. mellonella* for the control of *Phyllophaga* complex in citrus orchards where one larvae can kill a tree. The nematode is applied from July to December as a preventive way to protect plants that are up to 4 years old. The nematode is applied together with the *G. mellonella* cadavers by digging the cadaver around the tree at the dose of two cadavers per tree in regular areas and four in areas with hight humidity. Infected cadavers are produced by the company TicoFrut at the amount of 35.000 infected cadavers per day, 5 days a week since 2016.

18.3 In Vitro Production

Steinernema and *Heterorhabditis* started being studied for the control of insects from the 1930s. However, it was only in the late 1980s that commercial use for insect control was initiated due to the improvement of the mass production method by *vitro* process, in a monoxenic system, taking advantage of the symbiotic bacteria as a source of food (Ehlers 2001). EPNs, when needed on a large scale but with reasonable quality and cost, are produced by solid or liquid fermentation (Askary and Ahmad 2017). Both techniques are used to culture nematodes monoxenically to maintain consistency in quality and achieve predictablility (Lunau et al. 1993).

Solid-state media used for EPNs production started with agar media (House et al. 1965; Wouts 1981; Dunphy and Webster 1989), and now the three-dimensional media with nutrients in polyether-polyurethane sponge (Bedding 1981, 1984) are used for mass production. Regarding liquid culture, it provides economies of scale since the production units may be increased from a few mL to several m^3 with hardly any increase in labor cost. The nematodes are grown in a bioreactor composed of a stainless-steel tank with systems to stir and aerate the liquid culture, and with probes for various physical parameters and actors for controlling these parameters (Peters et al. 2017).

In Brazil, initiatives toward developing an EPN-based commercial product have been gaining attention in recent years. Leite (2006) highlights the development of media for the in vitro cultivation of nematodes and mass production techniques. A medium based on beef liver soaked in sponge provided high production rates of the nematode *Heterorhabditis* sp., with a yield of up to 180,000 juveniles per mL of medium. More recently, Leite et al. (2016a) tested a rich liquid medium based on egg yolk, white yolk, yeast extract, glucose, and peanut oil, obtaining more than 100,000 IJs mL of medium. The addition of 0.2% agar to the medium also increased the nematode production at 280 rpm and 25 °C (Leite et al. 2016a, b).

Currently, some companies in Brazil are producing EPNs by the in vitro solid-state process (Bedding 1981, 1984) to supply products for field tests and for commercial use. These companies have improved the solid-state process with funding from the Brazilian governments.

The solid-state process is based on monoxenic cultures where nematodes are added to a pure culture of their symbiotic bacteria in a nutritive medium. The in vitro solid-state production of EPNs involves four steps: preparation of culture media, culture and inoculation of symbiotic bacteria, inoculation with nematodes, and harvest, as described by Peters et al. (2017).

The symbiotic bacteria of different isolates of EPNs can be easily isolated from nematode-infected insect larvae or surface sterilized IJs. The symbiotic bacteria are well known to have two phase variants. The primary variant (Phase I) is conducive to growth of IJs and tends to be retained by them (Han and Ehlers 2001). Phase I variant can be selected on nutrient bromothymol blue agar (NBTA) (0.5% peptone, 0.3% beef extract, 1.5% agar, 0.025% bromothymol blue, and 0.004% 2,3,5-triphenyltetrazolium chloride) or MacConkey media (1.7% peptone17, 0.3% proteose peptone, 1% lactose, 0.15% bile salts, 0.5% sodium chloride, 0.003% neutral red, and 1.35% agar) (Akhurst 1980). Symbiotic bacteria can be cultured in lysogeny broth (LB) (0.5% NaCl 5 g, 1% tryptone, 0.5% yeast Extract) or yeast salts (YS) broth (0.05% $NH_4H_2PO_4$, 0.05% K_2HPO_4, 0.02% $MgSO_4$•$7H_2O$0.2 g, 0.5% NaCl and 0.5% yeast extract) (Hirao and Ehlers 2009). Symbiotic bacteria stock can be prepared by mixing fresh bacteria liquid culture with glycerol to a final concentration of 15% (v/v) and store the vials at −80 °C. A freshly grown bacterial culture should be inoculated into the solid media 1 to 4 days before nematode inoculation (Bedding 1981, 1984) as the symbiotic bacteria need time to convert the culture medium into a suitable medium for nematode development and reproduction (Forst et al. 2002). Bacterial inoculum size does not appear to be important in yield determination (Han et al. 1992, 1993).

Monoxenic cultures obtained as described above can be scaled-up by inoculation of agar with nematodes to the sponge media. Inoculation of monoxenic nematodes to new culture media can be accomplished either by transferring the sponge media with IJs directly or by washing the IJs out of the media before inoculation under aseptic conditions, depending on the different used culture containers. Containers used for EPN solid culture can be flasks or culture vessels comprising a tray with side walls and overlapping lids that allow gas exchange through a layer of foam (Bedding et al. 1991), or bags with gas permeable strips for ventilation (Gaugler and Han 2002).

Nematode inoculum concentration (IJs per unit of media) may impact yields of nematodes for some, but not all nematode species (Han et al. 1992, 1993; Leite et al. 2017b). However, increasing inoculum size may accelerate nematode propagation and decrease culture time.

After 2–5 weeks of culture, nematodes can be harvested (Bedding 1981, 1984). IJs can be extracted from solid media with centrifugal sifters or by washing nematodes out of the sponge in washing machines and then separating the IJs by sedimentation, centrifugation or sieving (Ehlers and Shapiro-Ilan 2005). Before formulation and storage, the separated IJs should be washed two or more times to remove impurities which would fuel the growth of undesired fungal or bacterial contaminants in the final product.

18.4 Formulation

Various studies have attempted to elucidate the most favorable conditions and formulations for the storage and application of EPNs. Various factors such as temperature and oxygen directly affect the survival and infectivity of IJs stored in aqueous suspension. Isolates of *Heterorhabditis amazonensis* Andalo, Nguyen, Moino remained viable for up to 6 months at a temperature of 16 °C with aeration but did not survive when kept at 8 °C and 28 °C (Andaló et al. 2006; Molina et al. 2006). IJs survival may be extended by the addition of certain products to IJs suspensions, such as glycerin, which help keep the *S. carpocapsae* IJs alive and infective for longer periods when compared to water (Andaló et al. 2008).

In addition, some substrates may assist and favor the storage of nematodes, such as the sponge that preserves *H. amazonensis* JPM4 IJs at 16 °C, with 50% viability after 6 months of storage (Andaló et al. 2010). An agar-based medium was also tested for the transport and storage of IJs. Laboratory and field studies evaluating the ability of nematodes to infect and kill *G. mellonella* larvae indicated that different agar concentrations (2% and 4%) remained effective for 7–45 days, respectively, for the storage, transport, and formulation of *H. baujardi* LPP7 IJs with no difference to water and with the advantage of using a lower volume (Dolinski et al. 2008).

When the application of EPNs occurs in the form of insect cadavers, the use of covering is recommended because the insect cadaver is often subject to physical damage during manipulation and application. For instance, on the ground, insect cadavers are subject to the action of organisms such as ants, which may alter the effectiveness of the application. Del Valle et al. (2009) determined the effects of different coverage of insect cadavers of *G. mellonella* infected by *H. baujardi* LPP7, and evaluated the behavior of *Ectatomma* spp. ants (Formicidae) toward the cadavers. The best coverage treatments were talc and gelatine capsules because they did not affect the emergence of IJs. However, when the formulated cadavers were placed close to the openings of ant nests, ants carried the insect cadavers with different coatings up to 20 cm away from the nest openings, with the exception of those formulated in gelatine capsules, which were not removed.

18.5 Final Considerations

Many nematode species are capable of parasitizing insects; however, only the so-called entomopathogenic – Steinernematidae and Heterorhabditidae – have been more widely used because of the possibility of commercial production and formulation of these organisms, and because of the wide range of hosts susceptible to them. In Brazil, neglected for decades, in the last years the subject finally began to gain the deserved prominence, apparently driven by several reasons such as (i) the growing market niche for the control of soil-dwelling insect pests in several countries and (ii) the accelerated growth of the agribusiness sector as a whole.

The extent of the use of entomopathogenic nematodes in biocontrol would not be thinkable without efficient mass-rearing techniques. The development and the increase in scale of the liquid culture resulted in reduction of production costs by more than 75% in the last 30 years and this process has not been finished yet (Toepfer et al. 2009; Töpfer et al. 2010).

Another important prerequisite for the success of entomopathogenic nematodes is the low level of regulation. Nematodes are macroorganisms and, like the widely used beneficial arthropods, do not need to be registered as plant protection agents in most countries. It is probably due to this liberal legislation that the nematode production business is versatile and various culture techniques coexist. This coexistence is important to sustain the innovative potential of EPNs (Peters et al. 2017). In Brazil, registration for macroorganism is required but it is easier in terms of application compared to registration for microorganisms. At least four companies are registering EPNs in Brazil.

The market for the use of nematode in Brazil is immense and there is a great need for alternative techniques to replace chemical insecticides for soil-dwelling pest control. Entomopathogenic nematodes have great potential for the control of various pests in different cultures and environments. For sugarcane, nematodes can be an important alternative considering the underground habit of most insect pests, and the very favorable and stable conditions offered by the crop, especially in mechanized harvest areas, where the soil remains covered with straw.

References

Akhurst RJ (1980) Morphological and functional dimorphism in *Xenorhabdus* spp., bacteria symbiotically associated with the insect pathogenic nematodes, *Neoaplectana* and *Heterorhabditis*. J Gen Microbiol 121:303–309

Andaló V, Cavalcanti RS, Molina JP et al (2006) Influência da aeração no armazenamento de nematodes entomopatogênicos (Rhabditida) em condições de laboratório. Nematol Bras 30:45–50

Andaló V, Cavalcanti RS, Molina JP et al (2008) Avaliação de substâncias na preservação de nematodes entomopatogênicos (Rhabditida: Heterorhabditidae, Steinernematidae) em diferentes temperaturas. Arq Inst Biol 75:301–312

Andaló V, Cavalcanti RS, Molina JP et al (2010) Substrates for storing entomopathogenic nematodes (Rhabditida: Steinernematidae, Heterorhabditidae). Sci Agric 67(3):342–347

Arahis CL, Ravelo HG, Hernández UA et al (2006) Susceptibilidad de *Homoeosoma electellum* (Hulst) a nematodos entomopatógenos. Cent Agríc 33(4):41–45

Askary TH, Ahmad MJ (2017) Entomopathogenic nematodes: mass production, formulation and application. In: Abd-Elgawad M, Askary TH, Coupland J (orgs) Biocontrol agents: entomopathogenic and slug parasitic nematodes. CABI Publishing, Boston, pp 348–361

Bedding RA (1981) Low cost *in vitro* mass production of *Neoaplectana* and *Heterorhabditis* species (Nematoda) for field control of insect pests. Nematologica 27(1):109–114

Bedding RA (1984) Large scale production, storage and transport of the insect parasitic nematodes *Neoaplectana* spp. and *Heterorhabditis* spp. Ann Appl Biol 104(1):117–120

Bedding RA, Stanfield MA, Crompton G (1991) Apparatus and method for rearing nematodes, fungi, tissue cultures and the like, and for harvesting nematodes. World Patent. WO 91/15569, 17 Jan 1991

Bernal S, González N, Andino M (1998a) Influencia de la temperatura del suelo en el parasitismo de Heterorhabditis spp. en suelos pardos con carbonatos. Cent Agríc 25(2):16–17

Bernal S, González N, Andino M (1998b) Influencia de la temperatura del suelo en el parasitismo de Heterorhabditis spp. en suelos fersialíticos. Cent Agríc 25(2):18–22

Del Valle EE, Dolinski C, Barreto ELS et al (2009) Effect of cadaver coatings on emergence and infectivity of the entomopathogenic nematode *Heterorhabditis baujardi* LPP7 (Rhabditida: Heterorhabditidae) and the removal of cadavers by ants. Biol Control 50(1):21–24

Dias PV, Dolinski C, Molina JP (2008) Influência da dose de juvenis infectantes e da massa de larvas de *Galleria mellonella* (Lepidoptera: Pyralidae) na produção in vivo de *Heterorhabditis baujardi* LPP7 (Rhabditida: Heterorhabditidae). Nematol Bras 32(4):317–321

Dolinski C, Barreto ELS, Del Valle EE et al (2008) Agar media: an alternative formulation for storage, transport and application of the entomopathogenic nematode *Heterorhabditis baujardi* LPP7 phan et al. (Nematoda: Heterorhabditidae). Bio Assay 3:1–5

Dunphy GB, Webster JM (1989) The monoxenic culture of *Neoaplectana carpocapsae* DD-136 and *Heterorhabditisheliothidis*. Rev Nématol 12:113–123

Dutky SR, Thompson JV, Cantwell GE (1964) A technique for the mass propagation of the DD-136 nematode. J Insect Pathol 6:417–422

Ehlers RU (2001) Mass production of entomopathogenic nematodes for plant protection. Appl Microbiol Biotechnol 56(5/6):623–633

Ehlers RU, Shapiro-Ilan DI (2005) Mass production. In: Grewal PS, Ehlers RU, Shapiro-Ilan DI (eds) Nematodes as biocontrol agents. CABI Publishing, Wallingford, pp 65–78

Evans G, Herrera RV, Morales MC et al (2009) Susceptibilidad de *Metamasius hemipterus sericeus* (L.) (Coleoptera; *Curculionidae)* a una cepa nativa de nematodos entomopatógenos. Cent Agríc 36(2):65–69

Ferraz LCCB (1998) Nematoides entomopatogênicos. In: Alves SB (ed) Controle microbiano de insetos. FEALQ, Piracicaba, pp 541–569

Folegatti MEG, Alves SB, Hawai PC et al (1988) Nova metodologia para produção in vivo de *Neoaplectana carpocapsae*. Nematol Bras 12:76–83

Forst S, Clarke D, Gaugler R (2002) Bacteria-nematode symbiosis. In: Gaugler R (ed) Entomopathogenic nematology. CABI Publishing, New York, pp 57–77

García M, Rodríguez Y, Cabrera D et al (2007) Producción de nematodos entomopatógenos en el centro nacional de referencia fitosanitaria para la montaña en Cuba. Rev Protección Veg 22(2):131–133

Gaugler R, Han R (2002) Production technology. In: Gaugler R (ed) Entomopathogenic nematology. CABI Publishing, Oxon, pp 289–310

Han R, Cao L, Liu X (1992) Relationship between medium composition, inoculum size, temperature and culture time in the yields of *Steinernema* and *Heterorhabditis* nematodes. Fundam Appl Nematol 15(3):223–229

Han R, Cao L, Liu X (1993) Effects of inoculum size, temperature and time on *in-vitro* production of *Steinernema carpocapsae* Agriotos. Nematologica 39(1):366–375

Han R, Ehlers RU (2001) Effect of *Photorhabdus luminescens* phase variants on the in vivo and *in vitro* development and reproduction of the entomopathogenic nematodes *Heterorhabditis bacteriophora* and *Steinernema carpocapsae*. FEMS Microbiol Ecol 35(3):239–247

Hirao A, Ehlers RU (2009) Influence of cell density and phase variants of bacterial symbionts (*Xenorhabdus* spp.) on dauer juvenile recovery and development of biocontrol nematodes *Steinernema carpocapsae* and *S. feltiae* (Nematoda: Rhabditida). Appl Microbiol Biotechnol 84(1):77–85

House HL, Welch HE, Cleugh TR (1965) Food medium of prepared dog biscuit for the mass-production of the nematode DD-136 (Nematoda: Steinernematidae). Nature 206(4986):847

Jackson TA (1999) Factors in the success and failure of microbial control agents for soil dwelling pests. J Integr Pest Manag Rev 4(4):281–285

Lacey LA (2008) Foreword. In: Alves SB, Lopes RB (eds) Controle microbiano de pragas na América Latina. FEALQ, Piracicaba, pp 19–20

Leite LG (2006) Nematoides entomopatogênicos. In: Batista Filho A (ed) Controle biológico de insetos e ácaros. Instituto Biológico, São Paulo, pp 42–51

Leite LG, Batista Filho A, Prada WLA (1990) Produção de Neoplectana glaseri em lagartas vivas e mortas de *Galleria mellonella*. Rev Agric 65:225–232

Leite LG, Shapiro-Ilan DI, Hazir S et al (2016a) A new medium for liquid fermentation of *Steinernema feltiae*: selection of lipid and protein sources. Nematropica 46(2):147–153

Leite LG, Shapiro-Ilan DI, Hazir S et al (2016b) The effects of nutriente concentration, addition of thickeners, and agitation speed on liquid fermentation of *Steinernema feltie*. J Nematol 48(2):126–133

Leite LG, Andaló V, Dolinski C et al. (2017a) Status of entomopathogenic nematodes in integrate pest management strategies in Brazil. In: ABD-Elgawad M, Askary TH, Coupland J (orgs) Biocontrol agents: entomopathogenic and slug parasitic nematodes. CABI Publishing, Boston, pp 348–361

Leite LG, Shapiro-Ilan DI, Hazir S et al (2017b) Effect of the inoculum age and phusical parameters on in vivo culture of the entomopathogenic nematode *Steinernema feltie*. J Helminthol 91(6):686–695

Liens BR, Andino M, Expósito I et al (1998) Permanencia de *Heterorhabditis* sp. en el suelo a diferentes temperaturas. Cent Agríc 25:43–45

Lucila G, Soler DM, Sánchez L (2001) Virulencia y potencial reproductivo de aislamientos cubanos de nematodos entomopatógenos. Rev Protección Veg 16(1):50–54

Lunau S, Stoessel S, Schmidt-Peisker AJ et al (1993) Establishment of monoxenic inocula for scaling-up *in vitro* cultures of the entomopathogenic nematodes *Steinernema* spp. and *Heterorhabditis* spp. Nematologica 39(1):385–399

Milstead JE, Poinar GO (1978) A new entomogenous nematode for pest management systems. Calif Agric 32:324–327

Molina JP, Oino AJ, Cavalcanti RS (2004) Produção in vivo de nematodes entomopatogênicos em diferentes insetos hospedeiros. Arq Inst Biol 71:347–354

Molina JP, Oino AJ, Cavalcanti RS et al (2006) Efecto da temperatura, concentración y tiempo de almacenamiento en la supervivencia de nematodos entomopatógenos. Rev Colomb Entomol 32(1):24–30

Peters A, Han R, Yan X et al (2017) Production of entomopathogenic nematodes. In: Lacey L (ed) Microbial control of insect and mite pests: from theory to practice. Academic Press, London, pp 157–172

Pozo VE, Herrera RVV, Morales MC et al (2007) Susceptibilidad de *Heliothis virescens* Fabr. (Lepidoptera; Noctuidae) a nematodos entomopatógenos. Cent Agríc 34:39–43

Pozo VE, Sisne ML, Aragón UR et al (2006) Susceptibilidad de escarabajos (Coleoptera; Scarabaeidae) presentes en la piña *Ananas comosus* (L.) Merril en Ciego de Ávila a nematodos entomopatógenos. Géneros *Phyllophaga* y *Cyclocephala*. Cent Agríc 33:83–86

Rodríguez I, Rodríguez MG, Sánchez L et al (1997) Efectividad de *Heterorhabditis bacteriophora* (Rhabditidae: Heteroderidae) sobre chinches harinosas del cafeto (Homoptera: Pseudococcidae). Rev Protección Veg 12(2):119–122

Sanchez L, Rodriguez MG (2008) Potenciales de *Heterorhabditis bacteriophora* poinar cepa HC1 para el manejo de *Hypothenemus hampei* Ferr: compatibilidad con *Beauveria bassiana* (Balsamo) vuillemin y endosulfan. Rev Protección Veg 23(2):104–111

Toepfer S, Burger S, Ehlers RU et al (2009) Controlling western corn rootworm larvae with entomopathogenic nematodes: effect of application techniques on plant-scale efficacy. J Appl Entomol 134:467–480

Töpfer K, Knuth P, Peters A et al (2010) Insektenpathogene nematoden gegen Wurzelbohrer. Mais 2:68–70

Valdés HR, Pérez EM, Vázquez MC et al (2005) Empleo de nematodos entomopatógenos (*Heterorhabditis* spp.) como contribución al manejo integrado de *Diaphania hyalinata* (L.) (Lepidoptera; Pyralidae) en el cultivo del pepino en sistemas de organopónicos. Cent Agríc 32:47–53

Wouts WM (1981) Mass production of the entomogenous nematode *Heterorhabditis heliothidis* (Nematoda: Heterorhabditidae) on artificial media. J Nematol 13(4):467–469

Chapter 19
Entomopathogenic Fungi

José Eduardo Marcondes Almeida, Luís Garrigós Leite, and Antonio Batista Filho

19.1 Introduction

Bioinsecticide production based on entomopathogenic fungi faces a series of challenges from isolates research to production due to aspects related to production, automation, formulation, legislation, and trading systems. Bioinsecticide manufacturers in Brazil are small and have little capital to invest in research, which delays the development of new products on a yearly basis. Multinationals have been turning their attention to bioinsecticides in the last years. Despite the faced difficulties, pest microbial control is one of the research fields that are mostly growing in entomology studies, and it influences the agricultural and industrial production sectors.

The global bioinsecticide market can reach US$ 3.4 billion up to 2017, and its main product is the bacterium *Bacillus thuringiensis* (Berliner), although fungi emerge as an important alternative for pest control. According to the Brazilian Association of Biological Control Companies (*Associação Brasileira das Empresas de Controle Biológico – ABCBio*), biological control agents may soon represent a US$11 billion market, which nowadays accounts for US$ 110 million. Microbial control in Latin America accounts for US$ 60 million (Barros and Ferreira 2014).

There are bioinsecticides in Brazil based on entomopathogenic fungi belonging to the species *Beauveria bassiana* (Bals.-Criv.) Vuill., *Metarhizium anisopliae* (Metchnikoff) Sorokin, and *Isaria fumosorosea* Wize, which have been produced in large scale. The objective of this chapter is to present the most commonly used entomopathogenic fungi production techniques in Latin America (LFA A et al. 2010; Alves and Lopes 2008; Batista Filho et al. 2006; Micothon 2012).

J. E. M. Almeida (✉) · L. G. Leite · A. Batista Filho
Instituto Biológico, Agência Paulista de Tecnologia dos Agronegócios (APTA), Campinas, SP, Brazil
e-mail: jemalmeida@biologico.sp.gov.br

B. Souza et al. (eds.), *Natural Enemies of Insect Pests in Neotropical Agroecosystems*, https://doi.org/10.1007/978-3-030-24733-1_19

19.2 Entomopathogenic Fungi

Fungi are able to attack a large number of insects in almost all development stages. Most fungi work by contact and ingestion, and their great genetic variability allows studies focused on the selection of more virulent strain and isolates for pest control purposes (Alves 1998).

The production processes of fungi pathogenic to invertebrates are reported based on the structure of the pathogen to be produced and used in the field. The different structures of the fungus to be used in pest control and the role they play in the natural cycle of the pathogen are (1) conidia, due to their reproduction and outspread; (2) blastospores, which outspread in the hemolymph of hosts; (3) mycelium, because it migrates to outside the host and allows fungal conidiogenesis; and (4) the resistance spores, which are responsible for fungus survival in the soil. Most hyphomycetes can be used in their conidia, blastopore, and mycelium forms; the two first forms are often the option of choice because both are host-infective. The choice made for the fungus form to be adopted will depend on the pathogen species and isolate, on how hard they are produced, on the environment where they will be applied in, and on the application method (Leite et al. 2003).

19.3 Entomopathogenic Fungi Production Through Solid Fermentation

The most produced and traded fungi structures are conidia generated in solid culture medium. This process has been adopted for the routine maintenance of isolates and for large-scale production aimed at field testing and trading. The fungus is produced on the surface of a solid medium inside different containers depending on the goal and on the production scale. There is no significant difference between small- and large-scale entomopathogenic fungi production concepts because all of them use isolates or strains and successive pricking in gelatinized medium such as agar or precooked rice, depending on the amount to be produced. They are usually produced in small-scale for research and laboratory or greenhouse testing, although the production principle is the same (Leite et al. 2003).

Low-cost vegetal products have been used for large-scale conidia production, mainly rice grains (Alves 1998; Leite et al. 2003; Michereff Filho et al. 2009).

The system for producing fungi by fermentation in solid culture medium is used worldwide, despite the advances of the system with liquid or semisolid culture medium. In Cuba, the *B. bassiana*, *M. anisopliae*, *Lecanicillium lecanii* (R. Zare and W. Gams), and *I. fumosorosea* fungi are produced in moist and sterilized rice or maize, using metal trays (Vazquez and Pérez 2017).

19.3.1 Strains and Isolates

The production process starts with the isolation of entomopathogenic fungi strains collected from insect corpses, soil, plants, and water; they are collected in the field, labeled and georeferenced. Next, they are taken to the laboratory to be isolated in general culture media such as potatoes – dextrose – agar (BDA) or agar – water, or even in more complex media such as mineral salts and nitrogen sources.

19.3.2 Isolates Selection

Isolate selection is an essential stage to get high-productivity virulent isolates in culture medium and to proceed with commercial-scale production. Molecular biology techniques help fungi isolates selection and shorten the time needed for such selection; therefore, they determine the groups of isolates and virulence that are similar to that of other insect species. Moreover, it is worth demystifying the idea of collecting isolates from the same region and plot since in most cases these isolates are endemic and of lower virulence because they are naturally generated. Therefore, chances are that exotic isolates become more virulent to the pest (Almeida and Batista Filho 2006).

19.3.3 Matrices Preparation

Matrices used in the solid fermentation system are alcoholic-yeast-like for liquid fermentation, since the process must start in Petri dishes or in culture tubes filled with BDA culture medium, which is used to grow the pure colonies of the fungus to be produced in large scale. These plates can be considered matrices for the beginning of large-scale production, or they can be pricked to Rolfs type, or similar, glass bottles filled with precooked rice sterilized in autoclave at 121 °C for 30 minutes (Leite et al. 2003).

19.3.4 Culture Medium Preparation for Large-Scale Production

Parboiled rice is firstly soaked in cold, or previously heated, water inside a container suitable for the volume to be sterilized per day. The aim is to get a rubbery consistency in cold water or in water at 70 °C, that is, 50 and 15 minutes are needed, respectively. Next, the precooked rice is placed in polypropylene bags or on autoclavable trays and autoclaved for 30 minutes at 121 °C. The substrate must cover at most half of the volume in the container in order to provide enough space and oxygen for the fungus to grow.

The medium is inoculated with conidia after cooling, and subsequently, it is incubated at room temperature in a room under positive pressure or in a biosecurity chamber. The use of autoclavable plastic bags is the mostly adopted method for fungi production.

Many other natural substrates were studied for fungi production through solid fermentation such as corn and wheat grains, wheat bran, corn meal, many rice types (1, 2, and *quirela*), and parboiled rice, among others. However, parboiled rice or type 1 rice were always the most efficient substrates, besides presenting the best cost-benefit (Ottati-de-Lima et al. 2010).

19.3.5 Inoculum Preparation

The inoculum for the small bags filled with sterilized rice can be liquid or solid. The liquid inoculum can be obtained by washing the conidia in the rice, filling the matrix bottles. The washing process must be done in water sterilized in autoclave at 121 °C for 30 minutes and cooled in biosecurity chamber under vertical airflow (Almeida et al. 2016).

19.3.6 Inoculation

Liquid inoculation happens in each small bag filled with sterile rice; 10–15 mL of this suspension is standardized with the aid of a sterile pipette. The ideal scenario is to have 10–15 mL of a suspension with 10^7–10^8 conidia/mL applied in each plastic bag (Almeida et al. 2016).

The solid solution is the simple application of one full ice cream spoon with rice + fungus taken straight from the matrix flask to each small bag. The entire inoculation process must be conducted in biosecurity chamber under vertical sterile air flow (Almeida et al. 2016)

19.3.7 Growth or Incubation

The inoculated bags filled with rice must be stored in incubator room at 24 ± 2 °C and relative humidity 60 ± 10% after inoculation. Sprinter-type air conditioners or cooling units, which are more efficient, are often used for the experiment.

Painted stainless shelves, or any other structure made of anything but wood, or a very porous material, can be used to store the polypropylene bags in the room. The room must have ceramic floor and walls covered with tiles or with epoxy painting in order to avoid the accumulation of microorganisms and dust, as well as to make cleaning easier (Almeida et al. 2016).

19.3.8 Drying Process

The bags are opened 10–15 days after incubation and their content is exposed in a room with slightly heated air flow (28 °C) for 48–120 hours to dry; the material will have 10–15% humidity. Drying can also be obtained in airflow drying tunnels at 28 °C, since they are more efficient to dry organic matter (Almeida et al. 2016).

19.3.9 Conidia Separation

Dry via conidia separation can be performed through sealed vibrating sieves or air mantle stirring and air cyclone. This equipment can be adapted for conidia extraction, although there are high conidia losses, from 5% to 10% (Leite et al. 2003).

Stirring with abrasive material such as clay and talc is another extraction option, although it will also depend on stirring or on closed systems, since it uses powder. Clay or talc extraction facilitates the formulation of dry powder or wettable powder (WP) and allows a more efficient rice-grain cleaning, depending on the fungus or isolate species, since *M. anisopliae* unleashes itself from the rice grain easier than *B. bassiana* (Almeida et al. 2007).

19.3.10 Formulation

There are formulations of the wettable powder type, although they regard a simply technical product composed of pure conidia or of ground rice mixed to conidia. Besides their low stability, these formulations lead to a wettability issue, that is, they are hard to be mixed in syrup or to be decanted on the bottom of the sprayer tank. They demand cooling in order to stabilize the propagules, although many manufacturers insist in selling the product without cooling, fact that does not assure fungi viability. The oil-based emulsifiable formulation (EO) with fungi propagules is also found in the Brazilian market. Despite its higher cost, the EO formulation enables obtaining a better tank mixture, and sometimes, it even allows product maintenance at room temperature, but never for longer than 30 days (Almeida et al. 2008).

With regard to entomopathogenic fungi, formulation is one of the greatest challenges, since advancements were accomplished in this field; therefore, any of the adopted cases still demand electricity-generated refrigeration. The development of bioinsecticides based on fungi must become an activity more and more absorbed by manufacturers, since it is an important field for technological innovation. However, along with the advantages from the formulations, cost must also be viable for pest biological control to remain as a broadly applied management technique (Almeida et al. 2008).

19.4 Entomopathogenic Fungi Production Through Liquid Fermentation

Entomopathogenic fungi production through liquid fermentation involves balancing the carbon–nitrogen (C:N) ratio in order to produce different structures of the main entomopathogenic fungi. Liquid culture media often have a carbon source, a nitrogen source, inorganic salts or organic substances (containing macro- and micronutrients), and possibly, some other growth sources (Leite et al. 2003).

The Biological Institute (*Instituto Biológico*) in Campinas, SP, Brazil, has been developing trials focused on standardizing carbon and nitrogen source substances to produce entomopathogenic fungi through liquid fermentation in order to search for conclusive outcomes in efficient and cheap culture medium. Tanzini and Batista Filho (1992) produced *Paecylomices fumosoroseus* (*I. fumosorosea*) in liquid culture media based on 20% beans, 20% beans +3% sucrose, 1% yeast extract, 1% yeast extract +3% sucrose, 10% soymilk, and 10% soymilk +3% sucrose. They found that the 20% beans macerate was the best one; it presented 97% viability and 1.41×10^8 con./mL.

Ottati-de-Lima et al. (2014) reported that the best media for *M. anisopliae* production are the ones presenting anhydrous D-glucose in their composition, whereas treatments with sucrose were more efficient for *B. bassiana* production.

Metarhizium anisopliae production advancements have been recorded all over the world, including in Brazil, since the country has a microbiological insecticide based on this fungus registered in the Agriculture, Livestock and Supply Ministry (*Ministério da Agricultura, Pecuária e Abastecimento* – MAPA) and produced through liquid fermentation. Furthermore, new research has reinforced the possibility of adopting liquid fermentation to produce and apply entomopathogenic fungi. Gotti (2016) verified that the culture medium containing albumin, sucrose, $NaNO_3$, and KCl was efficient for fungus development and led to 80% pathogenicity over the *Galleria mellonella* (L.) (Pyralidae) caterpillars used as test insect.

19.4.1 Stages of Entomopathogenic Fungi Production Through Liquid Fermentation

Similar to the production process through solid fermentation, it is necessary taking into consideration the isolation and selection of the fungi isolates to be produced, as well as the matrices – Petri dishes with purified fungus – to work as inoculum in the matrices.

19.4.2 Matrices Preparation

Liquid fermentation matrices are similar to the ones used for solid fermentation, that is, Petri dishes or culture tubes filled with BDA culture medium to grow pure colonies of the fungus to be produced (Leite et al. 2003).

19.4.3 Inoculum Preparation

The inoculum is prepared from small fractions of the liquid culture medium selected for the fungus to be produced. The containers used in the experiment can be Erlenmeyer or plastic or glass fermenting flasks, but they must allow sterilization by means of humid heat in autoclave at 121 °C for 30 minutes. These containers are inoculated with conidia of the fungus in the matrices. Subsequently, they are incubated for 2–8 days depending on the inoculum development.

19.4.4 Culture Medium Preparation for Large-Scale Production

The liquid fermentation process demands the culture medium to be prepared inside fermenters, which can have different sizes, from 1 liter to more than 1000 liters; only the sterilization system can change depending on the fermenter. In the case of glass or plastic gallons (1–20 liters), sterilization can be performed through humid heat in autoclave or through steam injection in the stainless-steel jacketing system. The ozone injection system can be used to sterilize the culture medium.

19.4.5 Inoculation

Inoculation is done by means of inoculum injection in the fermenter. This process poses high contamination risk; therefore, it is necessary following the flaming procedures and adopting the correct system to inject the inoculum in the fermenter (Almeida et al. 2016; Gotti 2016).

19.4.6 Incubation

Incubation changes according to fermenter size and its aeration system. The ideal is to have metal fins agitation and aeration via forced injection of compressed and filtered air in order to achieve constant culture medium aeration with fungus inoculum. Incubation time in the fermenter may last from 48 hours to 10 days, depending on the operational system, because the more constant the temperature (from 26 °C to 28 °C) and the aeration via agitation, the faster the fungus development (Almeida et al. 2016; Gotti 2016).

19.4.7 Formulation

Fungi harvest (after their growth) can be conducted through pipelines linked to the industrial fermenters or even through specific taps. It is already possible to find in Brazil a product registered in the Agriculture, Livestock and Supply Ministry (*Ministério da Agricultura, Pecuária e Abastecimento* – MAPA) based on *M. anisopliae* in liquid fermentation, which basically consists of culture medium and fungus with a stabilizing agent. According to Gotti (2016), blastospores are the infective structures.

It is possible to drain the culture from the fermenter to a continuous centrifuge to separate the blastospores, hyphal bodies, and/or conidia submerged in the culture medium, to cold drying and to produce formulations from this material, such as microsclerotia with clay or diatomaceous earth, or with other inert materials (Jackson and Jackson 2009).

19.5 Surface Cultivation

The surface cultivation system is old, and the fungus is produced on the surface of the liquid culture medium (incubated without agitation) inside special containers that enable maximizing the surface area of the culture medium. Kybal and Vlcek (1976) used polyethylene tubes (30 cm diameter) sealed in both ends and partially filled with nutritious broth (0.8% peptone +1% sorbitol) to produce *B. bassiana*. Sterile air was constantly pumped inside the tubes aiming at fungus oxygenation and gas exchange during incubation. Mycelial growth happens inside the liquid medium, whereas conidiogenesis takes place on its surface. The liquid is drained after incubation and the conidia are harvested. Samsináková et al. (1981) used this system 12 days after incubation and harvested 10^9 *B. bassiana* conidia per cm^2 from the medium surface.

Batista Filho et al. (1985) assessed the surface production of *B. bassiana* in liquid media and found that beans broth produced more conidia per cm^2 than other culture media.

Based on other methodologies, the fungus species *B. bassiana* and *M. anisopliae* were produced inside glass flasks partially filled with liquid medium, which was prepared with "guandú" beans grains (20%). After fungus inoculation, the flasks were subjected to hydrophobic cotton buffer and incubated at 27 °C. Fungus production reached the magnitude 7.5 × 10^9 and 6.25 × 10^9 conidia/cm^2, respectively (Alvarenga et al. 1988).

Despite the studies about the surface production of entomopathogenic fungi in liquid medium, this technique is not used in great industrial-scale since it demands more care against contamination and space for fungi development. Furthermore, it does not present the desired economic features.

19.6 Entomopathogenic Fungi Production Through Two-Phase System

Fungi production by the two-phase process associates the benefit from the high production of biomass resulting from cultivation inside liquid medium with the production of stable and hydrophobic conidia (lipophilics) in solid medium. This system is adopted to produce *Metarhizium flavoviride* (Gams and Roszypal) for grasshoppers' control in Western Africa. Initially, the fungus is produced in a simple liquid medium based on brewer's yeast + sucrose inside a flask, which is kept in a shaker. After 2–4 days, the produced biomass is inoculated in precooked rice and autoclaved. After being colonized by the fungus, the rice is transferred to plastic bowls (1 kg of rice/bowl). The bowls are unstacked after 10 days and kept in a dehumidified room for 48 hours in order to allow the fungus to dry. Conidia are extracted from the rice with the aid of a vacuum device, and subsequently, they are dried at 5% humidity inside a desiccator. Conidia can be stored as dry powder for 1–2 years in the freezer (Leite et al. 2003).

The two-phase process can also use inert substrates as the basis for fungus surface growth, as it was already presented. *Metarhizium rileyi* (Farlow) (Kepler, S.A. Rehner and Humber) production can be performed by using a medium composed of saboraud, maltose, and yeast extract to obtain biomass (which is inoculated in sugarcane bagasse) – approximately 8.7 × 10^8 conidia/5 g of medium (Leite et al. 2003).

The two-phase system is also little used, because it demands two production systems, fact that does not mean high investments unless the liquid production is only adopted for the inoculum of the plastic bags. This procedure needs smaller-sized equipment for production.

19.7 Quality Control

Quality control is essential and must start at isolate selection and featuring by means of molecular characterization, power rating, enzymes production, conidia production, and other parameters often set by research centers.

The most common tests conducted with bioinsecticides of fungal origin are conidia concentration and viability, purity, and virulence.

Virulence, which is characterized by Medium Lethal Dose CL_{50} and Mean Lethal Time TL_{50}, is often set by research centers. It is recommended to pass the agent on the natural host after 10 pathogens pricking in artificial medium in order to keep fungus virulence in a bioinsecticide industry. The fungus must be divided in the largest number of samples (matrices) as possible after it is passed on the host in order to have its total availability to start bioinsecticide production and to minimize the number of pricking in the artificial medium. The fungus must be stored under appropriate conditions such as cooling in artificial medium or passed on the natural host, whenever necessary.

19.8 Perspectives

The entomopathogenic fungi industry is a reality due to its annual growth resulting from the cane rootworm and spittlebugs, which demand biological control. The demand for such control generates job positions, more income, and boosts research in the entomopathogen field, besides its influence on the market of parasitoids and predators.

There are challenges to be faced such as improving the production methods to decrease costs and formulation facilitation, and the development of new fungi to control other pests and formulations. Register and trading are also challenges the Brazilian industry of entomopathogenic fungi needs to deal with. However, thanks to the incentive of businessmen committed to the environment and to the competence of the Brazilian research in this field, the bioinsecticide industry based on entomopathogenic fungi tends to grow and to gain new markets worldwide.

The industry of entomopathogenic bacteria decayed in the 1990s and in the first decade of the twenty-first century due to the emergence of new caterpillar species and to the changes from secondary to primary species such as the complex *Plusia*. However, importation became faster and easier to register, since there were registers of products based on *B. thuringiensis*.

References

Almeida JEM, Batista Filho A (2006) Controle biológico da cigarrinha-da-raiz da cana-de-açúcar com o fungo *Metarhizium anisopliae*. Bol Téc/Inst Biol 16:1–19

Almeida JEM, Batista Filho A, Alves SB et al (2008) Formulação de entomopatógenos na América Latina. In: Alves SB, Lopes RB (eds) Controle microbiano de pragas da América Latina. FEALQ, Piracicaba, pp 257–277

Almeida JEM, Batista Filho A, Leite LG (2016) Curso de controle microbiano de insetos – Fungos entomopatogênicos. Instituto Biológico, São Paulo

Almeida JEM, Rocha TC, Batista Filho A (2007) Desenvolvimento de método para extração física de conídios de *Metarhizium anisopliae* e *Beauveria bassiana* para formulação pó seco e molhável de bioinseticidas. Arq Inst Biol 74(4):369–371

Alvarenga ARM, Cruz BPB, Oliveira DA et al (1988) Novos testes de cultivo de fungos utilizados em controle biológico usando meios de cultura naturais líquidos. Arq Inst Biol 55(1/4):31–35

LFA A, PMJO N, Faria MR (eds) (2010) Recomendações para utilização de fungos entomopatogênicos no controle de pragas. CP 2: Rede Entomofungo, Piracicaba

Alves SB (1998) Controle microbiano de insetos. FEALQ, Piracicaba

Alves SB, Lopes RB (eds) (2008) Controle microbiano de pragas na América Latina. FEALQ, Piracicaba

Barros B, Ferreira C (2014) Microorganismos em alta no campo. Valor Econômico. http://www.valor.com.br/agro/3708192/microorganismos-em-alta-no-campo#ixzz3EF3P4QKs. Accessed 15 May 2017

Batista Filho A, Alves LFA, Almeida JEM et al (2006) Controle biológico de insetos e ácaros. Bol Téc/Inst Biol 15:1–85

Batista Filho A, Cruz BPB, Camargo LMPC et al (1985) Crescimento de *Beauveria* sp., isolado de bicudo do algodoeiro (*Anthonomus grandis* Boheman), em meios de cultura naturais, líquidos. O Biológico 51(1):17–21

Gotti IA (2016) Desenvolvimento de meio de cultura para a produção de *Metarhizium anisopliae* (Metsch.) Sorok. por fermentação líquida. Dissertação, Instituto Biológico de São Paulo

Jackson MA, Jackson ST (2009) Production of microesclerotia of the fungal entomopatogen *Metarhizium anisopliae* and their potential for use as a biocontrol agent for soil-inhabiting insects. Mycol Res 113(8):842–850

Kybal J, Vlcek V (1976) A simple device for stationary cultivation of microorganisms. Biotechnol Bioeng 18(12):1713–1718

Leite LG, Batista Filho A, Almeida JEM et al (eds) (2003) Produção de fungos entomopatogênicos. Sene, Ribeirão Preto

Michereff Filho M, Faria M, Whraight SP et al (2009) Micoinseticidas e micoacaricidas no Brasil: Como estamos após quatro décadas? Arq Inst Biol 76(4):769–779

Micothon (2012) Global biopesticides market to reach US$ 3,4 billion by 2017. http://micothon.nl/2012/03/global-biopesticides-market-to-reach-us3-4-billion-by-2017/. Accessed 15 May 2017

Ottati-de-Lima EL, Batista Filho A, Almeida JEM (2010) Produção semissólida de *Metarhizium anisopliae* e *Beauveria bassiana* em diferentes substratos e efeito da radiação ultravioleta e da temperatura sobre propágulos desses entomopatógenos. Arq Inst Biol 77(4):651–659

Ottati-de-Lima EL, Batista Filho A, Almeida JEM (2014) Liquid production of entomopatogenic fungi and ultraviolet radiation and temperature effects on produced propagules. Arq Inst Biol 81(4):342–350

Samsináková A, Kálalová S, Vlcek V et al (1981) Mass production of *Beauveria bassiana* for regulation for *Leptinotarsa decemlineata* populations. J Invertebr Pathol 38:169–174

Tanzini MR, Batista Filho A (1992) Produção de *Paecilomyces fumosoroseus* (Wise) Brown e Smith, em diferentes meios de cultura líquidos. Ecossistema 18:149–155

Vazquez LL, Pérez N (2017) El control biologico integrado al manejo territorial de plagas de insectos em Cuba. Agroecologia 12(1):39–46

Chapter 20
Entomophogenic Viruses

Fernando Hercos Valicente

20.1 Introduction

Although detailed descriptions of commercial production systems for baculovirus are scarce (Grzywacz et al. 2014), to achieve a successful in vivo baculovirus large-scale production, some important factors have to be taken into consideration. In general terms, baculovirus production can be divided into three parts:

1. Mass production of the healthy host.
2. Infection of the healthy host with baculovirus.
3. Harvest and/or storage and/or formulation of the infected host after its death.

These three steps may or may not be taken at the same time, depending on the facility production. It also depends on the production goal of the industry, that is, if it is a small-, medium-, or large-scale production system. Baculovirus production is usually used for many species of Lepidoptera. To assure aseptic conditions, the room used for rearing healthy hosts needs to be separated from the infection and incubation rooms. There is no standard facility design; however, these rooms must be physically separated. All materials used in this room must be sterilized and kept in the room. Separating the host artificial rearing is a very important caution in the larvae production process, since it avoids undesired contamination of the healthy insects and of the colony. UV (ultraviolet) light bulbs may be used in every room. Although this irradiation is only effective for surface sterilization, this method may help keep the rooms clean. In case of colony contamination, the colony must be restarted with healthy larvae and sterilization of the room. If the necessary measures are not taken, the yield in this step of the process will be compromised. To achieve a good mass production of baculovirus-based biopesticides for commercial purposes, some important factors of the healthy host and cannibalistic behavior, as well as laboratory conditions, should be taken into consideration.

F. H. Valicente (✉)
Embrapa Maize and Sorghum, Sete Lagoas, MG, Brazil
e-mail: fernando.valicente@embrapa.br

B. Souza et al. (eds.), *Natural Enemies of Insect Pests in Neotropical Agroecosystems*, https://doi.org/10.1007/978-3-030-24733-1_20

This chapter details the mass production of baculovirus using healthy hosts. These hosts present both cannibalistic and noncannibalistic behaviors. For this experiment, we infected healthy hosts with baculovirus and inoculated a large number of hosts with the baculovirus. It is worth mentioning that the use of an alternative host and of harvesting and storing dead hosts is essential. We also list the equipment and materials needed in a baculovirus biofactory, the formulation, and quality control.

20.2 Mass Production of the Healthy Host

Mass production of the healthy host is usually done using artificial diets and can be divided into two categories:

1. Hosts with cannibalistic behavior.
2. Hosts with noncannibalistic behavior.

If hosts, such as fall armyworm, *Spodoptera frugiperda* (J. E. Smith) (Noctuidae), and cotton bollworm, *Helicoverpa armigera* (Hübner) (Noctuidae), show a cannibalistic behavior, these larvae must be individualized at a specific age, in an acclimatized rearing room, as shown in Fig. 20.1.

When multiplying baculovirus for cannibalistic larvae in a large-scale production system, alternative hosts may be used. However, the baculovirus stock for large-scale production must be created in the original host. The mass production of the original and/or alternative host occurs artificially in acclimatized rearing rooms, under controlled and aseptic conditions (Fig. 20.2).

The room conditions in larvae-rearing laboratories must be controlled and monitored to ensure the uniformity of the host's size and weight prior to the infection. Three factors are the most important in a baculovirus large-scale production system: age of the larvae, temperature of the incubation room, and concentration of the baculovirus used to infect the larvae. Among the room conditions, temperature is one of the most important factors during the host production process. The ideal rearing and incubation temperature, for example, for the fall armyworm, is 25 °C ± 2 °C. Air conditioners and air circulators are used to maintain the room temperature. The air circulators help in the process of air homogenization inside the room, since the temperature in the superior part of the room is always higher than the inferior part. Another important factor is the uniformity of the caterpillars that will be infected so that their mortality may occur on the same day. This makes the management and use of both the rearing room and the incubation room easier. The relative air humidity must be maintained around 70% ± 10% to prevent the epidermis of the insects from going dry and contaminants from proliferating in the local area. The materials used for the artificial rearing of the healthy larvae are inexpensive, such as PVC cages for adults and plastic pots with plastic lids for larvae with artificial diet.

PVC cages are used for the oviposition of the adults (moths) of the fall army-

Fig. 20.1 Individualized cannibalistic larvae in an artificial rearing laboratory at IMAmt/Brazil. (Photo: Fernando H. Valicente)

Fig. 20.2 The mass production of the original and/or alternative host occurs artificially in acclimatized rearing rooms, under controlled and aseptic conditions. IMAmt/Brazil. (Photo: Fernando H. Valicente)

worm or any other lepidopterans. The cages should be maintained in ventilated areas, if possible with access to some sunlight (no sunlight directly onto the cages); however, we observed in our laboratory that when there is indirect sunlight into the room, the amount of eggs is much higher than when artificial lights are used (personal observation). Sheets of paper or napkins should be hanging on the sides of the cages in order to guarantee a higher amount of eggs and a lower mortality rate of the adults. The adults can be fed with a mixture of water, honey or sugar, and ascorbic acid (10%). The adults mate and lay eggs on the hanging sheets of paper. All the egg clusters are cut and each egg cluster is individually put inside a 50 mL plastic container containing artificial diet. Cannibalism among some larvae when they are 7 days old or third instar must be accounted for. At this moment, they must undergo an individualization process, in which the caterpillars are individually transferred to a plastic recipient that contains approximately 5 g of artificial diet. This amount of diet is enough for the caterpillars to grow into the pupal stage. As of the pupal stage, the cycle restarts with the adults. If larvae will be infected with baculovirus, 1/3 of the artificial diet may be used and it is sufficient for infection and death of the host.

When using the alternative host, the rearing follows the same standards; however, the caterpillars do not need to be kept individually at 7 days of age. In this case, they can remain together during the contamination and death caused by baculovirus.

20.3 Infection of the Healthy Host with the Biological Control Agent

One of the most important steps of the baculovirus production process is the infection of healthy hosts with the baculovirus. The baculovirus stock solution should be purified at first, because the stock of unpurified baculovirus should not be used in several generations in the large-scale production. Purification or semi-purification of the baculovirus is recommended at every 4–5 generations. The purification is made with a sucrose gradient (Valicente 1989). Hygienization and temperature control in the rearing room of the infected host are important, because, at this point, any chance of infection may compromise the entire batch being produced. The same caution should be considered when cleaning and controlling the physical area, as described in the previous item.

Healthy larvae can be inoculated in three different ways:

1. Inoculation of the baculovirus in an individual host.
2. Inoculation of the baculovirus in a large number of hosts.
3. Use of an alternative host.

20.4 Inoculation of the Baculovirus in Individual Caterpillars

The caterpillars used in the large-scale production must have uniform sizes and weights, preferably when they are, for example, 7 days old for fall armyworm, or third instar or fourth instar for soybean looper, in order to have a uniform and efficient infection. Our results confirm that the best temperature for artificially rearing most larvae (all of them being insect pests of major crops) and then infecting them with the baculovirus is 25 °C ± 2 °C.

In order to maximize the production system management, it is necessary to take notes of the dates, in order to monitor the ideal infection point starting from the day that the eggs hatch. The eggs must remain in a place with controlled temperature. Depending on the size of the biofactory and the necessity, eggs may be stored in a fridge at 4 °C, in order to postpone the hatching of the larvae. The caterpillars which ideal infection point was monitored produced a higher ratio of polyhedron/caterpillar, increasing the pib yield and lowering the production cost.

Caterpillars may be exposed to the baculovirus for a 24- to 48-hour period with contaminated artificial diet and/or leaves (Fig. 20.3).

The baculovirus suspension is sprayed on the leaves or artificial diet. However, the leaves should be previously washed with sodium hypochlorite (0.5%) and then washed with distilled water. The concentration of the baculovirus solution must be between 2×10^6 and 2×10^7 polyhedra/mL. If older caterpillars (8 days of age or third instar) are necessary, increasing the concentration of the baculovirus is recommended to assure the host's death. However, even using larger caterpillars, the rearing laboratory and the incubation room should remain at 25 °C ± 2 °C.

A B

Fig. 20.3 Caterpillars exposed to baculovirus in contaminated artificial diet (**a**) and leaves (**b**). (Photos: Fernando H. Valicente)

20.5 Inoculation of the Baculovirus in a Large/Massive Number of Caterpillars

This method is recommended because it provides the highest yield of dead caterpillars in regard to the production process. The healthy caterpillars are infected in groups that may vary from 100 to 600 caterpillars at once and not individually. In this process, the healthy caterpillars are exposed to corn leaves and/or artificial diet contaminated with baculovirus, as described previously, for a period of 24–48 hours, in larger plastic containers (Fig. 20.4). This process may also be automatic, in which the caterpillars and the diet contaminated with baculovirus are placed together in plastic recipients (Fig. 20.5). After the contamination period, the larvae may or may not be individualized into smaller plastic recipients. The advantage of using castor bean leaves (*Ricinus communis* L.), previously washed with sodium hypochlorite 0.5%, is the great reduction of cannibalism that occurs among caterpillars (Valicente et al. 2013). During the whole process, the temperature must remain at 25 °C ± 2 °C.

A B

Fig. 20.4 Larvae being infected with baculovirus using artificial diet (**a**) and leaves (**b**). (Photos: Fernando H. Valicente)

Fig. 20.5 Automated process where the artificial diet and the larvae are put into plastic recipients. (Photo: Fernando H. Valicente)

20.6 Use of the Alternative Host

The main objective of using an alternative host for multiplying the baculovirus is to avoid the cannibalism that occurs with caterpillars that have a cannibalistic behavior, such as fall armyworm, *S. frugiperda*, and cotton bollworm, *H. armigera*. Thus, a high yield in the large-scale production process is expected. The choice of the host should follow some steps. The alternative host cannot be cannibal and has to replicate the baculovirus like the original host. Our results indicate some excellent alternative hosts, such as *Spodoptera cosmioides* Walker (Noctuidae). When an alternative host is used, 50–600 caterpillars are inoculated into plastic containers and remain together until they die (Valicente et al. 2013). However, one must consider that most noncannibalistic caterpillars, when infected with baculovirus, change their behavior and become cannibals. If they stay together in the same plastic recipient during the incubation period, an average of cannibalism is observed (personal observation).

The cautions to avoid contamination must be the same as described before, and the incubation temperature must be kept around 25/26 °C. However, the entire baculovirus supply from the biofactory should be multiplied in *S. frugiperda* larvae. Only the first generation should be multiplied using *S. cosmioides*. This avoids a specificity of the baculovirus towards the alternative host. The baculovirus production should abide by the flowchart in Fig. 20.6.

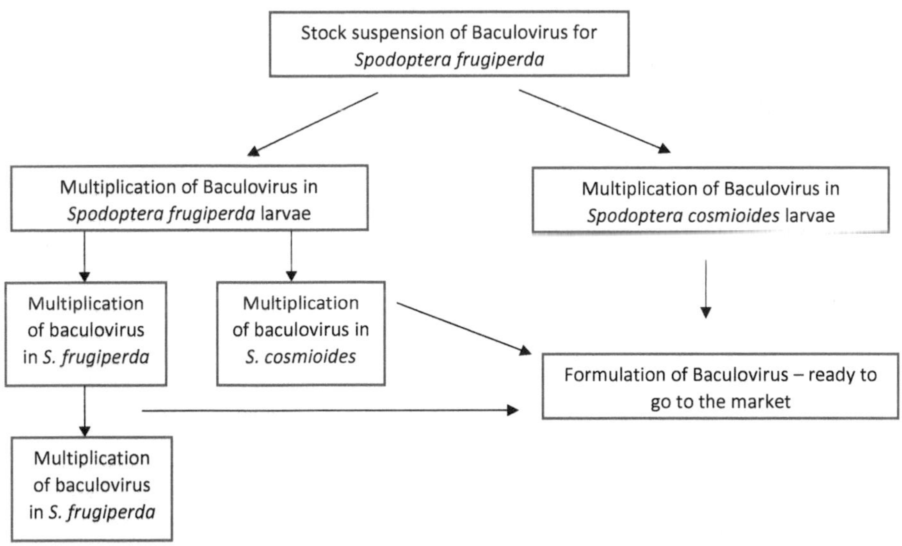

Fig. 20.6 Flowchart of the multiplication of the *Baculovirus spodoptera* in the original host, *Spodoptera frugiperda*, as well as in the alternative noncannibalistic host, *Spodoptera cosmioides*. (Figure: Fernando H. Valicente)

As displayed in the flowchart above, baculovirus should be multiplied only once in the alternative host and then be processed, formulated, and delivered to the market. One of the great advantages of using the alternative host is the better use of the laboratory space because this host is not cannibal and does not need individualization.

20.7 Harvesting and Storage of the Host After Its Death

The limiting factor for harvesting and storing of *baculovirus* when produced in large scale is the liquefaction of the caterpillar's tegument immediately after its death (Fig. 20.7a). The dead larvae liquefy due to tegument rupture, resulting in leakage of all the internal liquids, making the operation for gathering and storing baculovirus more difficult.

There is a new baculovirus that is effective against fall armyworm and presents a unique characteristic among the baculoviruses. This isolate does not cause the rupture of the insect's tegument immediately after its death (Fig. 20.7b) (Valicente et al. 2007; Vieira et al. 2012). This is a very important characteristic to reduce labor and production costs during the process of harvesting the infected larvae.

When this isolate is used, dead caterpillars with typical symptoms of baculovirus are collected using forceps/tweezers and put in plastic recipients. At this point of any production system, the dead caterpillars can be destined for processing and formulation or be frozen, preferably at −20 °C, for further processing. This factor depends on time and staff availability and on the output flow/manufacturing flow of the biofactory. The dead caterpillars infected with baculovirus remain viable for years when frozen at −20 °C.

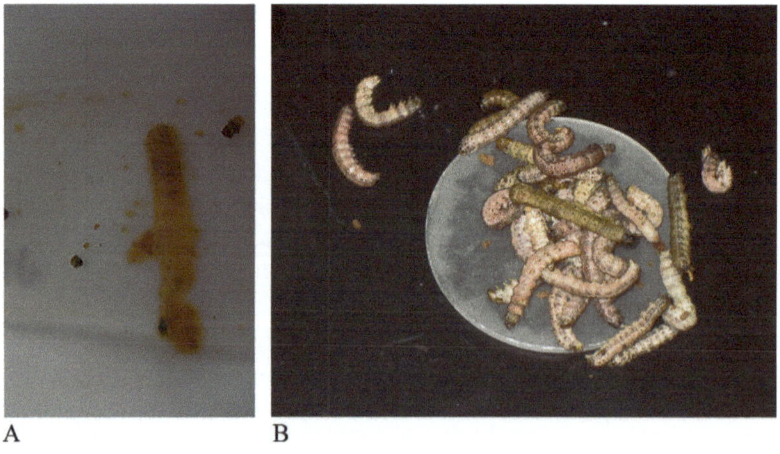

Fig. 20.7 *Spodoptera frugiperda* larvae dead by *Baculovirus spodoptera*. (**a**) Larva dead by the isolate that causes integument liquefaction; (**b**) infected and dead larvae by an isolate that does not cause immediate liquefaction of the integument. (Photos: Fernando H. Valicente)

20.8 Equipment and Materials Needed in a Baculovirus Biofactory

Some basic equipment is necessary for a small- to large-scale production laboratory. There should be a room with an autoclave for sterilizing the equipment, air conditioners and a digital scale, and a specific room for the preparation of the artificial diet. A laminar flow or hood is also recommended to use UV light for sterilizing the surface of some materials and ingredients, if necessary (Fig. 20.8). Some useful information on production and layout of an insectary may be found in Shapiro (1986), Shieh (1989), Hunter-Fujita et al. (1998), Grzywacz et al. (2004), and Grzywacz and Moore (2017). Some authors give valuable information on production systems (Van Beek and Davis 2009; Grzywacz et al. 2014).

20.9 Formulation and Quality Control

Formulation and quality control are key factors in the production system, and it is important to know the origin of the materials and reagents to be used in the production and formulation, the quality of the inoculum, and the quantification of the occlusion bodies. It is also important to check the activity of the baculovirus produced, the contaminants present in biological pesticide, as well as shelf life, that is, the longest period of time a product survives during storage, or the time between manufacture and use in the field (Grzywacz 2017). According to Jones and Burges (1998), viruses must remain viable during the shelf life, with minimal loss of their virulence and retention of formulation properties. These authors also suggest a minimum shelf life of 18 months.

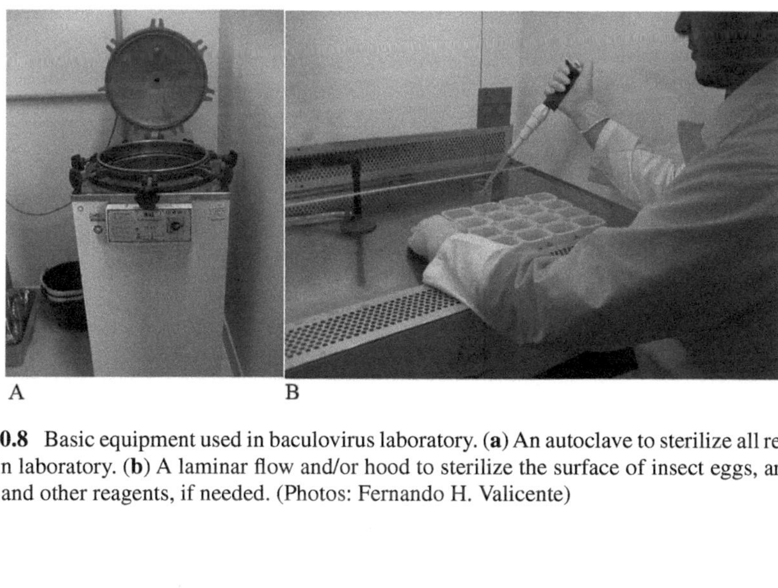

A B

Fig. 20.8 Basic equipment used in baculovirus laboratory. (**a**) An autoclave to sterilize all reagents used in laboratory. (**b**) A laminar flow and/or hood to sterilize the surface of insect eggs, artificial diets, and other reagents, if needed. (Photos: Fernando H. Valicente)

20.10 Final Remarks

To achieve a successful in vivo baculovirus large-scale production, some important factors have to be taken into consideration. Factors such as the healthy host and cannibalistic behavior, as well as laboratory conditions, should be considered. The room conditions in larvae-rearing laboratories must be controlled and monitored to assure the uniformity of the host size and weight prior to the infection. Three factors are the most important in a baculovirus large-scale production system: age of the larvae, temperature of the incubation room, and concentration of baculovirus used to infect larvae. Good formulation and quality control are also important as a final step.

References

Grzywacz D (2017) Basic and applied research: baculovirus. In: Lacey LA (ed) Microbial control of insect and mite pests: from theory to practice. Elsevier, Amsterdam, pp 27–46

Grzywacz D, Moore S (2017) Production, formulation, and bioassay of baculoviruses for pest control. In: Lacey LA (ed) Microbial control of insect and mite pests from theory to practice. Academic Press, Amsterdam, pp 109–124

Grzywacz D, Rabindra RJ, Brown M et al (2004) *Helicoverpa armigera* nucleopolyhedrovirus production manual. FAO, Genebra

Grzywacz D, Moore D, Rabindra RJ (2014) Mass production of entomopathogens in less industrialized countries. In: Morales-Ramos JA, Rojas MG, Shapiro-Ilan DI (eds) Mass production of beneficial organisms. Elsevier, Amsterdam, pp 519–553

Hunter-Fujita FR, Entwistle PF, Evans HF et al (1998) Insect viruses and pest management. Wiley, Chichester

Jones KA, Burges HD (1998) Technology of formulation and application. In: Burges HD (ed) Formulation of microbial biopesticides. Kluwer Academic Publishers, Dordrecht, pp 2–30

Shapiro M (1986) *In vivo* production of baculoviruses. In: Granados RR, Federici BA (eds) The biology of baculoviruses volume II: practical application for insect control. CRC Press, Boca Raton, pp 32–61

Shieh TR (1989) Industrial production of viral pesticides. Adv Virus Res 36:315–343

Valicente FH (1989) Levantamento dos inimigos naturais de *Spodoptera frugiperda* (J. E. Smith) (Lepidoptera: Noctuidae) em diferentes regiões do estado de Minas Gerais. An Soc Entomol Bras 18(1):119–130

Valicente FH, Tuelher ES, Pena RC et al (2007) The use of Baculovirus to control fall armyworm, *Spodoptera frugiperda*, in Brazil. In: Proceedings of the 16th international plant protection congress, British Crop Protection Council, Alton, 15–18 Oct 2007

Valicente FH, Tuelher ES, Pena RC et al (2013) Cannibalism and virus production in *Spodoptera frugiperda* (J.E. Smith) (Lepidoptera: Noctuidae) larvae fed with two leaf substrates inoculated with *Baculovirus spodoptera*. Neotrop Entomol 42(2):191–199

Van Beek N, Davis DC (2009) Baculovirus production in insect larvae. In: Murhammer DW (ed) Baculovirus and insect cell expression protocols. Humana Press, Towata, pp 367–378

Vieira CM, Tuelher ES, Valicente FH et al (2012) Characterization of a *Spodoptera frugiperda* multiple nucleopolyhedrovirus isolate that does not liquefy the integument of infected larvae. J Invertebr Pathol 111:189–192

Chapter 21
Bacillus thuringiensis

María Elena Márquez Gutiérrez, Deise Maria Fontana Capalbo, Regina de Oliveira Arruda, and Rodrigo de Oliveira Moraes

21.1 Introduction

There is a worldwide concern regarding the indiscriminate use of pesticides and their negative impacts, showing the need to develop environmentally safer methods for pest control. Even developing countries are interested in the implementation of integrated pest management (IPM) programs. As part of such programs, the use of microorganisms has shown some remarkable successes, including bacteria, fungi, and viruses as commercial products (Capalbo et al. 2008; Moraes et al. 2008a, b). Among the microorganisms, many spore-forming bacteria stands out as they are a source of distinct promissory molecules with antibiotic or insecticidal activities. Their main characteristic are explored for use as biopesticides. The endospore is resistant to stressing environmental conditions, but it will germinate again under specific circumstances. Those same characteristics of resistance are also important for industrial production.

Most of these spore-forming bacteria belong to the *Bacillaceae* group, including the *Bacillus*, *Sporolactobacillus*, *Clostridium*, *Desulfotomaculum*, and *Sporosarcina* genera (Moraes et al. 2001). The aerobic *Bacillus* genera is the most important for its bioinsecticide activity. The insecticides based on entomopathogenous are most

M. E. M. Gutiérrez (✉)
Havana University, Havana, Cuba

D. M. F. Capalbo
Embrapa Environment, Jaguariúna, SP, Brazil

R. de Oliveira Arruda
Probiom Technology Ltd, Campinas, Brazil

Universidade Guarulhos, Research and Development Department, Guarulhos, SP, Brazil

R. de Oliveira Moraes
Probiom Tecnologia - Pesquisa e Desenvolvimento Experimental em Ciências Físicas e Naturais Ltda., Research and Development Department, Campinas, SP, Brazil

B. Souza et al. (eds.), *Natural Enemies of Insect Pests in Neotropical Agroecosystems*, https://doi.org/10.1007/978-3-030-24733-1_21

often very specific and show low or no toxicity to vertebrates and beneficial insects (Capalbo et al. 2008). The maintenance of the pathogenicity of the strain is generally of vital importance, particularly for those pathogens that lose their virulence with the constant replications of the in vitro culture (Moraes 1973). The variability, mutability, and sometimes the ease of genetic manipulation are some advantages of insecticides based on entomopathogens compared to chemicals. The genetic manipulation of pathogens, promoting new potentially more active strains and the discovery of new species present more promising prospects in the area. At the same time, the development of new methods of preservation of microorganisms favors the maintenance of banks of this material for future researches and improvements. The preservation of the microorganisms of interest must be carried out through deposit in a recognized Collection of Cultures, such as The André Tosello Foundation that has the Tropical Cultures Collection, which meets the requirements and is registered in the World Federation Culture Collections (WFCC), as present in its online catalog: www.fat.org.br (Moraes et al. 2011, 2012).

Until the end of the twentieth century, the use of bioinsecticides reached 1–2% of the total insecticidal market (plant protection uses), but with the comprehension of the mode of action of *Bacillus thuringiensis* (Bt), the active ingredient of the most widely used biopesticide in the world was favored by the expansion of biological control methods, even in the area of new biotechnology and the use of Bt-plants (genetically modified plants expressing the Bt toxin). A recent increase in consumption of Bt-products is also attributed to novelties in the formulation and production processes, which in turn provide economic advantages.The aim of this chapter is to report the processes and parameters for the fermentation of Bt by industrial technologies together with other aspects of the formulation, quality control, and commercialization.

21.2 Commercial Production

Firstly, it was the fermentation companies that, in the search for new markets, embarked on the study of Bt production, which was viable for in vitro growth. Next, the chemical industries, already established in the production and sale of insecticides, showed an interest due to the potentiality of the bacterial insecticide represented in the production facility, viability, and effectiveness for insect control. As opposed to most chemical pesticides, the active ingredient of bacteria-based products is obtained directly from living organisms, which implies different stages of production and use than those routinely observed in chemical products (Bryant 1994; Moraes et al. 2008b). For the commercial production of microorganisms, there is a need to select a line well adapted to the fermentation process and variations are required in order to maximize production and perform growth under economical fermentation conditions. Regardless of the source of the lineage – isolated from natural sources such as soil or insects, or obtained by genetic manipulation or acquired from a collection of cultures – all new "materials" should follow optimization steps before they can be used in large commercial scale fermentations. These steps involve lineage stability and optimum production conditions.

Conventional Bt products have been targeted primarily against lepidopteran pests of agricultural and forestry crops; however, in recent years, *Bt* strains active against coleopteran pests have also been marketed. Strains of *Bt* active against dipteran vectors of parasiticand viral diseases are being used in public health programs. Commercial *Bt* formulations may be applied on foliage, in soil, in aquatic environments, or in food storage facilities. After the application of *Bt* to an ecosystem, the vegetative cells and spores may persist at concentrations that decrease gradually for weeks, months, or years, becoming a component of the natural microflora (Boucias and Pendland 1998).

Cuba has dedicated more than 20 years to the integration of chemical pesticides with biologicals as part of pest management programs, and there is a great experience in the massive production of Bt, which allows to protect, together with other biological control agents, more than one million hectares per year of crops of economic importance. There is a production infrastructure currently formed by 205 entomopathogens and entomophagues production centers (CREE) spread throughout the country and four plants of industrial production of biopesticides, in the west and center (Centro Nacional de Sanidad Vegetal 2016). Four lines of Bt bioproducts are obtained: THURISAVE 13, THURISAVE 24, THURISAVE 21, and THURISAVE 26 from stable and previously characterized strains. The results achieved mean that at present the demand for these products is much higher than the productive capacity of the country. Inundative applications made for preventive action (Murguido and Elizondo 2007) mainly against *Plutella xylostella* (L.) (Plutellidae) in cabbage and watercress; *Spodoptera frugiperda* (J.E. Smith) (Noctuidae) in corn, rice, and sorghum; *Trichoplusia ni* (Hübner) (Noctuidae), *Helicoverpa zea* (Boddie) (Noctuidae) in vegetables, roots and tubers, and cicer; *Diaphania* spp. (Crambidae) in cucurbitaceous crops; *Erinnyis ello* (L.) (Sphingidae); *Manduca sexta* (L.) (Sphingidae); *Herse cingulata* (Fabricius) (Sphingidae) in manihot, tobacco, papaya, and sweet potato; *Mocis latipes* (Guenée) (Noctuidae) in grasses and sugar cane; *Heliothis virescens* (Fabricius) (Noctuidae) in tobacco; *Mythimna unipuncta* Haworth (Noctuidae) in sugar cane; *Davara caricae* (Dyar) (Pyralidae) in papaya; *Omiodes indicata* (Fabricius) (Pyralidae) in beans; *Phyllocnistis citrella* Stainton (Gracillariidae) in citrus; and *Liriomyza trifolii* (Burgess) (Agromyzidae) in vegetables, grains, and root and tubers (Jiménez 1974; Jiménez and Vázquez 1990; Vázquez et al. 2010).

21.3 Fermentative Process

Fermentation processes involve several steps for obtaining large amounts of cells and/or their metabolites (Moraes et al. 1991, 2011). For Bt, the most common form is the submerged process in which a liquid nutrient medium is used to suspend and propagate the bacterial biomass (Moraes and Capalbo 1986; Moraes et al. 2018).

Each strain is kept as a stock culture in freeze-dried form or preserved in ultra-freezer at −80 °C. Whenever necessary, it is rehydrated and a small amount of this material is aseptically transferred to a propagation medium that could be an agar plate or a liquid one. This first stage of propagation (seed culture, seed, or starter) is used for purity tests that include, for example, serotyping, sensitivity to bacteriophages, observation of the presence of contaminants, and confirmation of insecticidal activity. A small aliquot of this parent culture is then transferred to a flask (up to 1 liter), which after a multiplication period under controlled conditions will be the inoculum of larger reactors (5 L, 10 L, 50 L, and more) until the desired final volume is reached. The volume of inoculum varies from 1% to 10% of the final volume according to other growing conditions. Aseptic conditions throughout the procedure should be observed to avoid contamination by other microorganisms (Moraes and Capalbo 1986; Capalbo et al. 2008; Moraes et al. 2018).

The fermentation parameters must be carefully monitored for safety in terms of progress and maintenance of productivity. During the growth period, the broth is constantly monitored for its physical and microbiological properties till the final desired volume is reached, which will reach up to 10,000 liters or more per reactor (Bernhard and Utz 1993). In the final stage of fermentation, the nutrient sources become limiting, which promotes the sporulation stage. Thus, a large number of crystals and spores are the final product, when the fermentation is concluded. The final broth of the reactor will contain a suspension of cells remaining in the process, small cell fragments (resulting from lysis of the cell wall), in addition to spores and crystals, which predominate (Moraes et al. 2001).

21.4 Fermentation Parameters

Growth Requirements All commercial products containing Bt produced until the end of the twentieth century were obtained by submerged fermentation, varying only the form of recovery of the spores and the final formulation. One of the keys to a successful commercial production of a bacterial insecticide is the development of the culture medium. The composition of the culture medium for fermentation should count basically on carbon, nitrogen, and traces of minerals.

Carbon, Nitrogen, and Minerals Fermentations involving Bt and most of the other bacterial biopesticides need high levels of carbon, nitrogen, and oxygen because:

(a) Aerobic microorganisms, such as Bt, present high air demand in the initial stages of fermentation. It is almost impossible to saturate with air broths of these cultures at the logarithmic growth stage, even with aeration of 2.0 vvm (volume of air/volume of medium/min^{-1}).
(b) High carbohydrate concentrations are generally achieved from dextrose or starch. However, from these carbohydrates, large amounts of organic acids can be pro-

duced, which will lower the pH of the broth to below 5.4, which in turn may cause decrease in bacterial growth. It can be generally overcome by pH neutralization.

(c) High levels of nitrogen (from proteins, hydrolysates, or maize wastewater) also stimulate growth and promote the release of organic bases, which will favor maintaining the pH of the broth at desirable levels.

Most of the successful attempts to produce bacterial biopesticides commercially employed totally natural products, mainly industrial by-products. Carbohydrates were supplied by tryptone, soy flour, partially defatted cotton seed cake, or cassava starch, while nitrogen was obtained from ammonium salts, corn steep liquor, residual brewer yeast, or soybean meal, for a liquid culture media (Moraes 1973; Fernández-Larrea et al. 2009; Capalbo et al. 2008). The levels and source of those elements also depend on the fermentation process. For example, a carbon source suitable for a semisolid fermentation (like rice bran or oilseed cake) may not be adequate for submerged fermentation, where beet molasses, sugarcane molasses, or starch could be more appropriate (Capalbo et al. 2008). Inorganic salts, such as calcium, zinc, manganese, and magnesium, are also required for the growth of microorganisms. The proper balance of mineral salts helps in the pH balance of the fermentation broth, which is of utmost importance in the production and subsequent recovery and stability of the desired toxin or end product. In the case of Bt, calcium is required for thermostability of its spores, while manganese is required for sporulation. Several authors studying the effect of some minerals on the growth of Bt found that the optimal concentrations of minerals for endotoxin growth and production are different.

For the Bt production in Brazil, it was interesting to use sugarcane molasses and corn steep liquor (Moraes 1973, 1976, 1985), as indicated by the cost analysis compared to the yield of endo and exotoxic fractions. The production in some other countries – Cuba (Márquez-Gutiérrez et al. 1999), Argentina, Colombia, Mexico, and Peru – varied from country to country and were summarized by Capalbo et al. (2008).

pH The balance between nitrogen and oxygen may have a great influence on pH during fermentation. The pH of the fermentation for Bt production can be controlled between 5.4 and 8.4 through the balance of nutrients, sources of protein that form acids and bases. The toxic activity of the final product in a fermentation process is usually dependent on the medium and the strain used. In this process, not only the number of crystals changes during growth in different media, but also the morphological characteristics, leading to changes in magnitude and specificity of the final product recovered, as presented in Table 21.1.

Temperature and Oxigen and Aeration Rates Another important parameter during fermentation is temperature. Dulmage (1993) studied the relationship of growth and yield on endotoxins for Bt and observed that growth and yield do not vary greatly between temperatures of 26 °C and 34 °C. There is no advantage in growing Bt at temperatures over 34 °C for commercial production. In addition to the risk of

Table 21.1 Main properties of the crystal produced by the Z-52 strain in different culture media (Moraes et al. 1998)

Culture media	HM[a] + 0,5% starch	HM + 1% starch	HM + 2% starch	AM[b]	YPM[c]
Crystal length (μm)	–[d]	1.38	1.43	1.91	1.51
Width of crystal base (μm))	–[d]	0.64	0.56	0.83	0.68
Dimension of the cube	–[d]	0.64	0.58	0.81	0.64
Volume of bipyramidal crystals (μm^3)	–[d]	0.19	0.15	0.44	0.23
Volume of cuboidal crystals (μm^3)	–[d]	0.26	0.20	0.53	0.26
Concentration of crystals (mg/ml)	0.5	1.2	1.9	2.9	3.1
Antibacterial specific activity of the crystal (U/ml)	2.3	2.2	2.0	9.0	6.7
General antibacterial activity of the crystal (U/ml)	1.2	2.6	3.8	26.0	20.8
Number of spores ($\times 10^9$/ml)	1.8	2.8	4.2	7.3	4.7
Insecticidal activity ($\times 10^6$ spores/ml)	–[d]	7,7	–[d]	1.7	2.4

[a]HM = yeast extract hydrolyzed (H_2SO_4) and different concentrations of starch (0.5–2.5%)
[b]AM = calcium acetate, yeast, corn flour
[c]YPM = 3% yeast extract, 1.5% corn flour
[d]= not measured

lower yield, there is the cost of energy to be added. Besides, at 37 °C microscopic examinations showed long strands of cells and low yields. Moraes et al. (1981) studied the influence of O_2 in *Bacillus thuringiensis* submerged fermentation.

Types of Fermentation Processes It can be done based on the kind of substrate (solid, semisolid, liquid, or submerged), on the flow process (batch, continuous, feed-batch, etc), or on the types of final product aimed (latic, for latic acid production, important for milk coagulation; alcoholic, for beverage production), among others. Most industrial processes used for the production of a microbial insecticide belong to the batch technique, also referred to as discontinuous (based on process flow). In this kind of process regime, once the culture medium is prepared and sterilized in the fermenter, it is inoculated and nothing else is added except oxygen (aerobic process), acid or base, (pH control) or antifoam. When reaching the end of the fermentation, the fermenter is unloaded and the final broth goes to the downstream steps (separation, concentration, formulation, and so on). The fermenter is washed, sterilized, and readied for the next batch. An example of such discontinuous fermentation under pilot plant conditions (7 and 100 L), using the LBT 25 strain of *Bt* active against *Meloidogyne incognita* (Márquez-Gutiérrez and Fernández 2006) reached sporulated stage in 26 hours at a concentration of 1×10^9 spores/mL, using vegetative cells as inoculum. However, using inocula containing mainly spores, the fermentation lasted for 30 hours with a final concentration of 2.45×10^9 spores/mL. The free pH curves decreased in 4 hours, reaching a minimum in the first 10–12 hours with a tendency to increase after 14 hours and remain constant until the end of the process. The highest glucose consumption was between 8 and 10 hours depending on the type of inoculum used.

Once the optimum conditions for a discontinuous process flow are known, they can be employed in a continuous process. In this process, the substrate is continuously added to the fermenter and at the same flow rate the final product is withdrawn, so that the concentrations are constant. There are difficulties in this process, but knowing the breakeven conditions of the system, they can be manipulated in a way that allows a deep study of the kinetics of the growth of the microorganism and the products of its metabolism. Capalbo (1982) found that for laboratory conditions, continuous fermentation with Bt is successful for systems with more than one stage, and the reduction of aeration of the last stage of the process is an important economic factor.

The semisolid or solid-state fermentation (SSF), as it is often called, is the production system for obtaining substances from microorganisms that develop on the surface of solid substrates. A huge range of products can be obtained (such as mushrooms, oriental foods, enzymes, etc.). Several characteristics intrinsic to this process have been extensively reviewed in Capalbo (1989). It is interesting to summarize that from the point of view of engineering process, the SSF offers attractive characteristics as an alternative process of submerged fermentation even for bacteria. In SSF, by-products of agricultural origin are generally used as support for microbial growth, with microorganisms growing internally on the substrate, on its surface, and in interstitial spaces. As the microorganism is closely linked to the support, it is often difficult to assess its mass directly. This is why microbial growth is often evaluated through O_2 consumption or CO_2 production. In a study on gas diffusion (CO_2 and O_2) in SSF (Auria et al 1992 apud Moraes et al. 2001), it emphasized the importance of the macroscopic variables (column size, shape, substrate particle size distribution, and reactor filling pressure) on diffusion rates and coefficients and, therefore, on biomass results. Another important factor is the heat transfer due to the increase in temperature resulting from the development of the microorganism.

21.5 Quality Control

In the production of Bt by any type of fermentation process, an efficient quality control is imperative to achieve consistency in production systems. A product with well-defined specifications and the consequent quality control procedures ensures its effectiveness and safety, promotes the standardization of production costs, and guarantees its stability in the market, which leads to the gain of confidence in the consumer. Quality not only refers to the final product but also to the evaluation of different parameters during the process (Elósegui et al. 2005, 2006; Márquez-Gutiérrez et al. 2010). The quality control begins with the strain and includes the maintenance and production stock; in both cases, it is necessary to establish conservation methods. The control of purity and the biological activity of the strains must be carried out as often as necessary to ensure that it maintains the conditions for which it has been selected. The preparation of the inoculum is one of the most important steps, and in addition to its purity, the cell concentration and viability need to be monitored. Sterility of the culture media should be checked: bottle not inoculated should be kept under the same conditions as the inoculated ones to confirm the absence of contaminants. If the production is done in industrialized fermentors, the samples are taken before inoculation and subjected to the same analyzes to determine the progress of the process and or the need or time to stop it. The tests that are mostly carried out to determine the quality of Bt biopesticide products are:

1. Concentration of infective units by direct counting in Neubauer chamber under optical microscope.
2. Viability test on Petri dishes of agar media (colony forming units/g or /ml of product).
3. Microbiological purity.
4. Bioassays in in vitro and in vivo conditions to check the virulence of the strain against the pest species of interest.

The specifications for the permissible level of contaminants in the final product will depend on the production methods and the processing of the product in the harvest, post-harvest, and formulation steps; usually the last stages are carried out in non-sterile environments. Therefore, a base level of pollutant input should be established through a well-controlled study, checking the controls in the early stages of the process to demonstrate that the level of contaminants in the final product is inevitably due to the later stages of the process.

Nardo et al. (1995) reviewed some standards regarding the active ingredient for registration of a biopesticide. They showed that in the United States, the use of international units (IU) was established in 1971 to express the potency of Bt-based products, a standard preparation being necessary. The first preparation provided by the Pasteur Institute in France was designated E-61, contained Bt var. *thuringiensis*, and its potency verified in bioassay against *T. ni* was assigned the arbitrary value 1000 IU/mg. This pattern exists today at the Pasteur Institute and was used to standardize the standards of S-1971 and S-1980, both obtained from Bt var. *kurstaki* and with potencies of 18,000 and 16,000 IU/mg, respectively. Bti IPS-82 was produced and assigned a power of 15,000 IU/mg. The standardization bioassays, in this case, were performed according to protocols of the World Health Organization (WHO) or the United States Department of Agriculture (USDA). Some authors also point out the impossibility of having an international standard insect due to the difficulties of the legislation of some quarantine countries. On the other hand, it is questionable whether it is better to use a test or multiple insect in order to get a useful complementation of information. In addition, in order to obtain reproductive results, it is necessary to have insects uniformity (age and weight), constancy of climatic factors and amount of food administered. Therefore, in the industrial routine, the constancy of product characteristics must be verified through a reference sample, and each producer can have its own quality standard (Capalbo et al. 2008). The topic of registration and standards for microbial control agents "Guidelines" published by the

Food and Agriculture Organization (FAO) and the WHO was made available by the end of 2017 (World Health Organization 2017).

21.6 Downstream Process and Formulation

Whichever the form of separation is desired or possible, it will involve purification and concentration steps of the fermented broth, as it contains water, solids, and dissolved material in the culture medium, cell fragments, spores, and crystals. This separation is another critical stage in the development of the productive process, because the power of the product can be reduced if the downstream design applied is not adequate (Bryant 1994). To process large volumes, the most applied method is centrifugation (Bernhard and Utz 1993). A continuous flow centrifuge can be used, discharging periodically the solids in which spores and toxic crystals are found. The solids obtained should not be kept at room temperature for more than 24 hours since can loose their activity and quality. It is recommended to dry the material, preferably in a spray drier; the wet solids from the centrifuge are sprayed into a large vessel through which hot air circulates. The water evaporates quickly and the small droplets become dust particles. In this way, the fine and stable powder obtained can be stored for later formulation. Before continuing with the formulation process, it is necessary to evaluate the insecticidal potency of the batch in question using a bioassay.

Moraes (1993) studied the thermobacteriological parameters aiming at the definition of a good drying process of a Bt biopesticide in pilot scale. With the centrifuged slurry obtained from a submerged fermentation and using clay as the inert material, the relative viability in a conventional dryer was maintained at 50 °C and 70 °C and was reduced at 90 °C. In a spray drier, relative viability only decreased when the temperature raised to 180 °C. Depending on the activity level of the powder obtained from Bt, it will be necessary to add more or less inert substances to ensure that the final product always has the same final power.

In addition, other additives such as dispersants, adhesives, UV protectants, and attractants can be supplied to the mixture to improve the physical properties and its effectiveness of the final commercial product (Rosas-García 2008). Other inerts such as talc, kaolin, lime, and sand can be found in Bt bioinsecticides, since they act directly on pests through various mechanisms; a toxic function proper of its chemical nature or size of its particles, and by the barrier effect that they provide in the surfaces where they are applied, constituting the first line of resistance to the plague is important. In this case, during the development of the Thurisave 25 nematicidal product, in the validation phase in Cuba, the centrifugation parameters were adjusted to obtain a cream with a concentration of 4.25×10^{10} spores/mL that was formulated with kaolin, which achieved an efficacy of 85% against *M. incognita.* The Standard Work Procedures of the Good Laboratory Practices system established in Probelte SA are followed, all of them based on the appropriate CIPAC methods. The selected final formula will also be subjected to a storage stability test for 1 year at 20–25 °C to verify that the loss of insecticidal potency does not exceed 10% the effectiveness of the UV protectors (Organización Mundial de la Salud 2002).

21.7 Marketing and Application

According to Mordor Intelligence,[1] the world market for biopesticides, in terms of value, is expected to reach US$ 4369.88 million by 2019, with an annual compound growth rate of 16.0% from 2014 to 2019. In 2006, Arysta Life Science, an international plant protection and life science company, has estimated the worldwide biopesticide market at approximately US$ 541 million. A 2008 study released by the Global Industry Analysts, Inc. (GIA) estimated that biopesticides accounted for about 3% (US$ 750 million) of the global pesticide market and were likely to reach the US$1 billion mark by 2010. This expectation agreed with the data of Damalas and Koutroubas (2018), which indicates a value of about US$3 billion worldwide market for 2017, accounting for just 5% of the total crop protection market.

The evolution of the biopesticide market is mainly triggered by the high prevalence of crop diseases and growing demand for healthy food in both developed and developing countries and also the benefits offered by biopesticides compared with the introduction of conventional pesticides, which are technologically advanced products. Others include the largest global investment in biopesticide research, the development, a better-established application of the IPM and Integrated Crop Management (ICM) concepts, and an increase in the organic production area. However, factors such as the lack of knowledge to farmers about the benefits of biopesticides and reluctance among farmers to change chemical protection practices restrict the growth of this market. Many market analysts and consultants[2] besides the International Biocontrol Manufactures Association (IBMA) (Mordor Inteligence 2014) indicate that the demand for volume and value for biopesticides is expected to outperform synthetic pesticides. A summary of information obtained from these sources is presented in Table 21.2, by region. While North America should maintain its dominance in terms of volume and value per demand for biopesticides, Asia-Pacific should outperform other global markets, progressing at a faster pace.

In Brazil, for example, aiming at advances in the production process and technology transfer of bioinsecticides, with a view to solving the problems caused by *Aedes aegypti* (L.) (Culicidae), two projects were developed with the support of FAPESP (Foundation to Support Research of the State of São Paulo[3]) by Probiom Technology Ltd. (www.probiom.com.br). In these projects, fermentation process and the formu-

[1] http://mordorintelligence.com

[2] Personal information. The authors consulted and or heard. Agrow; CPL Business Consultants; Markets & Markets; MicroMarketMonitor; BCCResearch; Research and Markets; DunhamTrimmer.

[3] FAPESP Processes 2006/06706-9 and 2009/52,990-9 Technological Innovation in the Formulation of Biological Insecticide Produced by *Bacillus thuringiensis israelensis*. Phase I and II.

Table 21.2 Volume of the global biopesticide market by main regions (2010–2020)

Regions	1000 tons					
	2010	2011	2012	2013	2014	2020
North America	43.46	46.46	50.24	54.89	60.6	135.77
Europe	40.23	43.24	46.97	51.48	57.06	128.97
Asia-Pacific	31.6	34.68	38.37	42.93	48.57	125.78
Latin America	13.1	14.35	15.93	17.83	20.17	51.76
Rest of the world	5.47	5.93	6.47	7.15	7.96	18.79

Source: Translated from Moraes et al. (2018)
Personal information. The authors consulted and or heard. Agrow; CPL Business Consultants; Markets & Markets; MicroMarketMonitor; BCCResearch; Research and Markets; DunhamTrimmer

lation were approached – selection of economical and available substrates and also formulation ingredients that are adequatee to the habits of *A. aegypti*. A 50 L and a 30 kg/batch was obtained, respectively, for submerged and semisolid process. Besides, a new bioreactor was developed. The technology is available to be used in the control for dengue fever. Other experimental products to control Lepidoptera (*H. zea*, among them) were also developed by Probiom. Other bioinsecticides are under development to control pests such as the horn fly *Haematobia irritans* (L.) (Muscidae), the stamen fly *Stomoxys calcitrans* (L.) (Muscidae), *Alphitobius diaperinus* Panzer (Tenebrionidae) pest of aviaries, and *Limnoperna fortunei* Dunker (Mytilidae) (mexilhão dourado).

The lack of adequate or specific standards for biological product evaluation in many countries reduces or hinders the registration process, and consequently restricts the market and this also occurs in Brazil (Nardo et al. 1995). Damalas and Koutroubas (2018) indicate the long and complex registration processes used in many countries such as the European Union (EU), as the cause for fewer biopesticide-active substances been registered. Cuban legislation, in harmony with international requirements, demands that microbial pesticides be subjected to detailed toxicological and environmental impact studies accordingly, the country has an official document for the registration of biopesticides, which includes the requirements for biosafety permits and environmental licensing, which favors the bidding of these products, one of the most limiting difficulties for their use and development worldwide (Fernández-Larrea 2013).

21.8 Major Advances and Limitations

Bt is the main focus of studies and research on microbial pesticides. From its discovery until about 1978, it was thought that its toxic activity was limited to the order Lepidoptera, and the varieties known until the time were the base of the known products of the great companies. By the 1980s, discoveries of lineages that act against nematodes, ectoparasites (such as mites), and endoparasites (protozoa), as well as discoveries of activity against Coleoptera and Diptera, have created a wider range of opportunities. New products are being marketed as patented processes, or featuring special aqueous formulations that allow use in conventional terrestrial and airborne applicators. There is widespread knowledge about its mass production and application, which favors its use and acceptance among farmers.

The use of genetic engineering and nonrecombinant techniques has generated varieties with greater or more extensive activity, as well as the possibility of new formulations. Products have already been released containing the Bt toxin internally to *Pseudomonas*, these being nonviable units, not presenting environmental concerns. Genetic recombination technology allowed the insertion of the gene responsible for the production of Bt toxin in different hosts, such as *Escherichia coli*, *Bacillus subtilis*, and even in algae. The Bt toxin gene has also been inserted into *Clavibacter xyli* var. *cynodontis* (endophytic bacterium in maize), which rapidly colonizes the roots, leaves and stalk of the corn where it remains for the entire life of the plant. It is also possible, by means of another vector (*Agrobacterium tumefasciens*), to insert Bt genes for the synthesis of δ-endotoxin in cotton, tomato, potato, tobacco, and other crops of economic importance.

Besides its new insecticidal action, it presents as an advantage to the environment the fact that it does not survive outside the plant. Transgenic plants containing Bt crystal also have limitations. One is the difficulty in producing cells in the crystal at levels that rapidly kill the larvae. Not all species of pest insects are equally sensitive to crystals, and often the level of crystals expressed in plants is not able to kill resistant insects. There is also the risk of exposing the larvae to very high doses of toxin, thus inducing resistance to the crystal. Resistance triggering may be particularly rapid for transgenic plants that produce their own toxic agent because they will exert an effect on population levels far below those that would occur with conventional, chemical, or biological pesticides, thus promoting a more rapid development of resistance. Strategies to delay the onset of resistance are under development.

In any of the mentioned examples, the importance of the fermentative process in the mass production of the new microorganisms, both those containing or expressing the Bt toxin, and those with potential for biocontrol, whether obtained through genetic manipulation or not, is indisputable. For all of them, the variables and parameters of production and recovery noted in this chapter are valid and should be carefully evaluated to obtain a quality product. The possibilities of enhancement of the biocontrol potential presented by a microorganism can be extended, provided that the production process is carefully optimized. However, in practice, even enhancing the development of this biological control agent with improvements in selection procedures or inoculation production and formulation methods, it will still require the support of other control measures for the sustainable management of the majority of the pests. Another challenge in the successful exploitation of microbial control with Bt, is the necessary support to local producers through agricultural extension programs.

21.9 Final Remarks

Studies with the entomopathogenic bacterium *Bacillus thuringiensis* have led to new multidisciplinary approaches. The progress achieved shows the great impact of these bacteria as biological control agents. The widespread adoption of concepts provided by integrated pest management with the integration of bioinsecticides is a well-supported policy in itself, which can contribute to a greater mass use of these microorganisms.

Many of the factors that determine the viability of microbial insecticides focus on product formulation and application methods. Product formulations should ideally be compatible with popular application methods or be inexpensive and easy to apply by some other method. Producers will need to be more effective in demonstrating and communicating their benefits.

Bt is a good example of biopesticide that has maintained acceptably – priced products through large-scale production and large-volume (and primarily high-value) markets. The obvious benefits of nontarget and environmental safety for utilizing control strategies based on this bacterium provide compelling support to continue research and development in this category of biological control.

The use of Bt-based products in Latin American countries depends more on political and economic aspects than on technological approaches. The importance of government agencies in relation to marketing opportunities and the development of incentives should not be overlooked. Pressures on governments and food producers to reduce the use of pesticides can favor the success of Bt. The ability to formulate these products on a commercial scale should be considered by the governments of Latin American countries and/or private industries. The decrease in the cost to the farmer would encourage the use of these microbial products and increase the benefits of the producers, since the "products of biological origin" or those grown without pesticides have a better acceptance in the market in several countries.

References

Bernhard K, Utz R (1993) Production of Bacillus thuringiensis insecticides for experimental and commercial uses. In: Entwistle PF, Cory JS, Bailey MJ et al (eds) Bacillus thuringiensis: an environmental biopesticide: theory and practice. John Wiley, Chichester, pp 255–267

Boucias DG, Pendland JC (1998) Principles of insect pathology. Klumer Academic Publishers, Norwell

Bryant JE (1994) Commercial production and formulation of *Bacillus thuringiensis*. Agric Ecosyst Environm 49(1):31–35

Capalbo DMF (1982) Contribuição ao estudo de fermentação de *Bacillus thuringiensis*. Dissertação, Universidade Estadual de Campinas

Capalbo DMF (1989) Desenvolvimento de processo de fermentação semi-sólida para obtenção de *Bacillus thuringiensis* Berliner. Tese, Universidade Estadual de Campinas

Capalbo DMF, Moraes IO, Arruda ROM et al (2008) Impact assessment of the production and utilization of entomopathogens and its effects on public policies. In: Crop improvement and biotechnology. Taylor & Francis, New Delhi, pp 223–229

Centro Nacional de Sanidad Vegetal (2016) Análisis del cumplimiento del programa de producción e medios biológicos. Ministerio de la Agricultura, La Habana

Damalas CA, Koutroubas SD (2018) Current status and recent developments in biopesticide use. Agriculture 8(1):13

Dulmage HT (1993) Development of isolates of *Bacillus thuringiensis* and similar aerobic microbes for use in developing countries. In: Salama HS, Morris ON, Rached E (eds) The biopesticide *Bacillus thuringiensis* and its applications in developing countries. National Research Centre, Cairo, pp 15–42

Elósegui O, Carr A, Fernández-Larrea O (2005) Influencia de la carga microbiana contaminante inicial delsustratoenlacalidad final de biopreparados de *Trichoderma harzianum* Rifai y *Beauveria bassiana* Balsamo Vuillemin. Fitosanidad 9(1):51–55

Elósegui O, Fernandez-Larrea O, López O (2006) Control de lacalidad de biopreparados de hongos. In: Carreras B (ed) Memorias del curso internacional: laproducción de microorganismosentomopatógenos y antagonistas para elcontrol de plagas. CIDISAV, La Habana

Fernández-Larrea O (2013) Programa para la recuperación de bioplaguicidas, biofertilizantes y bioestimulantes en Cuba. Agricultura Orgánica 19(2):2–5

Fernández-Larrea O, Valencia AC, Gutiérrez MEM (2009) Influencia del medio de cultivo, el tipo de inóculo y las condiciones de agitación-aireación en la reproducción de dos cepas de *Bacillust huringiensis* (Berl.) en fermentadores de diferentes volúmenes. Fitosanidad 13(3):201–209

Jiménez J (1974) Integración del control biológico de *Heliothisvirescens* (F.) en el cultivo del tabaco. Agrotecnia de Cuba 6:25–31

Jiménez J, Vázquez M (1990) Efectividad de *Bacillus thuringiensis* Berl. contra *Mocislatipes* (Guenée), en áreas de pastos. INISAV, La Habana

Márquez-Gutiérrez ME, Fernández E (2006) Selección de cepas de *Bacillus thuringiensis* con efecto nematicida. Manejo Integr Plagas Agroecol 78:63–69

Márquez-Gutiérrez ME, Fernández-Larrea O, Díaz A (1999) Influencia de la composición del medio de cultivo en los rendimientos de esporas y cristales de *Bacillus thuringiensis*. Fitosanidad 3:31–34

Márquez-Gutiérrez ME, Fernández-Larrea O, Jiménez-Ramos J et al (2010) Producción masiva de bioplaguicidas para el control biológico de plagas y enfermedades. CIDISAV, La Habana

Moraes IO (1973) Obtenção de inseticidas bacterianos por fementação submersa Dissertação, Universidade Estadual de Campinas

Moraes IO (1976) Ensaios de fermentação submersa para produção de inseticida bacteriano em minifermentador. Tese, Universidade Estadual de Campinas

Moraes IO (1985) Processo de produção de toxina termoestável de *Bacillus thuringiensis*. BR Patente 8500663, 16 Sept 1985

Moraes RO (1993) Determinação da viabilidade de Bacillusthuringiensis após processo de secagem. Dissertação, Universidade Estadual de Campinas

Moraes IO, Capalbo DMF (1986) Produção de bactérias entomopatogênicas. In: Alves SB (coord) Controle microbiano de insetos. Manole, São Paulo, pp 296–310

Moraes IO, Santana MHA, Hokka CO (1981) The influenceofoxygenconcentrationon microbial insecticideproduction. In: Young MM, Campbell WR, Vezina C (eds) Advances in biotechnology: scientific and engineering principles, vol 1. Elsevier, New York, pp 75–79

Moraes IO, Capalbo DMF, Moraes RO (1991) Multiplicação de agentes de controle biológico. In: Bettiol W (coord) Controle biológico de doenças de plantas. Embrapa, Brasília, pp 253–272

Moraes IO, Capalbo DMF, Arruda ROM (1998) Produção de bactérias entomopatogenicas. In: Alves SB (ed) Controle microbiano de insetos, vol 4. FEALQ, Piracicaba, pp 815–843

Moraes IO, Capalbo DMF, Arruda ROM (2001) Produção de bioinseticidas. In: Lima UA, Aquarone E, Borzani W et al (coords) Biotecnologia industrial: processos fermentativos e enzimáticos, vol 3. Edgard Blücher, Porto Alegre, pp 245–265

Moraes IO, Capalbo DMF, Arantes O et al (2008a) Produção de bactérias entomopatogências na América Latina. In: Alves SB, Lopes RB (orgs) Controle microbiano de pragas na América Latina: avanços e desafios, vol 1. FEALQ, Piracicaba, pp 239–256

Moraes IO, Capalbo DMF, Valicente FH et al (2008b) Case studies of *Bacillus thuringiensis* production and biocontrol applications. In: Thangadurai D, Tripathi L, Vasanthaiah HKN et al (eds) Crop improvement and biotechnology. Bioscience, Puliyur, pp 229–241

Moraes IO, Baró YR, Marquez-Gutiérrez ME et al (2011) Characterization of *Bacillus thuringiensis* strains under the umbrela of Brazil-Cuba cooperation on bioinsecticides. In: Mendez-Vilas A (ed) Microbes in applied research: current advances and challenges. Formatex Research Center, Malaga, pp 505–509

Moraes IO, Moraes RO, Arruda ROM (2012) Brazilian biodiversity in a culture collection devoted to the identification and preservation of microorganisms useful in environmental, industrial and applied microbiology. In: Mendez-Vilas A (ed) Microbes in applied research: current advances and challenges. Formatex Research Center, Malaga, pp 367–371

Moraes IO, Capalbo DMF, Arruda ROM (2018) Produção de bioinseticidas. In: Alterthun F, Lima UA, Moraes IO et al (coords) Série biotecnologia industrial. Blucher, Porto Alegre

Mordor Inteligence (2014) Global biopesticides market-growth, trends and forecasts. http://mordorintelligence.com/product/global-biopesticides-market-growth-trends-and-forecasts-2014-2019. Accessed 15 Aug 2017

Murguido CA, Elizondo AI (2007) El manejo integrado de plagas de insectosen Cuba. Fitosanidad 11(3):23–28

Nardo EAB, Capalbo DMF, Moraes GJ et al (1995) Requisitos para a análise de risco de produto contendo agentes microbianos de controle de organismos nocivos: uma proposta para os órgãos federais registrantes. Embrapa-CNPMA, Jaguariúna

Organización Mundial de la Salud (2002) Manual sobre elaboración y empleo de lãs especificaciones de la FAO y de la OMS para Plaguicidas: especificaciones de plaguicidas. FAO, Rome

Rosas-García NM (2008) Avances en el desarrollo de formulaciones insecticidas a base de *Bacillus thuringiensis*. Rev Colomb Biotecnol 10(1):49–63

Vázquez LL, Caballero S, Carr A et al (2010) Diagnóstico de la utilización de entomófagos y entomopatógenos para el control biológico de insectos por los agricultores en Cuba. Fitosanidad 14:159–169

World Health Organization (2017) International code of conduct on pesticide management: guidelines for the registration of microbial, botanical and semiochemical pest control agents for plant protection and public health uses. FAO, Rome

Part IV
Biological Control in Major Crops, Forests, Pasture, Weeds and Plant Diseases in the Neotropical Region

Chapter 22
Beans

Eliane Dias Quintela, Tiago Cardoso da Costa-Lima, and Alexandre José Ferreira Diniz

22.1 Introduction

Common bean (*Phaseolus vulgaris* L.) and cowpea (*Vigna unguiculata* L.) are the most commonly consumed legume worldwide, with a commercial value exceeding that of all other legume crops combined (Porch et al. 2013). The FAO reported that total bean production in 2017 was over 31.4 million tons (FAO 2019). Beans are nearly "perfect" food: nutritionally rich, good source of protein, folic acid, iron, dietary fiber, and complex carbohydrates (Silochi et al. 2016). Today beans are grown in more than 120 countries. The top bean producers of the world are Brazil, Myanmar, India, and China, who together account for more than 50% of world production (Harvest Choice 2017). Brazil is the world's largest producer and consumer of common bean and cowpea, producing around 3.12 million metric tons of grain on 3.1 million ha in 2017 (Conab 2019).

In many regions where beans are grown, full yield potentials are not met due to several factors, including inadequate soil fertility, insect predation, diseases, high temperatures, and drought (Oyewale and Bamaiyi 2013). There are several serious insect pests that attack the common bean, depending on the geographic location, but predation by a wide range of arthropods – aphids, beetles, caterpillars, leafhoppers, whiteflies, mites, and thrips – is seen in the Neotropical region (Cardona 1989; Quintela et al. 1991; Quintela and Barbosa 2015). Typically, chemical pesticides are used more commonly in the commercial production setting, rather than by

E. D. Quintela (✉)
Empresa Brasileira de Pesquisa Agropecuária (Embrapa Arroz e Feijão), Goiânia, GO, Brazil
e-mail: eliane.quintela@embrapa.br

T. C. da Costa-Lima
Empresa Brasileira de Pesquisa Agropecuária (Embrapa Semiárido), Petrolina, PE, Brazil

A. J. F. Diniz
Faculdade de Ensino Superior Santa Bárbara (FAESB), Tatuí, SP, Brazil

B. Souza et al. (eds.), *Natural Enemies of Insect Pests in Neotropical Agroecosystems*, https://doi.org/10.1007/978-3-030-24733-1_22

smallholder farmers (FAO 2019). Subsistence farmers with little access to costly farm inputs such as pesticides and biological control agents must rely on natural pest regulation and cultural controls. Therefore, most of the applied biological control has been used by medium- and large-scale farmers in Brazil and other Latin American countries. Several examples of natural enemies of bean pest have been cited in the literature. Some of them are available as commercial product in Brazil (ABCBio 2019). Parasitoid as *Trichogramma* spp. and pathogens as nucleopolyhedrovirus of *Helicoverpa armigera* (Hübner) (Noctuidae), *Chrysodeixis includens* (Walker) (Noctuidae) and *Spodoptera frugiperda* (Smith) (Noctuidae), *Bacillus thuringiensis*, and *Beauveria bassiana* (Bals.) Vuill. (Cordycipitaceae) have been the most used although in small scale by bean farmers.

The research on insect pathogens of bean/cowpea in Brazil had great improvement after a project was initiated with Bean/Cowpea Collaborative Research Support Program (Bean/Cowpea CRSP) supported by Boyce Thompson Institute (BTI) in collaboration with the Brazilian Agricultural Research Corporation (Embrapa Rice and Beans) from 1981 through 1993. Besides the basic and applied research, training and germplasm conservation of entomopathogenic fungi had also been an important element of the project. Almost 300 strains have been collected and stored in liquid nitrogen at Invertebrate Fungal Collection at Embrapa Genetic Resources and Biotechnology (Brasília, DF, Brazil) and the Agricultural Research Service Collection of Entomopathogenic Fungal Cultures (Ithaca, New York) (Roberts and Quintela 1990).

A review of the most important pest of beans considering the ones that has more information about their natural enemies will be addressed in this chapter.

22.1.1 Soil Pests

Bean crops are attacked by a series of soil pests, as Diptera, Coleoptera, and Lepidoptera larvae, mainly in the initial growing stage. Predators that habit the soil can have an important role in their control. Hong et al. (2011) identified six taxonomic groups feeding on *Agrotis ipsilon* (Hufnagel) (Noctuidae); among those, 70% was represented by the ant *Solenopsis invicta* Buren (Formicidae), following by other natural enemies, as beetles (Carabidae, Cicindelidae, and Staphylinidae), assassin bugs (Reduviidae), and spiders. These same insect groups were found to be the most important predators of *Elasmopalpus lignosellus* (Zeller) (Pyralidae), and the study also showed that their population was not influenced by the use of mulching (Gill et al. 2010). Pereira et al. (2010) studied the same soil fauna in conventional and no tillage bean planting systems. The mite *Cheyletus* sp. (Cheyletidae) was found as the most abundant predator. The planting system showed an influence over ant predators that had a sevenfold higher abundance under no-tillage system, with the retention of crop residue.

Many of the principal pests of beans and cowpeas pass all of part of their larval stages within the soil environment. *Chalcodermus* (Curculionidae) weevil larvae, after completing development in cowpea seeds, exit the pods and enter the soil to pupate. Soil applications of *Metarhizium anisopliae* (Metschn.), Sorokin (Clavicipitaceae), and *B. bassiana* killed 30–50% of *Chalcodermus bimaculatus* Boheman larvae entering the soil over a 1–2-week period in three field trials conducted at central and northeast of Brazil (Quintela et al. 1990, 1992b, 1994; Quintela and Roberts 1992). *B. bassiana* applied to cowpea seeds resulted in 73% of confirmed mortality of first-instar *Cerotoma arcuata* (Olivier) (Chrysomelidae) larvae (Roberts and Quintela 1990). Pianoski et al. (1990) tested *B. bassiana* strains in combination with fertilizer treatments for control of *Diabrotica speciosa* (Germar) (Chrysomelidae) in bean crops. In the field, *B. bassiana* had the strongest effect when plants were treated with excess nitrogen, but in laboratory trials, *B. bassiana* was most effective without fertilizers. Applications of conidia of *M. anisopliae* and *B. bassiana* to the stems and soil surrounding the stems of cowpea seedlings caused greater than 85% of neonate *E. lignosellus* larvae (Roberts and Quintela 1990).

22.1.2 Leaf Miners

Crops with low insecticide impact, generally, do not show economic damage by leaf miners outbreaks, due to the high prevalence of parasitoids and predators.

Leaf miner is one of the guilds with more parasitoids richness. Only the *Liriomyza* genus (Agromyzidae) has more than 140 parasitoid species related (Liu et al. 2009), mainly associated with the main species, *Liriomyza sativae* (Blanchard), *Liriomyza trifolii* (Burgess), and *Liriomyza huidobrensis* (Blanchard). The *Liriomyza* parasitoids commercialization is a reality since the 1980s in Europe and North America (Van Lenteren 2012). *Diglyphus isaea* (Walker) (Eulophidae) and *Dacnusa sibirica* Telenga (Eulophidae) are the major species used, and more recently, these species are also applied in Mexico and Costa Rica. In Brazil, a leaf miner parasitoid is expected to be commercialized in the next years, mainly for melon crops.

The leaf miners parasitoids are idiobionts or koinobionts. The first ones are mainly from the Eulophidae family. They parasitize and paralyze the *Liriomyza* larvae, while their larvae consume the host and pupate in the tunnel inside the leaf. The koinobiont parasitizes the larvae and emerges from the pupa that generally occurs in the soil. The braconids from the *Opius* genus are the most common associated to beans and cowpea in Brazil. There are reports of *Opius scabriventris* (=*Phaedrotoma*) Nixon (Braconidae), *Chrysocharis bedius* Walker (Eulophidae), *Chrysocharis vonones* (Walker), and *Zaeucoila unicarinata* Ashmead (Figitidae) (Campos et al. 1984; Gallardo et al. 2010; Costa-Lima et al. 2014).

The main predator species of larvae and pupae of *Liriomyza* spp. are the bugs *Cyrtopeltis modestus* (Distant) (Miridae), *Dicyphus cerastii* Wagner (Miridae), *Dicyphus tamaninii* Wagner e *Macrolophus caliginosus* Wagner (Miridae) (Castane et al. 2004). Nymphs and adults of these predators can prey on leaf miner larvae and pupae. Ants also can prey on *Liriomyza* spp. prepupae and pupae (Xavier 2013). Flies of the families Cecidomyiidae, Empididae, Muscidae, and Dolichopodidae were observed preying on adults of leaf miners (Van Lenteren et al. 1979; Rauf et al. 2000).

22.1.3 Whiteflies

Considering the importance of *Bemisia tabaci* (Gennadius) (Aleyrodidae) to a wide range of crops, the natural enemy complex of this pest is well studied. In the last review of *B. tabaci* natural enemies, Gerling et al. (2001) reported 55 parasitoid species. However, certainly that many species are still unknown. Only in China, Li et al. (2011) identified 56 whitefly parasitoid species. Over 90% of these species are represented by two genera, *Encarsia* and *Eretmocerus*, both in Aphelinidae family. Eight whiteflies parasitoid species were already commercialized in Europe, the United States, Canada, and Mexico (Van Lenteren 2012). The first one was *Encarsia formosa* (Gahan) in 1926 (Van Lenteren 2012). *E. formosa* parasitizes all immature stages older than settled first-instar nymph; however, it prefers to oviposit in third- and fourth-instar nymphs (Hoddle et al. 1998). In Brazil, Lourenção et al. (2014) reviewed the occurrence of ten *Encarsia* species associated to *B. tabaci* and reported also the finding of *Eretmocerus mundus* Mercet (Aphelinidae). Santos (2017) evaluated the influence of the edges between native vegetation and agricultural crops on the population dynamics of whitefly and the patterns of parasitism in 20 fields in the center west of Brazil. The edge adjacent to the native vegetation did not affect the natural parasitism, but it was correlated positively with higher whitefly host density. The species of greatest abundance was *Encarsia lutea* Masi, which was found in approximately 60% of parasitized nymphs. Other three species, *Encarsia porteri* (Mercet), *Encarsia nigricephala* Dozier, and *Eretmocerus mundus* Mercet were also identified. The rate of parasitism ranged from 6.2% to 46.5% in 2013/2014 and from 3.6% to 50% in the next season.

The *B. tabaci* ability to transmit the *bean gold mosaic virus* (BGMV) and the *cowpea mild mottle virus* (CPMMV) in common beans causes the producers to apply chemicals preventively. The high spraying frequency limits the parasitoid field abundance. For example, Quintela et al. (1992a) found high level of nymphs parasitized by *Encarsia* sp. on natural vegetation (at the edge, adjacent to crops) when compared with planted crops. The adoption of insecticides with higher selectivity to natural enemies can aid the parasitoid increment and achieve higher efficacy control.

There are several predators being commercialized aiming the whiteflies control in different countries. Among these biological control agents are mite predators (*Amblyseius swirskii* Athias-Henriot, Phytoseiidae, and *Amblydromalus limonicus* Garman and McGregor, Phytoseiidae), Coccinellidae (*Delphastus catalinae* Horn), and Miridae bug (*Macrolophus pygmaeus*, Rambur). A whitefly natural enemies' survey in Brazil conducted in eight crops, including beans, detected seven

Coccinellidae species, four Chrysopidae species, and three Syrphidae species (Oliveira et al. 2003). Considering the bugs, the families Miridae and Anthocoridae were the most registered attacking whiteflies. *Macrolophus caliginosus* (Wagner) preys upon *B. tabaci* eggs and nymphs, preferring the latter one (Bonato et al. 2006). In Brazil, *Chrysoperla externa* Hagen and *Ceraeochrysa cincta* (Schneider) are commonly found attacking whiteflies (Oliveira et al. 2003). Among the mite predators, *A. swirskii* has been successfully commercialized mainly in Europe (Calvo et al. 2015), and more recently in countries in Latin America, as Mexico and Ecuador (Koppert Biological Systems 2019). This mite preys on eggs and nymphs of *B. tabaci*. In Brazil, studies showed another species of Phytoseiidae mite, *Amblyseius tamatavensis* Blommers, as potential candidate for biological control of *B. tabaci* (Cavalcante et al. 2017).

Entomopathogenic fungi have been recognized as important agents of biological control of whiteflies and are the only pathogens group capable of infecting the cuticle of their hosts, different from viruses and bacteria that must be ingested (Faria and Wraight 2001). The fungi most often observed as biocontrol agents of *B. tabaci* are *Isaria* spp. [proposed new name *Cordyceps* (Kepler et al. 2017)], *Lecanicillium* spp., and *Aschersonia* spp. (Hypocreales) (Faria and Wraight 2001, 2007; Lacey et al. 2008; Mascarin et al. 2013; Quintela et al. 2016) because they naturally infect and thus regulate populations of whiteflies (Faria and Wraight 2001). Most of the research involving fungi for control of *B. tabaci* has focused on the use of *B. bassiana* mainly on protected crops (Wraight et al. 2000; Faria and Wraight 2001), and 13 products are registered in Brazil for whitefly control (ABCBio 2019). Although *B. bassiana* has never been isolated from whitefly under natural conditions, the nymphal stage is highly susceptible to some isolates of this pathogen (Vicentini et al. 2001; Mascarin et al. 2013). However, at the field level, *B. bassiana* efficiency for whitefly control has not been very promising (Azevedo et al. 2005).

On the other hand, *Cordyceps javanica* (formerly *Isaria*) (Friedrichs & Bally) Samson and Hywel-Jones (Cordycipitaceae) has been considered a promising biocontrol candidate for whiteflies mainly due to its ability to cause epizootics, high virulence to *B. tabaci* life stages, and easy mass production (Cabanillas and Jones 2009; Mascarin et al. 2013, 2014). Epizootics of *C. javanica* were observed on whitefly nymphs and adults in common bean, soybean, cotton, corn, guava, and tomato crops, in the 2012/2013 growing season at Federal District and Goiás state in Brazil (Quintela et al. 2016). Three of ten of these isolates collected were highly virulent to *B. tabaci* nymphs at laboratory and screenhouse conditions (Mascarin et al. 2018, Quintela ED, unpublished data). At three field trials conducted in January to April 2018 at Santo Antonio de Goiás, GO, Brazil, the efficiency of uncoated and formulated conidia of *C. javanica* for the control of *B. tabaci* was greater than 82.9% (considering the natural of parasitism of 30–50% of nymphs by *Encarsia* sp.) (Boaventura H et al, unpublished data). In all experiments, high amount of conidia has been produced from cadavers of nymphs and adults as well as mycelial growth from infected insects. These results confirm the great capacity of multiplication and dispersion of this fungus that has favored the epizootics to the field, showing its potential as a mycoinsecticide.

22.1.4 Leafhoppers

Leafhoppers from the *Empoasca* genus are important bean pest, as an example of *Empoasca kraemeri* Ross and Moore (Cicadellidae) in South America and *Empoasca fabae* Harrisin (Cicadellidae) in North America. Egg parasitoids of leafhoppers are common in Mymaridae and Trichogrammatidae (Chalcidoidea) families (Freytag 1985). In Brazil, *E. kraemeri* eggs were parasitized by *Anagrus flaveolus* Waterhouse (Mymaridae) and *Aphelinoidea plutella* Girault (Trichogrammatidae) reaching upon 19% and 31% of parasitism, respectively, in *P. vulgaris* (Pizzamiglio 1979). However, more surveys are needed to better characterize these important group of natural enemies.

On bean crops, *E. kraemeri* populations were more correlated to predators from the genus *Orius* (Anthocoridae). Other predators registered were *Franklinothrips* sp. and spiders (Fernandes et al. 2010). Comparing three *E. fabae* predators, *O. insidiosus* Say, *Coleomegilla maculata* (De Geer) (Coccinellidae), and *C. carnea,* the latter one showed the highest nymph and adult preying rate (Erlandson and Obrycki 2010). Some predators can also prey upon *E. fabae* eggs, as *O. insidiosus* and *Nabis americoferus* Carayon (Nabidae) (Martinez and Pienkowski 1982).

Zoophthora radicans (Brefeld) Batko (Entomophthorales) is an important natural enemy of leafhoppers, and its development for biological control has been pursued for many years. Mass production had been a hindrance for the use of this fungus, since it produces fragile, thin-walled spores that are difficult to mass produce, harvest, and formulate on an industrial scale. An alternative approach, which circumvented this problem, involved production of vegetative mycelium. USDA scientists have developed and patented a process for formulation of *Z. radicans* mycelium into dry granules that can be stored for many months under refrigeration. At field trials, the sporulation behavior on leafhopper cadavers of the formulated fungus was similar to that of the wild fungus. After being applied in the crop in only five location per ha, the mycelium produces spores that infect the target pests and the infected pest spread the disease. Information about the main research on *Z. radicans* for *Empoasca* control in Brazil can be found in Leite et al. (1991, 1996), Magalhães et al. (1988a, 1990, 1991a, b, c, 1996), Wraight et al. (2003), Sandskär and Magalhães (1994), and Galaini-Wraight et al. (1991, 1992).

Another fungus, *Hirsutella guyana* Minter and B.L. Brady (Ophiocordycipitaceae), collected from epizootic outbreaks on *E. kraemeri*, in Caucaia, CE, also has great potential as biopesticide. Field trials on cowpea showed that an application of the fungus as dry mycelium on leaves underwent sporulation, producing a dense mat of conidiophores and conidia within 2 or 3 days (Roberts and Quintela 1990).

22.1.5 Chrysomelid Beetles

The cucurbit beetles, *D. speciosa* and *C. arcuata,* are one the main bean pests. The adult feeds on the leaves and the larvae attacks the roots. The South American *Celatoria bosqi* Blanchard (Tachinidae) was originally described from *D. speciosa* (Blanchard 1937). Since then, it has been collected from *D. speciosa* (up to 18% parasitism) (Cabrera Walsh 2004). The species of *Celatoria* have also been frequently described parasitizing *Cerotoma* sp. (Chrysomelidae). Danielson et al. (2000) found adults of *Cerotoma trifurcata* (Forster) parasitized by *Celatoria* sp. in Nebraska (North America). Herzog (1977) and Marrone et al. (1983) found adults of *C. trifurcata* parasitized by *Celatoria diabroticae* Shimer. In Brazil, 32.2% parasitism were observed on *C. arcuata* (Olivier) (Chrysomelidae) by *C. bosqi*, in cowpea (Magalhães and Quintela 1987).

Insects of 35 families from 12 orders are known as predators of Diabroticina species (Toepfer et al. 2009). All are generalists, with no information about predation rate in the field (Walsh et al. 2016). Three groups of Diabroticina predators have been most studied: mites (Acari), ants (Formicidae), and carabid beetles (Carabidae) (Toepfer et al. 2009). Among those, the most important are beetles (Carabidae and Coccinellidae), earwigs (Dermaptera), true bugs (Anthocoridae and Nabidae), lacewings (Chrysopidae), and wasps (Vespidae) (Milanez 1984; Hohmann and Carvalho 1989). The low number of *D. speciosa* predators is related to the capability of this species to store cucurbitacin, that is obtained feeding from plants rich is this compound (Metcalf 1994). Thus, the insect shows a deterrent effect over natural enemies.

Diabroticina beetles are susceptible to infection by a broad range of microbial agents including viruses, bacteria, and fungi. Toepfer et al. (2009) published an extensive review of the natural enemies of *Cerotoma* and *Diabrotica*, including the pathogens. Some of these studies were done in Brazil on cowpea and commons beans (Magalhães et al. 1986, 1988a, b; Lord et al. 1987, 1988).

22.1.6 Lepidoptera

Bean crops present several Lepidoptera pests that can cause defoliation; others are stem borers or attack the pods. In Brazil, some beans producers are using the egg parasitoid, *Trichogramma pretiosum* Riley (Trichogrammatidae), to control mainly *C. includens* and, eventually, *H. armigera*. Actually, there are four *T. pretiosum* biodefensives registered for caterpillars (ABCBio 2019). The current recommendation is to conduct three releases per cycle, with 100.000 parasitoids/hectare (personal communication, Bruno M. Arroyo – Koppert).

Among the pathogens, the viruses, mainly members of the Baculoviridae (nucleopolyhedrovirus and granulovirus) offer the greatest potential for development as microbial-based insecticides. In Brazil, many baculovirus-based biological insecticides are registered and used for inundative biological control of lepidopterans, including *H. armigera*, *S. frugiperda*, and *C. includens* (ABCBio 2019). Recently, a product based in a mixture of nuclear polyhedrosis virus (NPV) of *H. armigera* and *C. includens* was registered by AgBiTech.

22.2 Final Remarks

Arthropod pests are major constraints to production of cowpeas and beans in Brazil and several other Latin American with large population of subsistence farmers. Several natural enemy species appear to be promising candidates for control strategies for the different species. An IPM approach by incorporation of pest-monitoring systems, application of biological control products, crop rotation, and modification of cultural techniques (i.e., reduced tillage, reduced weed control, cover crops, diversified crop rotations, or soil amendments) must be pursued to enhance the natural enemy populations. The implementation of these techniques is not an easy task, mainly for subsistence farmers. Coordinated actions among researchers and public and private extension institutions are necessary to implement the IPM techniques and explore the full potential of the natural enemies of bean and cowpea pests.

References

ABCBio (2019) Biodefensivos registrados. http://www.abcbio.org.br/. Accessed 10 Mar 2019

Azevedo FR, Guimarães JA, Braga Sobrinho R, Lima MAA (2005) Eficiência de produtos naturais para o controle de *Bemisia tabaci* biótipo b (Hemiptera: Aleyrodidae) em meloeiro. Arq Inst Biol 72:73–79

Blanchard EE (1937) New or little known Argentine Diptera. Revista de la Sociedad Entomologica Argentina 9:35–58

Bonato O, Couton L, Fargues J (2006) Feeding preference of *Macrolophus caliginosus* (Heteroptera: Miridae) on *Bemisia tabaci* and *Trialeurodes vaporariorum* (Homoptera: Aleyrodidae). J Econ Entomol 99:1143–1151

Cabanillas HE, Jones WA (2009) Pathogenicity of *Isaria* sp. (Hypocreales: Clavicipitaceae) against the sweet potato whitefly B biotype, *Bemisia tabaci* (Hemiptera: Aleyrodidae). Crop Prot 28:333–337. https://doi.org/10.1016/j.cropro.2008.11.015

Cabrera Walsh G (2004) Distribution, host specificity, and overwintering of *Celatoria bosqi* Blanchard (Diptera: Tachinidae), a South American parasitoid of *Diabrotica* spp. (Coleoptera: Chrysomelidae: Galerucinae). Biol Control 29:427–434. https://doi.org/10.1016/j.biocontrol.2003.08.003

Calvo JF, Knapp M, Van Houten YM, Hoogerbrugge H, Belda JE (2015) *Amblyseius swirskii:* what made this predatory mite such a successful biocontrol agent? Exp Appl Acarol 65:419–433. https://doi.org/10.1007/s10493-014-9873-0

Campos TB, Takematsu AP, Rodrigues Neto SM, Bitran EA, Oliveira DA, Chiba S (1984) Controle químico do díptero minador *Liriomyza* sp. em feijoeiro (Phaseolus vulgaris L.). Biológico 50:33–38
Cardona C (1989) Insects and other invertebrate bean pests in Latin America. In: Schwartz HF, Pastor-Corrales MA (eds) Bean production problems in the tropics. Centro Internacional de Agricultura Tropical, Cali, pp 505–570
Castane C, Alomar O, Goula M, Gabarra R (2004) Colonization of tomato greenhouses by the predatory mirid bugs *Macrolophus caliginosus* and *Dicyphus tamaninii*. Biol Control 30:591–597. https://doi.org/10.1016/j.biocontrol.2004.02.012
Cavalcante ACC, Mandro MEA, Paes ER, Moraes GJ (2017) *Amblyseius tamatavensis* Blommers (Acari: Phytoseiidae) a candidate for biological control of *Bemisia tabaci* (Gennadius) biotype B (Hemiptera: Aleyrodidae) in Brazil. Int J Acarol 43:10–15. https://doi.org/10.1080/01647954.2016.1225816
Conab (2019) Safra brasileira de grãos. https://www.conab.gov.br/info-agro/safras/graos. Accessed 9 Jan 2019
Costa-Lima TC, Chagas MCM, Parra JR (2014) Temperature-dependent development of two Neotropical parasitoids of Liriomyza sativae (Diptera: Agromyzidae). J Insect Sci 14:245. https://doi.org/10.1093/jisesa/ieu107
Danielson SD, Brandle JR, Hodges L (2000) Parasitoids of the bean leaf beetle (Coleoptera: Chrysomelidae) found in Nebraska soybean fields. J Kans Entomol Soc 73:242–244
Erlandson LAW, Obrycki JJ (2010) Predation of immature and adult *Empoasca fabae* (Harris) (Hemiptera: Cicadellidae) by three species of predatory insects. J Kans Entomol Soc 83:1–6
FAO (2019) Faostst – food and agriculture data statistical. http://www.fao.org/faostat/en/#data/QC. Accessed 9 Jan 2019
Faria M, Wraight SP (2001) Biological control of *Bemisia tabaci* with fungi. Crop Prot 20:767–778. https://doi.org/10.1016/S0261-2194(01)00110-7
Faria MR, Wraight SP (2007) Mycoinsecticides and mycoacaricides: a comprehensive list with worldwide coverage and international classification of formulation types. Biol Control 43:237–256. https://doi.org/10.1016/j.biocontrol.2007.08.001
Fernandes FL, Picanço MC, Fernandes MLS, Xavier VM, Martins JC, Silva VF (2010) Controle biológico natural de pragas e interações ecológicas com predadores e parasitoides em feijoeiro. Biosci J 26:6–14
Freytag PH (1985) The insect parasites of leafhoppers, and related groups. In: Nault LR, Rodriguez J (eds) The leafhoppers and planthoppers. Wiley, New York, pp 423–467
Galaini-Wraight S, Wraight SP, Carruthers RI, Magalhães BP, Roberts DW (1991) Description of a *Zoophthora radicans* (Zygomycetes: Entomophthoraceae) epizootic in a population of *Empoasca kraemeri* (Homoptera: Cicadellidae) on beans in Central Brazil. J Invert Pathol 58:311–326
Galaini-Wraight S, Wraight SP, Carruthers RI, Roberts DW (1992) Temperature-dependent germination and host penetration of the entomophthoralean fungus *Zoophthora radicans* on the leafhopper *Empoasca kraemeri*. Mycol Res 96:38–48. https://doi.org/10.1016/S0953-7562(09)80994-5
Gallardo FE, Diaz NB, Reche V, Costa-Lima, TC (2010) Lista de Eucoilinae (Hymenoptera, Cynipoidea, Figitidae) asociados a Diptera Agromyzidae en la región Neotropical. In: IV Reunión Argentina de Parasitoidólogos, 2010, Concórdia, Argentina
Gerling D, Alomar O, Arno J (2001) Biological control of *Bemisia tabaci* using predators and parasitoids. Crop Prot 20:779–799. https://doi.org/10.1016/S0261-2194(01)00111-9
Gill HK, McSorley R, Goyal G, Webb SE (2010) Mulch as a potential management strategy for lesser cornstalk borer, *Elasmopalpus lignosellus* (Insecta: Lepidoptera: Pyralidae), in bush bean (*Phaseolus vulgaris*). Fla Entomol 93:183–190. https://doi.org/10.1653/024.093.0206
Harvest Choice (2017) Bean. http://harvestchoice.org/commodities/bean. Accessed 25 Nov 2017
Herzog DC (1977) Bean leaf beetle parasitism by *Celatoria diabroticae* (Shimer) and *Hyalomyodes triangulifer* (Leow.). J Georgia Entomol Soc 12:64–68

Hoddle MS, Van Driesche RG, Sanderson JP (1998) Biology and use of the whitefly parasitoid *Encarsia formosa*. Annu Rev Entomol 43:645–669. https://doi.org/10.1146/annurev.ento.43.1.645

Hohmann CL, Carvalho SM (1989) Pragas e seu controle. In: O feijão no Paraná. Iapar, Londrina, pp 217–246

Hong SC, Held DW, Williamson RC (2011) Generalist predators and predation of black cutworm *Agrotis ipsilon* larvae in close mown creeping bentgrass. Fla Entomol 94:714–715. https://doi.org/10.1653/024.094.0345

Kepler RM, Luangsa-Ard JJ, Hywel-Jones NL, Quandt CA, Sung GH, Rehner SA, Aime MC, Henkel TW, Sanjuan T, Zare R, Chen M, Li Z, Rossman AY, Spatafora JW, Shrestha B (2017) A phylogenetically-based nomenclature for Cordycipitaceae (Hypocreales). IMA Fungus 8:335–353. https://doi.org/10.5598/imafungus.2017.08.02.08

Koppert Biological Systems (2019) Swirski-mite: predatory mite Amblyseius swirskii. https://www.koppert.com/swirski-mite/. Accessed 16 Jan 2019

Lacey LA, Wraight SP, Kirk AA (2008) Entomopathogenic fungi for control of *Bemisia tabaci* Biotype B: foreign exploration, research and implementation. In: Gould J, Hoelmer K, Goolsby J (eds) Classical biological control of *Bemisia tabaci* in the United States – a review of interagency research and implementation. Springer, Netherlands, pp 33–69

Leite LG, Batista Filho A, Cruz BPB (1991) Crescimento vegetativo do fungo *Zoophthora radicans* (Brefest) humber, em diferentes meios de cultura naturais. Arq Inst Biol 58:21–24

Leite LG, Alves SB, Wraight SP, Galaini-Wraight S, Roberts DW, Magalhães BP (1996) Conidiogênese de *Zoophthora radicans* (Brefeld) Batko sobre *Empoasca kraemeri* (Roos & Moore), à diferentes temperaturas. Arq Inst Biol 63:47–53

Li SJ, Xue X, Ahmed MZ, Ren SX, Du YZ, Wu JH, Cuthbertson AGS, Qiu BL (2011) Host plants and natural enemies of *Bemisia tabaci* (Hemiptera: Aleyrodidae) in China. Insect Sci 18:101–120. https://doi.org/10.1111/j.1744-7917.2010.01395.x

Liu TX, Kang L, Heinz KM, Trumble J (2009) Biological control of *Liriomyza* leafminers: progress and perspective. In: CAB Reviews: Perspectives in Agriculture, Veterinary Science, Nutrition and Natural Resources, 4, p 16

Lord JC, Magalhães BP, Roberts DW (1987) Effects of the fungus *Beauveria bassiana* (Bal.) Vuill, behavior, oviposition, and susceptibility to secondary infections of adult *Cerotoma arcuata*. An Soc Entomol Brasil 16:187–197

Lord JC, Magalhães BP, Roberts DW (1988) Biological control of cowpea pests in Brazil. In: Watt EE, Araújo JPP (eds) Cowpea research in Brazil. IITA/Embrapa, Brasília, pp 301–313

Lourenção AL, Costa VA, Silveira Pereira L, Cardoso Prado J (2014) Occurrence of *Eretmocerus mundus* Mercet (Hymenoptera: Aphelinidae) parasitizing *Bemisia tabaci* (Genn.) biotype B (Hemiptera: Aleyrodidae) in Brazil. Bragantia 73:160–162. https://doi.org/10.1590/brag.2014.025

Magalhães BP, Quintela, ED (1987) Níveis de parasitismo de *Urosigalphus chalcodermi* Wilkinson sobre *Chalcodermus bimaculatus* Fiedier e de *Celatoria bosqi* Blanchard sobre *Cerotoma arcuata* Olivier em caupi (Vigna unguiculata (L.) Walp.) em Goiás. An Soc Entomol Brasil 16:235–238

Magalhães BP, Lord JD, Roberts DW, Daoust RA (1986) Efeito de *Beauveria bassiana* sobre *Cerotoma arcuata* (Olivier), apósexposição de esporos do fungo a tuberculos de Taiuia. An Soc Entomol Brasil 15:327–334

Magalhães BP, Lord JC, Wraight SP, Daoust RA, Roberts DW (1988a) Pathogenicity of *Beauveria bassiana* and *Zoophthora radicans* to the coccinellid predators *Coleomegilla maculata* and *Eriopis connexa*. J Invert Pathol 52:471–473. https://doi.org/10.1016/0022-2011(88)90060-2

Magalhães BP, Lord JC, Roberts DW (1988b) Controle biológico de pragas do caupi no Brasil. In: Araújo JPP, Watt EE (eds) O caupi no Brasil. IITA/Embrapa, Brasília, pp 659–673

Magalhães BP, Butt TM, Humber RA, Shields EJ, Roberts DW (1990) Formation of appressoria in vitro by the entomopathogenic fungus *Zoophthora radicans* (Zygomycetes:Entomophthorales). J Invert Pathol 55:284–288. https://doi.org/10.1016/0022-2011(90)90065-E

Magalhães BP, Leger RS, Humber RA, Allee L, Shields EJ, Roberts DW (1991a) Nuclear events during germination and appressorial formation of the entomopathogenic fungus *Zoophthora radicans* (Zygomycetes: Entomophthorales). J Invertebr Pathol 57:43–49. https://doi.org/10.1016/0022-2011(91)90039-S

Magalhães BP, Wayne R, Humber RA, Shields EJ, Roberts DW (1991b) Calcium-regulated appressorium formation of the entomopathogenic fungus *Zoophthora radicans*. Protoplasma 160:77–88

Magalhães BP, Humber RA, Shields EJ, Roberts DW (1991c) Effects of environment and nutrition on conidium germination and appressorium formation by *Zoophthora radicans* (Zygomycetes: Entomophthorales): a pathogen of the potato leafhopper (Homoptera: Cicadellidae). Environ Entomol 20:1460–1468. https://doi.org/10.1093/ee/20.5.1460

Magalhães BP, Leite LG, Alves SB, Wraight SP, Galaini-Wraight SG, Roberts DW (1996) Conidiogenese de *Zoophthora radicans* (Brefel) Batko sobre *Empoasca kraemeri* (Ross). Arq Inst Biol 63:47–53

Marrone PG, Brooks WM, Stinner RE (1983) The incidence of tachinid parasites and pathogens in adult populations of the bean leaf beetle, *Cerotoma trifurcata* (Forster) (Coleoptera: Chrysomelidae) in North Carolina. J Georgia Entomol Soc 18:363–370

Martinez DG, Pienkowski RL (1982) Laboratory studies on insect predators of potato leafhopper eggs, nymphs, and adults. Environ Entomol 11:361–362. https://doi.org/10.1093/ee/11.2.361

Mascarin GM, Kobori NN, Quintela ED, Delalibera Junior I (2013) The virulence of entomopathogenic fungi against *Bemisia tabaci* biotype B (Hemiptera: Aleyrodidae) and their conidial production using solid substrate fermentation. Biol Control 66:209–218. https://doi.org/10.1016/j.biocontrol.2013.05.001

Mascarin GM, Kobori NN, Quintela ED, Arthurs SP, Delalibera Junior I (2014) Toxicity of non-ionic surfactants and interactions with fungal entomopathogens toward *Bemisia tabaci* biotype B. BioControl 59:111–123. https://doi.org/10.1007/s10526-013-9543-x

Mascarin GM, Pereira-Junior RA, Fernandes EKK, Quintela ED, Dunlap CA, Arthurs SP (2018) Phenotype responses to abiotic stresses, asexual reproduction and virulence among isolates of the entomopathogenic fungus *Cordyceps javanica* (Hypocreales: Cordycipitaceae). Microbiol Res 216:12–22. https://doi.org/10.1016/j.micres.2018.08.002

Metcalf RL (1994) Chemical ecology of Diabroticites. In: Jovilet PH, Cox ML, Petitpierre E (eds) Novel aspects of the biology of Chrysomelidae. Kluwer, Netherlands, pp 153–169

Milanez JM (1984) Ocorrência de artrópodes em consórcio-milho feijão, comparado com as respectivas monoculturas. In: IX Congresso Brasileiro de Entomologia, Londrina

Oliveira MRV, Amancio E, Laumann RA, Gomes LO (2003) Natural enemies of *Bemisia tabaci* (Gennadius) B biotype and *Trialeurodes vaporariorum* (Westwood) (Hemiptera: Aleyrodidae) in Brasília, Brazil. Neotrop Entomol 32:151 154. https://doi.org/10.1590/S1519-566X2003000100023

Oyewale RO, Bamaiyi LJ (2013) Management of cowpea insect pests. Sch Acad J Biosci 1:217–226

Pereira JL, Picanço MC, Pereira EJG, Silva AA, Jakelaitis A, Pereira RR, Xavier VM (2010) Influence of crop management practices on bean foliage arthropods. Bull Entomol Res 100:679–688. https://doi.org/10.1017/S0007485310000039

Pianoski J, Bertucci E, Capassi MC, Cirelli EA, Calafiori MH, Teixeira NT (1990) Efficiency of *Beauveria bassiana* (Bals) Vuill for the control of *Diabrotica speciosa* (Germar, 1824.) on beans *Phaseolus vulgaris* L. with different fertilizer treatments. Ecossistema 15:24–35

Pizzamiglio MA (1979) Aspectos da biologia de *Empoasca kraemeri* (Ross & Moore, 1957) (Homoptera: Cicadellidae) em *Phaseoulus vulgaris* (Linnaeus 1753) e ocorrência de parasitismo de ovos. An Soc Entomol Brasil 8:369–372

Porch T, Beaver JS, Debouck DG, Jackson SA, Kelly JD, Dempewolf H (2013) Use of wild relatives and closely related species to adapt common bean to climate change. Agronomy 3:433–461. https://doi.org/10.3390/agronomy3020433

Quintela ED, Barbosa FR (2015) Manual de identificação de insetos e outros invertebrados pragas do feijoeiro, 2nd edn. Embrapa, Santo Antônio de Goiás, p 91

Quintela ED, Roberts DW (1992) Controle de *Chalcodermus bimaculatus* (Boheman) (Coleoptera: Curculionidae) no solo com *Beauveria bassiana* (Bals.) Vuillemin e *Metarhizium anisopliae* (Metsch) Sorokin. Pesqui Agropecu Bras 27:95–105

Quintela ED, Lord JC, Wraight SP, Alves SB, Roberts DW (1990) Pathogenicity of *Beauveria Bassiana* (Hyphomycetes: Moniliales) to larval and adult *Chalcodermus Binaculatus*(Coleoptera: Curculionidae). J Econ Entomol 83:1276–1279. https://doi.org/10.1093/jee/83.4.1276

Quintela ED, Neves BP, Quinderé MAW, Roberts DW (1991) Principal insect pests of cowpea in Brazil. Embrapa, Goiânia, p 37

Quintela ED, Lord JC, Alves SB, Roberts DW (1992a) Persistência de *Beauveria bassiana* em solo do cerrado e sua interação com microorganismos do solo. An Soc Entomol Brasil 21:69–82

Quintela ED, Sanchez E, Yokoyama M (1992b) Parasitismo de *Encarsia* sp (Hymenoptera: Aphelinidae) em Bemisia tabaci (Gennadius, 1989) (Homoptera: Aleyrodidae). An Soc Entomol Brasil 21:471–475

Quintela ED, Wraight SP, Quinderé MW, Roberts DW (1994) Aplicação de conídios de *Beauveria Bassiana* (Bals.) Vuill. e *Metarhizium Anisopliae* (Metsch.) Sor. para controle de larvas de *Chalcodermus bimaculatus* Boh. (Coleoptera: Curculionidae) no solo. An Soc Entomol Brasil 23:367–377

Quintela ED, Abreu AG, Lima JF, Mascarin GM, Santos JB, Brown JK (2016) Reproduction of the whitefly *Bemisia tabaci* (Hemiptera: Aleyrodidae) B biotype in maize fields (*Zea mays* L.) in Brazil. Pest Manag Sci 72:2181–2187. https://doi.org/10.1002/ps.4259

Rauf A, Shepard BM, Johnson MW (2000) Leafminers in vegetables, ornamental plants and weeds in Indonesia: surveys of host plants, species composition and parasitoids. Int J Pest Manag 46:257–266. https://doi.org/10.1080/09670870050206028

Roberts DW, Quintela E (1990) 10 years of collaborative research on beans and cowpeas: insect pathogens in cowpea pest management systems for developing nations. Proceedings of the Michigan State University, pp 7–13

Sandskär B, Magalhães BP (1994) Cryopreservation of *Zoophthora radicans* (Zygomycetes, Entomophthorales) in liquid nitrogen. Cryobiology 31:206–213. https://doi.org/10.1006/cryo.1994.1025

Santos JB (2017) Estudo da ecologia de paisagem e comportamento sazonal da mosca-branca, *Bemisia tabaci* biótipo B. (Gennadius, 1889) e de seus parasitoides em soja no Brasil central. Tese Doutorado. Universidade Federal de Goiás, Goiânia, p 92

Silochi RMHQ, Coelho SRM, Bischoff TZ, Cassol FDR, Prado NV, Bassinello PZ (2016) Nutritional technological characterization and secondary metabolites in stored carioca bean cultivars. Afr J Agric Res 11:2102–2111. https://doi.org/10.5897/AJAR2016.11013

Toepfer S, Haye T, Erlandson M, Goettel M, Lundgren JG, Kleespies RG, Weber DC, Walsh GC, Peters A, Ehlers RU, Strasser H, Moore D, Keller S, Vidal S, Kuhlmann U (2009) A review of the natural enemies of beetles in the subtribe Diabroticina (Coleoptera: Chrysomelidae): implications for sustainable pest management. Biocontrol Sci Tech 19:1–65. https://doi.org/10.1080/09583150802524727

Van Lenteren JC (2012) The state of commercial augmentative biological control: plenty of natural enemies, but a frustrating lack of uptake. BioControl 57:1–20. https://doi.org/10.1007/s10526-011-9395-1

Van Lenteren JC, Ramakers PMJ, Woets J (1979) The biological control situation in Dutch glasshouses; problems with *Trialeurodes vaporariorum* (Westwood), *Liriomyza bryoniae* Kalt. and *Myzus persicae* Sulz. Med Fac Landbouw Rijksuniv Gent 44:117–125

Vicentini S, Faria M, Oliveira RVM (2001) Screening of *Beauveria bassiana* (Deuteromycotina: Hyphomycetes) isolates against nymphs of *Bemisia tabaci* (Genn.) biotype B (Hemiptera: Aleyrodidae) with description of a bioassay method. Neotrop Entomol 30:97–103. https://doi.org/10.1590/S1519-566X2001000100015

Walsh GC, Pinto AS, Nava DE (2016) Controle bológico de *Diabrotica speciosa*: parasitoides e predadores. In: Nava DE, Ávila CJ, Pinto AS (eds) Diabrotica speciosa. Occasio, Piracicaba, pp 105–117

Wraight S, Carruthers RI, Jaronski ST, Bradley CA, Garza CJ, Galaini-Wraight S (2000) Evaluation of the entomopathogenic fungi *Beauveria bassiana* and *Paecilomyces fumosoroseus* for microbial control of the silverleaf whitefly, *Bemisia argentifolii*. Biol Control 17:203–217. https://doi.org/10.1006/bcon.1999.0799

Wraight SP, Galaini-Wraight S, Carruthers RI, Roberts DW (2003) *Zoophthora radicans* (Zygomycetes: Entomophthorales) spore production from naturally-infected *Empoasca kraemeri* cadavers and dry-formulated mycelium under laboratory and field conditions. Biol Control 28(1):60–77. https://doi.org/10.1016/S1049-9644(03)00035-5

Xavier VM (2013) Variação sazonal dos fatores de mortalidade natural e limiares térmicos para *Liriomyza huidodrensis*. Tese Doutorado. Universidade Federal de Viçosa, Viçosa, p 52

Chapter 23
Coffee

César Freire Carvalho, Stephan Malfitano Carvalho, and Brígida Souza

23.1 Introduction

Coffee production represents one of the most important agricultural activities, both socially and economically, in several Latin-American countries, such as Brazil, Colombia, Ecuador, Peru, Costa Rica, Honduras, Guatemala, El Salvador, and Mexico.

Some of the traditional production practices in Brazil involved the use of diverse systems, consisting of native trees that provided shade for coffee plants. With the modifications implemented to crop management throughout the years, coffee crops became optimal environments for protection, food source, reproduction, and development of many phytophagous arthropods. In large crops, these agents may reach high population densities, demanding the use of chemical products to control them. However, this practice destabilizes the agricultural system due to the adverse effects of certain compounds that act indistinctly against pests and their natural enemies.

More recently, because of a new perspective for the agricultural ecosystem, different coffee crop systems have been adopted, not only in Brazil but also in many other Latin-American countries. Some of these systems are the organomineral (without the application of phytosanitary products), natural, and organic, which may be shaded coffee cultivation system or full sun-grown system. Despite its small percentage when compared to conventional crops, the organic system is an activity with potential to implement environmental preservation, create social and economic valuation in an area, reduce social inequality, and bring opportunities for small producers (Fantini 2014). A fourth type of crop system is characterized by intercropping (conventional or organic). In some regions, coffee producers grow their crops

C. F. Carvalho (Deceased)

S. M. Carvalho · B. Souza (✉)
Department of Entomology, Federal University of Lavras (UFLA), Lavras, MG, Brazil
e-mail: brgsouza@ufla.br

B. Souza et al. (eds.), *Natural Enemies of Insect Pests in Neotropical Agroecosystems*, https://doi.org/10.1007/978-3-030-24733-1_23

in association with some other vegetal species, such as annual crops, fruit trees, Fabaceae plants, forest species, among others, to reduce costs, explore agricultural activities (other than coffee), and improve the physicochemical conditions of the soil. Northern Latin-American countries commonly intercrop several fruit species with coffee, creating a multi-stratified system, in which coffee plants form the shrub layer (Perfecto et al. 1996).

The practices adopted for coffee crops in Latin America and in other producing countries have received attention because of the important role this agroecosystem plays in the maintenance of the local and global biodiversity and in the provision of ecological services (Jha et al. 2014). In Mato Grosso do Sul, Brazil, for example, the most diverse groups of macroinvertebrates were observed in systems with organic coffee enriched with tree species, as opposed to conventional organic coffee systems, without any other botanical species (Portilho and Silva 2008). In Colombia, Armbrecht et al. (2005) observed that not only the number of ant species (Formicidae) decreased in less-shaded crops but also their ecological relations. In nonshaded crops, *Solenopsis picea* (Forel) (Formicidae) and *Pheidole radoszkowski* Mayr (Formicidae) usually compete and eliminate other ant species. However, in a research carried out in coffee cooperatives in El Salvador, Méndez et al. (2009) showed that, despite the potential for better provisions of ecosystem services and more income for producers, the highest diversification of trees promoted the greatest reduction in coffee yield. In Chiapas, Mexico, Soto-Pinto et al. (2000) observed the positive effect of canopy shade (23–38%), which kept production at 48%; however, the authors reported that yield may decrease under shading darker that 50%. In southern Minas Gerais, Brazil, there is a great diversity of arthropods associated with coffee crops under full sun that receive either mechanical or chemical hoeing (César F Carvalho, personal communication 2018). This observation confirms the research on the importance of coffee ecosystems as "ecological corridors" (Cooper 2007; Jha et al. 2014), which are a means to preserve biodiversity. These results are evidence of a conflict regarding the diversification of services provided by shading and the yield of the main product, which should be assessed following criteria according to the producer's needs.

Pretensions aside, many studies show that ecosystem services may be reduced in agricultural systems that are more intensely managed. Thus, despite the diversity of natural enemies associated with coffee pests—to be discussed in this chapter—the simplification of the ecosystem is believed to reduce the survival and the permanence of these agents in the environment. The lack of food for adults, the scarcity of hosts/alternative prey, and the asynchrony of life cycles are some factors that affect the stability of natural enemies and interfere with the regulation of insect-prey populations (Landis et al. 2000).

Some of the phytophagous organisms that are potential key coffee pests are the coffee leaf miner *Leucoptera coffeella* (Guérin-Mèneville & Perrottet) (Lyonetiidae), the coffee berry borer *Hypothenemus hampei* (Ferrari) (Curculionidae), several species of cochineal (coccids, pseudococcids, and diaspidids) (Hemiptera), cicadas (Cicadidae), and mites (Acari). The phytophagous insects and mites associated with coffee plants, as well as any other organism, have natural enemies that feed on them.

Regarding biological control, the natural enemies (entomophagous and entomopathogens) in an agroecosystem use phytophagous for food and reproduction, providing an important ecological service for the reduction of population density of arthropod pests and the promotion of natural biological control.

Many studies show evidence on the positive association between the increase in biodiversity and the increase in number of natural enemies in coffee plants; however, not many of them consider the reduction in damage caused by pests as a response to the increase in the affluence of natural enemies in the diversified area. This chapter aims at presenting the main natural enemies of the most important coffee pests in Latin-American countries, emphasizing the effects of vegetal diversification on these natural enemies.

23.2 Coffee Insect Pests

23.2.1 Coffee Leaf Miner

Surveys on pests that directly affect coffee plants point out *L. coffeella* as one of the most important ones in all Neotropical countries, causing up to 80% of loss. The intensity of the damage caused by larvae is associated to a series of factors, such as cultivars, crop management, climate, altitude, season, indiscriminate use of phytosanitary products, and presence of adventitious and associated plants, among others. Chemical control has been the most widely used method; however, the cultural method, varietal resistance, and biological control may be used to maintain the level of economic damage according to the place and period of the year (Tuelher et al. 2003; Pereira et al. 2007; David-Rueda et al. 2016).

Another relevant aspect was characterized in a research on seasonal factors that affect the mortality of *L. coffeella* during rainy and dry seasons. Mortality at egg, larvae, and pupae stages reached 94.3% in the rainy season and 89% during the dry season. Furthermore, only 57 adults that emerged from 1000 eggs were collected. These observations demonstrate that factors intrinsic to the plants, in addition to rain (46%), inviability of eggs (33%), anomalies during tegument changes (60%), natural enemies (50%), and local climate variations, are determinant for the populational dynamics of the leaf miner (Pereira et al. 2007).

With respect to natural enemies of *L. coffeella*, studies carried out in Latin-American countries that produce coffee demonstrate that parasitoids and predators represent the most important control agents of the coffee leaf miner. In western and southern Bahia, Brazil, four species of Eulophidae were identified [*Cirrospilus neotropicus* Diez & Fidalgo, *Closteroscerus coffeella* (Ihering), *Horismenus aeneicollis* Ashmead, and *Neochrysocharis coffeae* (Ihering)], along with two other species of *Stiropius* (Braconidae) in association with the leaf miner in organic and conventional systems (Melo et al. 2007). In a crop located at an altitude of 1000 m, in Santo Antônio do Amparo, MG, Brazil, and conducted under the same crop systems, the predominant parasitoids of larvae of *L. coffeella* were *Orgilus niger* Penteado Dias, *Centistidea striata* Penteado Dias, *Stiropius reticulatus* Penteado Dias (Braconidae), and *Horismenus* sp. (Ecole et al. 2013). In Soconusco, Chiapas, Mexico, 22 species of parasitoids and 17 species of predators were found in association with *L. coffeella* (Lomeli-Flores et al. 2009). In Colombia, sporadic damages caused by the leaf miner have been registered; in 2013 and 2014, in the Department of Antioquia, approximately 2% of loss and 58–89% of parasitism were observed in association with seven species of Eulophidae. The main one was *C. coffeella*, in addition to *Closterocerus lividus* (Ashmead), *Zagrammosoma multilineatum* (Ashmead), *Pnigalio sarasolai* De Santis, *Horismenus* sp., *Horismenus* n. sp., and *Apleurotropis* n. sp. In this case, conservative biological control is recommended for pest population decrease (David-Rueda et al. 2016).

Parasitoid species of *L. coffeella* differ according to the region. Species of Braconidae and Eulophidae, as identified by Melo et al. (2007), in coffee crops in western and southern Bahia, Brazil, diverged from the ones found in crops in Zona da Mata and Campo das Vertentes, MG, Brazil, where the braconids *C. striata* and *O. niger* were collected, along with the eulophids *Horismenus* sp. and *Proacrias coffeae* Inhering, which were responsible for 26.5% of mortality. The highest mortality percentage of *L. coffeella* occurred during egg, larvae, and pupae stages (Pereira et al. 2007). Studies on these natural enemies characterize them as relevant at natural control and maintenance of pest population densities, with parasitism at a level of 16–30% (Parra and Reis 2013).

Practices, such as weed management and vegetal diversification, affect the community of natural enemies in coffee crops. A study conducted in Machado and Poço Fundo, southern Minas Gerais, Brazil, showed that the natural agroforest system (without the use of fertilizers or phytosanitary products, but with promotion of vegetal diversity) provided more richness and diversity of insects, when compared to the other crop systems (organic, without phytosanitary products and conventional in full sun) and evidenced that the richness of hymenopterans that parasitize leaf miners and the percentage of parasitism were higher in natural agroforest systems than in other systems (Fernandes 2013). In Chiapas, Mexico, Pak et al. (2015) identified, among 27 species of natural enemies, specimens of Encyrtidae, Ichneumonidae, and Vespidae as the most abundant agents in coffee crops and highlighted the importance of botanical composition in the diversification of the habitat as well as the provision of shade for these organisms. The parasitoids associated with larvae of coffee leaf miners in a crop associated with five vegetal species, in Heliodora, MG, Brazil, were the braconids *O. niger, S. reticulatus, C. striata*, and *Horismenus* sp. However, in this case, plant diversity did not affect the total number of mined leaves, nor did it influence the proportion of parasitized miners (Amaral et al. 2010). Rezende et al. (2014) mentioned the presence of parasitoids (16.7%) and predators (59.9%) (53.3% of which were ants) in extrafloral nectaries of *Inga* sp. (Fabaceae) and considered them as important regulators for the population density of coffee leaf miner and the coffee borer beetle.

Research aimed at knowing the dynamics of *L. coffeella* populations associated with the action of parasitoid wasps in conventional and agroecological transition crops (conventional to agroecological) in southern Minas Gerais evidenced endemism of leaf miners in both systems, though without causing economic damages to the crop. The regulation of leaf miner populations was related to the action of parasitoids and to climate factors in the region (Marques 2017).

The larvae of leaf miners, despite being restricted to the leaf parenchyma, are targeted by predators. The association between wasps and the coffee leaf miner is evidenced in several studies. The evaluation of the influence of seasons and altitude on the occurrence of leaf miners and their natural enemies in coffee crops in Viçosa, MG, Brazil, showed that the occurrence of predator wasps varies according to the seasons and the pest population density, having been registered a predation of 50–90%. Wasps did not occur during rainy season, but during dry season they were the most expressive mortality factor of the pest (Pereira et al. 2007). The highest mortality percentage of *L. coffeella* by these predators occurred during egg, larva, and pupa phases, which were considered critical stages for the reduction of population density and dynamics of the coffee leaf miner (Pereira et al. 2007). The space-time distribution of leaf miners and predator wasps in organic coffee plants, in Santo Antonio do Amparo, MG, Brazil, evidenced a positive correlation between the number of wasps and the larvae mortality. Representatives of 14 species of Vespidae were collected. *Brachygastra augusti* (Saussure), *Polybia occidentalis* Olivier, and *Polybia paulista* von Ihering interacted negatively with the pest, being responsible for 25–50% of larval mortality (Scalon et al. 2011).

Ants are also important natural enemies of larvae and pupae of *L. coffeella*, affecting their population dynamics. In coffee crops in Chiapas, Mexico, these predators were represented by 14 species, which were considered the most important mortality factor of the pest (58% of the total) (Lomelí-Flores et al. 2010). Some studies also report the predation of leaf miners by larvae of *Chrysoperla externa* (Hagen) (Chrysopidae). According to Ecole et al. (2002), this green lacewing may contribute to the regulation of leaf miner populations by preying pre-pupae and pupae in coffee plants.

Regarding the effect of vegetal diversification on the fauna of leaf miner predators, Amaral et al. (2010) observed that the association of coffee plants with five botanical species in Heliodora, MG, Brazil, did not affect the infestation by the pest; however, the diversity of plants positively influenced the number of predated mines, and Vespidae was the most important group in coffee plants in a shaded cultivation system. In Nicaragua, the number of insects collected in agroforest systems was also larger than in full-sun cultivation (Guharay et al. 2001). In coffee plants intercropped with five plant species in Machado, MG, Brazil, 15 species of wasps representatives of six genera were identified, the most abundant being *Polistes* (41%), *Polybia* (36%), and *Angiopolybia* (18%). The highest diversity of wasps occurred in areas intercropped with *Mucuna pruriens* (L.) (Fabaceae) and *Crotalaria juncea* (L.) (Fabaceae), which should be more precisely assessed in conservative biological control programs (Tomazella VB – personal communication 2018).

Gusmão et al. (2000), in tests for evaluation of phytosanitary products against wasps *Apoica pallens* (Fabricius), *Brachygastra lecheguana* (Latreille), and *Polistes versicolor versicolor* (Olivier), and Fragoso et al. (2001), in evaluations for *B. lecheguana*,

P. paulista, and *Protopolybia exigua* (Saussure), mentioned *Protonectarina sylveirae* (Saussure), *Polybia scutellaris* (White), *P. occidentalis*, and *Synoeca surinama cyanea* (L.) as important predators of coffee leaf miners.

23.2.2 Cochineals

Cochineals belonging to several families, especially Coccidae, Pseudococcidae, and Diaspididae, occur naturally in coffee plants in tropical regions (Costa et al. 2009, 2016; Fornazier et al. 2017; Santa-Cecília et al. 2017) and oftentimes cause significant damages. These insects occur in branches, leaves, flowers, fruits, and root system of the plant, and the damages may vary according to the cultivar, location, crop system, climate conditions, crop management, and associated and interrow plants.

When referring to natural enemies of cochineals, it must be considered that several species produce a waxy shell that protects them; others produce a whitish secretion with the same purpose. Insects that feed on cochineals find these characteristics as barriers; however, the ladybugs *Azya luteipes* Mulsant, *Pentilia egena* Mulsant, *Coccidophilus citricola* Brèthes, *Cryptolaemus montrouzieri* Mulsant, *Hyperaspis* sp., *Scymnus* sp. and *Calloeneis* sp. (Coccinellidae), *C. externa,* and *Ceraeochrysa cubana* (Hagen) (Chrysopidae) are mentioned as the most important predators of nymphs and adults (Santa-Cecília et al. 2007; Costa et al. 2009, 2014).

One of the most common coccinellids in coffee crops in Brazil is *A. luteipes*, which is found in every coffee-producing region. In Rondônia state, for example, larvae and adults of this ladybug were observed predating *C. viridis* and *Planococcus citri* (Risso) (Pseudococcidae) in *Conillon* coffee plants (Costa et al. 2014). Also in Rondônia, the following parasitoids of *P. citri* were registered: *Leptomastix dactylopii* Howard, *Aphycus alboclavatus* Ishii, *Anagyrus pseudococci* (Girault), *Anagyrus coccidovorus* Dozier, *Leptomastidea abnormis* (Girault) (Encyrtidae), *Apanteles paraguayensis* Brèthes (Braconidae), *Coccophagus caridei* (Brèthes) (Aphelinidae), *Chartocerus niger* (Ashmead) [= *Thysanus niger* (Ashmead)] (Signiphoridae), and *Pachyneuron* sp. (Pteromalidae).

Research on the antagonism and dominance between the ladybug *Azya orbigera* Mulsant and the fungus *Lecanicillium lecanii* (Zimmermann) R. Zare & W. Gams (= *Verticillium lecanii*) on the green coffee scale, *C. viridis*, in organic coffee in Chiapas, Mexico, showed that fungal infection was reduced in presence of the coccinellid and that simultaneous occurrence of both natural enemies favors the biological control only when the predator is dominant (Ong and Vandermeer 2014).

Beside *L. lecanii*, other infectious entomopathogenic fungi of scale insects are *Uredinella* sp. and *Myriangium duriaei* Mont. & Berk., which, in addition to occurring in coffee, are associated with the same insect pests in citrus crops. Species of *Coccus* genus are also parasitized by *Acrostalagmus albus* Preuss [= *Verticillium album* (Preuss) Pidopl.]. The mealybugs *Planococcus* spp. can be infected by *L. lecanii* and *Neozygites fumosa* (Speare) Remaud. & S. Keller (= *Empusa fumosa* Speare) (Costa et al. 2009). The fungi *Fusarium coccophilum* (Desm.) Wollenw. &

Reink, *Tetracrium cocciolum* Hönell, *Aschersonia* sp., *M. duriaei,* and *L. lecanii* are mentioned infecting the diaspidid *Pinnaspis aspidistrae* (Signoret), which occurs in coffee plants and citrus (Prade et al. 2007).

Fungi and nematodes highlight among the natural enemies of cochineals that occur in the root system of the coffee plant. For the former, *Beauveria bassiana* (Bals.-Criv.) Vuill., *Metarhizium anisopliae* (Metchnikoff) Sorokin, *L. lecanii,* and *Paecilomyces* sp. are mentioned. The most important nematodes belong the families Steinernematidae and Heterorhabditidae, from which *Heterorhabditis* sp. and *Steinernema carpocapsae* (Weiser) stand out as the most efficient ones in the control of these insects in coffee roots (Santa-Cecília et al. 2007).

23.2.3 Cicadas

Coffee cicadas (Cicadidae) have been considered as crop pests for over a century. Because they are necessarily associated with the root system of the plant and always protected in some way, controlling these insects is not easy. There are eight species most commonly associated with coffee, and several control methods are used to minimize crop damage, such as the cultural, mechanical, chemical, and biological methods (Parra and Reis 2013).

The fungus *M. anisopliae* is mentioned as one of the natural mortality agents of cicadas; however, much research has revealed its inefficiency in terms of population control of the pest. In laboratory, Cintra et al. (2013) observed this fungus causing mortality in *Fidicinoides pronoe* (Walker) and being compatible with some phytosanitary products used in coffee crops. Other fungi are *Massospora spinosa* Cif., A.A. Machado & Vittal, *Massospora dorisiana* R.S. Soper (= *M. dorisianae*, valid name), *Massospora carineta* R.S. Soper (= *M. carineta*, valid name), and *Massospora diminuta* R.S. Soper, which are found colonizing nymphs of some species of cicadas in Brazil (Silva 2011).

Similar to the cochineals by their cryptic habit, the control of *Quesada gigas* (Olivier) has been tested by using the nematodes *Heterorhabditis* sp. and *Steinernema riobrave* Cabanillas, Poinar & Raulston, which seems to be an alternative to coffee producers (Silva 2011).

23.2.4 Mites

Phytophagous mites, mainly represented by *Oligonychus ilicis* (McGregor) (Tetranychidae), *Polyphagotarsonemus latus* (Banks) (Tarsonemidae), and *Brevipalpus phoenicis* (Geijskes) (Tenuipalpidae), constitute a group of coffee crop pests because they hinder the development of foliar tissues and fruits and cause a typical necrosis. Their main natural enemies are predatory mites and the fungus *Beauveria* spp. Martins et al. (2016) selected and characterized some isolates of this fungus to control *P. latus* in laboratory, greenhouse, and field environments. Although the research was conducted using another host plant, the results varied from 66% to 70% of mortality, confirming the efficiency of this entomopathogen as a natural enemy. Regarding predatory mites, the main families are Phytoseiidae and Stigmaeidae, which have numerous species. The coffee agroecosystem generally holds an abundant fauna of these predators, and 134 species of phytoseiids were observed, primarily *Iphiseiodes zuluagai* Denmark & Muma, *Ambleyseius acalyphus* Denmark & Muma, *Ambleyseius herbicolus* (Chant), and *Euseius alatus* DeLeon (Silva et al. 2010). Experimentally, *I. zuluagai, E. alatus,* and *A. herbicolus* promoted population reduction of *B. phoenicis*, a mite responsible for transmitting citrus leprosis virus and coffee ringspot virus (Reis et al. 2003, 2007).

The fauna of predatory mite species may differ according to the geographical region in which they occur. In coffee crops in the Cerrado region of Minas Gerais, Brazil, among the 1,299 specimens of 12 families of mites selected as samples, three were phytophagous, four predators (Phytoseiidae, Stigmaeidae, Bdellidae, and Ascidae), and five generalists. *Euseius citrifolius* Denmark & Muma was the most frequent species, as well as *A. herbicolus* and *Euseius concordis* (Chant) (Spongoski et al. 2005). Mineiro et al. (2008b) reported *E. citrifolius* and *E. concordis* as the most frequent *B. phoenicis* predatory mites in coffee crops in Garça, SP, Brazil. Similarly, Mineiro et al. (2008a) observed the phytophagous *B. phoenicis* and *Oligonychus yothersi* (McGregor) and the predators *I. zuluagai* and *Bdella* sp. in Catuai coffee crop, in Atibaia, SP, Brazil. Mineiro et al. (2009) reported 39 species, the most common being the phytoseiids *E. citrifolius*, *E. concordis,* and *I. zuluagai*. A study carried out by Lara et al. (2010) on the diversity of Hemerobiidae (Neuroptera) in coffee plants and their relation with several pests revealed a negative correlation between the population density of *Sympherobius miranda* (Navás) and *O. ilicis* and between the occurrence of *Nusalala tessellata* (Gerstaecker) and the reduction in the population of this mite.

In relation to the changes to the structure and composition of the mite communities in function of the coffee crop management system, the natural coffee crop system in southern Minas Gerais, Brazil, hosts a greater number of predatory and generalist mites, as opposed to organic and conventional systems, and systems without phytosanitary products (Fernandes 2013). According to Teodoro et al. (2008), the vegetative characteristics of the coffee-growing agroforest system also contribute to the reduction of arthropod-pest populations, and the population density of mites and coffee borer beetles was lower in structurally more complex agroforests.

The management of spontaneous plants in the crop is another important factor that contributes to the structure of the mite communities associated with coffee. In organic crops, practices such as weeding and mowing (in total area or alternate rows) promoted a population increase of *E. citrifolius*, but non-management was beneficial to predators such as *E. concordis* and *Neoseiulus affs. mumai* (Denmark) (Pedro Neto 2009). In Ecuador, Teodoro et al. (2009) also observed a negative response of these predators as response by the intensification of crop management. In São Sebastião do Paraíso, MG, spontaneous plants (especially broad-leafed ones) in coffee crops host predatory mites, even though they may also host pest mites

(Fernanda A Abreu – personal communication 2013). Consortiating these plants between the rows of *C. arabica* provides shelter and food for predatory mites and contributes to their permanence in the area; therefore, they should be properly managed.

The diversification of coffee production areas through the introduction of beneficial plants has been adopted by farmers as green manures or cover crops to increase organic matter, soil protection, nitrogen fixation, nutrient cycling, or spontaneous plant management (Amaral et al. 2010). In Machado, MG, Brazil, the community of phytophagous, predatory, and generalist mites was evaluated in diversified coffee crops by intercropping with popular tropical green manures: pigeon pea (*Cajanus cajan*) (Fabaceae), rattlepod (*Crotalaria spectabilis*) (Fabaceae), velvet bean (*Mucuna deeringiana*) (Fabaceae), Congo grass (*Brachiaria ruziziensis*) (Poaceae), and marigold (*Tagetes erecta*) (Asteraceae) associated with spontaneous plants. The composition of the predatory mite fauna was affected by botanical species consortiated; there was a greater abundance in the system with rattlepod (30.6%) and Congo grass (21.8%). The mite community was also affected by the phenological stage of the plant, showing a higher density of predatory species and lower density of phytophagous during the full flowering period of the intercropped plants (Fernandes 2013).

23.2.5 Coffee Berry Borer

The coffee berry borer *Hypothenemus hampei* (Ferrari) (Curculionidae) is the most important insect pest of coffee beans in the world (Morris and Perfecto 2016). The arrival of *H. hampei* in the American continent occurred in Brazil, probably in 1913, with the introduction of infested coffee beans from Africa (Laurentino and Costa 2004). Since then, this insect is considered one of the main pests, being found in every crop in several producing regions of the country. A similar situation occurred in Mexico in 1978, Colombia in 1988, Costa Rica in 2000, and in other Latin-American countries in decades of 1970s, 1980s, and 1990s when the borer was detected (Baker 1984; Kein-Koch 1990; Bustillo et al. 1998; Rojas 2012). Although their importance has been reduced due to the changes in crop management, this insect has caused considerable damage to coffee berry. The biotic and abiotic differences in crops with and without shading are examples of crop management, impacting the severity of damage, distribution, and reproduction of insects. Shading, besides affecting the phytophagous agents, favors the occurrence of natural enemies, such as ants and entomopathogens (Mariño et al. 2016). For example, the presence of *C. viridis* and its mutualistic relationship with the ant *Azteca instabilis* (Smith, F.) caused a positive indirect effect on the coffee plant by reducing the number of several insect pests, including coffee berry borer (Perfecto and Vandermeer 2006).

There is a considerable number of studies about the natural enemies of the coffee berry borer that occur naturally in all regions where the plant is cultivated, including fungi, bacteria, protozoa, nematodes, parasitoids, and predators. The infection of the coffee borer by the entomopathogenic fungus *B. bassiana* occurs naturally in several countries, including Brazil (Villacorta 1984), Colombia (Vélez-Arango and Benavides-Gómez 1990), and Puerto Rico (Gallardo-Covas et al. 2010), among others. However, the natural occurrence of *B. bassiana* on adults of the borer is variable, with infection rates being below 1% in Brazil (Costa et al. 2002) and higher in other countries, such as up to 44% in Nicaragua (Monzón et al. 2008) and up to 70.6% in Cameroon (Mbang et al. 2012). The effectiveness of *B. bassiana* under field conditions depends on several factors, such as strain, concentration, virulence, climate conditions and application efficiency (Posada et al. 2003; Agegnehu et al. 2015).

Classical biological control is evaluated in Latin America, as well as in other coffee-growing regions. In Brazil, the introduction of the parasitoid *Prorops nasuta* Waterston (Bethylidae) occurred in March 1929, followed by a wide campaign to raise awareness among farmers, being reared and distributed in the field. This was considered the first biological pest control program implemented in Brazil and the main program developed by researchers from the Biological Institute, São Paulo. It was proposed a protocol to ensure the best way to control the pest, but even with all the technical–scientific effort made at the time, the wasp could not settle in Brazilian territory as supposed. In other countries, a similar fact happened. In Mexico, where the occurrence of the pest was registered in 1978 in Chiapas state, the African parasitoids *P. nasuta* and *Cephalonomia stephanoderis* Betrem (Bethylidae) were introduced, which were not successful in controlling the borer. In addition to these parasitoids, *Cephalonomia hyalinipennis* Ashmead is also mentioned, which occurs naturally in North America and Europe (Batchelor et al. 2006). Infante et al. (2013) reported that the eulophid *Phymastichus coffea* LaSalle, native to Africa and introduced in Mexico, has a potential to control the borer in the Latin-American coffee agroecosystems; however, this species does not occur in Brazil. Laurentino and Costa (2004) registered *Heterospillus coffeicolla* Schmiedknecht (Braconidae) in the state of Rondônia, Brazil, whose larvae feed on eggs, larvae, and pupae of the borer. Also in Rondônia, *Polynema* sp. (Mymaridae), an egg endoparasite, is a potential candidate for the control of the borer in its egg stage.

In Colombia, although coffee borer was introduced at the end of the 1990s, several natural enemies were recorded, including five species of fungi, represented by *B. bassiana, Hirsutella eleutheratorum* (Nex ex Gray) Petch, *M. anisopliae, Fusarium oxysporum* Schlecht emend. Snyder & Hansen, and *Paecilomyces lilacinus* (Thom.) Samson. *Isaria fumosorosea* (Wise) is a soil fungus and may be used to control insects, such as the borer, that occur in fruits fallen on the ground (Bustillo et al. 2002). In Ecuador, *B. bassiana* and fruit infection by bacteria and fungi that destroy endosperms are the most important natural factors that may reduce *H. hampei* populations (Kein-Koch 1990).

With respect to the coffee borer predators, the most important ones are ants (Formicidae), hemipterous of the Anthocoridae family, and coleopterous of the Cucujidae family. However, regardless of the coffee-growing region, ants are the most efficient natural enemies. In Brazil, *Crematogaster curvispinosus* Mayr (Reis and Souza 1998) is recorded, and in Colombia, the genera *Solenopsis*, *Pheidole*, *Wasmannia*, *Paratrechina*, *Crematogaster*, *Brachymyrmex*, and *Prenolepis* are registered (Bustillo et al., 2002; Armbrecht et al. 2005). The predatory potential of *Wasmannia auropunctata* (Roger) and *Solenopsis picea* Emery was evaluated against immature forms of the borer, supporting them as efficient predators (Morris and Perfecto 2016). Studies conducted in the laboratory by Larsen and Philpott (2010) have shown that several ant species nesting on tree branches are effective predators of borer both outside and inside grains, such as *Pseudomyrmex ejectus* (Smith, F.), *Pseudomyrmex simplex* (Smith, F.), and *Pseudomyrmex* PSW-53. In the field, the abundance of colonies of these ants was influenced by shade management techniques, with a higher number of colonies in places where trees were pruned. However, the richness and diversity of these predators were less affected by shading, since these ants can colonize coffee branches in shade and in full sun (Larsen and Philpott 2010).

H. hampei is a widespread insect in all regions where coffee is grown, and other natural enemies are also found, such as *Karnyothripes flavipes* (Jones) (Phlaeothripidae) in Kenya, Africa. The adults of *K. flavipes* are predators and feed on borer eggs (Jaramillo et al. 2010). The occurrence of this thrips species is also documented in other countries and throughout Latin America (Vega and Jaramillo 2010). Some nematode species may also be mentioned, for example, *Heterorhabditis* sp., *Heterorhabditis bacteriophora* Poinar, *Steinernema feltiae* (Filipjev), *Sphaerulariopsis* sp. (Tylenchidae), *Metaparasitylenchus hypothenemi* Poinar Jr., Vega, Castillo, Chavez & Infante (Allantonematidae), infecting larvae and adults of the borer on fallen fruits or interfering on the female fecundity (Jaramillo et al. 2006).

23.3 Final Remarks

Coffee crops (specially arabica or robusta) are explored in most Latin-American countries distributed predominantly between latitudes 25°N and 30°S. The great latitudinal extension, different altitudinal gradients, and peculiar edaphic and climatic characteristics of each region allow a diverse range of environments explored by numerous organisms, many of which are associated with coffee plants. Thus, coffee crop management assumes an important role not only in relation to the production itself but also in the maintenance of biodiversity. The restructuring of coffee crops by intercropping with other plants, maintenance of spontaneous plants, or growth of other related botanical species, in or around the cultivated area, are biodiversity management strategies, which can benefit the regulation of pest populations aimed by increasing populations of natural enemies and the ecological balance of the coffee system.

References

Abreu FA (2013) Manejo de plantas adventícias em cafezal e seu efeito sobre ácaros fitófagos e predadores. Ed. UFLA, Lavras

Agegnehu E, Thakur A, Mulualem T (2015) Potential impact of climate change on dynamics of coffee berry borer (*Hypothenemus hampei* Ferrari) in Ethiopia. Open Access Libr J 2(1):1–11

Amaral DS, Venzon M, Pallini A et al (2010) A diversificação da vegetação reduz o ataque do bicho-mineiro-do-cafeeiro *Leucoptera coffeella* (Guérin-Mèneville) (Lepidoptera: Lyonetiidae)? Neotrop Entomol 39(4):543–548

Armbrecht I, Rivera L, Perfecto I (2005) Reduced diversity and complexity in the leaf-litter ant assemblage of Colombian coffee plantations. Conserv Biol 19(3):897–907

Baker PS (1984) Some aspects of the behavior of the coffee berry borer in relation to its control in southern Mexico (Coleoptera: Scolytidae). Folia Entomol Mex 62:9–24

Batchelor TP, Hardy ICW, Barrera JF (2006) Interactions among bethylid parasitoid species attacking the coffee berry borer, *Hypothenemus hampei* (Coleoptera: Scolytidae). Biol Control 36(1):106–118

Bustillo AE, Cárdenas R, Villalba D et al (1998) Manejo integrado de la broca del café, *Hypothenemus hampei* (Ferrari) en Colombia. Cenicafé, Chinchiná

Bustillo AE, Cárdenas R, Posada FJ (2002) Natural enemies and competitors of *Hypothenemus hampei* (Ferrari) (Coleoptera: Scolytidae) in Colombia. Neotrop Entomol 31(4):635–639

Cintra ERR, Loureiro ES, Almeida JEM et al (2013) Patogenicidade de *Metarhizium anisopliae* à cigarra do café *Fidicinoides pronoe* (Hemiptera: Cicadidae) e sua compatibilidade com agrotóxicos utilizados na cultura do cafeeiro. Biológico 75(1):63–70

Cooper DS (2007) Ecological assessment of five coffee farms in north-central. Cooper Ecological Monitoring, Nicaragua

Costa JNM, Silva RB, Ribeiro PA et al (2002) Ocorrência de *Beauveria bassiana* (Bals.) Vuill. em broca-do-café (*Hypothenemus hampei*, Ferrari) no estado de Rondônia, Brasil. Acta Amaz 32(3):517–519

Costa JNM, Teixeira CAD, Sallet LAP et al (2009) Cochonilhas ocorrentes em cafezais de Rondônia. Embrapa, Porto Velho. Circular Técnica 110

Costa JNM, Uchoa TL, Silva TT et al (2014) Ocorrência de joaninha *Azya luteipes* Mulsant, 1850 (Coleoptera: Coccinellidae), agente de controle biológico de cochonilhas do cafeeiro, em Porto Velho, Rondônia. Embrapa, Porto Velho. Comunicado Técnico, 395

Costa MB, Souza BS, Santa-Cecília LVC et al (2016) Tabela de vida de fertilidade de *Planococcus citri* (Risso) e *Planococcus minor* (Maskel) (Hemiptera: Pseudococcidae) em cafeeiro. Coffee Sci 11(2):204–210

David-Rueda G, Constantino LM, Montoya EC et al (2016) Diagnóstico de *Leucoptera coffeella* (Lepidoptera: Lyonetiidae) y sus parasitoides en el departamento de Antioquia, Colombia. Rev Colomb Entomol 42(1):4–11

Ecole CC, Silva RA, Louzada JNC et al (2002) Predação de ovos, larvas e pupas do bicho-mineiro do cafeeiro, *Leucoptera coffeella* (Guérin-Mèneville & Perrottet, 1842) (Lepidoptera: Lyonetiidae) por *Chrysoperla externa* (Hagen, 1861) (Neuroptera: Chrysopidae). Ciênc Agrotec 26(2):318–324

Ecole CC, Vilela M, Moraes JC et al (2013) Dinâmica populacional do bicho-mineiro e seus inimigos naturais em cafeeiro orgânico e convencional adensados. Ver Agric 88(1):44–52

Fantini C (2014) Tudo sobre o cultivo do café orgânico. Saber na Rede. http://www.sabernarede.com.br/tudo-sobre-o-cultivo-do-cafe-organico/. Accessed 12 Abr 2017

Fernandes LG (2013) Diversidade de inimigos naturais de pragas do cafeeiro em diferentes sistemas de cultivo. Tese, Universidade Federal de Lavras

Fornazier MJ, Martins DS, Willink MCG et al (2017) Scale insects (Hemiptera: Coccoidea) associated with arabic coffee and geographical distribution in the Neotropical region. An Acad Bras Ciênc 89(4):3083–3092

Fragoso DB, Jusselino-Filho P, Guedes RNC et al (2001) Seletividade de inseticidas a vespas predadoras de *Leucoptera coffeella* (Guér.-Mènev.) (Lepidoptera: Lyonetiidae). Neotrop Entomol 30(1):139–144

Gallardo-Covas F, Hernández E, Pagán J (2010) Presencia natural del hongo *Beauveria bassiana* (Bals.) Vuill. en la broca del fruto del café *Hypothenemus hampei* (Ferrari) en Puerto Rico. J Agric Univ P R 94:195–198

Guharay F, Monterosso D, Staver C (2001) El diseño y manejo de la sombra para la supresión de plagas en cafetales de América Central. Agrofor Am 8:22–29

Gusmão MR, Picanço M, Gonring AHR et al (2000) Seletividade fisiológica de inseticidas a Vespidae predadores do bicho-mineiro-do-cafeeiro. Pesq Agropec Bras 35(4):681–686

Infante F, Castillo A, Pérez J et al (2013) Field-cage evaluation of the parasitoid *Phymastichus coffea* as a natural enemy of the coffee berry borer, *Hypothenemus hampei*. Biol Control 67(3):446–450

Jaramillo J, Borgemeister C, Baker P (2006) Coffee berry borer *Hypothenemus hampei* (Coleoptera: Curculionidae): searching for sustainable control strategies. Bull Entomol Res 96(3):223–233

Jaramillo J, Chapman EG, Veja FE et al (2010) Molecular diagnosis of a previouly unreported predator-prey association in coffee: *Karnyothrips flavipes* Jones (Thysanoptera: Phlaeothripidae) predation on the coffee berry borer. Naturwissenschaften 97(3):291–298

Jha S, Bacon CM, Philpott SM et al (2014) Shade coffee: update on a disappearing refuge for biodiversity. Bioscience 64(5):416–428

Kein-Koch C (1990) Natural regulation factors and classical biological control of the coffee berry borer (*Hypothenemus hampei* Ferrari) in Ecuador. In: Proceedings of the international DLG-symposium on integrated pest management in tropical and subtropical systems, [s.n.], Bad Dürkheim, 8–15 Feb. 1990

Landis DA, Wratten SD, Gurr GM (2000) Habitat management to conserve natural enemies of arthropod pests in agriculture. Annu Rev Entomol 45:175–201

Lara RIR, Perioto NW, Freitas S (2010) Diversidade de hemerobiídeos (Neuroptera) e suas associações com presas em cafeeiros. Pesq Agropec Bras 45(2):115–123

Larsen A, Philpott SM (2010) Twig-nesting ants: the hidden predators of the coffee berry borer in Chiapas, México. Biotropica 42(3):342–347

Laurentino E, Costa JNM (2004) Descrição e caracterização da broca-do-café *Hypothenemus hampei* (Ferrari) no estado de Rondônia. Embrapa, Porto Velho

Lomelí-Flores JR, Barrera JF, Bernal JS (2009) Impact of natural enemies on coffee leafminer *Leucoptera coffeella* (Lepidoptera: Lyonetiidae) population dynamics in Chiapas, México. Biol Control 51(1):51–60

Lomelí-Flores JR, Barrera JF, Bernal JS (2010) Impact of weather, shade cover and elevation on coffee leafminer *Leucoptera coffeella* (Lepidoptera: Lyonetiidae) population dynamics and natural enemies. Crop Prot 29(6):1039–1048

Mariño YA, Pérez M-E, Gallardo F et al (2016) Sun vs. Shade affects infestation, total population and sex ratio of the coffee berry borer (*Hypothenemus hampei*) in Puerto Rico. Agric Ecosyst Environ 222:258–266

Marques KBSC (2017) Infestação e parasitismo de *Leucoptera coffeella* (Guérin-Mèneville & Perrottet, 1842) em cafeeiros em transição agroecológica. Dissertação, Universidade Federal de Lavras

Martins CC, Alves LFA, Mamprim AP et al (2016) Selection and characterization of *Beauveria* sp. isolates to control of broad mite *Polyphagotarsonemus latus* (Banks, 1904) (Acari: Tarsonemidae). Braz J Biol 76(3):629–637

Mbang JA, Mounjouenpou P, Mahob RJ et al (2012) Evaluaation naturelle de l'impact de *Beauveria bassiana*: Champignon entomopathogène dans la dynamique de population de *Hypothenemus hampei*, scolyte de baies des cerises de *Coffea canefora*. Afr Crop Sci J 20:443–451

Melo TL, Castellani MA, Nascimento ML et al (2007) Comunidades de parasitoides de *Leucoptera coffeella* (Guérin-Méneville & Perrotet) (Lepidoptera: Lyonetiidae) em cafeeiros nas regiões Oeste e Sudeste da Bahia. Ciênc Agrotec 31(4):966–972

Méndez EV, Shapiro EN, Gilbert GS (2009) Cooperative management and its effects on shade tree diversity, soil properties and ecosystem services of coffee plantations in western El Salvador. Agrofor Syst 76(1):111–126

Mineiro JLC, Sato ME, Raga A et al (2008a) Incidência de ácaros em cafeeiro cv. Catuaí amarelo. Bragantia 67(1):197–201

Mineiro JLC, Sato ME, Raga A et al (2008b) Population dynamics of phytophagous and predaceous mites on coffee in Brasil, with emphasis on *Brevipalpus phoenicis* (Acari: Tenuipalpidae). Exp Appl Acarol 44(4):277–291

Mineiro JLC, Raga A, Sato ME et al (2009) Ácaros associados ao cafeeiro (*Coffea* spp.) no estado de São Paulo, Brasil. Parte I Mesostigmata. Biota Neotrop 9(1):37–46

Monzón AJ, Guharay F, Klingen I (2008) Natural occurrence of *Beauveria bassiana* in *Hypothenemus hampei* (Coleoptera: Curculionidae) population in unsprayed coffee fields. J Invertebr Pathol 97(2):134–141

Morris JR, Perfecto I (2016) Testing the potential for ant predation of immature coffee berry borer (*Hypothenemus hampei*) life stages. Agric Ecosyst Environ 233:224–228

Ong TW, Vandermeer JH (2014) Antagonism between two natural enemies improves biological control of a coffee pest: the importance of dominance hierarchies. Biol Control 76:107–113

Pak D, Iverson AL, Ennis KK et al (2015) Parasitoid wasps benefit from shade tree size and landscape complexity in Mexico coffee agroecosystems. Agric Ecosyst Environ 206:21–32

Parra JRP, Reis PR (2013) Manejo integrado para as principais pragas da cafeicultura, no Brasil. Visão Agríc 12:47–50

Pedro Neto M (2009) Influência de cobertura vegetal do solo e da precipitação pluvial na população de ácaros-praga e ácaros predadores em cafeeiro orgânico e convencional. Tese, Universidade Federal de Lavras

Pereira EJG, Picanço MC, Bacci L et al (2007) Seasonal mortality factors of the coffee leafminer, *Leucoptera coffeella*. Bull Entomol Res 97(4):421–432

Perfecto I, Vandermeer J (2006) The effect of an ant-hemipteran mutualism on the coffee berry borer (*Hypothenemus hampei*) in southern Mexico. Agric Ecosyst Environ 117(2/3):218–221

Perfecto I, Rice RA, Greenberg R et al (1996) Shade coffee: a disappearing refuge for biodiversity. Bioscience 46(8):598–608

Portilho II, Silva RF (2008) Macrofauna epigéica em diferentes sistemas de manejo de café orgânico em Mato Grosso do Sul. Rev Bras Agroecol 3(1):63–66

Posada FFJ, Salazar EHM, Aristizábal ALF et al (2003) Taller de evaluación de *Beauveria bassiana* con caficultores experimentadores para el control de *Hypothenemus hampei* (Coleoptera: Scolytidae). Ver Colomb Entomol 29(1):63–68

Prade CA, Dal Soglio F, Wolf VRS et al (2007) Ocorrência de fungos filamentosos associados às cochonilhas com escudo (Hemiptera: Diaspididae) em pomar de citros no município de Taquari-RS. Biociencias 15(1):68–72

Reis PR, Souza JC (1998) Manejo integrado das pragas do cafeeiro em Minas Gerais. Inf Agropecu 19(193):17–25

Reis PR, Sousa EO, Teodoro AV et al (2003) Effect of prey density on the functional and numerical responses of two species of predaceous mites (Acari: Phytoseiidae). Neotrop Entomol 32(3):461–467

Reis PR, Teodoro AV, Pedro Neto M et al (2007) Life history of *Amblyseius herbicolus* (Chant) (Acari: Phytoseiidae) on coffee plants. Neotrop Entomol 36(2):282–287

Rezende MQ, Venzon M, Perez AL et al (2014) Extrafloral nectaries of associated tres can enhance natural pest control. Agric Ecosyst Environ 188:198–203

Rojas M (2012) Manejo sostenible de la broca del café (*Hypothenemus hampei*) mediante poda sistemática del cafeto en Costa Rica. Agronomía Costarricense 36(2):71–79

Santa-Cecília LVC, Souza B, Souza JC et al (2007) Cochonilhas-farinhentas em cafeeiros: bioecologia, danos e métodos de controle. Epamig, Belo Horizonte. Boletim Técnico, 79

Santa-Cecília LVC, Prado E, Pereira AB et al (2017) *Planococcus citri* (Risso) (Hemiptera: Pseudococcidae): relações macho-fêmea e influências na biologia reprodutiva. Coffee Sci 12(3):336–343

Scalon JD, Avelar MBL, Alves GF et al (2011) Spatial and temporal dynamics of coffee leafminer and predatory wasps in organic coffee field in formation. Ciênc Rural 41(4):646–652

Silva MAT (2011) Controle de *Quesada gigas* (Hemiptera: Cicadidae) pela aplicação de nematoides entomopatogênicos e compatibilidade com alguns produtos fitossanitários em cafeeiro. Tese, Universidade Federal de Lavras

Silva EA, Reis PR, Zacarias MS et al (2010) Fitoseídeos (Acari: Phytoseiidae) associados a cafezais e fragmentos florestais vizinhos. Ciênc Agrotec 34(5):1146–1153

Soto-Pinto L, Perfecto I, Castillo-Hernandez J et al (2000) Shade effect on coffee production at the northern Tzeltal zone of the state of Chiapas, Mexico. Agric Ecosyst Environ 80(1/2):61–69

Spongoski S, Reis PR, Zacarias MS (2005) Acarofauna da cafeicultura de cerrado em Patrocínio, Minas Gerais. Ciênc Agrotec 29(1):9–17

Teodoro AV, Klein A-M, Tscharntke T (2008) Environmentally mediated coffee pest densities in relation to agroforestry management, using hierarchical partitioning analyses. Agric Ecosyst Environ 125(1/4):120–126

Teodoro AV, Klein A-M, Tscharntke T (2009) Temporally mediated responses of the diversity of coffee mites to agroforestry management. J Appl Entomol 133(9/10):659–665

Tuelher ES, Oliveira EE, Guedes RNC et al (2003) Ocorrência de bicho-mineiro do cafeeiro (*Leucoptera coffeella*) influenciada pelo período estacional e pela altitude. Acta Sci Agron 25(1):119–124

Vega FE, Jaramillo J (2010) Apredator of the coffee berry borer: is it present in your country? International Coffee Organization, London

Vélez-Arango PE, Benavides-Gómez M (1990) Registro e identificación de *Beauveria bassiana* en *Hypothenemus hampei* en Ancuya, Departamento de Nariño, Colombia. Cenicafé 41(2):50–57

Villacorta A (1984) Ocorrência de *Beauveria* sp. infectando a broca-do-café, *Hypothenemus hampei* (Ferrari, 1867) (Coleoptera: Scolytidae) em lavouras no estado do Paraná. An Soc Entomol Bras 13:177–178

Chapter 24
Cotton

Cecilia Czepak, Karina Cordeiro Albernaz Godinho, Pablo da Costa Gontijo, and Janayne Maria Rezende

24.1 Introduction

Several biotic and abiotic factors can affect the growth, development, and productivity of agricultural systems. Biotic factors such as the attack of arthropod pests are mostly fostered by simultaneous or successive crops, such as soybean, corn, and cotton. In these crops, problems related to plant health become more frequent each season. An example of this occurs in the cotton crop, which permanently hosts a complex of arthropod pests that might attack part or the entire plant from germination to harvest. These pests have high reproductive capacity and wide dispersion. They can rapidly infest crops causing direct and indirect damage, either by reducing productivity or by affecting important characteristics of seeds and fibers.

Although many arthropod pests are associated with cotton crops, their agroecosystem also has a high diversity of natural enemies (Ali et al. 2016). These natural enemies are important biological control agents that contribute to regulate pest populations at below economic threshold levels (Symondson et al. 2002). Eilenberg et al. (2001) defined biological control as the use of biological agents to suppress specific populations of pests to reduce their abundance or the damage they cause. Despite the presence of many natural enemies in the cotton agroecosystem, the performance of these biological agents, either by increase and/or conservation, is limited by the excessive pesticide applications in the crop. Nevertheless, a way to establish a balance in this agroecosystem is the implementation of Integrated Pest Management (IPM) programs with the association of different control tactics, in

C. Czepak (✉) · K. C. A. Godinho · P. da Costa Gontijo
Universidade Federal de Goiás (UFG) – Escola de Agronomia, Goiânia, GO, Brazil

J. M. Rezende
Agbitech Controles Biológicos Ltda, São Paulo, SP, Brazil

B. Souza et al. (eds.), *Natural Enemies of Insect Pests in Neotropical Agroecosystems*, https://doi.org/10.1007/978-3-030-24733-1_24

which natural enemies play a fundamental role. IPM supports that the coexistence between arthropods that are pests and those that provide benefits to the crop must be maintained in a harmonic way, always taking into account the costs and benefits of this practice.

Thus, the objective of this chapter is to provide information on biological control agents (predators, parasitoids, and entomopathogens) that are associated with the main arthropod pests of cotton crops, especially in Brazil. However, firstly, the history of the different phases of IPM implementation in cotton crop in Brazil is presented.

24.2 The Caterpillar Phase

This phase occurred in the early 1980s, a period known for ecological disasters due to the numerous applications of pesticides in the fields, such as those in Santa Helena de Goiás and Rio Verde (Goiás State, Brazil). At that time, cotton fields received more than 40 pesticide sprayings in just one crop. Due to this excessive spraying and with the recurrent pest problems, the concepts of IPM began to be implemented in the culture. The main strategy was to avoid initial sprayings for aphid control when their populations were not causing damage. The aim was for the aphids to serve as alternative food for natural enemies, especially predators and parasitoids, which would act to control intermediate and late pests such as caterpillars. With the adoption of this strategy, pesticide use was reduced by more than 90% for the control of arthropod pests in cotton. This period was considered the golden phase of cotton IPM in Brazil.

24.3 The Boll Weevil Phase

By this phase, IPM concepts in the cotton crop were already producing good results and were being disseminated by the Coordenadoria de Assistência Técnica Integral (CATI) and by the Universidade Estadual Paulista (UNESP) throughout the State of São Paulo. However, the appearance of cotton boll weevil *Anthonomus grandis* Boheman (Curculionidae) increased the number of sprayings in the crop because the natural biological control of this pest was not very effective. With the occurrence of boll weevil, the number of pesticide sprayings in cotton went from one application to 12 or 15 applications per season.

24.4 The Aphid Phase

In this phase, cotton varieties with longer fibers were brought to Brazil from the United States and Australia. These varieties were not resistant to the virus transmitted by the aphid *Aphis gossypii* Glover (Aphididae), unlike those developed by the

Instituto Agronômico de Campinas (IAC). Thus, the early spraying, which had been reduced, became unavoidable to control aphids. From two to three sprayings, cotton began to receive more than ten sprayings per season. However, in this period, new pest management strategies were developed within the IPM concepts, which helped to reduce the number of sprayings to five to six per season.

24.5 Current Phase

Currently, most cotton grown throughout the world uses genetically modified plants, which express insecticidal proteins (Cry) derived from the bacterium *Bacillus thuringiensis* (Bt). The objective of these plants is resistance to arthropod pests, especially lepidopteran. However, the adoption of Bt cultivars has modified the entire pest management scenario in the cotton crop, directly or indirectly influencing the occurrence of other pests, such as whitefly *Bemisia tabaci* (Gennadius) MEAM1 (Middle East Asia Minor 1) (Aleyrodidae), which has recently become important in crops. In addition, Bt technology does not efficiently control all species, such as *Spodoptera frugiperda* and *Helicoverpa armigera* (Noctuidae).

24.6 Major Natural Enemies of Arthropod Pests in Cotton

24.6.1 Predators

Approximately, 45 species of predators are associated with arthropod pests of cotton (Pallini et al. 2006). However, detailed information on the biology, ecology, and behavior of many of these species are still rare.

The most common predators related to pest control in cotton crop include ants of the genus *Solenopsis* (Formicidae); hemipteran of the genera *Geocoris* (Lygaeidae), *Nabis* (Nabidae), *Zelus* (Reduviidae), and *Podisus* (Pentatomidae); lady beetles of the genera *Cycloneda*, *Scymnus*, *Eriopis*, *Coleomegilla*, *Harmonia*, *Hippodamia*, and *Diomus* (Coccinellidae); and lacewings *Chrysoperla* (Chrysopidae), as well as several species of spiders (Gravena and Cunha 1991; Fitt 1994; Luttrell et al. 1994). As an example of the predation potential of some of these organisms, *Chrysoperla externa* (Hagen) (Chrysopidae) species can consume an average of up to 800 aphids and 250 nymphs of whitefly during its life cycle (Costa et al. 1999). Adults of *Podisus nigrispinus* (Dallas) (Pentatomidae) can prey on up to two small caterpillars of *Alabama argillacea* (Hübner) (Noctuidae) per day under laboratory conditions (Oliveira et al. 2005). As for mites, Phytoseiidae of the species *Amblyseius operculatus* (DeLeon) and *Iphizeiodes zuluagai* Denmark & Muma (Phytoseiidae) are the most frequently species found associated with phytophagous cotton mites (Evangelista-Jr et al. 2006).

Studies conducted in Brazil have not revealed drastic changes in the arthropod community present in areas of Bt and non-Bt cotton (Nunes 2010). Thomazoni et al. (2010) observed that there were no differences in larval and adult abundance of predators *Cycloneda sanguinea* (Linnaeus) *Scymnus* sp., *Geocoris* sp., and *Chrysoperla* sp. recorded in Bt and non-Bt cultivars. Similarly, Naranjo (2005) did not record difference in the density of the generalist bug *Geocoris* sp. in Bt cotton cultivars relative to non-Bt cultivars. However, according to Liu et al. (2005), Bt cotton can indirectly affect specialist natural enemies by removing lepidopteran eggs, larvae, and pupae that serve as food source and oviposition substrate for natural enemies.

24.6.2 Parasitoids

Behind predators, parasitoids are the second largest group of natural enemies used to control invertebrates. The parasitoids are usually highly specialized with the different stages of the development of the host and act as natural biological control agents and can also be used in applied biological control of the main cotton pests. The main orders of parasitoids are Hymenoptera and Diptera; however, the orders Coleoptera, Lepidoptera, and Neuroptera have families with rare species of parasitoids (Hajek 2004). The order Hymenoptera is estimated to contain more than 65,000 species of parasitoids (Gordh and Caltagirone 1999). Braconidae, Ichneumonidae, Chalcididae, Pteromalidae, Encyrtidae, Aphelinidae, and Trichogrammatidae are the main families used in biological control (Parra et al. 2002; Hajek 2004). In the order Diptera, the family that has the greatest diversity of parasitoids is Tachinidae (Greathead 1986).

The cotton agroecosystem contains high diversity and abundance of parasitoids, which help regulate populations of insect pests in the crop. In a fauna survey conducted in cotton fields around the region of Ribeirão Preto, SP, Brazil, more than 16,000 parasitoid Hymenoptera, belonging to 22 families and 8 superfamilies, were collected. The most abundant families were Encyrtidae (45.1%), Trichogrammatidae (19.1%), Mymaridae (14.3%), and Scelionidae (6.6%) (Perioto et al. 2002). Diptera parasitoids of the family Tachinidae are also frequently found in areas with cotton, with the main species being *Hyalomyodes brasiliensis* Townsend, *Patelloa* spp., *Euphorocera floridensis* Townsend, *Eutrichopodopsis nitens* Blanchard, *Trichopoda* sp., *Archytas* spp., *Peleteria robusta* (Wiedman), *Winthemia quadripustulata* (Fabricius), and *Winthemia* sp. (Bastos and Torres 2005).

In Brazil, according to Ramalho and Wanderley (1996), approximately 13 species of parasitoids are associated with the immature stages of the cotton boll weevil *A. grandis*, and evaluations in the northern region of the country indicated that the natural mortality of this pest due to parasitoid attack was over 10% (Evangelista Jr. et al. 2006). The main parasitoids of boll weevil in Brazil are endoparasitoid

Bracon vulgaris Asmead (Braconidae) and the ectoparasitoid *Catolaccus grandis* (Burks) (Pteromalidae), whose origin is southeast Mexico and northeast Central America (Evangelista-Jr et al. 2006). However, the most common species is *B. vulgaris*, which in addition to parasitizing *A. grandis*, also attacks the pink bollworm *Pectinophora gossypiella* (Saunders) (Gelechiidae). In addition to these parasitoids, *Chelonus* sp. (Braconidae) was found parasitizing *A. grandis* in the Brazilian Cerrado region, mainly in flower buds in the soil (Nunes and Fernandes 2000).

Although the parasitoid *C. grandis* is found in cotton fields less frequently than *B. vulgaris*, it has great potential for use in applied biological control programs because it can be easily reared in the laboratory with its alternative hosts including *Euscepes postfasciatus* Fairmaire (Curculionidae) (Dias et al. 1998) and cowpea weevil *Callosobruchus maculatus* (Fabricius) (Bruchidae) (Ramalho and Dias 2003). Recently, associations of the Brazilian cotton producers partnered with research institutions to enable the use of *C. grandis* in the applied biological control of the boll weevil. Research on releases of *C. grandis* found that this parasitoid prefers third instar larvae of *A. grandis* and provided efficient control (Bastos and Torres 2005).

Parasitoids are also associated with the control of the lepidopteran pest complex that attacks cotton crop. The parasitoids of the genus *Trichogramma* (Trichogrammatidae) are the main egg parasitoids used in the biological control applied in several agriculturally important crops (Gardner et al. 2011). Approximately 210 *Trichogramma* species have been described worldwide (Querino et al. 2010). The efficiency of *Trichogramma* parasitism on the genus *Helicoverpa* caterpillars in cotton is 70–80% and up to 50% for the pink bollworm *P. gossypiella*. To control the cotton leafworm *A. argillacea*, research demonstrated that the level of nonaction is 70% of the plants infested with the presence of 70% of the eggs parasitized by *Trichogramma* (Bastos and Torres 2005). Parasitoids of the genus *Trichogramma* also present high rates of egg parasitism in the larva of *Chloridea virescens* (Fabricius) (Noctuidae) (Soares et al. 2008). However, the cotton cultivar may influence the rate of parasitism of different *Trichogramma* species, which would explain varied results in different areas (Andrade et al. 2009).

Parasitoids are also widely used as classical biological control in cotton areas in Latin America, especially for exotic pest control. As an example, several species of parasitoids, mainly from India, were introduced in Mexico to control *P. gossypiella* (Bastos and Torres 2005). In addition, parasitoids are associated with the natural control of *H. armigera* in areas of structured refuge of *Bt* cotton, without the application of insecticides. In these areas, the main *H. armigera* parasitoids found in the western region of Bahia, Brazil, were *Campoletis sonorensis* (Cameron) (Ichneumonidae), *Archytas marmoratus* (Townsend), and *Archytas incertus* (Marcquart) (Tachinidae), with a parasitism rate higher than 40% (Luz et al. 2018).

Populations of sucking insects such as the cotton aphid *A. gossypii* and whitefly *B. tabaci* are also regulated by the action of parasitoids. In Brazil, the main parasitoids of cotton aphid are *Aphidius colemani* Viereck and *Lysiphlebus testaceipes* (Cresson) (Braconidae), which present a parasitism rate ranging from 40 to 76% (Bastos and Torres 2005). The occurrence of *Aphelinus gossypii* Timberlake (Aphelinidae) in cotton fields has also been reported in northern Brazil (Fernandes et al. 2000). The parasitoid *L. testaceipes* has high parasitism capacity, rapidly reducing the aphid population. Bedoya et al. (2018) verified high parasitoid richness around cotton fields in the region of Montería (Colombia) and this richness was higher than that found near rice and corn fields. For whitefly, more than 50 parasitoid species are known worldwide. Among these species, the most important are species of the genera *Encarsia* and *Eretmocerus* (Aphelinidae), with the species *Encarsia formosa* Gahan as the natural enemy most employed for whitefly control in the world, with commercial production (Hoddle et al. 1998). The whitefly stages most susceptible to the parasite *E. formosa* are third and fourth instar nymphs (puparium), which a few days after the parasitism become dark. This is not observed in nymphs parasitized by *Eretmocerus*.

24.6.3 Entomopathogens

In addition to beneficial insects, the presence of entomopathogenic microorganisms is a natural form of pest management. Numerous entomopathogenic agents occur in the cotton fields. For example, the population of *Chrysodeixis includens* (Noctuidae) is frequently regulated by epizootics of the fungi *Metarhizium rileyi, Cordyceps* sp. (formerly in *Isaria* sp.), and the baculovirus ChinNPV (*Chrysodeixis includens nucleopolyhedrovirus*) (Alexandre et al. 2010). The natural occurrence of *Beauveria bassiana, Metarhizium anisopliae*, and *Cordyceps* sp. has been frequently recorded enzootically or causing epizootics in populations of cotton boll weevil (*Anthonomus grandis,* Curculionidae), which is also the target of bacterial diseases (Andrade et al. 1984; Pierozzi Jr. and Habib 1993; Almeida 1998). Whitefly (*B. tabaci*) infected by *Cordyceps* sp. was collected in cotton leaves in Bahia, Brazil (Tamai et al. 2015).

However, with the predominant dynamics of pest control with conventional chemical insecticides and fungicides, the natural occurrence of entomopathogens has not been able to keep pest populations below the level of economic damage. Therefore, application of microbial control against diverse pests has significantly increased in several crops, including cotton. Market demand for more sustainably managed cotton and the increasing pressure of insect pests and resistance problems on insecticides in cotton crop has led the growers to pay more attention to microbial control. The use of entomopathogenic agents is an effective method to control pest arthropods and they do not harm the environment, animals, or human; therefore, they have been used as an alternative to conventional chemical insecticides. These agents include bacteria, fungi, and viruses.

24.6.3.1 Bacteria

Firstly discovered in 1900, First discovered in the early 1900s, *Bacillus thuringiensis* (Bt) is an entomopathogenic bacterium that affects insects, mainly insects of orders Lepidoptera and Coleoptera.

Brazil currently boasts 21 commercial products containing Bt as an active ingredient, which mainly target the control of *S. frugiperda*, *H. armigera*, *Helicoverpa zea* (Boddie), and *Chloridea virescens* (Fabricius) in cotton. However, the product must be applied when the caterpillars are in their early stages of life.

The use of foliar applications of commercial Bt-based products in cotton cultivation has significantly reduced since the arrival of genetically modified cotton that produces *B. thuringiensis* insecticidal proteins.

At present, on-farm production of *B. thuringiensis* is widely practiced in Brazil. However, this production does not always comply with the necessary production quality controls, which may lead to contamination and a product that may not be safe for the applicator and the fauna. Instead, it could be extremely toxic depending on the contamination in the production environment.

24.6.3.2 Fungi

Entomopathogenic fungi have a major advantage over the other microorganisms that cause diseases in insects because fungi infect their hosts through the cuticle and do not need to be ingested. Bioprospecting of the entomopathogenic fungi for the cotton crop has focused on the control of Coleoptera such as *A. grandis* and sucking insects such as aphids and whitefly with fungi such as *B. bassiana, M. anisopliae,* and *Cordyceps* sp.

The efficiency of *Cordyceps* spp. in the biological control of *B. tabaci* was observed in several studies (Faria and Wraight 2001; Lacey et al. 2008; Mascarin et al. 2013). Nymphs and adult whiteflies treated with *C. fumosorosea* reached mortality rates above 70% and 90%, respectively (Mascarin et al. 2013). However, the most used fungus to control *B. tabaci* in cotton in Brazil is currently *B. bassiana* because there is still no microbial insecticide based on *Cordyceps* spp. registered for this pest. Formulated bioinsecticides containing *Cordyceps* spp. are in the process of being registered and may soon contribute to the biological control of whitefly in various crops.

24.6.3.3 Virus

The use of baculovirus, an important category of entomopathogen, has great potential for use in Brazilian cotton growing. Baculoviruses fill the basic requirements as an alternative and/or complement to other tactics for regulation of lepidopteran populations. In Brazil, the use of baculovirus as biological insecticides began in the 1970s, in velvetbean caterpillar [*Anticarsia gemmatalis* (Hübner) (Erebidae)] management programs. The first baculovirus-based bioinsecticide (HzSNPV) was commercially registered in 1975 (Virion-H, Biocontrol-VHZ, Elcar) and used on a large scale to control the Heliothinae family species complex, such as *H. zea* and *C. virescens*, key pests of the cotton crop in the United States. Currently, approximately 50 baculovirus-based bioinsecticides are used in different parts of the world (Czepak et al. 2018).

In Brazil, the baculoviruses *Helicoverpa armigera nucleopolyhedrovirus* (HearNPV), *Chrysodeixis includens nucleopolyhedrovirus* (ChinNPV), and *Spodoptera frugiperda multiple nucleopolyhedrovirus* (SfMNPV) have been registered with the Brazilian Ministry of Agriculture, Livestock, and Food Supply. They can be used to control caterpillars of the *Heliothinae* and *Plusiinae* subfamilies, and *S. frugiperda*, which are Lepidoptera key pest of cotton.

In cotton crop, sequential spraying of HearNPV and SfMNPV is recommended at the beginning of flowering when the first appearance of the target pest(s) is observed, as illustrated in Fig. 24.1. It is important to start the applications at the beginning of the infestation by the target pest (3% incidence), reapplying at intervals of 5–10 days according to the monitoring. The baculovirus HearNPV and SfMNPV are effective on first to third instar caterpillars (≤8 mm in length).

The adoption of SfMNPV for cotton production program increased the efficacy against *S. frugiperda* caterpillars and reduced damage to reproductive structures (Fig. 24.2).

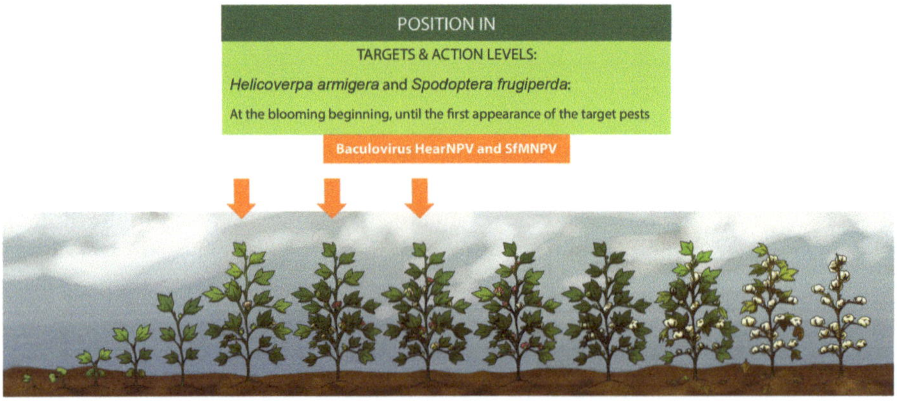

Fig. 24.1 Position to place baculovirus HearNPV and SfMNPV in the cotton crop to control *Helicoverpa armigera* and *Spodoptera frugiperda* (AgBiTech 2018)

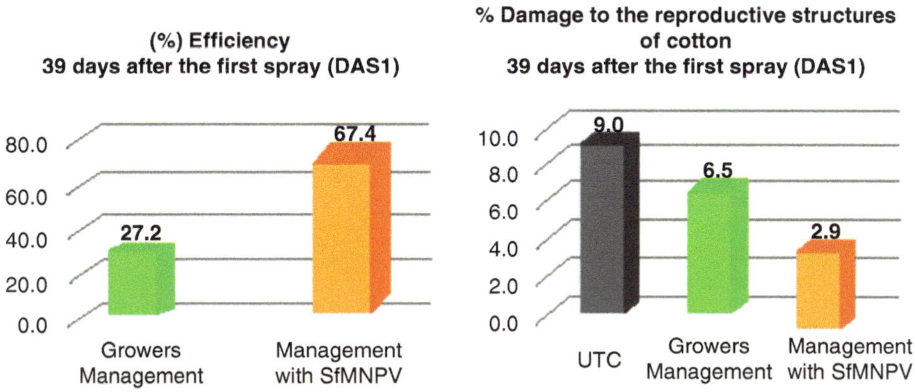

Fig. 24.2 Management of *Spodoptera frugiperda* in cotton, in Luis Eduardo Magalhães, Bahia, Brazil, with baculovirus SfMNPV: (**a**) efficiency of SfMNPV to control caterpillars and (**b**) damage in reproductive structures of cotton. *Management with SfMNPV (orange): 6 applications of SfMNPV (100 ml/ha). Growers Producer Management (green): 3 applications (1) thiodicarb (0.48 kg/ha); (2) emamectin benzoate (0.015 kg/ha); (3) emamectin benzoate (0.015 k/ha). Untreated check (UTC): grey(AgBiTech 2018)

24.7 Final Considerations

Although many natural enemies are present in the cotton agroecosystem, they are not always sufficient to stop pests from causing economic damages. This is because some pests attack the reproductive parts of plants as well as because of excessive use of nonselective pesticides. Thus, knowledge about the main species of natural enemies associated with cotton and basic research on the biology and behavior of these species are essential for the development and success of natural or applied biological control in the cotton crop.

Acknowledgments This research was funded by the National Council for Scientific and Technological Development (CNPq), the Coordination for Improvement of Higher Education Personnel (CAPES), the Foundation for Research Support of the State of Goiás (FAPEG), and collaborations.

References

AgBiTech (2018) Vai mais longe por mais tempo. BolTéc 1:11–12

Alexandre TM, Ribeiro ZMA, Craveiro SR et al (2010) Evaluation of seven viral isolates as potential biocontrol agents against *Pseudoplusia includens* (Lepidoptera: Noctuidae) caterpillars. J Invertebr Pathol 105(1):98–104

Ali A, Desneux N, Lu Y et al (2016) Characterization of the natural enemy community attacking cotton aphid in the *Bt* cotton ecosystem in Northern China. Sci Rep 6(24273):1–9

Almeida RP (1998) Seletividade de inseticidas ao fungo entomopatogênico Metarrhizium anisopliae (Merch.) Sorokin. Embrapa – CNPA, Campina Grande. (Circular Técnica, 19)

Andrade CFS, Pierozzi Júnior I, MEM H (1984) Ocorrência natural de doenças infecciosas em populações do "bicudo", *Anthonomus grandis* Boheman, 1843. In: Resumos do 9° congresso brasileiro de entomologia, SEB, Londrina, 22-27 jul

Andrade GS, Pratissoli D, Torres JB et al (2009) Parasitismo de ovos de *Heliothis virescens* por *Trichogramma* spp. pode ser afetado por cultivares de algodão. Acta Sci Agron 31(4):569–573

Bastos CS, Torres JB (2005) Controle biológico e o manejo de pragas do algodoeiro. Embrapa Algodão, Campina Grande

Bedoya A, Herrera CF, García KDP (2018) Entomofauna diversity associated to sorrounding vegetation on rice, corn and cotton crops. Temas Agrarios 23(2):107–120

Costa RIF, Almeida SA, Guerra CL et al (1999) Consumo de *Bemisia argentifolii* (Bellows & Perring) (Hemiptera: Aleyrodidae) e *Aphis gossypii* (Glover, 1877) (Homoptera: Aphididae) por *Chrysoperla externa* (Hagen, 1861) (Neuroptera: Chrysopidae). In: Resumos do 2° congresso brasileiro de algodão, Embrapa Algodão, Campina Grande, 5-10 set

Czepak C, Marcon PG, Lima M et al (2018) Alternativo e complementar. Cultiv Gd Cult 229:8–11

Dias JM, Lemos WP, Medeiros RS et al (1998) Reprodução de *Catolaccus grandis* (Burks) (Hymenoptera: Pteromalidae), parasitóide do bicudo do algodoeiro, tendo como hospedeiro larvas de *Euscepes postfasciatus* (Fairmaire) (Coleoptera: Curculionidae). In: Resumos do 17° congresso brasileiro de entomologia, UFRRJ, Rio de Janeiro, 9-14 jun

Eilenberg J, Hajek A, Lomer C (2001) Suggestions for unifying the terminology in biological control. BioControl 46(4):387–400

Evangelista Júnior WS, Zanuncio Júnior SZ, Zanuncio JC (2006) Controle biológico de artrópodes pragas do algodoeiro com predadores e parasitoides. Rev Bras Ol Fibros 10(3):1147–1165

Faria M, Wraight SP (2001) Biological control of *Bemisia tabaci* with fungi. Crop Prot 20(9):767–778

Fernandes AMV, Farias AMI, Faria CA et al (2000) Ocorrência de *Aphelinus gossypii* Timberlake (Hymenoptera: Aphelinidae) parasitando *Aphis gossypii* Glover (Hemiptera: Aphididae) em algodão no estado de Pernambuco. An Soc Entomol Bras 29(4):831–834

Fitt GP (1994) Cotton pest management: part 3. An Australian perspective. Annu Rev Entomol 39:543–562

Gardner J, Hoffmann MP, Pitcher SA et al (2011) Integrating insecticides and Trichogramma ostriniae to control European corn borer in sweet corn: economic analysis. Biol Control 56(1):9–16

Gordh GEF, Caltagirone LE (1999) Biology of parasitic Hymenoptera. In: Bellows TS, Fisher TW (eds) Handbook of biological control. Academic Press, New York, pp 355–371

Gravena S, Cunha HF (1991) Artrópodes predadores na cultura algodoeira. CEMIP/Departamento de Entomologia e Nematologia, Jaboticabal

Greathead DJ (1986) Parasitoids in classical biological control. In: Greathead D (ed) Insect parasitoids. Academic Press, Orlando, pp 289–318

Hajek A (2004) Natural enemies: an introduction to biological control. Cambridge University Press, New York

Hoddle MS, Van Driesche RG, Sanderson JP (1998) Biology and use of the whitefly parasitoid *Encarsia formosa*. Ann Rev Entomol 43:645–669

Lacey LA, Wright SP, Kirk AA (2008) Entomopathogenic fungi for control of *Bemisia tabaci* biotype B: foreign exploration, research and implementation. In: Gould J, Hoelmer K, Goolsby J (eds) Classical biological control of *Bemisia tabaci* in the United States – a review of interagency research and implementation, 4a edn. Springer Netherlands, Cham, pp 33–70

Liu XD, Zhai BP, Zhang XX et al (2005) Impact of transgenic cotton plants on a non-target pest, *Aphis gossypii* Glover. Ecol Entomol 30(3):307–315

Luttrell RG, Fitt GP, Ramalho FS et al (1994) Cotton pest management: part 1. A worldwide perspective. Annu Rev Entomol 39:517–526

Luz PMC, Paula-Moraes SV, López JMP et al (2018) Parasitoides associados com *Helicoverpa armigera* em áreas de refúgio de algodão, no Oeste da Bahia, Brasil. Cienc Rural 48:1):1–1):4

Mascarin GM, Kobori NN, Quintela ED et al (2013) The virulence of entomopathogenic fungi against *Bemisia tabaci* biotype B (Hemiptera: Aleyrodidae) and their conidial production using solid substrate fermentation. Biol Control 66(3):209–218

Naranjo SE (2005) Long-term assessment of the effects of transgenic *Bt* cotton on the abundance of nontarget arthropod natural enemies. Environ Entomol 34(5):1193–1210

Nunes DH (2010) Efeito do algodoeiro geneticamente modificado (Bollgard) em organismo não alvo. Tese, Escola Superior de Agricultura "Luiz de Queiroz"

Nunes JCS, Fernandes PM (2000) Parasitismo do bicudo do algodoeiro (*Anthonomus grandis*) em botões florais do algodoeiro, no município de Goiânia-GO. Pesqui Agropecu Trop 30(2):13–15

Oliveira EE, Oliveira CL, Sarmento RA et al (2005) Aspectos biológicos do predador *Cycloneda sanguinea* (Coleoptera: Coccinellidae) alimentado com *Tetranychus evansi* (Acari: Tetranychidae) e *Macrosiphum euphorbiae* (Homoptera: Aphididae). Biosci J 21(6):33–39

Pallini A, Silvie P, Monnerat RG et al (2006) Non-target and biodiversity impactson parasitoids. In: Hilbeck A, Andow DA, Fontes EMG (eds) Environmental risk assessment of genetically modified organisms: methodologies for assessing *Bt* cotton in Brazil. CABI, Wallingford, pp 200–224

Parra JRP, Botelho PSM, Corrêa-Ferreira S et al (2002) Controle biológico no Brasil: parasitoides e predadores. Manole, São Paulo

Perioto NW, Lara RIR, Santos JCC et al (2002) Himenópteros parasitoides (Insecta, Hymenoptera) coletados em cultura de algodão (*Gossypium hirsutum* L.) (Malvaceae), no município de Ribeirão Preto, SP, Brasil. Rev Bras Entomol 46(2):165–168

Pierozzi Júnior I, Habib MEM (1993) Identificação de fatores de mortalidade natural dos estágios imaturos de *Anthonomus grandis* Boh. (Coleoptera: Curculionidae), na região de Campinas, São Paulo. An Soc Entomol Bras 22(2):326–329

Querino RB, Zucchi RA, Pinto JD (2010) Systematics of the Trichogrammatidae (Hymenoptera: Chalcidoidea) with a focus on the genera attacking Lepidoptera. In: Cônsoli FL, Parra JRP, Zucchi RA (eds) Egg parasitoids in agroecosystems with emphasis on *Trichogramma*. Springer, Dordrecht, pp 191–218

Ramalho FS, Dias JM (2003) Efeitos de hospedeiros alternativos na biologia de *Catolaccus grandis* (Burks) (Hymenoptera: Pteromalidae), parasitóide de *Anthonomus grandis* Boheman (Coleoptera: Curculionidae). Neotrop Entomol 32(2):305–310

Ramalho FS, Wanderley PA (1996) Ecology and management of cotton boll weevil in South America cotton. Am Entomol 42:41–47

Soares JJ, Nascimento ARB, Silva MV (2008) Predadores e parasitoides chaves e seletividade de inseticidas na cultura algodoeira. Embrapa Algodão, Campina Grande

Symondson WOC, Sunderland KD, Greenstone MH (2002) Can generalist predators be effective biocontrol agents? Annu Rev Entomol 47:561–594

Tamai M, Ghazzaoui HAN, Sampaio MAS et al (2015) Epizootia de *Isaria* sp. em população de *Bemisia tabaci* (Gennadius) na cultura do algodão no cerrado do oeste da Bahia, safra 2014/15. In: Resumos do 10° Congresso Brasileiro do Algodão, Abrapa, Foz do Iguaçu, 1-4 set

Thomazoni D, Degrande PE, Silvie P et al (2010) Impact of Bollgard genetically modified cotton on the biodiversity of arthropods under practical field conditions in Brazil. Afr J Biotechnol 9(37):6167–6176

Chapter 25
Forests

Leonardo Rodrigues Barbosa, Juliana Mendonça Campos, Carlos Frederico Wilcken, and José Cola Zanuncio

25.1 Introduction

Forests provide environmental, economic, social, and aesthetic benefits (Miura et al. 2015) as sources of products such as wood, firewood, fiber, and other nontimber products (Food and Agriculture Organization of the United Nations 2014). Forests cover approximately 31% of the terrestrial area (Keenan et al. 2015) and contain the largest number of species in the terrestrial ecosystem (Matthews et al. 2000), accounting for about 80% of species diversity (Carnus et al. 2006).

Forest benefits include maintaining biodiversity (Brockerhoff et al. 2013) and ecosystem processes and services (Thompson et al. 2011) such as food, soil erosion prevention, increased pollination, carbon sequestration, climate regulation, recreation, and pest reduction (Diaz et al. 2005; Nadrowski et al. 2010; Cardinale et al. 2011).

Insects are important in the forest ecosystem (Yang and Gratton 2014) for pest control, pollination (Kremen and Chaplin-Kramer 2007), herbivory (Jactel and Brockerhoff 2007), secondary seed dispersal (Slade et al. 2011), and nutrient recycling (Freymann et al. 2008). Insect population increases can cause negative impacts on trees and even the destruction of native or planted forests (Wingfield et al. 2008).

L. R. Barbosa (✉)
Empresa Brasileira de Pesquisa Agropecuária - Embrapa Florestas, Colombo, Paraná, Brazil
e-mail: leonardo.r.barbosa@embrapa.br

J. M. Campos
Departamento Fitotecnia, Universidade Federal de Viçosa, Viçosa, Minas Gerais, Brazil

C. F. Wilcken
Departamento de Proteção Vegetal, Faculdade de Ciências Agronômicas, Botucatu, SP, Brazil

J. C. Zanuncio
Departamento de Entomologia/BIOAGRO, Universidade Federal de Viçosa, Viçosa, Minas Gerais, Brazil

B. Souza et al. (eds.), *Natural Enemies of Insect Pests in Neotropical Agroecosystems*, https://doi.org/10.1007/978-3-030-24733-1_25

According to the World Food and Agriculture Organization (FAO) on Global Assessment of Forest Resources (Food and Agriculture Organization of the United Nations 2015), in 2015, more than 1.1 million hectares of forest were damaged by insect pests in South America (van Lierop et al. 2015).

Biodiversity can affect herbivory by mechanisms such as reducing forest pests (Tylianakis et al. 2008) through biological control (Bridgeland et al. 2010). The annual value of the biological control, pollination, decomposition, and food services offered by wild animals was estimated at $ 57 billion in the United States and represents a small fraction of the services provided by insects (Losey and Vaughan 2006).

A combination of biotic and abiotic factors, including natural enemies, regulates forest insect pests (Wylie and Speight 2012). These organisms can act without human intervention as natural biological control or used to reduce the population density of a pest when applied as biological control (Bale et al. 2008). This chapter provides an overview of the natural enemy benefits to forest ecosystems, especially for predators, parasitoids, and pathogens on pest insects in planted forests. Successful cases of biological control programs in planted forests in Latin America are discussed.

25.2 Biological Control of Forest Insect Pests in Latin America

Pest insects in native forests are regulated without human intervention, mainly via natural control with a wide variety of natural enemies. Classic and augmentative biological control strategies with inoculative releases are used in forest plantations to allow interactions between natural enemies and pests over a long period (Bale et al. 2008). The more extensive use of biological control in forest areas is also because pest control in these systems is not necessarily immediate and the economic damage level is higher than in agriculture.

Forest certification systems represent an initiative for environmental conservation and sustainable forest development, influencing forest pest management. The Forest Stewardship Council (FSC) is the leading nongovernmental organization responsible for forest certification and is committed to promoting the use of nonchemical methods of pest control. Restrictions on the use of chemical insecticides in the integrated management of forest pests favor the adoption of biological control.

The main natural enemies used in biological control of forest pests include entomophagous agents (predatory insects and parasitoids) and entomopathogens (bacteria, fungi, nematodes, and viruses). Endemic or introduced populations of natural enemies can regulate pest abundance in native and cultivated forest plantations. *Thaumastocoris peregrinus* Carpintero & Dellapé (Thaumastocoridae) x *Cleruchoides noackae* Lin & Huber (Mymaridae), *Sirex noctilio* Fabricius (Siricidae) x *Deladenus siricidicola* Bedding (Nematoda: Neotylenchidae), *Thyrinteina arnobia* (Stoll) (Geometridae), *Psorocampa denticulata* Schaus (Notodontidae),

Eupseudosoma aberrans Schaus (Arctiidae), *Sarsina violascens* (Herrich-Schäffer) (Lymantriidae) x predatory bedbugs (Pentatomidae: Asopinae): *Podisus nigrispinus* (Dallas), *Supputius cincticeps* (Stäl), and *Brontocoris tabidus* (Signoret) are examples of successful biological control programs in Latin America.

25.3 Principal Species of Natural Enemies of Forest Insect Pests

1. *Predators*

Predatory insects of forest pests are found mainly in the Carabidae, Chrysomelidae and Coccinellidae, Pentatomidae and Reduviidae, and Chrysopidae families.

Insects of the Pentatomidae family, subfamily Asopinae [*Alcaeorrhynchus grandis* Dallas, *B. tabidus, Podisus distinctus* (Stäl), *P. nigrispinus, S. cincticeps, Thynacantha marginata* (Dallas)] have a high fauna diversity (Pires et al. 2015) preying on eggs, larvae, pupae, and adults of Coleoptera and Lepidoptera (Torres et al. 2006; Menezes et al. 2012) agricultural and forest pests (Zanuncio et al. 2006, 2014). These insects can be released in areas with low caterpillar infestation or in larger numbers when the population density of these pests reaches outbreak levels (Molina-Rugama et al. 1997; Zanuncio et al. 1998; De Bortoli et al. 2011). In Latin America, the genera *Podisus* and *Brontocoris* have more aggressive species (van Lenteren 2012) and those of the genera *Alcaeorrhychus* and *Supputius* are less common.

Insects of the Neuroptera order are used in augmentative biological control. These insects are common in native and planted forests, feeding on a wide range of insects. Species of the genera *Chrysoperla, Ceraeochrysa, Chrysopodes, Plesiochrysa,* and *Leucochrysa* are among the Neotropical Chrysopidae with potential to prey on arthropod pests. *Chrysoperla externa* (Hagen) is a high-potential biological control agent in Central and South America for being a widespread species well suited to that region (Albuquerque et al. 1994). In Brazil, this species preys on *T. peregrinus* nymphs (Barbosa et al. 2017). *Ceraeochrysa cincta* (Schneider), *Ceraeochrysa cubana* (Hagen), and *Ceraeochrysa smithi* (Navás) present excellent opportunities for increasing and diversifying the arsenal of biological control agents that are monitored and manipulated for integrated pest management in Mexico and in Central and South America (Lopez-Arroyo et al. 1999).

2. *Parasitoids*

The main parasitoid insects associated with forest insect pests are of the orders Hymenoptera and Diptera. *Cleruchoides noackae* Lin and Huber is a solitary egg micro-hymenopteran endoparasitoid of the bronzed bug *T. peregrinus*. The embryonic (egg) and post-embryonic (larva and pupa) development of this parasitoid occurs in the host egg within approximately 15 days. Adults are about 0.5 mm long, with a longevity of 6 days at 25 °C feeding mainly on host eggs (Mutitu et al. 2013). *Cleruchoides noackae* was reported in Australia in 2007 parasitizing *T. peregrinus* eggs (Lin et al. 2007), with 21–25% parasitism rate (Cross 2009). Introduced in Brazil in 2012, since then, it has been bred and released in eucalyptus plantations in several states of Brazil, a parasite rate of 50% in the field and laboratory (Barbosa et al. 2017). This parasitoid has already been introduced and released in Chile for the control of *T. peregrinus*, but reports of its efficiency are still unknown (Nadel and Noack 2012).

Anaphes nitens Girault (Mymaridae) is an egg parasitoid with excellent localization capacity and host discrimination (Carbone and Rivera 2003; Carbone et al. 2004). This natural enemy parasitizes eggs of *Gonipturs platensis* (Marelli) and *Gonipterus pulverulentus* Lea (Curculionidae), Until the publication of Mapondera et al. (2012), clarifying the taxonomy of the genus Gonipterus, these species have been confused in literature. Both G. platensis and the undescribed weevil were referred to as G. scutellatus, and G. pulverulentus was referred to as G. gibberus (EPPO 2005; Mapondera et al. 2012) known as eucalyptus weevils, which are Australian insects and the main species of eucalyptus defoliating beetles (Oliveira 2006). *Anaphes nitens* larvae hatch, feed, and pupate within the host egg until adult emergence (Sanches 2000). The biological cycle lasts 17–33 days (Kevan 1946). *Anaphes nitens* maintained population equilibrium of the *G. platensis* in eucalyptus plantations in the southern region and in the south of the State of São Paulo, Brazil, over a decade (Wilcken et al. 2008a). In 1998, this parasitoid was introduced in Chile and appeared successful, with up to 98% parasitism achieved 1 year after the release of the parasitoid (Lanfranco and Dungey 2001).

The genus *Psyllaephagus* (Encyrtidae) has 237 species described (Noyes 2015), with the greatest taxonomic diversity in Australia (Noyes and Hanson 1996) and 12 species in the Neotropical region (Singh 2016). Most species of this genus attack Psylloidea nymphs, but some are hyperparasitoids of *Psyllaephagus* species (Riek 1962). The effectiveness and specificity of *Psyllaephagus bliteus* Riek and *Psyllaephagus pilosus* Noyes have facilitated their use in classical biological control programs of eucalyptus psyllids (Paine et al. 2011). *P. bliteus* and *P. pilosus* were successfully introduced in California, USA, to control *Glycaspis brimblecombei* Moore (Psyllidae) (Daane et al. 2005) and *Ctenarytaina eucalypti* (Maskell) (Psyllidae) (Dahlsten et al. 1998), respectively. Adults of these parasitoids deposit their eggs in the abdomen of psyllid nymphs of all instars (Daane et al. 2005). *P. bliteus* specimens were registered in Brazil in 2003 (Berti Filho et al. 2003). *Psyllaephagus pilosus* was introduced in Chile to control *C. eucalypti* (Rodríguez and Sáiz 2006) and spread to Argentina (Fidalgo et al. 2005) and Brazil (Kurylo 2008). The parasitism of this pest by this natural enemy can reach 94% in eucalyptus plantations in Brazil (Wilcken et al. 2008b). In Mexico, 80% of the *G. brimblecombei* nymphs were parasitized after the first release of this parasitoid (Herrera 2001).

The genus *Trichogramma* (Trichogrammatidae) stands out due to its ease of breeding in alternative hosts (Haji et al. 1998) and high parasitism in pest insect eggs (Botelho 2010). Species of this genus are exclusively endoparasitoids of eggs, mainly hosts of the order Lepidoptera (Pinto 2006). *Trichogramma minutum* Riley is the most efficient parasitoid of lepidopteran defoliation eggs such as *Actebia fennica* Tauscher (Noctuidae), *Lambdina fiscellaria* Guenee (Geometridae), *Choristoneura fumiferana* (Clemens) and *Choristoneura occidentalis* Freeman (Tortricidae), *Oryia leucostimata* Smith, and *Lymantria dispar* L. (Erebidae)

(Bai et al. 1995). In Brazil, many of the major lepidopteran defoliator species are parasitized by *Trichogramma* (Moraes et al. 1983; Oliveira et al. 2000): *Trichogramma maxacalii* Voegelé & Pointel in *Euselasia euploea eucerus* (Hewitson) (Riodinidae) (Voegelé and Pointel 1980), *Trichogramma manicobai* Brun, Moraes & Soares and *Trichogramma caiaposi* Voegelé & Pointel in eggs of *Erinnyis ello* (L.) (Sphingidae), and *Trichogramma acacioi* Brun, Moraes & Soares in *Psorocampa denticulata* (Schaus) (Notodontidae) (Brun et al. 1984). In Chile, since 1970, a large augmentative project has been implemented for the control of *Rhyacionia buoliana* (Denis & Schiffermüller) (Tortricidae), with releases of the parasitoids *Trichogramma nerudai* Pintureau & Gerding; Lepidoptera (defoliating caterpillars) are under augmentative biological control in large forest areas with *Trichogramma* spp. in Colombia (Bueno and van Lenteren 2002). *Trichogramma* spp. are also used to control *Rhyacionia frustrana* (Comstock) (Tortricidae) and *Mocis latipes* (Guennée) (Noctuidae) in Cuba (Vázquez et al. 2010).

Palmistichus elaeisis Delvare & LaSalle (Eulophidae), an endoparasitoid of gregarious habit, preferably of Lepidoptera, and reported in more than 20 species, is a promising biological control agent (Andrade et al. 2010; Menezes et al. 2014). This insect oviposits inside the pupae, where its larvae hatch and feed on host body tissues until adulthood (Soares et al. 2009). *P. elaeisis* parasitized Lepidoptera defoliators such as *Eupseudosoma involuta* (Sepp) (Arctiidae), *Euselasia eucerus* (Hewitson) (Riodinidae) (Delvare and Lasalle 1993), *Sabulodes* sp., *T. arnobia*, *Thyrinteina leucocerae* Rindge (Geometridae) (Pereira et al. 2008), and *S. violascens* (Zaché et al. 2012). In addition, it developed in *Tenebrio molitor* L. (Tenebrionidae) (Zanuncio et al. 2008) and *Bombyx mori* L (Bombycidae) pupae (Pereira et al. 2010).

Species of *Rhyssa* (Ichneumonidae) are siricid ectoparasitoid with potential for the biological control of *S. noctilio* Fabricius (Coyle and Gandhi 2012). Some species of *Rhyssa* genus, such as *R. alaskensis* Ashmead, *R. crevieri* (Provancher), *R. hoferi* Rohwer, *R. howdenorum* (Townes), *R. persuasoria* (L.), and *R. ponderosae* Townes, are used in the biological control of *S. noctilio* and *Urocerus* spp. (Siricidae) in North America (Coyle and Gandhi 2012). *Rhyssa persuasoria* was introduced in Brazil in 1996, 1997, and 2003 from Australia, but its establishment was not confirmed until 2010 for the control of *S. noctilio* (Iede et al. 2000). This species was also introduced in Argentina, Chile, and Uruguay for the control *S. noctilio* (Slippers et al. 2012).

In the genus *Megarhyssa* (Ichneumonidae), *M. nortoni* (Cresso), *M. atrata* (Fabricius), *M. greenei* Viereck, and *M. macrurus* (L.) are abundant in forest areas presenting potential biological control of Siricidae species (Coyle and Gandhi 2012). *Megarhyssa nortoni* was introduced in Brazil in 1996, 1997, and 2003 from Tasmania, Australia, through a project between Embrapa Forests, CSIRO, International Institute of Biological Control CABI-Bioscience and the USDA Forest Service (Iede et al. 2000). Parasitoids of this genus were also introduced in Argentina, Chile, and Uruguay to control *S. noctilio* populations (Slippers et al. 2012).

Species of *Ibalia* (Ibaliidae) are egg and first instar larvae endoparasitoids with a 3-year life cycle (Middlekauff 1960). In the last instars of their host though, they become ectoparasites. *Ibalia* species respond to the volatiles of *Amylostereum*, symbiotic fungi of the wood wasp (Spradbery and Kirk 1978). *Ibalia anceps*, *I. montana*, *I. rufipes rufipes*, *I. kirki*, *I. arizonica* and *I. ruficollis* are used in the biological control of *S. noctilio* in North America (Coyle and Gandhi 2012). The parasitism of the wood wasp by *I. leucospoides* (Hochenwarth) was 23% with a high establishment capacity (Iede et al. 2000). The parasitism of *S. noctilio* was 20% in Argentina (Klasmer et al. 1998), 24% in Uruguay (Rebuffo 1988), and 4% (Carvalho 1993) and 24% (Iede and Zanetti 2007) in Brazil. This parasitoid is established in Brazil; companies breed this natural enemy in the laboratory and release it in nonoccurrence areas or to increase parasitism rates (Iede et al. 2000).

Selitrichodes neseri Kelly & La Salle (Eulophidae) is an ectoparasitoid native to Australia, introduced to Brazil in 2015 as an alternative to control larvae and prepupae of *Leptocybe invasa* Fisher & La Salle (Eulophidae) and can be bred in the laboratory (Dittrich-Schröder et al. 2014; Souza et al. 2015). This parasitoid was also introduced in Argentina (Aquino et al. 2011), Chile (Servicio Agrícola y Ganadero 2014), and Uruguay (Jorge et al. 2016) to control gall wasps (*L. invasa*).

3. *Entomopathogens*

Entomopathogenic microorganisms such as fungi, bacteria, viruses, and nematodes have been successfully used for biological control in agriculture, forests, and urban habitats, and against insects of veterinary importance (Lacey et al. 2015).

25.3.1 Fungi

Entomopathogenic fungi attack Coleoptera, Diptera, Hemiptera, Hymenoptera, Lepidoptera, and Orthoptera insect pests. Some species of the Clavicipitacea family have a broader spectrum than others, such as Entomophthorales, which are more specific pathogens. The natural occurrence of fungi such as *Metarhizium* and *Beauveria* was recorded on the herb borer *Hedypathes betulinus* (Klug) (Cerambycidae) (Dalla Santa et al. 2009) and subsequently isolated from their hosts and used in pest control. Some strains of *Beauveria bassiana* (Bals.-Criv.) Vuill. with high virulence were isolated from *Oligonychus yothersi* (McGregor) (Acari: Tetranychidae) (Oliveira et al. 2004) and *T. peregrinus* (Lorencetti et al. 2011). In general in Latin-American countries, the practical application of these agents in the biological control of forest pests is limited. *Metarhizium anisopliae* (Metch.) Sorokin and *B. bassiana* are the most commonly used commercial mycoinsecticides in forest areas (Wright 1993; Alves 1998). *Beauveria bassiana* has a wide host range, broad action spectrum, various infection pathways (oral, tracheal, or tegumentary), and great genetic variability and durability in the environment (Lorencetti et al. 2011). In Cuba, *M. anisopliae* is used to control *Hypsipyla grandella* (Zeller) (Pyralidae) and *R. frustrana*; *B. bassiana* is used to control *Phyllophaga* spp. (Scarabaeidae), *Ips* spp. (Curculionidae), *Atta insularis* (Formicidae), and *H. grandella* (Vázquez et al. 1999, 2010).

Metarhizium anisopliae naturally attacks more than 300 insect species. When infected, the hosts are mummified and take on a light to dark green, grayish, or whitish coloration (Alves 1998). Commercial products based on *M. anisopliae* were effective against nymphs and adults of the shell psyllid *G. brimblecombei* (Wilcken et al. 2015). Entomophthorales fungi (Entomophthora, Zoophthora, and Neozygites) can maintain insect populations in equilibrium (Sosa-Gómez et al. 2003) and a field epizootiae of *Zoophthora radicans* (Brefeld) (Entomophthorales: Entomophthoraceae) has been reported on the bronze bug (Soliman 2010).

25.3.2 Bacteria

The main bacteria used in biological control are *Bacillus thuringiensis* (various varieties), such as kurstaki that attacks caterpillars (Lepidoptera), israelensis that infects Diptera larvae (mosquitoes and black flies), tenebrionis that infects Coleoptera, and *Bacillus sphaericus* that infects Diptera larvae.

Bacillus thuringiensis is a bacterium widely used to control *E. involuta* (Arctiidae), *S. violascens* (Lymantriidae), and *T. arnobia* (Geometridae) (Zanuncio et al. 1993). In Mexico and South America, damage caused by *L. dispar* (Erebidae) is controlled by products based on *B. thuringiensis* (Demir et al. 2012; Praciak 2013). In Cuba, this bacterium is used to control *M. latipes* (Vázquez et al. 1999, 2010).

25.3.3 Virus

The lepidopterans species *C. fumiferana*, *C. occidentalis*, *Choristoneura pinus* Freeman (Tortricidae), and *Hyphantria cunea* Drury (Erebidae), *L. dispar*, *Orgyia pseudotsugata* McDunn (Lymantriidae), and *Panolis flammea* (Denis & Schiffermüller) (Noctuidae) are forest plant defoliators susceptible to baculovirus (Moscardi 1999). In Brazil, there is a baculovirus-based product registered for *Condylorrhiza vestigialis* (Guenée) (Crambidae) (Haase et al. 2015). The use of TaV (*Thyrinteina arnobia* virus) as a bioinsecticide is an alternative for the biological control of this pest in Latin-American countries with reports of natural occurrence in the field.

25.3.4 Nematodes

The specie *Deladenus* (= *Beddingia*) *siricidicola* (Nematoda: Neotylenchidae) was introduced in Argentina, Brazil, Chile, and Uruguay to control *Sirex noctilio* (Siricidae) (Hurley et al. 2007). This nematode presents a free life cycle, feeding on the fungus *Amylostereum areolatum* (Chaillet ex Fr.) Boidin and can be easily bred in the laboratory and released in the field. When applied to trees attacked by *S. noctilio*, it can achieve near 100% parasitism levels of this pest (Iede et al. 1998). In Cuba, there is a program to use nematodes to control *Ips* spp. (Vázquez et al. 2010).

25.4 Final Remarks

The use of natural enemies is among the most effective and promising methods to control forest pests. The success and development of a biological program requires planning and knowledge of several factors, including the biology of natural enemies and the pests to be controlled. Biodiversity and climate favor the biological control in Latin America. However, its implementation is limited due to the need for technical assistance, more knowledge, and lack of public policies that contribute to the development and understanding of the process and its benefits. Among the biological agents, parasitoids have the greater potential for commercial use since they have wide global distribution, are highly specialized and efficient, and occur with greater abundance in plantations due to their accidental and planned introductions. Currently, biological control with native and introduced natural enemies is an integral part of any forest pest-management program.

Acknowledgments The authors thank the "Conselho Nacional de Desenvolvimento Cientifico e Tecnológico (CNPq), Coordenação de Aperfeiçoamento de Pessoal de Nível Superior (CAPES), Fundação de Amparo à Pesquisa do Estado de Minas Gerais (FAPEMIG)," and "Programa Cooperativo de Proteção Florestal/PROTEF of the Instituto de Pesquisas e Estudos Florestais/IPEF." Dr. Phillip John Villani (University of Melbourne, Australia) revised and corrected the English language in this manuscript.

References

Albuquerque GS, Tauber CA, Tauber MJ (1994) *Chrysoperla externa* (Neuroptera: Chrysopidae): life history and potential for biological control in Central and South America. Biol Control 4(1):8–13

Alves SB (1998) Controle microbiano de insetos, 2nd edn. FEALQ, Piracicaba

Andrade GS, Serrão JE, Zanuncio JC, Zanuncio TV, Leite GLD (2010) Immunity of an alternative host can be overcome by higher densities of its parasitoids *Palmistichus elaeisis* and *Trichospilus diatraeae*. PLoS One 5(10):e13231

Aquino DA, Botto EN, Loiácono MS, Pathauer P (2011) Avispa de la agalla del eucalipto", *Leptocybe invasa* Fischer & La Salle (Hymenoptera: Eulophidae: Tetrastichinae) en Argentina. RIA 37(2):159–164

Bai B, Çobanoĝlu S, Smith SM (1995) Assessment of *Trichogramma* species for biological control of forest lepidopteran defoliators. Entomol Exp Appl 75(2):135–143

Bale J, van Lenteren JC, Bigler F (2008) Biological control and sustainable food production. Philos Trans R Soc Lond B 363(1492):761–776

Barbosa LR, Rodrigues AP, Soler LS, Fernandes BV, Castro BMC, Wilcken CF, Zanuncio JC (2017) Establishment in the field of *Cleruchoides noackae* (Hymenoptera: Mymaridae), an exotic egg parasitoid of *Thaumastocoris peregrinus* (Hemiptera: Thaumastocoridae). Fla Entomol 100(2):372–374

Berti Filho E, Costa VA, Zuparko RL, Lasalle J (2003) Ocorrência de *Psyllaephagus bliteus* Riek (Hymenoptera: Encyrtidae) no Brasil. Rev Agric 78(3):304

Botelho PSM (2010) Eficiência de *Trichogramma* em campo. In: Parra JRP, ZucchI RA (eds) *Trichogramma* e o controle biológico aplicado. FEALQ, Piracicaba, pp 303–318

Bridgeland WT, Beier P, Kolb T, Whitham TG (2010) A conditional trophic cascade: birds benefit faster growing trees with strong links between predators and plants. Ecology 91(1):73–84

Brockerhoff EG, Jactel H, Parrotta JA, Ferraz SFB (2013) Role of eucalypt and other planted forests in biodiversity conservation and the provision of biodiversity: related ecosystem services. For Ecol Manag 301:43–50

Brun PG, Moraes GWG, Soares LA (1984) Três espécies novas de Trichogrammatidae parasitoides de lepidópteros desfolhadores de mandioca e eucalipto. Pesq Agropec Bras 19(7):805–810

Bueno VHP, van Lenteren JC (2002) The popularity of augmentative biological control in Latin America: history and state of affairs. In: Proceedings of the 1st symposium on biological control of Arthropods, Honolulu, Hawaii, pp 180–184

Carbone SS, Rivera AC (2003) Egg load and adaptive superparasitism in *Anaphes nitens*, an egg parasitoid of the Eucalyptus snout-beetle *Gonipterus scutellatus*. Entomol Exp Appl 106(2):127–134

Carbone SS, Rodriguez-Ilamola A, Rivera AC (2004) Host finding and host discrimination ability in *Anaphes nitens* Girault, an egg parasitoid of the Eucalyptus snout-beetle *Gonipterus scutellatus* Gyllenhal. Biol Control 29(1):24–33

Cardinale BJ, Matulich KL, Hooper DU, Byrnes JE, Duffy E, Gamfeldt L, Gonzalez A (2011) The functional role of producer diversity in ecosystems. Am J Bot 98(3):572–592

Carnus JM, Parrotta J, Brockerhoff EG, Arbez M, Jactel H, Kremer A, Lamb D, O'Hara K, Walters B (2006) Planted forests and biodiversity. J For 104(2):65–77

Carvalho AG (1993) Aspectos biológicos de *Ibalia leucospoides* (Hockewarth), Hymenoptera, Ibaliidae. In: Anais da Conferencia Regional da vespa da madeira, *Sirex noctilio*, na America do Sul, Florianópolis, pp 111–120

Coyle DR, Gandhi KJ (2012) The ecology, behavior, and biological control potential of hymenopteran parasitoids of woodwasps (Hymenoptera: Siricidae) in North America. Environ Entomol 41(4):731–749

Cross D (2009) The parasitoids of *Thaumastocoris* spp. in the Sydney region. Thesis, Department of Food and Natural Resources, University of Sydney

Daane KM, Sime KR, Dahlsten DL, Andrews JW, Zuparko RL (2005) The biology of *Psyllaephagus bliteus* Riek (Hymenoptera: Encyrtidae), a parasitoid of the red gum lerp psyllid (Hemiptera: Psylloidea). Biol Control 32(2):228–235

Dahlsten DL, Rowney DL, Copper WA, Tassan RL, Chaney WE, Rob KL, Tjosvold S, Bianchi M, Lane P (1998) Parasitoid wasp controls blue gum psyllid. Calif Agric 52(1):31–34

Dalla Santa HS, Sousa NJ, Pittner E, Dalla Santa OR, Soccol CR (2009) Controle biológico em pragas de *Ilex paraguariensis* (A. St.-Hil.) com fungo Beauveria sp. Rev Floresta 39(1):67–76

De Bortoli SA, Otuka AK, Vacari AM, Martins MI, Volpe HX (2011) Comparative biology and production costs of *Podisus nigrispinus* (Hemiptera: Pentatomidae) when fed different types of prey. Biol Control 58(2):127–132

Delvare G, Lasalle JA (1993) New genus of Tetrastichinae (Hymenoptera: Eulophidae) from the Neotropical region, with the description of a new species parasitica on key pests of oil palm. J Nat Hist 27(2):435–444

Demir İ, Eryüzlü E, Demirbağ Z (2012) A study on the characterization and pathogenicity of the bacteria from *Lymantria dispar* L. (Lepidoptera: Lymantriidae). Turk J Biol 36(4):459–468

Diaz SD, Tilman J, Fargione FS, Chapin R, Dirzo T, Kitzberger B, Gemmill M, Zobel M, Vila C, Mitchell A, Wilby GC, Daly M, Galetti WF, Laurence J, Pretty R, Naylor A, Power D, Harvell

S, Potts C, Kremen T, Griswold C (2005) Biodiversity regulation of ecosystem services. In: Hassan R, Scholes R, Ash N (eds) Ecosystems and human well-being: current state and trends, millennium ecosystem assessment, vol 1. Island Press, Washington, pp 297–329

Dittrich-Schröder G, Harley M, Neser S, Joffe T, Bush S, Hurley BP, Wingfield MJ, Slippers B (2014) Biology and host preference of *Selitrichodes neseri:* a potential biological control agent of the *Eucalyptus* gall wasp, *Leptocybe invasa*. Biol Control 78:33–41

Fidalgo P, Olivares TS, Haro AM, De Baranao JJ (2005) Control biologico de *Ctenarytaina eucalypti* (Hem.: Psyllidae), plaga del eucalipto en Argentina. Bosque 26(2):91–93

Food and Agricultural Organization of the United Nations (2014) State of the World's Forests: enhancing the socioeconomic benefits from forests. Rome

Food and Agricultural Organization of the United Nations (2015) Terms and definitions (Forest Resources Assessment Working Paper, vol 180). Rome. http://www.fao.org/docrep/017/ap862e/ap862e00.pdf. Accessed 10 Sept 2016

Freymann B, Buitenwerf R, Souza O, Olff H (2008) The importance of termites (Isoptera) for the recycling of herbivore dung in tropical ecosystems: a review. Eur J Entomol 105(2):165–173

Haase S, Sciocco-Cap A, Romanowski V (2015) Baculovirus Insecticides in Latin America: historical overview, current status and future perspectives. Viruses 7(5):2230–2267

Haji FND, Jimenez Velasquez J, Bleicher E, Alencar JA, Haji AT, Diniz RS (1998) Tecnologia de produção massal de *Trichogramma* spp. Embrapa-CPATSA, Petrolina

Herrera IG (2001) Avances del control biológico de la conchuela del eucalipto *Glycaspis brimblecombei* y dispersión del parasitóide *Psyllaephagus bliteus*. Tu Bosque 26(1):5–6

Hurley BP, Slippers B, Wingfield MJ (2007) A comparison of control results for the alien invasive woodwasp, *Sirex noctilio*, in the southern hemisphere. Agric For Entomol 9(3):159–171

Iede ET, Zanetti R (2007) Ocorrência e recomendações para o manejo de *Sirex noctilio* Fabricius (Hymenoptera, Siricidae) em plantios de *Pinus patula* (Pinaceae) em Minas Gerais, Brasil. Rev Bras Entomol 51(4):529–531

Iede ET, Penteado SRC, Leite MSP (1998) Utilização do nematóide *Deladenus siricidicola* (Nematoda: Neotylenchidae) no controle biológico de *Sirex noctilio* (Hymenoptera: siricidae), praga de Pinus spp. In: Anais do I. Congresso Latino-Americano IUFRO, Valdívia

Iede ET, Penteado SRC, Reis Filho W, Schaitza EG (2000) Updated situation of Sirex noctilio integrated management program in Brazil. Série Tecnica IPEF 13:11–20

Jactel H, Brockerhoff EG (2007) Tree diversity reduces herbivory by forest insects. Ecol Lett 10(9):835–848

Jorge C, Martínez G, Gómez D, Bollazzi M (2016) First record of the eucalypt gallwasp *Leptocybe invasa* (Hymenoptera: Eulophidae) from Uruguay. Bosque 37(3):631–636

Keenan RJ, Reams GA, Achard F, Freitas JV, Grainger A, Lindquist E (2015) Dynamics of global forest area: results from the FAO Global Forest Resources Assessment. For Ecol Manag 352:9–20

Kevan DKMCE (1946) The *Eucalyptus weevil* in East Africa. E Afr Agric J 12(1):40–44

Klasmer P, Fritz G, Corley J, Botto E (1998) Current status of research on *Sirex noctilio* F. in the Andean-Patagonian region in Argentina. In: Proceedings of a conference: training in the control of *Sirex noctilio* by the use of natural enemies, Colombo, pp 89–90

Kremen C, Chaplin-Kramer R (2007) Insects as providers of ecosystem services: crop pollination and pest control. In: Proceedings of the Royal Entomological Society's 23rd symposium, insect conservation biology. University of California, Berkeley, pp 349–382

Kurylo CL (2008) Ocorrência e bionomia de *Ctenarytaina eucalypti* (Maskell) (Hemiptera: Psyllidae) e seu inimigo natural *Psyllaephagus pilosus* Noyes (Hymenoptera: Encyrtidae) em plantas de *Eucalyptus globulus*. Dissertação, Universidade Federal de Pelotas

Lacey LA, Grzywacz D, Shapiro-Ilan DI, Frutos R, Brownbridge M, Goettel MS (2015) Insect pathogens as biological control agents: back to the future. J Inverteb Pathol 132:1–41

Lanfranco D, Dungey HS (2001) Insect damage in Eucalyptus: a review of plantations in Chile. Aust Ecol 26(5):477–481

Lin NQ, Huber JT, La Salle J (2007) The Australian genera of Mymaridae (Hymenoptera: Chalcidoidea). Zootaxa 1596(1):1–111

Lopez-Arroyo JI, Tauber CA, Tauber MJ (1999) Comparative life histories of the predators *Ceraeochrysa cincta, C. cubana*, and *C. smithi* (Neuroptera: Chrysopidae). Ann Entomol Soc Am 92(2):208–217

Lorencetti GAT, Menezes MJS, Jung PH, Gonçalves TE, Barbosa LR, Potrich M, Mazaro SM, Silva ERL (2011) Análise do potencial de isolados de *Beauveria bassiana* Vuill. (Ascomycetes: Clavicipitaceae) para o controle de *Thaumastocoris peregrinus* Carpintero & Dellapé (Hemiptera: Thaumastocoridae). In: Anais do Congresso de Ciência e Tecnologia da UTFPR. Universidade Tecnológica Federal do Paraná, Dois Vizinhos, pp 94–97

Losey JE, Vaughan M (2006) The economic value of ecological services provided by insects. Bioscience 56(4):311–323

Matthews E, Payne R, Rohweder M, Murray S (2000) Pilot Analysis of Global Ecosystems (PAGE): forest ecosystems. World Resources Institute, Washington, DC

Menezes CWG, Soares MA, Santos JB, Assis SL Jr, Fonseca AJ, Zanuncio JC (2012) Reproductive and toxicological impacts of herbicides used in *Eucalyptus* culture in Brazil on the parasitoid *Palmistichus elaeisis* (Hymenoptera: Eulophidae). Weed Res 52(6):520–525

Menezes CW, Soares MA, Fonseca AJ, Santos JB, Camilo SDS, Zanuncio JC (2014) *Palmistichus elaeisis* (Hymenoptera: Eulophidae) as an indicator of toxicity of herbicides registered for corn in Brazil. Chil J Agric Res 74(3):361–365

Middlekauff WW (1960) The siricid woodwasps of California (Hymenoptera: Symphyta). Bull Calif Insect Surv 6(4):59–77

Miura S, Amacher M, Hofer T, San-Miguel-Ayanz J, Thackway R (2015) Protective functions and ecosystem services of global forests in the past quarter-century. For Ecol Manag 352:35–46

Molina-Rugama AJ, Zanuncio JC, Torres JB, Zanuncio TV (1997) Longevidad y fecundidad de *Podisus nigrispinus* (Heteroptera: Pentatomidae) alimentado con *Musca domestica* (Diptera: Muscidae) y frijol. Rev Biol Trop 45(3):1125–1130

Moraes GWG, Brun PG, Soares LA (1983) O controle biológico dos lepidópteros desfolhadores de eucaliptos em Minas Gerais. Inf Agrop 9(104):23–30

Moscardi F (1999) Assessment of the application of baculoviruses for control of Lepidoptera. Annu Rev Entomol 44(1):257–289

Mutitu K, Garnas JR, Hurley BP, Wingfield MJ, Harney M, Bush SJ, Slippers B (2013) Biology and rearing of *Cleruchoides noackae* (Hymenoptera: Mymaridae), an egg parasitoid for the biological control of *Thaumastocoris peregrinus* (Hemiptera: Thaumastocoridae). J Econ Entomol 106(5):1979–1985

Nadel RL, Noack AE (2012) Current understanding of the biology of *Thaumastocoris peregrines* in the quest for a management strategy. Int J Pest Manag 58(3):257–266

Nadrowski K, Wirth C, Scherer-Lorenzen M (2010) Is forest diversity driving ecosystem function and service? Curr Opin Environ Sustain 2(1–2):75–79

Noyes JS (2015) Universal Chalcidoidea catabase. http://www.nhm.ac.uk/chalcidoids Accessed 10 Sept 2016

Noyes JS, Hanson P (1996) Encyrtidae (Hymenoptera: Chalcidoidea) of Costa Rica: the genera and species associated with jumping plant-lice (Homoptera: Psylloidea). Bull Natl Hist Mus 65:105–164

Oliveira NC (2006) Biologia de *Gonipterus scutellatus* (Coleoptera: Curculionidae) em *Eucalyptus* spp. em diferentes temperaturas. Tese, Universidade Estadual Paulista "Júlio de Mesquita Filho"

Oliveira HN, Zanuncio JC, Pratissoli D, Cruz I (2000) Parasitism rate and viability of *Trichogramma maxacalii* (Hym.: Trichogrammatidae) parasitoid of the Eucalptus. For Ecol Manag 130(1–3):1–6

Oliveira RC, Neves PMOJ, Alves LFA (2004) Seleção de fungos entomopatogênicos para o controle de *Oligonychus yothersi* (McGregor) (Acari: Tetranychidae), na cultura da erva-mate (*Ilex paraguariensis* St. Hill.). Neotrop Entomol 33(3):347–351

Paine TD, Steinbauer MJ, Lawson SA (2011) Native and exotic pests of Eucalyptus: a worldwide perspective. Annu Rev Entomol 56(1):181–201

Pereira FF, Zanuncio JC, Tavares MT, Pastori PL, Jacques GC, Vilela EF (2008) New record of *Trichospilus diatraeae* as a parasitoid of the eucalypt defoliator *Thyrinteina arnobia* in Brazil. Phytoparasitica 36(3):304–306

Pereira FF, Zanuncio JC, Serrão JE, Zanuncio TV, Pratissoli D, Pastori PL (2010) The density of females of *Palmistichus elaeisis* Delvare and LaSalle (Hymenoptera: Eulophidae) affects their reproductive performance on pupae of *Bombyx mori* L. (Lepidoptera: Bombycidae). An Acad Bras Ciênc 82(2):323–331

Pinto JDA (2006) Review of the new world genera of Trichogrammatidae (Hymenoptera). J Hymenopt Res 15(1):38–163

Pires EM, Soares MA, Nogueira RM, Zanuncio JC, Moreira PSA, Oliveira MA (2015) Seven decades of studies with Asopinae predators in Brazil (1933–2014). Biosci J 31(5):1530–1547

Praciak A (2013) The CABI encyclopedia of forest trees. CAB International, Wallingford

Rebuffo S (1988) La "Avispa de la Madera" *Sirex noctilio* F. en el Uruguay. Ministério e Ganaderia, Agricultura y Pesca, Dirección Forestal, Montevideo

Riek EF (1962) The Australian species of *Psyllaephagus* (Hymenoptera: Encyrtidae), parasites of Psyllids (Homoptera). Aust J Zool 10(4):684–757

Rodríguez AF, Sáiz GF (2006) Parasitoidismo de *Psyllaephagus pilosus* Noyes (Hym.: Encyrtidae) sobre el psílido del eucalipto *Ctenarytaina eucalypti* (Maskell) (Hem.: Psyllidae) en plantaciones de eucaliptos en la V Región. Agric Téc 66(4):342–351

Sanches MA (2000) Parasitismo de ovos de *Gonipterus scutellatus* Gyllenhal, 1833 e *Gonipterus gibberus* Boisduval, 1835 (Coleoptera, Curculionidae) por *Anaphes nitens* (Girault, 1928) (Hymenoptera, Mymaridae) em Colombo (Paraná, Brasil). Arq Inst Biol 67(1):77–82

Servicio Agrícola y Ganadero (2014) *Leptocybe invasa* Fischer et Lasalle (Hymenoptera: Eulophidae) Microavispa formadora de agallas en eucalipto. Servicio Agrícola y Ganadero, Santiago. http://www.sag.cl/ambitos-de-accion/leptocybe-invasa-fischer-lasalle-hymenoptera-eulophidae-microavispa-formadora-de. Accessed 18 Mar. 2018

Singh S (2016) Description of a new species of *Psyllaephagus Ashmead* (Hymenoptera: Encyrtidae) parasitizing pit gall forming *Psyllid trioza fletcheri* (Hemiptera: Psylloidea: Triozidae) on *Terminalia arjuna* from India. Ann Zool 66(3):393–402

Slade EM, Mann DJ, Lewis OT (2011) Biodiversity and ecosystem function of tropical forest dung beetles under contrasting logging regimes. Biol Conserv 144(1):166–174

Slippers B, De Groot P, Wingfield MJ (2012) The sirex woodwasp and its fungal symbiont: research and management of a worldwide invasive pest. Springer, New York

Soares MA, Gutierrez CT, Zanuncio JC, Pedrosa ARP, Lorenzon AS (2009) Superparasitismo de *Palmistichus elaeisis* (Hymenoptera: Eulophidae) y comportamiento de defensa de dos hospederos. Rev Colomb Entomol 35(1):62–65

Soliman EP (2010) Bioecologia do percevejo bronzeado *Thaumastocoris peregrinus* Carpintero & Dellapé (Hemiptera: Thaumastocoridae) em eucalipto e prospecção de inimigos naturais. Dissertação, Faculdade de Ciências Agronômicas, Universidade Estadual Paulista "Júlio de Mesquita Filho"

Sosa-Gómez DR, Delpin KE, Moscardi F, Nozaki MH (2003) The impact of fungicides on *Nomuraea rileyi* (Farlow) Samson epizootics and on populations of *Anticarsia gemmatalis* Hübner (Lepidoptera: Noctuidae), on soybean. Neotrop Entomol 32(2):287–291

Souza AR, Puretz BO, Souza NM, Becchi LK, Veloso SGM, Wilcken CF (2015) Controle biológico de pragas exóticas de *Eucalyptus* spp. In: Baldin ELL, Kronka AZ, Fujihara RT (eds) Proteção vegetal. Fepaf, Botucatu, pp 71–84

Spradbery JP, Kirk AA (1978) Aspects of the ecology of siricid woodwasps (Hymenoptera: Siricidae) in Europe, North Africa and Turkey with special reference to the biological control of *Sirex noctilio* F. in Australia. Bull Entomol Res 68(3):341–359

Thompson ID, Okabe K, Tylianakis JM, Kumar P, Brockerhoff EG, Schellhorn NA, Parrotta JÁ, Nasi R (2011) Forest biodiversity and the delivery of ecosystem goods and services: translating science into policy. Bioscience 61(12):972–981

Torres JB, Zanuncio JC, Moura MA (2006) The predatory stinkbug *Podisus nigrispinus*: biology, ecology and augmentative releases for lepidoperan larval control in Eucalyptus in Brazil. Biocontrol News Info 27(1):1–18

Tylianakis JM, Rand TA, Kahmen A, Klein AM, Buchmann N, Perner J, Tscharntke T (2008) Resource heterogeneity moderates the biodiversity-function relationship in real world ecosystems. PLoS Biol 6(5):947–956

van Lenteren JC (2012) The state of commercial augmentative biological control: plenty of natural enemies, but a frustrating lack of uptake. BioControl 57(1):1–20

van Lierop PE, Lindquist S, Sathyapala G, Franceschini G (2015) Global forest area disturbance from fire, insect pests, diseases and severe weather events. For Ecol Manag 352:78–88

Vázquez Moreno LL, Caballero S, Carr A, Gil J, Armas JL (2010) Diagnóstico de la utilización de entomófagos y entomopatógenos para el control biológico de insectos por los agricultores en Cuba. Fitosanidad 14(3):159–169

Vázquez LL, Menéndez JM, López R (1999) Manejo de insectos de importância forestal en Cuba. Rev Man Integr Plagas 54:4–14

Voegelé JJ, Pointel JG (1980) Une nouvelle espèces de Trichogramme, *Trichogramma maxacalii* (Hymenoptera: Trichogrammatidae). Ann Soc Entomol France 16:599–603

Wilcken CF, Oliveira NC, Sartório RC, Loureiro EB, Bezerra N Jr, Rosado Neto GH (2008a) Ocorrência de *Gonipterus scutellatus* Gyllenhal (Coleoptera: Curculionidae) em plantações de eucalipto no Estado do Espírirto Santo. Arq Inst Biol 75(1):113–115

Wilcken CF, Sá LAN, Berti-Filho E, Ferreira-Filho PJ, Oliveira NC, Dal Pogetto MHFA, Soliman EP (2008b) Plagas exóticas de importancia en Eucalyptus en Brasil. In: Anales de 23. Jornadas Forestales de Entre Rios, Concórdia, pp 1–5

Wilcken CF, Firmino-Winckler DC, Dal Pogetto MHFA, Dias TKR, Lima ACV, Sá LD, Ferreira Filho PJ (2015) Psilídeo-de-concha-do-eucalipto, *Glycaspis brimblecombei* Moore. In: Vilela Filho E, Zucchl RA (eds) Pragas introduzidas no Brasil: insetos e ácaros. FEALQ, Piracicaba, pp 883–897

Wingfield MJ, Slippers B, Hurley BP, Coutinho TA, Wingfield BD, Roux J (2008) Eucalypt pests and diseases: growing threats to plantation productivity. South For 70(2):139–144

Wright JE (1993) Control of the boll weevil (Coleoptera: Curculionidae) with Naturalis-L: a mycoinsecticide. J Econ Entomol 86(5):1355–1358

Wylie FR, Speight MR (2012) Insect pests in tropical forestry, 2nd edn. CABI, Wallingford

Yang LH, Gratton C (2014) Insects as drivers of ecosystem processes. Curr Opin Insect Sci 2:26–32

Zaché B, Zaché RRC, Wilcken CF (2012) Evaluation of *Palmistichus elaeisis* Delvare & LaSalle (Hymenoptera: Eulophidae) as parasitoid of the *Sarsina violascens* Herrich-Schaeffer (Lepidoptera: Lymantriidae). J Plant Stud 1(1):85–89

Zanuncio JC, Santana DL, Nascimento ECD, Santos GP, Alves JB, Sartório RC, Zanuncio TV, Alves JB, Zanuncio TV (1993) Descrição das lagartas desfolhadoras. In: Manual de pragas em florestas-lepidoptera desfolhadoras de eucalipto: biologia, ecologia e controle. IPEF, Piracicaba, pp 75–77

Zanuncio TV, Zanuncio JC, Miranda MMM, Medeiros AGB (1998) Effect of plantation age on diversity and population fluctuation of Lepidoptera collected in Eucalyptus plantations in Brazil. For Eco Manage 108:91

Zanuncio JC, Lemos WP, Lacerda MC, Zanuncio TV, Serrão JE, Bauce E (2006) Age-dependent fecundity and fertility life tables of the predator *Brontocoris tabidus* (Heteroptera: Pentatomidae) under field conditions. J Econ Entomol 99(2):401–407

Zanuncio JC, Pereira FF, Jacques GC, Tavares MT, Serrão JE (2008) *Tenebrio molitor* Linnaeus (Coleoptera: Tenebrionidae), a new alternative host to rear the pupae parasitoid *Palmistichus elaeisis* Delvare & LaSalle (Hymenoptera: Eulophidae). Coleopt Bull 62(1):64–66

Zanuncio JC, Tavares WS, Fernandes BV, Wilcken CF, Zanuncio TV (2014) Production and use of Heteroptera predators for the biological control of Eucalyptus pests in Brazil. Ekoloji 23(91):98–104

Chapter 26
Fruit Crops

Lenira V. C. Santa-Cecília, Brígida Souza, Kethullyn H. Silva, and Ernesto Prado

26.1 Introduction

When thinking on biological control, we often associate it with man's intervention in the agroecosystems. As a larger concept, we can state that natural enemies, predators, parasites, entomopathogens, among others, are almost always present acting or regulating the insect populations in nature. It is another topic if they fill our expectations in agriculture keeping pests under the tolerance level. Biological control is not an alternative to the use of insecticides, just the opposite; the use of insecticides is an alternative when natural enemies are not able to keep pests under the tolerance level and is perhaps the last tool to be considered in pest control.

Phytophagous insects becoming pests are minimal considering the great number of insect consuming plants. Most of them are probably kept at low level by natural enemies or physical constraints. Nevertheless, even considering the relative few cases of pest insects, they are enough injurious to reduce food production.

Natural enemies' action can be improved by means of plant diversification or by releasing a large amount of natural enemies, usually when those naturally present in the agroecosystem are insufficient to regulate pest population. An increase in plant diversity is commonly named as Ecological Strategy for Pest Management. A diverse and heterogeneous environment enables the attraction and maintenance of diverse organisms, increasing their interaction. For example, predation and parasitism are favored in a more complex system because more food resources are available

L. V. C. Santa-Cecília (✉)
IMA/EPAMIG Sul, Lavras, MG, Brazil
e-mail: scecilia@epamig.ufla.br

B. Souza
Department of Entomology, Federal University of Lavras (UFLA), Lavras, MG, Brazil

K. H. Silva · E. Prado
EPAMIG Sul, Lavras, MG, Brazil

B. Souza et al. (eds.), *Natural Enemies of Insect Pests in Neotropical Agroecosystems*, https://doi.org/10.1007/978-3-030-24733-1_26

(Bugg and Waddington 1994; Altieri 1999). Parasitoids are provided mainly with refuge, nectar, and pollen to increase fecundity or additional hosts when their range is broader. Predators, less specific than parasitoids, benefit from refuge, extra preys, nectar, and pollen. Syrphids and chrysopids, for example, two important groups of predators, need pollen and nectar to survive (Venzon et al. 2006; Oliveira et al. 2009; Amaral et al. 2013; Resende et al. 2015). Natural enemies conservation appears especially important in those species that commonly do not reach the economic injury level (EIL) and natural control must be preserved. In this way, biodiversity management is an ecological strategy within the Integrated Pest Management (IPM).

However, when we deal with biodiversity and biological control, our initial question would be whether a diverse agroecosystem is a source of pests or natural enemies that could diminish insect population (Andow 1991; Gurr et al. 1998; Landis et al. 2000). Cover crops, for example, can influence many aspects of an orchard culture by affecting directly and indirectly both beneficial and pests organisms. Thus, the choice of plant species to be incorporated in the agroecosystem requires detailed knowledge to avoid an opposite effect, increasing pests (Colley and Luna 2000; Lavandero et al. 2006). Preferably, companion plants should not be host for the same herbivores than the crop plant, or compete with the crop and they have to be well adapted to common agricultural practices.

Environmental manipulation research to improve microbial control of pests has also been limited; however, many environmental conditions are known to improve epizooties. High humidity and protecting fungi from solar radiation, for example, increase the probability of spreading infection inside insect population (Fuxa 1998; Wilson 1998).

Plant management contributes to the ecological strategy by using different approaches such as cash cropping, multicropping, side cropping, cover crop, intracropping, or intercropping. The latter comprises management of naturally occurring plants or the introduction of new plants that can even be used as green manure (Gliessman 2001; Gurr et al. 2003). Environmental manipulation can occur at different levels, namely, within-crop, within-farm, or modifying landscape (Landis et al. 2000). Trap crops, used as intracropping or border crop, can attract, divert, intercept, and/or retain insects, protecting the main crop from damaging (Shelton and Badenes-Perez 2006). It appears to be an efficient practice in addition to localized insecticide applications or genetically modified plants.

In addition to aid in pest control, plants can improve the physical, chemical, and biological characteristics of soil. This strategy of pest control by excluding synthetic insecticides is a response to the increasing demand of consumers for organic products without using chemicals and in agreement to government guidelines. In Cuba, for example, the Urban and Familiar Agriculture Proposal focuses on producing about 180 fruit crops with sustained agricultural practices without using pesticides and trusting in the plant diversity to prevent pests and diseases (Nodals and Pérez 2005).

This chapter deals with research—private and from governmental initiatives—in some Latin American countries focusing on the sustainability of fruit crop production, stressing yet the lack of available information.

26.2 Biodiversity in Fruits Crops

Most studies related to biodiversity in Latin American agroecosystems have focused on vegetables such as tomato, beans, and maize (Venzon et al. 2011). However, orchards, as a perennial system, are especially suitable to respond to a more diverse environment, which could increase the natural enemies' efficacy (Fig. 26.1).

However, cover crops, despite their benefits, are also associated with detrimental effects on the economic crops by competing for water and nutrients, allelopathic effects, changing the microenvironment, interfering with management, eventually providing refuge for pests and diseases, or being a source for insect pests. In consequence, a proper cover crop must be chosen with care in order to avoid or minimize these deleterious effects (Bugg and Waddington 1994; Smith and Mulder Jr. 2007).

Even considering that a high biodiversity could reduce the use of pesticides, practical research is still scarce or not documented, especially in Latin America. Bugg and Waddington (1994) presented a detailed review about cover crops and pest management in the United States showing opportunities and challenges, mainly regarding temperate crops. They recognized at that time the limited experimental success of cover crops for pest management explaining the meagre adoption in

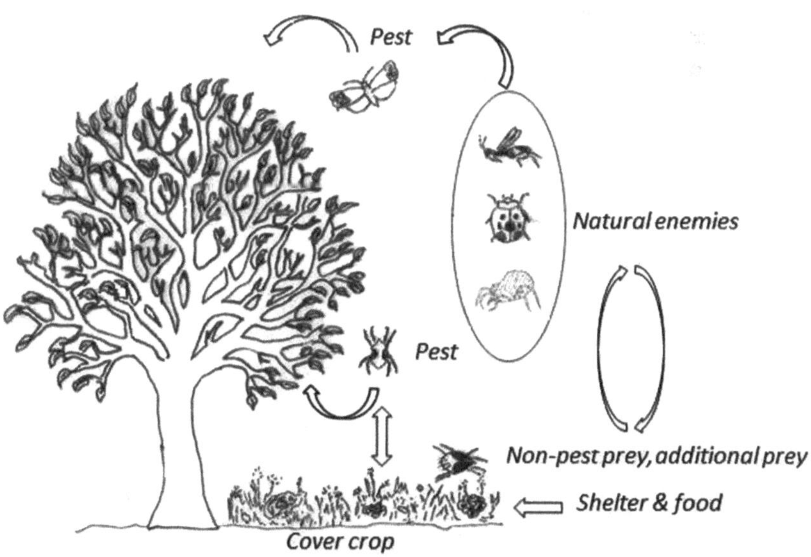

Fig. 26.1 Interactions between cover crop, fruit tree, natural enemies, and pests. (Figure: Ernesto Prado)

practical situations. However, a more recent review (Letourneau et al. 2011) throughout a meta-analysis of a number of papers related to biodiversity reported encouraging results of crop diversification. Significant natural enemies' enhancement and herbivore suppression were obtained in most of the analyzed papers; however, generally, crop yield increase was absent.

Cover crops could harbor generalist pests that can infest crops or alternatively divert these insects. Contrarily, cover crops are not sources for specialist pests and, therefore, should provide good conditions for natural enemies, reducing the pest population. On the other hand, cover crops can protect pupal stages and overwintering stages of specialists, increasing survival rate and infestation.

Mites are generally good candidates to manage with cover crops because they can provide alternative preys for phytoseid mites that can migrate to fruit trees to control mite pests. Even controversial, dust has been blamed to produce spider mite outbreaks by increasing leaf temperature and water stress favoring mite reproduction and development (Demirel and Çabuk 2008). Cover crops can also reduce dust avoiding mite outbreaks. In the State of Minas Gerais, Brazil, predatory mites such as *Phytoseius intermedius* Evans & MacFarlane (Phytoseiidae) and *Amblyseius herbicolus* (Chant) (Phytoseiidae) are favored by the natural vegetation and plant diversity in litchi orchards, both predators of *Aceria litchii* (Keifer) (Eriophyidae) (Venzon et al. 2016) (Fig. 26.2). Mite population has also been reduced by increasing predatory mites such as *Neoseiulus californicus* (McGregor) (Phytoseiidae) in spontaneous vegetation in apple orchards in Santa Catarina, Brazil (Monteiro et al. 2002).

Cover crops composed of white clover (*Trifolium repens* L.), ryegrass (*Lolium multiflorum* Lam.), common sow thistle (*Sonchus oleraceus* L.), Spanish needle (*Bidens pilosa* L.), dock (*Rumex* sp.), and dandelion (*Taraxacum officinalis* Weber)

Fig. 26.2 Weeds in a litchi orchard (*Litchia chinensis*). (Photo: C.S. Ferraz)

reduced the population of the ground pearl *Eurhizococcus brasiliensis* (Hempel) (Margarodidae) in vineyards in South Brazil. On the other hand, the velvet bean, *Stizolobium aterrimum* Piper and Tracy, did not affect the ground pearl population and represent a threat to the crop because it is the host of the insect (Botton et al. 2010). Grapes produced in South of Minas State in Brazil associated with oat (*Avena* spp.) and legume plants of the genus *Vicia* resulted in a higher diversity and abundance of natural enemies; however, their impact on the pest was not evaluated (Fadini et al. 2001).

Cover crops in passion fruit, *Passiflora edulis* Sims, also showed an increased biodiversity and beneficial establishment (Fancelli et al. 2016). Similar effects have been obtained in strawberries, *Fragaria* sp., where plants of the family Asteraceae such as the Spanish needle increased the abundance of *Orius* sp., a natural enemy of thrips (Rocha et al. 2017).

The adverse effect of guava trees, *Psidium guajava* L., on the Asian citrus psyllid, *Diaphorina citri* Kuwayama (Liviidae), and consequently on greening disease attack, has been documented in Vietnam (Beattie et al. 2006). The repellent effect of sesquiterpens volatiles of the guava fruit has been reported in Brazil using olfactory tests (Miranda et al. 2017). Recently, Tomaseto et al. (2019) reported the attractiveness of orange jasmine (*Murraya paniculata*) to *Diaphorina* and its potential to be used as border trap to avoid settlement and dispersal.

Aromatic plants grown in small areas of fruit orchards attract predators and parasitoids to improve pest control; for example, mint (*Mentha canadensis* L.) have been used in orchards to attract green lacewings (Chrysopidae) (Song et al. 2014). Greenhouse experiments in Brazil also showed the attractiveness of sweet basil (*Ocimum basilicum* L.) (Batista et al. 2017) and coriander (*Coriandrum sativum* L.) in association with roses (Salamanca et al. 2015) to green lacewings, opening a possibility to be used on fruit crops. Tests done in Brazil showed that flowers of dill, *Anethum graveolens* L., cilantro, *Coriandrum sativum* (L.), and fennel, *Foeniculum vulgare* Mill, improve survival and reproduction of the green lacewings *Chrysoperla externa* (Hagen) under laboratory conditions. These plants could be added to the list of potentially valuable species as insectary plants for habitat management and conservation biological control programs (Resende et al. 2015, 2017).

The increase in the biodiversity by using associated plants within the crop and outside it have been reported as responsible for pest reduction in vineyards in Berissio, Argentine, by protecting and increasing the number of ground beetles (Carabidae) (Paleologos et al. 2007). Similar results were obtained in Mexico, referring to fruit flies of genre *Anastrepha* (Tephritidae) by limiting adult dispersion (Montoya et al. 2000).

Native parasitoids of fruit fly *Anastrepha fraterculus* (Wiedemann) are well established in wild fruits in North of Argentine and should be preserved as natural reservoirs. Entomopathogens are also abundant in tropical forests in the same region and could aid to control diverse fruit flies species (Ovruski et al. 2004; Schliserman et al. 2010).

Natural reservoirs, nurseries, and reforestation programs have been proposed as part of an integrated pest management of fruit flies (Aluja 1999; Carvalho et al.

2010) considering that no specific control measure is enough to keep fruit fly population in an acceptable level.

The black scale, *Saissetia oleae* (Olivier) (Coccidae), is a primary pest on olives and other crops throughout the world. Natural enemies control the scale partially and insecticide sprayings are often necessary. Severe black scale outbreaks occur if natural enemies, especially parasitoids, are eliminated by using non-specific insecticides. Cover crops composed of bird's-foot trefoil, *Lotus corniculatus* L., in association with *Festuca* sp., have been used to provide additional food and refuge to scale parasitoids to improve scale control. This technic has been successfully used in irrigated orchards in Mendoza, Argentine (Fig. 26.3). *Lotus* plants have about 30 cm deep roots that do not compete with olives roots. Plants with deep roots such as lucerne, *Medicago sativa* L., are not suitable as cover crop as they compete for water and nutrients.

Fig. 26.3 Olives trees associated with bird's-foot trefoil focusing to improve black scale control (Mendoza, Argentine, 2017). (Photo: Ernesto Prado)

Fig. 26.4 Table grapes associated with white clover or *Lotus* focusing to improve mealybug parasitism (Mendoza, Argentine, 2017). (Photo: Ernesto Prado)

Grapes in association with white clover (*Trifolium repens* L.) and *Lotus* plants are also used in Mendoza to improve the action of natural enemies, mainly focusing on mealybugs (Fig. 26.4).

26.3 Final Remarks

Monoculture represents, by definition, a little diversified system too exposed to extended attacks of diseases and pests. Benefits and disadvantages of a multicrop system must be evaluated for each particular case and it is not a generalized model to be applied without experimental verification. Increasing market requirements forced researchers to search for new strategies of pest control, and multicrop system is one of them. Unfortunately, few examples of multicrop system have experimental support in Latin America, and more research are needed.

Crop management could change drastically when introducing any system to increase biological diversity, and this is a challenge to face with patience and determination because new problems will probably arise when implementing.

The concept of *functional diversity* has gained space in the past decades, despite the wide range of definitions. For example, a part of the biodiversity of an ecosystem, here an agroecosystem, aims to know the role or action of each organism inside a biological community, also called a trait-based approach (Petchey and Gaston 2006; Laureto et al. 2015). Much more progress has been achieved to distangle plant and insects traits and impacts to be included in a model. However, much work has to be done especially under field conditions. Studies under controlled circumstances have limitations and sometimes difficult to transfer to field.

The "use" of biodiversity (increase, conservation, modification, etc.) in an agroecosystem should aim to produce food in quantity and quality, under economic and sustainable ways. Other effects such as decreased loss in the production and improved water quality, soil improvement, or environment conservation are also impacts to be considered. Finally, studies should be conducted considering an applied and replicated technology approach to be implemented by farmers; otherwise it will remain as an academic exercise.

Acknowledgments The authors thank the Consortium of Coffee Research and Research Support Foundation of the State of Minas Gerais (FAPEMIG) for the scholarships granted to the last two authors.

References

Altieri MA (1999) Applying agroecology to enhance the productivity of pleasant farming systems in Latin America. Environ Dev Sustain 1(3/4):197–217

Aluja M (1999) Fruit fly (Diptera: Tephritidae) research in Latin America: myths, realities and dreams. An Soc Entomol Bras 28(4):565–594

Amaral DSSL, Venzon M, Duarte MVA et al (2013) Non-crop vegetation associated with chili pepper agroecosystems promote the abundance and survival of aphid predators. Biol Control 64(3):338–346

Andow DA (1991) Vegetational diversity and arthropod population response. Annu Rev Entomol 36:561–586

Batista MC, Fonseca MCM, Teodoro AV et al (2017) Basil (*Ocimum basilicum* L.) attracts and benefits the green lacewing *Ceraeochrysa cubana* Hagen. Biol Control 110:98–106

Beattie GAC, Holford P, Mabberley DJ et al (2006) Aspects and insights of Australia – Asia collaborative research on huanglongbing. In: Proceedings of the international workshop for prevention of citrus greening diseases in severely infested areas. Multilateral Research Network for Food and Agricultural Safety, Japan, 7–9 Dec 2006

Botton M, Melo GWB, Oliveira OLP et al (2010) Efeito da cobertura vegetal sobre a pérola-da-terra (Hemiptera: Margarodidae) na cultura da videira. Acta Sci Agron 32(4):681–684

Bugg RL, Waddington C (1994) Using cover crops to manage arthropod pests in orchards: a review. Agric Ecosyst Environ 50(1):11–28

Carvalho RS, Soares Filho WS, Ritzinger R (2010) Umbu-cajá como repositório natural de parasitoide nativo de moscas-das-frutas. Pesqui Agropecu Bras 45(10):1222–1225

Colley MR, Luna JM (2000) Relative attractiveness of potential beneficial insectary plants to aphidophagous hoverflies (Diptera: Syrphidae). Environ Entomol 29(5):1054–1059

Demirel N, Çabuk F (2008) Population trends of two spotted spider mite, *Tetranychus urticae* Koch (Acari: Tetranychidae) on cotton nearby soil and asphalt road. J Entomol 5(2):122–127

Fadini MAM, Regina MA, Fraguas JC et al (2001) Efeito da cobertura vegetal sobre a abundância e diversidade de inimigos naturais de pragas em Vinhedos. Rev Bras Frutic 23(3):573–576

Fancelli M, Carvalho RS, Machado CF et al (2016) Pragas do maracujazeiro. Inf Agropecu 37:44–55

Fuxa JR (1998) Environmental manipulation for microbial control of insects. In: Barbosa P (ed) Conservation biological control. Academic Press, San Diego, pp 255–268

Gliessman SR (2001) Agroecologia: processos ecológicos em agricultura sustentável. UFRGS, Porto Alegre
Gurr GM, Van Emden HF, Wratten SD (1998) Habitat manipulation and natural enemy efficiency: implications for the control of pests. In: Barbosa P (ed) Conservation biological control. Academic Press, San Diego, pp 155–183
Gurr GM, Wratten SD, Luna JM (2003) Multi-function agricultural biodiversity: pest management and other benefits. Basic Appl Ecol 4(2):107–116
Landis DA, Wratten SD, Gurr GM (2000) Habitat management to conserve natural enemies of arthropod pests in agriculture. Annu Rev Entomol 45:175–201
Laureto LMO, Cianciaruso MV, Samia DSM (2015) Functional diversity: an overview of its history and applicability. Nat Conserv 13(2):112–116
Lavandero B, Wratten SD, Didham RK et al (2006) Increasing floral diversity for selective enhancement of biological control agents: a double-edged sward? Basic Appl Ecol 7:236–243
Letourneau DK, Armbrecht I, Rivera BS et al (2011) Does plant diversity benefit agroecosystems? A synthetic review. Ecol Appl 21(1):9–21
Miranda MP, Volpe HXL, Veiga AC et al (2017) Manejo de insetos e ácaros vetores de fitopatógenos nos citros. Inf Agropecu 38:83–93
Monteiro LB, Belli L, Souza A et al (2002) Efeito do manejo de plantas daninhas sobre *Neoseiulus californicus* (Acari: Phytoseiidae) em pomar de macieira. Rev Bras Frutic 24(3):680–682
Montoya P, Liedo P, Benrey B et al (2000) Biological control of *Anastrepha* spp. (Diptera: Tephritidae) in mango orchards through augmentative releases of *Diachasmimorpha longicaudata* (Ashmead) (Hymenoptera: Braconidae). Biol Control 18(3):216–224
Nodals AAR, Pérez PS (2005) Especies de frutales cultivadas en Cuba en la agricultura urbana. Agrifor, La Habana
Oliveira SA, Auad AM, Souza B et al (2009) Benefícios do mel e pólen de forrageiras nos parâmetros biológicos de *Chrysoperla externa* (Hagen, 1861) (Neuroptera: Chrysopidae). Arq Inst Biol 76(4):583–588
Ovruski SM, Schliserman P, Aluja M (2004) Indigenous parasitoids (Hymenoptera) attacking *Anastrepha fraterculus* and *Ceratitis capitata* (Diptera: Tephritidae) in native and exotic host plants in Northwestern Argentina. Biol Control 29(1):43–57
Paleologos MF, Bonicatto MM, Marasas ME et al (2007) Abundancia de la coleopterofauna edáfica asociada a la cobertura vegetal y al monte cercano en viñedos tradicionales de la costa de Berisso, Buenos Aires, Argentina. Rev Bras Agroecol 2(1):373–377
Petchey OL, Gaston KJ (2006) Functional diversity: back to basics and looking forward. Ecol Lett 9(6):741–758
Resende ALS, Ferreira RB, Souza B (2015) Atratividade de adultos de *Chrysoperla externa* (Hagen, 1861) aos compostos voláteis de coentro, endro e erva-doce (Apiaceae) em condições de laboratório. Rev Ceres 62(1):37–43
Resende ALS, Souza B, Ferreira RB et al (2017) Flowers of Apiaceous species as sources of pollen for adults of *Chrysoperla externa* (Hagen) (Neuroptera). Biol Control 106:40–44
Rocha LCD, Costa R, Vasconcelos Sobrinho VL (2017) Principais artrópodes-praga na cultura do morangueiro: estratégias de manejo sustentável. Inf Agropecu 38(297):33–40
Salamanca J, Pareja M, Rodriguez-Saona C et al (2015) Behavioral responses of adult lacewings, *Chrysoperla externa*, to a rose–aphid–coriander complex. Biol Control 80:103–112
Schliserman P, Ovruski S, De Coll OR et al (2010) Diversity and abundance of Hymenopterous parasitoids associated with *Anastrepha fraterculus* (Diptera: Tephritidae) in native and exotic host plants in Misiones, northeastern Argentina. Fla Entomol 93(2):175–182
Shelton AM, Badenes-Perez FR (2006) Concepts and applications of trap cropping in pest management. Annu Rev Entomol 51:285–308
Smith MW, Mulder Jr PG (2007) Role of cover crops in the management of arthropod pest in orchard. In: Koul O, Cuperus GW (eds) Ecologically based integrated pest management. CABI Publishing, Wallingford, pp 56–732

Song B, Jiao H, Tang G et al (2014) Combining repellent and attractive aromatic plants to enhance biological control of three Tortricid species(Lepidoptera: Tortricidae) in an apple orchard. Fla Entomol 97(4):1679–1689

Tomaseto AF, Marques RN, Fereres A et al (2019) Orange jasmine as a trap crop to control *Diaphorina citri*. Sci Rep 9:2070

Venzon M, Rosado MC, Euzébio DE et al (2006) Suitability of leguminous cover crop pollens as food source for the green lacewing *Chrysoperla externa* (Hagen) (Neuroptera: Chrysopidae). Neotrop Entomol 35(3):371–376

Venzon M, Amaral DSSL, Togni PHB et al (2011) Manejo de pragas na agricultura orgânica. In: Lima PC, Moura WM, Venzon M et al (eds) Tecnologias para produção orgânica. EPAMIG Zona da Mata, Viçosa, pp 107–128

Venzon M, Diez-Rodriguez GI, Ferraz CS et al (2016) Manejo agroecológico das pragas das fruteiras. Inf Agropecu 37(293):94–103

Wilson CL (1998) Conserving epiphytic microorganisms on fruits and vegetables for biological control. In: Barbosa P (ed) Conservation biological control. Academic Press, San Diego, pp 335–350

Chapter 27
Maize

Rosangela C. Marucci, Simone Martins Mendes, and Ivana Lemos Souza

27.1 Introduction

Maize (*Zea mays* L.) is profoundly and extensively relevant for Latin American countries due to its historical, social, and sustainability values. The crop is produced within the most several systems, ranging between family agriculture and organic cultivation and integrated systems of production. Brazil, Argentina, and Mexico stand out as the main producing countries of Latin America, with production of 96,000, 41,000, and 26,000 tons, respectively (Agrianual 2019).

However, pest infestation in maize plantation is an important issue for farmers and producers. Pest insects may cause losses close to 7% of production in Brazil, or approximately two billion dollars (Oliveira et al. 2014). Although most insects may damage maize plantations and frequently harm the crop since sowing to harvest, the fall armyworm, *Spodoptera frugiperda* (Smith) (Noctuidae) is a key pest in cornfield.

Spodoptera frugiperda is native to the Americas and is distributed from Uruguay and Northern Argentina to Florida and Texas in southern United States, including Central America and the Caribbean Islands (Sparks 1979; Ashley et al. 1989). The fall armyworm is a leaf-chewing caterpillar that reduces the stand of recently emergent plants. At first, the recently hatched larvae usually eat the green tissue from one side of the leaf, leaving the membranous epidermis on the other side intact; after the second or third instar, they penetrate the corn ear and consume voraciously the leaf

R. C. Marucci (✉)
Department of Entomology, Federal University of Lavras (UFLA), Lavras, MG, Brazil
e-mail: rosangelac.marucci@ufla.br

S. M. Mendes
Lab de Entomologia, Embrapa Milho e Sorgo, Sete Lagoas, MG, Brazil

I. L. Souza
Department of Entomology, Federal University of Lavras (UFLA), Lavras, MG, Brazil

B. Souza et al. (eds.), *Natural Enemies of Insect Pests in Neotropical Agroecosystems*, https://doi.org/10.1007/978-3-030-24733-1_27

area. Armyworms may attack the stem, the tassel, the spike's insertion point, and the grain-in-formation within the spike, with direct and indirect injuries, owing to its auxiliary stance in the penetration and contamination of several microorganisms (mycotoxins) (Cruz 2008).

Furthermore, besides the biological imbalance within the maize production system, the *S. frugiperda* control by insecticides becomes crucial when the attack occurs during the plant's more advanced stages, due to the lack of means to enter the plantation with the normal equipment (Cruz 1988). Efficient low-cost alternatives are required, such as conservation biological control, which ensures the maintenance of natural enemies in the area, and augmentative control, which increases the number of control agents through inundative releases.

Nevertheless, several factors restrict the biological control use, particularly, excessive insecticides, which, in their turn, negatively affect natural enemies due to their high costs and decreasing efficaciousness due to the selection of resistant insect populations (Van Lenteren and Bueno 2003; Wyckhuys et al. 2013; Van Lenteren et al. 2017). On the other hand, huge resources are available to farmers for safe control of pests. Consequently, the inherent effects of agroecosystems may be enhanced to cope with phytosanitary issues when several measures, such as the proper management of the habitat and the conservation of natural enemies, are taken (Van Lenteren et al. 2017).

The natural enemy complex attacking fall armyworm consists of a variety of insect parasitoids (Hoballah et al. 2004; Cruz 2009; Vázquez 2009; Figueiredo et al. 2015; Cortez-Mondaca et al. 2012); arthropod predators (Cruz 2009; Vázquez 2009; Sueldo et al. 2010), including social wasps and ants (Hoballah et al. 2004; Wyckhuys and O'Neil 2006); and nematodes and enthomopathogens such as fungi, bacteria, and viruses (Valicente and Barreto 1999; Polanczyk et al. 2000; Molina-Ochoa et al. 2003b; Vázquez 2009).

The maize plant presents, as a differential, the production of large amounts of pollen, which is consumed by a number of insects, including many predatory species, such as *Doru luteipes* (Scudder) (Forficulidae) (Fig. 27.1a), *Harmonia axyridis* (Pallas) (Fig. 27.1b) and *Coleomegilla maculate* (De Geer) (Coccinellidae), *Orius insidiosus* (Say) (Anthocoridae) (Fig. 27.1c), and *Chrysoperla carnea* (Stephens) (Chrysopidae) (Pilcher et al. 1997; Lundgren et al. 2004; Lundgren and Wiedenmann 2004; Marucci et al. 2019). Corn pollen is abundant during anthesis and may allow entomophagous insects survive periods of low prey densities or provide them with critical or extra nutrients necessary for egg production or overwintering.

Furthermore, most predators are generalists and are included in the list of insects regulating aphids, thrips, and tiny arthropods populations, which enhance the maintenance and increase of predators on the field.

Therefore, maize pollen and diversified crop system (Fig. 27.1d) may be employed in the strategy to attract and maintain entomophagous insects and entomopathogens in the area, especially as *S. frugiperda* regulators. The current study analyzes the predominance of *S. frugiperda* predators and parasitoids in diversified systems and the entomopathogens used in maize production regions in Latin America.

Fig. 27.1 Predators that consume maize pollen: (**a**) *Doru luteipes*, (**b**) Coccinellidae, (**c**) *Orius insidious*, and (**d**) maize in diversified system. Lavras, MG, Brazil, 2017. (Photos: Rosangela C. Marucci)

27.2 Natural Enemies in Maize Culture

More than 50 natural enemies of *S. frugiperda* have been listed in the literature worldwide (Cruz 2009; Vázquez 2009), however few of them have been researches focus, mainly in diversified systems.

27.2.1 Predators

Wyckhuys and O'Neil (2007) quantified the influence of extra-field characteristics on fall armyworm populations and their main natural enemies in subsistence maize fields in Honduras. The authors recorded low fall armyworm infestation levels in maize fields bordered by habitats featuring high floral richness and cover. In-field abundance of social wasps was related to floral cover, while earwigs were associated with grass cover in habitats situated beyond the field border. Within-field density of the fire ant, *Solenopsis geminate* (Fabricius) (Formicidae), has been associated with its presence in the surrounding agro-landscape. At a larger spatial scale, they measured relationships between natural enemy abundance and spatial cover of habitats that dominated the extra-field environment. This analysis showed that abundance of spiders and ground beetles was highest in environments dominated by coffee plantations or mid-successional habitats.

Doru taeniatum (Dorhn) (Forficulidae) is the most abundant predator in Honduras, coupled to ants (Formicidae), social wasps (Vespidae), beetles (Carabidae), bugs (Pentatomidae), and spiders (Araneae) (Wyckhuys and O'Neil 2006, 2010). *D. taeniatum* has been shown to be an efficacious predator of *S. frugiperda* in all Central America countries (Jones et al. 1988). There are two more important Dermaptera predator species in Brazil: *D. luteipes*, which is not only abundant but also the most successful predator of *S. frugiperda* and *Helicoverpa zea* (Boddie) (Cruz 2009), and *Doru lineare* (Eschs) (Forficulidae) which is acknowledged to be a predator of other Lepidoptera species. Both, *D. luteipes* and *D. lineare* species, have been also found in maize crops in northern Argentina (Sueldo et al. 2010).

Doru luteipes is a predator constantly found in Brazilian regions in maize succession crops. It preys on egg masses and newly emerged larvae of *S. frugiperda*, the average consumption ranges from 12.5 to 21 eggs or small caterpillars, respectively, in the nymphal and adult phases of the predator (Cruz 2009).

Hoballah et al. (2004) observed several predators preying on *S. frugiperda* in fields in Mexico. The most common predators were the true bugs *Castolus* sp. (Reduviidae), *Podisus sagitta* (Fabricius) (Pentatomidae), *Zelus longipes* (Linnaeus) (Reduviidae), predominantly attacking larger *S. frugiperda* larvae, and the coccinellid *Coleomegilla* sp., the forficulid *Doru* sp., and the bug *Orius* sp., which attack newly emerged larvae.

In Brazil, research with predators has emphasized the species: *D. luteipes*, *Euborellia annulipes* (Lucas) (Anisolabididae), *Eriopis connexa* (German), and *Hippodamia convergens* Guérin-Meneville (Coccinellidae) (Cruz 2009).

Preying insects in organic maize plantations in Sete Lagoas, MG, Brazil, comprised *D. luteipes*, *Orius* sp., and coccinellids, which feed on eggs and small caterpillars of *S. frugiperda* (Figueiredo et al. 2015). Similarly, *Orius* sp. and *D. lineare* were the predominant predators in plantations in Londrina, PR, Brazil, with or without supplementary feed applied with sugared solutions to attract natural enemies. They made up 80% of predator abundance (Bortolotto et al. 2014).

The diversity of natural enemies of the phytophagous insects of maize in the agricultural systems in Cuba is not sufficiently known. In the case of predators, 30 species of 9 arthropods orders have been detected, with a greater representation of the families Syrphidae (Diptera) and Coccinellidae (Coleoptera), of which 23.3% consumes immature *S. frugiperda* (Vázquez 2009).

Predators need nectar, pollen, honeydew, alternative preys or hosts, adequate sites for the diapause phase, and other types of shelter so that they remain in the area. The maizes' own pollen is an alternative resource to maintain the predators *C. maculata*, *H. axyridis*, *C. externa*, and *Orius* sp. Consequently, management practices in the predators conservation by the provision of supplementary resources by cover plants, diversified culture, row plantation, adequate management or harvest remains, maintenance of plant communities at the edges, and provision of resources should be given priority.

The preys or maintenance of phytophagous insects with low damage capacity to plants is also an adequate strategy for the attraction and conservation of generalist predators on the field. The maintenance of small colonies of aphids, *Rhopalosiphum maidis* (Fitch) or *Schizaphis graminum* (Rondani) (Aphididae), is also suitable to maintain predators *Orius* spp. or *Chrysoperla* spp. (Maia et al. 2004).

27.2.2 Parasitoids

Molina-Ochoa et al. (2003a) prepared an inventory of parasitoids and parasites of the fall armyworm, *S. frugiperda*, using references describing parasitized *S. frugiperda* eggs, larvae, pupae, and adults collected from different crops or habitats throughout the Americas and the Caribbean. According to these authors, the largest number of parasite and parasitoid species attacking *S. frugiperda* was collected from corn (134), sorghum (40), cotton (28), peanut (24), alfalfa (17), rice (13), and bermuda grass (11). The largest number of parasitoid species attacking *S. frugiperda* larvae was reported in the United States (75), followed by Brazil (45), Honduras (43), and Nicaragua (42). Parasitoid species that attack *S. frugiperda* eggs were more prevalent in Nicaragua (11), Brazil (8), Barbados (7), and Mexico (6). Parasitoid species that attacked *S. frugiperda* pupae were reported from only four countries, with the greatest number of these species collected from the United States (6).

Cortez-Mondaca et al. (2012) estimated natural parasitism of seven parasite species of *S. frugiperda* in cultivated and volunteer maize and volunteer sorghum in México, namely, *Chelonus insularis* (Cresson) (Braconidae), *Campoletis sonorensis* (Cameron) (Ichneumonidae), *Pristomerus spinator* (Fabricius) (Ichneumonidae), *Cotesia marginiventris* (Creson) (Braconidae), *Meteorus* sp. (Braconidae), *Ophion flavidus* Brullé (Ichneumonidae), *Lespesia* sp. (Tachinidae). *C. insularis*, and *C. marginiventris* were the most abundant. According to these authors, although *C. insularis* may have a high possibility of being a candidate for augmentative biological control of *S. frugiperda*, care should be taken in in vitro breeding of the species. In fact, the percentage of females may decrease and production may become unfeasible.

In Cuba, the parasitoids that regulate populations of phytophagous insects in maize are also composed of twenty-nine species, in case, from the Diptera and Hymenoptera orders, 69% of these, parasitized immature *S. frugiperda* (Vázquez 2009).

In Brazil, *Trichogramma atopovirilia* Oatman & Platner (Trichogrammatidae), *Telenomus remus* Nixon (Platygastridae), *C. insularis*, *Campoletis flavicincta* (Ashmead) (Ichneumonidae), and *Exasticolus fuscicornis* (Cameron) (Braconidae) (Cruz 2009) may be mentioned among the most researched parasitoids.

Chelonus insularis is very common in several Brazilian regions as *S. frugiperda* parasitoid. Females lay their eggs within the pest's eggs. Caterpillars are hatched but fail to have normal development. They die when the larvae perforate the caterpillar's abdomen to become a cocoon in the external environment. By the end of the cycle, the authors observed 89.2% weight decrease of parasitized caterpillars compared to non-parasitized ones (Cruz 2009).

Penagos et al. (2003) registered that *C. insularis* featured 92–94% of all parasitism by *S. frugiperda* larvae collected in all plots with or without volunteer plant control. Other species comprised *Eiphosoma vitticole* Cresson (Ichneumonidae), *Meteorus* sp., *P. spinator*, and *Lespesia archippivora* (Riley) (Tachinidae), although in smaller numbers.

In subsistence maize areas in Honduras, Wyckhuys and O'Neil (2006) recorded that parasitism rates of *S. frugiperda* varied more between communities than between years. In 2002, parasitism rates were 2.7% and 11.5% for El Retiro and Lavanderos, respectively, while in 2003, the rates were 1.0% and 11%, respectively. The most abundant species were *Aleiodes laphygmae* (Viereck) (Braconidae) and *C. sonorensis*. The low level of parasitism (1–11.5%) was possibly due to the high elevation of the study sites and the associated relatively low number of hosts.

In 2001, a 34.8% parasitism rate was detected in *C. sonorensis* emerging from 23.1% of the recovered *S. frugiperda* caterpillars collected by Hoballah et al. (2004) in Mexico. *C. marginiventris* (6.2%) ranked second as the most abundant parasitoid, followed by *Meteorus laphygmae* Viereck (Ichneumonidae) (1.9%), *P. spinator* and *C. insularis* (1.7% each), *Euplectrus plathypenae* Howard (Eulophidae) (0.2%), and *A. laphygmae* (0.1%). *Trichogramma atopovirilia* emerged from several *S. frugiperda* egg batches collected.

Figueiredo et al. (2015) naturally detected *E. laphygmae*, *C. insularis*, *C. flavicincta*, *E. fuscicornis*, and Diptera in cultures of organic maize.

A great abundance of *C. flavicincta* (33.33%) was reported, followed by the flies *Winthemia trinitatis* Thompson (14.29%) and *Exorista* sp. (Meigen) (Tachnidae) (11.23%) in maize plots with or without the application of sugary solution as feed supplementation to attract natural enemies (Bortolotto et al. 2014).

Several interactions may occur in agroecosystems. It is highly important to note that *C. insularis* competes with *Trichogramma pretiosum* Riley (Trichogrammatidae) for parasitized eggs (Cabello et al. 2011). Several species of predators feed on parasitized and on non-parasitized eggs. Their occurrence within the production system may cause intra-guild competition (Figueiredo et al. 2015), with highly complex interactions.

It should also be underscored that undesirable effects may occur to entomophagous agents within diversified systems. Penagos et al. (2003) reported that parasitized egg mass of *S. frugiperda* was lower in weedy plots. The egg masses were placed at a height of approximately 30 cm above the soil surface, at a similar level to the non-crop vegetation and may have been less apparent to searching parastoids. On the other hand, egg mass within the plots without these plants was more visible to the parasitoids females. The specimens individually started parasitism as soon as the egg mass was distributed. Even if only *S. frugiperda* is analyzed, the interactions may be highly diversified due to landscape structure, host plants, and entomophagous species involved.

Figueiredo et al. (2015) tested the application of one, two, or three releases of *T. pretiosum* in organic corn. Results showed that 79.2% of *S. frugiperda* egg masses were parasitized or, rather, 19.4% gain of productivity and US$96.5 gain per hectare. Therefore, biological control with egg parasitoids is a promising alternative to control *S. frugiperda* in organic maize.

Since very few papers provide data on relevant field experiments showing the influence of released natural enemies on corn pest reduction or yield increase, a critical evaluation of the biological control programs results in Latin America is difficult. Mass breeding and release of *Trichogramma* sp. for the control of Lepidoptera in maize are predominant in Mexico, Ecuador, and Nicaragua; *Trichogramma* and *Telenomus* are predominant in Peru; *T. pretiosum* with *T. remus* is predominant in Venezuela (Van Lenteren and Bueno 2003).

27.2.3 Entomopathogens

In subsistence maize areas in Honduras, Wyckhuys and O'Neil (2006) recorded that insect-parasitic nematodes were entirely composed of *Hexamermis* spp. (Mermithidae). Entomopathogenic fungi and parasitic nematodes were responsible for very little larval mortality. Among the pathogens that parasitize maize phytophages in agroecosystems in Cuba, Vázquez (2009) reports that the entomopathogenic fungus *Metarhizium rileyi* (Farl.) Kepler, S.A. Rehner & Humber comb. nov. (= *Nomuraea rileyi* [Farl.] Samson; Clavicipitaceae) causes natural epizootics in populations of *S. frugiperda*. Alves et al. (2008) reported efficiency of 75–85% by using *M. rileyi* for the control of *S. frugiperda* in Cuba maizes and the commercial production of the entomopathogen in Venezuela. According to Lezama-Gutiérrez et al. (2001), the most widely distributed microbial control agent on *S. frugiperda* larvae in the Western Coast of Mexico was the fungus *M. rileyi* and, from soil, were the bacterium *Bacillus thuringiensis* (Berliner) (Bacillaceae) and steinernematid nematodes. In Chihuahua, Mexico, in 2014, two species of entomopathogenic fungi were found: *M. rileyi* and *Beauveria bassiana* (Balsamo) Vuillemin (Cordycipitaceae), with incidences of 8.6% and 0.65%, respectively. Forty-nine nucleopolyhedrovirus (Baculoviridae) isolates were obtained, corresponding to an incidence of 0.8% (Ordonez-Garcia et al. 2015).

In Brazil, *Hexamermis* spp. nematodes, *M. rileyi* fungus, and granulovirus and nucleopolyhedrovirus (Valicente 1986, 1989; Valicente and Barreto 1999) were also found.

Several factors positively affect results in the microorganisms use for pest control in Latin America, particularly climatic conditions favorable to infection development, the establishment of pathogens on the field, and the great diversity in most countries of the region.

Nucleopolyhedrovirus and granulovirus are the main viruses produced and used in Latin America for the *S. frugiperda* control in maize. They are produced and applied in Argentina, Brazil, and Nicaragua (Sosa-Gómez et al. 2008). *S. frugiperda* multiple nucleopolyhedrovirus (SfMNPV) is already available on the Brazilian market (Brasil 2016).

Although American countries feature 50% of consumption of *B. thuringiensis*, Latin America uses between 8% and 10% of production (Guerra et al. 2001). There are reports of its use for the regulation of *S. frugiperda* in Argentina, Brazil, Colombia, Cuba, Mexico, Peru, and Panama (Polanczyk et al. 2008). In Brazil, there are five Bt-based commercial products for the control of *S. fugiperda* (Brasil 2016).

27.3 Population Regulation of Pest Insects in Maize Within Diversified Systems

Through vegetation manipulation, conservation biological control can provide both primary and secondary pests control, while reducing the likelihood of pest outbreaks and resurgences (Naranjo and Ellsworth 2009).

Pest populations, in most agricultural systems, may be decreased when crops are developed with volunteer plants or intercalated with other cultures. Monoculture maize had highest infestation rates by *S. frugiperda* caterpillars in Colombia or in the United States when compared to rates in crops with volunteer plants or intercropped with beans (Altieri 1980). This comes down to the fact that the crops diversification is highly beneficial and enhances the biological control of *S. frugiperda*, even though it is greatly influenced by landscape and management (Wyckhuys and O'Neil 2010). Furthermore, in diversified surroundings, there is a greater availability of shelter, alternative preys (aphids, thrips, and tiny arthropods), or supplementary feed (pollen and nectar) (Penagos et al. 2003).

Several research works have demonstrated the role of diversification in reducing populations of maize phytophagous insects. Bastos et al. (2003) assessed the occurrence of phytophagous insects and predators in maize and beans cultivated in exclusive and polyculture systems, there was lower herbivores population density, which attacks preferentially maize such as *Dalbulus maidis* (Delong & Wolcott) (Cicadellidae) and *S. frugiperda*.

Besides the intercrop method, the several vegetal species use in succession with maize reduces *S. frugiperda* population. Some cover plants used in no-tillage systems may affect the bioecology parameters of *S. frugiperda*. In Brazil, Dias et al. (2016) have shown that *S. frugiperda* showed less adaptability to sunn hemp

(*Crotalaria juncea* L.). The authors revealed the negative influence of sunn hemp when provided to *S. frugiperda*, due to lower larval biomass, longer duration of larval stage in the laboratory, and lower biomass after 21 days in the greenhouse. Besides these data, sunn hemp plants showed the lowest damage rate in the greenhouse. McSorley et al. (2009) insisted that factors such as the branches architecture and the sunn hemp density, associated with increased humidity and temperature reduction, favored areas of refuge and shelter for parasitoids and predators. Consequently, they may be used in crop systems to attract or repel pests, natural enemies, and other organisms (Santos et al. 2008).

It becomes highly relevant to select plant species that may be employed in intercrop systems and/or in succession with maize for the decrease of pest insects, particularly *S. frugiperda*, and the maintenance of natural enemies.

27.4 Final Considerations

Pest management strategies in maize crops are rather complex, especially when biological control is involved. Knowledge on the system, insect-plant interaction, and bioecology of the biological control agents are crucial. Interaction complexity comprises direct (provision of resources and shelter) and indirect effects (modification of the vegetal community and changing of habitats). Effects should be known and the combination or arrangement should be selected that would reduce *S. frugiperda* populations. At the same time, predators, parasitoids, and entomopathogens populations should be maintained in the area so that biological control can be effective and accomplish its ecosystem role.

References

Agrianual (2019) Anuário da agricultura brasileira, 24th edn. Informa Economics FNP, São Paulo

Altieri MA (1980) Diversification of corn agrosystems as a means of regulating fall armyworm populations. Fla Entomol 63(4):450–456

Alves SB, Lopes RB, Vieira AS et al (2008) Fungos entomopatogênicos usados no controle de pragas na América Latina. In: Alves SB, Lopes RB (eds) Controle microbiano de pragas na América Latina: avanços e desafios. FEALQ, Piracicaba, pp 69–110

Ashley TR, Wiseman BR, Davis FM et al (1989) The fall armyworm: a bibliography. Fla Entomol 72(1):152–200

Bastos CS, Galvão JCC, Picanço MC et al (2003) Incidência de insetos fitófagos e de predadores no milho e no feijão cultivados em sistema exclusivo e consorciado. Cienc Rural 33(3):391–397

Bortolotto OC, Menezes Júnior AO, Hoshino AT et al (2014) Sugar solution treatment to attract natural enemies and its impact on fall armyworm *Spodoptera frugiperda* in maize fields. Interciencia 39(6):416–421

Brasil. Ministério da Agricultura Pecuária e Abastecimento (2016) Agrofit: sistema de agrotóxicos fitossanitários. http://agrofit.agricultura.gov.br/agrofit_cons/principal_agrofit_cons. Accessed 20 Sept 2017

Cabello T, Gámez M, Torres A et al (2011) Possible effects of inter-specific competition on the coexistence of two parasitoid species: *Trichogramma brassicae* Bezdenko and *Chelonus oculator* (F.) (Hymenoptera: Trichogrammatidae, Braconidae). Community Ecol 12:78–88

Cortez-Mondaca E, Pérez-Márquez J, Bahena-Juárez F (2012) Control biológico natural de gusano cogollero (Lepidoptera: Noctuidade) em maíz y en sorgo, en el norte de Sinaloa, México. Southwest Entomol 37(3):423–428

Cruz I (1988) Influência do equipamento de aplicação e do estágio de desenvolvimento da planta na eficiência de inseticidas no controle de lagartas de *Spodoptera frugiperda*. Embrapa/CNPMS, Brasília. (Pesquisa em andamento, 30)

Cruz I (2008) Manejo de pragas na cultura do milho. In: Cruz JC, Karam D, Monteiro MAR et al (eds) A cultura do milho. Embrapa Milho e Sorgo, Sete Lagoas, pp 303–362

Cruz I (2009) Métodos de criação de agentes entomófagos de *Spodoptera frugiperda*. In: Bueno VHP (ed) Controle biológico de pragas: produção massal e controle de qualidade. UFLA, Lavras, pp 237–275

Dias AS, Marucci RC, Mendes SM et al (2016) Bioecology of *Spodoptera frugiperda* (Smith, 1757) in different cover crops. Biosci J 32(2):337–345

Figueiredo MLC, Cruz I, Silva RB et al (2015) Biological control with *Trichogramma pretiosum* increases organic maize productivityby 19.4%. Agron Sustain Dev 35(3):1175–1183

Guerra PT, Wong LJG, Roldán HM et al (2001) Bioinseticidas: su empleo, production y comercialización em México. Cienc UANL 2:143–152

Hoballah ME, Degen T, Bergvinson D et al (2004) Occurrence and direct control potential of parasitoids and predators of the fall armyworm (Lepidoptera: Noctuidae) on maize in the subtropical lowlands of Mexico. Agric For Entomol 6(1):83–88

Jones R, Gislstrap F, Andrews K (1988) Biology and life table for the predaceous earwig, *Doru taeniatum* (Dermaptera: Forficulidae). Entomophaga 33(1):43–54

Lezama-Gutiérrez R, Hamm JJ, Molina-Ochoa J et al (2001) Occurrence of entomopathogens of *Spodoptera frugiperda* (Lepidoptera: Noctuidae) in the Mexican States of Michoacán, Colima, Jalisco and Tamaulipas. Fla Entomol 84(1):23–30

Lundgren JG, Wiedenmann RN (2004) Nutritional suitability of corn pollen for the predator *Coleomegilla maculate* (Coleoptera: Coccinellidae). J Insect Physiol 50(6):567–575

Lundgren JG, Razzak AA, Wiedenmann RN (2004) Population responses and food consumption by predators *Coleomegilla maculata* and *Harmonia axyridis* (Coleoptera: Coccinellidae) during anthesis in an Illinois cornfield. Environ Entomol 33:958–963

Maia WJMS, Carvalho CF, Souza B et al (2004) Capacidade predatória e aspectos biológicos de *Chrysoperla externa* (Hagen, 1861) (Neuroptera: Chrysopidae) alimentada com *Rhopalosiphum maidis* (Fitch, 1856) (Hemiptera: Aphididae). Ciênc Agrotec 28(6):1259–1268

Marucci RC, Souza IL, Silva LO, Auad AM, Mendes SM (2019) Pollen as a component of the diet of the predator *Doru luteipes* (Scudder) (Dermaptera: Forficulidae). Braz J Biol 79:584–588

Mcsorley R, Seal DR, Klassen W et al (2009) Non-target effects of sunn hemp and marigold cover crops on the soil invertebrate community. Nematropica 39(2):235–245

Molina-Ochoa J, Carpenter JE, Heinrichs EA et al (2003a) Parasitoids and parasites of *Spodoptera frugiperda* (Lepidoptera: Noctuidae) in the Americas and Caribbean Basin: an inventory. Fla Entomol 86:254–289

Molina-Ochoa J, Lezama-Guttierez R, Gonzalez-Ramirez M et al (2003b) Pathogens and parasitic nematodes associated with populations of fall armyworm (Lepidoptera: Noctuidae) larvae in Mexico. Fla Entomol 86(3):244–253

Naranjo SE, Ellsworth PC (2009) Fifty years of the integrated control concept: moving the model and implementation forward in Arizona. Pest Manag Sci 65(12):1267–1286

Oliveira CM, Auad AM, Mendes SM et al (2014) Crop losses and the economic impact of insect pests on Brazilian agriculture. Crop Prot 56:50–54

Ordonez-Garcia M, Rios-Velasco C, Berlanga-Reyes DI et al (2015) Occurrence of natural enemies of *Spodoptera frugiperda* (Lepidoptera: Noctuidae) in Chihuahua, Mexico. Fla Entomol 98(3):843–847

Penagos DI, Magallanes R, Valle J et al (2003) Effect of weeds on insect pests of maize and their natural enemies in Southern Mexico. Int J Pest Manag 49(2):155–161

Pilcher CD, Obrycki JJ, Rice ME et al (1997) Preimaginal development, survival, and field abundance of insect predators on transgenic *Bacillus thuringiensis* corn. Environ Entomol 26(2):446–454

Polanczyk RA, Pires RFS, Fiuza LM (2000) Effectiveness of *Bacillus thuringiensis* strains against *Spodoptera frugiperda* (Lepidoptera: Noctuidae). Braz J Microbiol 31(3):165–167

Polanczyk RA, Valicente FH, Barreto MR (2008) Utilização de *Bacillus thuringiensis* no controle de pragas agrícolas na América Latina. In: Alves SB, Lopes RB (eds) Controle microbiano de pragas na América Latina: avanços e desafios. FEALQ, Piracicaba, pp 111–136

Santos GG, Silveira PMD, Marchão RL et al (2008) Soil macrofauna communities and cover crops in a Cerrado Oxisolunder no tillage. Pesqui Agropec Bras 43(1):115–122

Sosa-Gómez DR, Moscardi F, Santos B et al (2008) Produção e uso de vírus para o controle de pragas na América Latina. In: Alves SB, Lopes RB (eds) Controle microbiano de pragas na América Latina: avanços e desafios. FEALQ, Piracicaba, pp 49–68

Sparks A (1979) A review of the biology of the fall armyworm. Fla Entomol 62(2):82–87

Sueldo MR, Bruzzone OA, Virla EG (2010) Characterization of the earwig, *Doru lineare*, as a predator of larvae of the fall armyworm, *Spodoptera frugiperda*: a functional response study. J Insect Sci 10:1–10

Valicente FH (1986) Ocorrência de nematoides mermitídeos em lagartas de *Spodoptera frugiperda* (J. E. Smith, 1797) (Lepidoptera: Noctuidae), em Sete Lagoas, MG. An Soc Entomol Bras 15:393–395

Valicente FH (1989) Levantamento dosinimigos naturais de *Spodoptera frugiperda* (J. E. Smith, 1797) (Lepidoptera: Noctuidae) em diferentes regiões do estado de Minas Gerais. An Soc Entomol Bras 18(1):119–130

Valicente FH, Barreto MR (1999) Survey of natural enemies of the fall armyworm, *Spodoptera frugiperda* (J. E. Smith) (Lepidoptera: Noctuidae), in Cascavel region, PR, Brazil. An Soc Entomol Bras 28(2):333–337

Van Lenteren JC, Bueno VHP (2003) Augmentative biological control of arthropods in Latin America. Biol Control 48(2):123–139

Van Lenteren JC, Bolckmans K, Köhl J et al (2017) Biological control using invertebrates and microorganisms: plenty of new opportunities. Biol Control 58:1–21

Vázquez LL (2009) Una visión entomológica de la introducción del maíz transgénico FR-Bt1 en Cuba. In: Funes-Monzote FR, Roach EFF (eds) Transgénicos: ¿qué se gana? ¿qué se pierde? textos para un debate en Cuba. Acuario, Centro Félix Varela, La Habana, pp 159–188

Wyckhuys KAG, O'Neil RJ (2006) Population dynamics of *Spodoptera frugiperda* Smith (Lepidoptera: Noctuidae) and associated arthropod natural enemies in Honduran subsistence maize. Crop Prot 25(11):1180–1190

Wyckhuys KAG, O'Neil RJ (2007) Influence of extra-field characteristics to abundance of key natural enemies of *Spodoptera frugiperda* Smith (Lepidoptera: Noctuidae) in subsistence maize production. Int J Pest Manag 53(2):89–99

Wyckhuys KAG, O'Neil RJ (2010) Social and ecological facets of pest management in Honduras subsistence agriculture: implications for IPM extension and natural resource management. Environ Dev Sustain 12(3):297–311

Wyckhuys KAG, Lu Y, Morales H et al (2013) Current status and potential of conservation biological control for agriculture in the developing world. Biol Control 65(1):152–167

Chapter 28
Oleraceous

Alessandra de Carvalho Silva, Carolina Rodrigues de Araújo, and Luis L. Vázquez

28.1 Introduction

Vegetables are short-cycle crops characterized by the need of high plant turnover and intensive management in production areas. Only a limited number of vegetal crops have high economic value. In urban agriculture, large species diversity is cultivated in small spaces through spatial arrangements of the crops.

Although numerous insects inhabit these crops, only a few of them eventually become a pest, with the exception of those that are virus vectors, which demand constant care. The severity of damage varies according to the geographic location, production system, and practices used to maintain the production process. In general, insects of economic importance to vegetable crops can be described as aphids, whiteflies, thrips, mites, leafminers, defoliators, and borers. Moreover, there are many ways to cultivate vegetables: organic or conventional management, small- or large-scale, monoculture, open field or at greenhouses.

The fact that vegetables are usually consumed freshly or with minimal processing requires greater care in their production, avoiding the presence of chemical residues in the product to be consumed. In this scenario, biological control plays an important role in the management of phytophagous insects. This practice leaves no residues on the plants or the environment, does not harm the health of the applicators and consumers, and it is very effective. In vegetable crops, biological control using entomophagous insects can be used in augmentative or conservative approaches, with the latter by the diversification of the productive system.

A. de Carvalho Silva (✉) · C. R. de Araújo
Empresa Brasileira de Pesquisa Agropecuária - EMBRAPA, Embrapa Agrobiologia, Seropédica, Brazil
e-mail: alessandra.carvalho@embrapa.br

L. L. Vázquez
Instituto de Investigaciones de Sanidad Vegetal (INISAV), Miramar, La Habana, Cuba

B. Souza et al. (eds.), *Natural Enemies of Insect Pests in Neotropical Agroecosystems*, https://doi.org/10.1007/978-3-030-24733-1_28

The use of natural enemies with inundative releases in vegetables is conditioned by the local commercial supply, which is generally low. Conservative biological control (CBC) has a wide variety of native natural enemy species that can be used in phytosanitary managements. In this chapter, advances on plants with the potential to provide the necessary resources for the survival of natural enemies targeting the CBC of the main pests of vegetables are reported. The main predators, parasitoids, and entomophagous insects used in the biological control of phytophagous insects, which are recognized as potentially limiting for the production of vegetables in Latin-American countries, are discussed below.

28.2 Predators

28.2.1 Coccinellidae

Much of the contribution of ladybugs, which is conditioned to the supply of floral resources for their survival, in vegetable crops occurs by their natural occurrence, often through intercropping. In organic cabbage crops, in Brazil, intercropping with dwarf mucuna (*Mucuna deeringiana* Holland) and crotalaria (*Crotalaria spectabilis* Roth) promoted the occurrence of 17 species of these predators (Resende et al. 2006), which also benefit from sorghum [*Sorghum bicolor* (L.) Moench] or pigeon bean [*Cajanus cajan* (L.) Mills] intercropping (Ramos 2015). Intercropping of vegetables with coriander (*Coriandrum sativum* L.) has also been shown to be advantageous, since it attracts ladybugs that control aphids on cabbage (Resende et al. 2011) and tomato (Togni et al. 2010).

As for aphidophagous species, the presence of coccinelids may be conditioned to the occurrence of prey, as observed in tomato plants, in *Bidens pilosa* L. (Fonseca et al. 2017) and in coriander, which function only as a mating area for adults (Sacramento et al. 2013). Vegetables such as lettuce can also be used as substrate, since these insects avoid placing their eggs on hairy surfaces, a common characteristic of weeds (Sengonca et al. 2002).

There are many examples on the contribution of ladybugs on pest control practices. In pepper crops, they are considered natural enemies that are particularly important for the reduction of aphid populations that cause direct and indirect crop damage (Fonseca et al. 2017). Several species of Coccinellidae were found near the colonies or feeding on *Aphis gossypii* Glover (Aphididae) in okra [*Albelmoschus esculentus* (L.) Moench]; this species also contributed to the reduction in the number of eggs of Chrysomelidae beetles and lepidopterans. In addition, the fungivorous species *Psyllobora confluens* (Fabricius) might also contribute to the crop by feeding on and controlling the fungus *Oidium* (Leite et al. 2007; Santos-Cividanes et al. 2010).

Ladybugs are generally good controllers of aphids and have little specificity for this phytophagous group. However, specificity has been observed between *Harmonia axyridis* (Pallas) and *Lipaphis erysimi* (Kaltenbach) (Aphididae) and between *Hippodamia convergens* Guérin-Méneville and *Brevicoryne brassicae* (L.) (Aphididae) (Ramos 2015).

Despite the known contribution to pest control, the use of Coccinellidae species in augmentative biological control (ABC) in Latin America is still limited. In this part of the American continent, four species are most commonly traded: *Adalia* spp. (control of aphids), *Eriopis connexa* (Germar) (control of scale insects and aphids), *Olla v-nigrum* (Mulsant) [= *Olla abdominalis* (Say)] (control of aphids), and *Stethorus* sp. (control of mites) (Van Lenteren et al. 2018). In addition, in Cuba, *Coleomegilla cubensis* (Casey), *Cycloneda sanguinea* (L.), and *Cryptolaemus montrouzieri* Mulsant are produced and released to control pests in vegetable crops (Moreno et al. 2010). In Mexico, species *H. convergens* and *C. sanguinea* are produced for aphids controlling (Bernal 2016).

28.2.2 Syrphidae

Brazil and Colombia have the highest diversity of syrphid fly species in the Neotropical region (Montoya et al. 2012). The action of the larvae of the predatory species is responsible for the reduction of the population of several vegetable crops pests, mainly aphids, both in open and in protected cultivation systems. Adults are considered floral visitors associated with different plants and therefore are selected for CBC. The population of Syrphidae was lower in broccoli monoculture production when compared to when it is intercropped with *Vicia sativa* L., which presents extrafloral nectaries (Garcia and Altieri 1993), as well as tomatoes intercropped with coriander, in which syrphid *Allograpta* sp. is a common species (Togni et al. 2010). In Buenos Aires, Argentina, these predators were found mainly on the edges of crops. A greater diversity of plants was found, being characterized by herbaceous vegetation with predominance of the families Poaceae and Asteraceae (Berensztein et al. 2017).

In Brazil, in Rio Grande do Sul State, plants of the families Apiaceae and Asteraceae presented the highest number of visiting syrphid flies, especially *Palpada urotaenia* (Curran) (Morales and Köhler 2008). In the Southeast region, adult syrphid flies are common in "gallant soldier" flowers (*Galinsoga parviflora* Cav.), growing on the edges of the vegetable (personal observation) and on chili pepper (*Capsicum frutescens* L.); also adult syrphid flies were frequently associated with the spontaneous plants *Digitaria* sp., *B. pilosa*, and *Ageratum conyzoides* L. (Amaral et al. 2013).

In the absence of floral resources, adult syrphid flies also use honeydew that is released by aphids, which promotes increased survival of these predators, as observed with *B. brassicae* in cabbage (Berensztein et al. 2017). When choosing the place to lay their eggs, female syrphid flies prefer colonies of larger aphids and not necessarily the plant that has more aphids, as observed in nine different brassicas infested by *B. brassicae* (Jankowska 2005).

28.2.3 Hemipteran Predators

28.2.3.1 Anthocoridae

The *Orius* species has showed preference for thrips (Thysanoptera), which are considered critical pests of vegetables due to direct damage and transmission of tospovirus. These Hemiptera predate nymphs and adults, being effective in the control of *Frankliniella occidentalis* (Pergande) and onion thrips, *Thrips tabaci* Lindeman. *Orius insidiosus* (Say) is also important in the control of mite *Tetranychus urticae* Koch (Tetranychidae) (Kingsley and Harrington 1981).

Anthocoridae are best established in crops that have pollen abundance, such as peppers. In vegetables such as cucumbers, where pollen is practically absent in most commercial varieties, these predators are established only when there is a high density of prey; however, in this case the application of natural control may be late. It is recommended the association of vegetables with plants that produce pollen, such as the marigold flower *Tagetes erecta* L. (Asteraceae), which associated with organic cultivation of onion (*Allium cepa* L.) hosted nearly 10 *O. insidiosus*/m^2 (Silveira et al. 2009).

The presence of prey is also essential for the maintenance of *Orius* in vegetable crops and, in this sense, weeds are important. In a horticultural region of La Plata, Argentina, these predators use *Alternanthera philoxeroides* (Mart.) Griseb., red clover (*Trifolium pratense* L.), and white clover (*Trifolium repens* L.) as alternative vegetable hosts, since they are also hosts of the TSWV virus vectors (*tomato spotted wilt virus*) (Juranovic and Ojeda 2017). In Brazil, many weeds are hosts of thrips and *Orius* spp., for example, amaranth (*Amaranthus* sp.), parthenium weed (*Parthenium hysterophorus* L.), and Joseph's coat (*Alternanthera ficoidea* L.) (Silveira et al. 2003).

Although 16 *Orius* species occur in Latin America (Carpintero 2002) and some of them are widely used in releases in protected vegetable crops worldwide, in this region, there are few records of commercial production. Nevertheless, in the tropics, the species of *O. insidiosus* and *O. thyestes* present an advantage, since they do not enter into diapause as in temperate climate, which favors their use throughout the year (Silveira and Bueno 2003; Carvalho et al. 2006). In Cuba, these insects are produced to control *Thrips palmi* Karny in potatoes, peppers, and beans (Moreno et al. 2010). In Colombia, *O. insidiosus* are massively reared to control thrips in vegetables and ornamental plants (Tinjacá 2016), mainly in integrated pest management (IPM) practices.

28.2.3.2 Miridae

These predators are associated with natural occurrences in vegetable crops. The use of banker plants is a measure to increase their colonization capacity, causing the ABC to be adopted only eventually. Although some mirids are zoophytophagous, these bugs are used in the control of vegetable pests, mainly in Europe.

In Latin America, the species *Nesidiocoris tenuis* (Reuter) has been recorded in Colombia, Cuba, Nicaragua, Puerto Rico, Haiti, Venezuela, and Mexico (Pineda et al. 2017). This species frequently appears in high populations in tomato crops and shows a certain correlation with the populations of immature insects of *Bemisia tabaci* (Gennadius) (Aleyrodidae) and with the period of flowering-fructification of this crop, where it can manifest its zoophytophagous habit, feeding on immature white fly and tomato flowers.

In Cuba, in addition to *N. tenuis*, species such as *Macrolophus praeclarus* (Distant) and *Engytatus varians* (Distant) have been studied for controlling tomato and tobacco pests, despite their dual predator and phytophagous condition (Martínez et al. 2014). In Brazil, the native species *Macrolophus basicornis* (Stål), *Campyloneuropsis infumatus* (Carvalho), and *E. varians* showed high rates of *T. absoluta* eggs and larvae predation in the laboratory (Bueno et al. 2013). In Uruguay, the native species *Tupiocoris cucurbitaceus* (Spinola) showed potential for controlling *T. vaporariorum* (Burla et al. 2014).

The damage that mirids causes in vegetables such as zucchini, cucumber, tomato, and melon due to phytophagy was observed only in high densities of predators and low availability of prey. *N. tenuis* populations decrease as prey decreases, migrating to other crops. However, when migration is blocked in covered greenhouses and without ventilation openings, the risk of crop damage increases.

28.2.3.3 Chrysopidae

Chrysopidae larvae, known as "leones de áfidos" in Latin-American countries, can be used to control pests through the release or attraction/conservation of natural populations. In this sense, rotation systems with Poaceae species, including weeds, favor the inclusion of these natural enemies in the productive system (Freitas 2002). In Cuba, species of the genus *Chrysopa* are massively reared and released in an inoculative manner against aphids and whiteflies (Moreno et al. 2010). In other Latin-American countries, *Chrysoperla asoralis* (Banks), *Ceraeochrysa cincta* (Schneider), and *Ceraeochrysa smithi* (Navas) are released for the control of aphids (Van Lenteren et al. 2018).

In Brazil, the population density of *B. tabaci* biotype B nymphs was reduced by 50% in semi-field conditions when larvae of *Chrysoperla externa* (Hagen) were released in tomato plants, in predator/prey ratios close to 1:40; for instar nymphs and "pupae," the ratio 1:30 resulted in 40% control of the pest (Auad et al. 2007). In Colombia, under laboratory conditions and under protected cultivation, *C. externa* exerted a regulatory effect on populations of *T. vaporariorum* in its three nymphal stages. In Argentina, a satisfactory reduction in the population of this pest in greenhouse cultivation of sweet pepper (*Capsicum annuum* L.) was achieved, with releases of the native species *Chrysoperla argentina* González Olazo and Reguiló in subtropical regions to the north of the country. Because it has no preference for the development of whitefly, the pest exerts an intense predation, establishing itself in the surrounding vegetation (Flores et al. 2015).

In addition to the whitefly, in Brazil, the first instar larvae of *C. externa* also promoted reductions in the population of *Myzus persicae* (Sulzer) (Aphididae) in sweet pepper and the predator was more efficient in ratios of 1:5 and 1:10 (Barbosa et al. 2008).

28.2.4 Other Predators

Although neglected as natural pest controllers, flies of the genus *Condylostylus* (Dolichopodidae) have been considered frequent and constant in organic systems for the production of vegetables or without the use of insecticides. They predate the leafminer *Liriomyza huidobrensis* (Blanchard) (Agromyzidae) and the whitefly *Bemisia argentifolli* Bellow & Perring (Aleyrodidae), in potato plantations in Peru (Junchaya 2010). These flies are also common in mixed organic vegetable crops (Sacramento et al. 2013) in Brazil, which highlights their preference to diversified systems.

Spiders are also predators of great importance for regulating pest populations in vegetable crops. Although most spiders can capture taxonomically diverse preys, some species have a restricted range of prey. For the establishment of an assembly of spiders, the environment must have gradients of microhabitats and refuges. Amaral et al. (2013) observed that weeds could contribute in this sense by increasing the number of these predators in pepper crops, mainly by offering shelter and prey.

Predatory mites are important in controlling vegetable pests and are widely used in augmentative biological control, such as phytophagous mite controllers, thrips nymphs, and other small insects. For several decades, biological control with predatory mites in greenhouse was emphasized with the use of the most specialist species, believing that success would be greater. However, successful examples with generalist mites of the Phytoseiidae family have been shown (Janssen and Sabelis 2015). The most used species belong to the genus *Amblyseius, Neoseiulus, Phytoseiulus, Typhlodromus,* and *Stratiolaelaps*.

28.3 Parasitoids

In Latin America, there are few species that are massively produced for biological control of pests of vegetable crops. In this sense, species of the genus *Trichogramma* (Trichogrammatidae) are the most used, especially in tomato cultivation, although parasitoids of leafminer, whitefly, aphids, and scale insects are also produced (Van Lenteren et al. 2018). As a result of the innovation processes carried out in different territories of the country, Cuba has expanded the use of parasitoids through mass production: nine species of the families Trichogrammatidae, Braconidae, Eulophidae, Scelionidae, and Aphelinidae have been used to produce cucurbitaceae, tomato, potato, sweet potato, and onion (Moreno et al. 2010).

The main parasitoids of pests of vegetables are listed below.

28.3.1 Trichogrammatidae

The parasitoids of the genus *Trichogramma* contribute to the control of many horticultural pests in an applied or natural way, with emphasis on *Trichogramma pretiosum* Riley and *Trichogramma atopovirilia* Oatman & Planter. An advantageous feature of the use of parasitoids of this genus is the lower host specificity, since many species are considered generalists (Rukmowati-Brotodjojo and Walter 2006) and can control a large number of species occurring in the field. In addition, the use of Trichogramma is generally compatible with the use of biological insecticides.

Trichogramma pretiosum, T. atopovirilia, Trichogramma evanescens Westwood, and *Trichogramma chilonis* Ishii have been emphasized as the most promising species for the management of *Plutella xylostella* L. (Plutellidae), being *T. pretiosum* the most studied species, reaching parasitism rates of 80% (Meira et al. 2011). For the control of *Helicoverpa zea* (Boddie) (Noctuidae), Velásquez and Gerding (2006) recommended *T. pretiosum* strains for superior parasitism efficiency in relation to other species.

The cabbage looper, *Trichoplusia ni* (Hübner) (Noctuidae), a pest of various vegetables, can also be controlled by *Trichogramma*. In the laboratory, the species *T. pretiosum, T. atopovirilia*, and *Trichogramma acacioi* Brun, Morais & Soares reached a parasitism rate of 90.3% in eggs of this host, with viability of 99.5% and a mean of 1.7 parasitoids emerged by egg. This parameter is important for the permanence of this natural enemy in the field at the time of release (Krechemer and Foerster 2015).

In tomato, *T. pretiosum* has been used in the control of *T. absoluta, Neoleucinodes elegantalis* (Guenée) (Crambidae), and *H. zea* with weekly releases associated with applications of the biological insecticide *Bacillus thuringiensis* (Berliner) or chemical insecticides with high selectivity. This strategy has ensured the efficiency and lower cost of control, obtaining productions with less than 2% of damaged fruits (Silva et al. 2006). Studies on the biology of *Trichogramma* and the mastery of techniques of creation and mass production of these insects significantly stimulated the commercialization and acceptance of this natural enemy by the producers of vegetable crops, mainly in tomato.

For the control of *Diaphania hyalinata* (L.) (Crambidae) in melons and cucumbers, mass releases of *Trichogramma pintoi* Voegelé (Trichogrammatidae) have been satisfactory in Cuba (Grillo and Pozo 1998; Moreno et al. 2010).

Studies on other species of Trichogrammatidae for biological control are still scarce and considered a challenge because of the great economic and ecological potential that this group presents.

28.3.2 Aphelinidae

Wasps of the family Aphelinidae have received attention due to their importance in biological control, especially those of the genus *Aphytis, Aphelinus, Encarsia,* and *Eretmocerus*. Most wasps are parasitoids of Sternorrhyncha nymphs, especially scale insects of the superfamily Cocooidea, Aleyrodidae, and Aphidoidea (Hanson and Gauld 2006).

Encarsia is one of the most important genera for the control of whiteflies and scale insects. *Encarsia formosa* Gahan is the most known species in the history of biological control of *T. vaporariorum* and *B. tabaci* in Europe and has a wide range of hosts. Takahashi (2003) found a faster egg-adult development of *E. formosa* and a larger number of *B. tabaci* parasitized when fed on cabbage in relation to soybean and tomato crops, suggesting greater suitability of the cabbage in this trophic interaction.

28.3.3 Braconidae

Braconid wasps can control many insect pests, preferentially parasitizing nymphs of Hemiptera and larvae and pupae of Lepidoptera. The main species that parasitize larvae and pupae belong to the genera *Cotesia, Apanteles, Bracon, Opius, Chelonus, Rogas, Microplitis,* and *Meteorus*. The specie *Cotesia glomerata* L. the main natural control agent of cabbage, *Ascia monuste orseis* (Godart) (Pieridae), in Europe, which parasitizes primarily first and second instar larvae (Mattiacci and Dicke 1995). Populations of *Spodoptera exigua* (Hübner) (Noctuidae) on onion were equilibrated through the action of the parasites *Cotesia marginiventris* Cresson, *Chelonus insularis* Cresson and *Rogas* sp. in Havana, Cuba (Ayala et al. 1991). Another example is the use of *Opius* (*Gastrosema*) *scabriventis* Nixon, a parasitoid of larvae of the genus *Liriomyza* (Agromyzidae) in melon, which has shown satisfactory results in the field (Lima et al. 2017). Onody (2009) mentioned braconid wasps of the genus *Hypomicrogaster* (Microgastrinae) parasitizing larvae of *Hellula phidilealis* (Walker) (Crambidae) in organic vegetable gardens in Brazil.

The braconids of the subfamily Aphidiinae constitute the group of natural enemies most used in programs of biological control of aphids, especially *Diaeretiella rapae* (McIntosh), *Lysiphlebus testaceipes* (Cresson), *Aphidius* spp., and *Adialytus* spp. The development of the parasitoid causes mummification of the host (it becomes immobile, with a globular external appearance and a golden or black color), which can be recognized by the rural producer with a field magnifier or by direct observation.

The *D. rapae* wasp is a specialist associated with brassicas (Blande et al. 2004), being attracted by the odor of these plants. In Brazil, *D. rapae* was indicated as the main parasitoid of *M. persicae, L. erysimi*, and *B. brassicae* (Resende et al. 2006). In Minas Gerais, Brazil, this species was the most abundant in cabbage cultivation, parasitizing *B. brassicae* (93.2%) in the field (Vaz et al. 2004).

One of the most common species in South America is *L. testaceipes*, which has great potential in biological control programs. It is responsible for parasitism rates above 50% on *A. gossypii* (Rodrigues et al. 2004). The species *L. testaceipes* and *D. rapae* are also commonly associated in vegetables grown within urban agriculture in Havana (Martínez et al. 2013). The *D. rapae* species has caused parasitism of 28% on *B. brassicae* and *L. testaceipes* and presented indices of 56% on *A. gossypii* and 68% on *B. brassicae* (Vázquez et al. 2005).

The genus *Aphidius* has several endoparasitoid species. Their adults have high mobility and the host search for oviposition causes mechanical disturbance of the colonies, causing drop and death of many specimens. The species *Aphidius colemani* Viereck (which may have been mistaken for *Aphidius plantensis* Brethes) is mentioned as the most efficient in the control of *M. persicae* and *A. gossypii* when compared to *Aphidius matricariae* Haliday and *L. testaceipes* (Van Steenis 1993). Recently, *A. plantensis* had its survival and nutrient reserves increased when *T. erecta* (Souza et al. 2018) were present, which positively affects the behavior of this parasitoid and develops well on species of the genus *Aphis* and *Schizaphis*.

28.3.4 Ichneumonidae

Several species of ichneumon wasps parasitize pest species and are generally associated with the larvae and the pupae of Lepidoptera. The genera *Campoletis*, *Eiphosoma*, *Ophion*, and *Diapetimorpha* frequently parasitize larvae of the family Noctuidae, such as *Spodoptera*; and the parasitoids of the genera *Diadegma* and *Diadromus* are generally associated with the control of *P. xylostella*. The abundance of parasitoids of the Ichneumonidae family in organic gardens was demonstrated by Onody et al. (2012), who found 14 species of the genus *Eiphosoma* in São Paulo, Brazil. The authors also reported the parasitism of *D. hyalinata* by species of this genus and mentioned *Campoletis flavicincta* (Ashmead), *Eiphosoma dentator* (Fabricius), and *Colpotrochia* sp. as important parasitoids of species of the *Spodoptera* complex.

28.4 Entomopathogens

Microorganisms have significant potential as bioinsecticides in horticulture. The use of fungi, bacteria, viruses, and entomopathogenic nematodes is another way of using biological control in vegetable crops. They are usually used through formulated products applied by means of leaf spraying or on the soil. However, the technical knowledge on the efficiency of these microorganisms in the control of pests in vegetables is limited to few species and there are a limited number of products registered for use in vegetables crops. On the other hand, they are naturally found in the productive system, and under certain management conditions, they can manifest on phytophagous insects, causing diseases and consequent death.

The bacterium *B. thuringiensis* (Bt) is one of the most widely used entomopathogenic microorganisms in the control of agricultural pests in Latin America. Several Bt-based products are marketed, expanding their adoption in the field for control of caterpillars. Its efficiency is commonly cited in the control of cruciferous (*P. xylostella* and *T. ni*) and cucurbit (*D. hyalinata*) pests as observed in Cuba (Moreno et al. 2010). In Brazil, the use of Bt-based bioinsecticides in the control of *P. xylostella* in cabbage resulted in 85–100% of marketable heads, having minimal effect on parasitoids (Monnerat et al. 2000).

The entomopathogenic fungus *Beauveria bassiana* (Bals.-Criv.) Vuill. has also become an option for the control of many insects harmful to vegetables, such as whitefly, mites, thrips, aphids, and beetles (Van Lenteren et al. 2018). Another fungus that is being used in this environment is *Metarhizium anisopliae* (Metchnikoff) Sorokin, whose efficiency on *F. occidentalis* in hydroponic lettuce cultivation reached 60% (Lopes et al. 2000). In addition to these insects, *B. bassiana* and *M. anisopliae* fungi are highly efficient on aphids *A. gossypii* and *M. persicae*, whose mortality reached 100% on the seventh day after inoculation (Loureiro and Moino Jr 2006). In Cuba, besides these fungi, genera *Paecilomyces*, *Lecanicillium*, *Aschersonia*, and *Cladosporium* are found on the whitefly *B. tabaci* (Moreno 2002). Natural epizootics of the fungus *Lecanicillium lecanii* (Zimm.) Zare & W. Gams are also observed in populations of aphids, mealybugs, and whiteflies in tropical and subtropical regions (Shah and Pell 2003); in Cuba, it is used for the control of *B. tabaci* in tomato, beans, and cucumber; against *B. brassicae* and *L. erysimi* aphids in cabbage; and against the aphids *M. persicae* and *A. gossypii* in different vegetables (Moreno et al. 2010).

The use of entomopathogenic nematodes to control pests in vegetables is not a reality in all Latin-American countries, but in some Central American countries, they are mainly used in urban agriculture to control *P. xylostella* in cabbage and *Agrotis* sp. in vegetables (Moreno et al. 2010). It is also used for the control of thrips in sweet pepper through species of the genera *Steinernema* and *Heterorhabditis* in export production systems (Diaz 2017).

The use of commercial products based on nuclear polyhedrosis viruses in horticulture has increased in Brazil. However, regarding vegetable crops, they are currently restricted to the control of *Helicoverpa* caterpillars such as *H. armigera* (Hübner) and *H. zea* (Boddie).

28.5 Final Considerations

There are many ways to grow vegetables, and the characteristics of the crops can influence the outcome of the biological control. However, in general, when it comes to ABC, there is currently a worldwide tendency to prioritize the use of native species instead of exotic natural enemies. Within this same trend, CBC has been more studied, aiming at the diversification of vegetable-growing areas with plants that offer resources to the natural control agents in the open field or protected cultivation, with the latter by the adoption of banker plants. In horticulture, pest regulation functions that are achieved by crop border manipulation, polyculture, and auxiliary plant integration are considered practices that increase the ability to regulate insect pest populations by their natural enemies.

References

Amaral DSSL, Venzon M, Duarte MVA et al (2013) Non-crop vegetation associated with chili pepper agroecosystems promote the abundance and survival of aphid predators. Biol Control 64(3):338–346

Auad AM, Carvalho CF, Souza B et al (2007) Potential of *Chrysoperla externa* (Hagen) at B-biotype *Bemisia tabaci* (Gennadius) nymphs control in tomato plants. Acta Sci Agron 29:29–32

Ayala JL, Rodríguez T, Borges M et al (1991) Himenópteros parásitos larvales de *Spodoptera exigua* (Hubner) en Guira de Melena, provincia Habana. Rev Ctro Agr 18:89–91

Barbosa LR, Carvalho CF, Souza B et al (2008) Efficiency of *Chrysoperla externa* (Hagen, 1861) (Neuroptera: Chrysopidae) in the *Myzus persi*cae (Sulzer, 1776) (Hemiptera: Aphididae) population reduction in sweet pepper (*Capsicum annum* L.). Ciênc Agrotec 32(4):1113–1119

Berensztein ND, Ricci M, Polack LA et al (2017) Control biológico por conservación: evaluación de los enemigos naturales de *Brevicoryne brassicae* (Hemiptera: Aphididae) en un manejo agroecológico de producción al aire libre de repollo (*Brassica oleracea*) del Cinturón Hortícola de La Plata, Buenos Aires. Rev Fac Agron 116(1):141–154

Bernal HCA (2016) Estatus del control biológico en México. In: Anales 43° Congreso Sociedad Colombiana de Entomología, Sociedad Colombiana de Entomología, Manizales, 28–29 Jul 2016

Blande JD, Pickett JA, Poppy GM (2004) Attack rate and success of the parasitoid *Diaeretiella rapae* on specialist and generalist feeding aphids. J Chem Ecol 30(9):1781–1795

Bueno VHP, Lenteren JC, Lins JC et al (2013) New records of *Tuta absoluta* (Meyrick) (Lepidoptera: Gelechiidae) predation by Brazilian Hemipteran predatory bugs. J Appl Entomol 137(1/2):29–34

Burla JP, Grille G, Lorenzo ME et al (2014) Effect of different diets on the development, mortality, survival, food uptake and fecundity of *Tupiocoris cucurbitaceus* (Hemiptera: Miridae). Fla Entomol 97(4):1816–1824

Carpintero DL (2002) Catalogue of the Neotropical Anthocoridae (Heteroptera). Rev Soc Entomol Argent 61(1/2):25–44

Carvalho AR, Bueno VH, Pedroso EC et al (2006) Influence of photoperiod on *Orius thyestes* Herring (Hemiptera: Anthocoridae) reproduction and longevity. Neotrop Entomol 35(4):489–492

Diaz DMB (2017) Integración del control biológico en sistemas de producción agrícola con fines de exportación en Centroamérica y Colombia. In: Anales 44° Congreso Sociedad Colombiana de Entomología, Simposio 6, Sociedad Colombiana de Entomología, Bogotá, 5–7 Jul 2017

Flores GC, Reguilón C, Alderete GL et al (2015) Liberación de *Chrysoperla argentina* (Neuroptera: Chrysopidae) para el control de *Trialeurodes vaporariorum* (Westwood) (Hemiptera, Aleyrodidae) en invernáculo de pimiento en Tucumán, Argentina. Intropica 10:28–36

Fonseca MM, Lima E, Lemos F et al (2017) Non-crop plant to attract and conserve an aphid predator (Coleoptera: Coccinellidae) in tomato. Biol Control 115:129–134

Freitas S (2002) O uso de crisopídeos no controle biológico de pragas. In: Parra JRP, Botelho PSM, Corrêa-Ferreira BS et al (eds) Controle biológico no Brasil: parasitoides e predadores, Manole, São Paulo, pp 209–224

Garcia MA, Altieri MA (1993) Comunidades de artrópodos em sistemas simples e diversificados: efeito do consórcio brócolos-leguminosas portadoras de nectários extra-florais. In: Anais do 14° Congresso Brasileiro de Entomologia, Sociedade Entomológica do Brasil, Piracicaba, 9–4 jul 1993

Grillo H, Pozo H (1998) Complejo de enemigos naturales de *Diaphania hyalinata* (L.) (Lepidoptera: Pyralidae) en la región central de Cuba. Rev Ctro Agr 25:22–24

Hanson PE, Gauld I (2006) Hymenoptera de la region neotropical. American Entomological Institute, Utah

Jankowska B (2005) Predatory syrphids (Diptera, Syrphidae) occurring in the cabbage aphid (*Brevicoryne brassicae* L.) colonies on the different cabbage vegetables. J Plant Prot Res 45(1):9–16

Janssen A, Sabelis MW (2015) Alternative food and biological control by generalist predatory mites: the case of *Amblyseius swirskii*. Exp Appl Acarol 65(4):413–418

Junchaya MAD (2010) Diagnóstico y desarrollo de componentes para un plan de manejo integrado de plagas y enfermedades en la pequeña y mediana agricultura de la costa peruana. Pueblo Cont 21(1):221–237

Juranovic V, Ojeda JP (2017) Relevamiento de flora espontánea asociada a cultivos hortícolas de La Plata como posibles hospederos de enemigos naturales con especial interés en el género *Orius* (Wolff). Tesis, Universidad Nacional de La Plata

Kingsley PC, Harrington BJ (1981) Development of *Orius insidiosus* (Hemiptera: Anthocoridae) in relation to temperature. Great Lakes Entomol 14(3):141–144

Krechemer FS, Foerster LA (2015) Temperature effects on the development and reproduction of three *Trichogramma* (Hymenoptera: Trichogrammatidae) species reared on *Trichoplusia ni* (Lepidoptera: Noctuidae) eggs. J Insect Sci 15(1):1–6

Leite GLD, Picanço M, Zanuncio JC et al (2007) Factors affecting colonization abundance of *Aphis gossypii* Glöer (Hemiptera: Aphididae) on okra plantations. Cienc Agrotec 31(2):337–343

Lima TCC, Geremias LD, Begiato AM et al (2017) Sistema de criação de parasitoide de mosca-minadora. Embrapa Semiárido, Petrolina

Lopes RB, Alves SB, Tamai MA (2000) Fungo *Metarhizium anisopliae* e o controle de *Frankliniella occidentalis* em alface hidropônico. Sci Agric 57(2):239–243

Loureiro ES, Oino AJ (2006) Pathogenicity of hyphomycet fungi to aphids *Aphis gossypii* Glover and *Myzus persicae* (Sulzer) (Hemiptera: Aphididae). Neotrop Entomol 35(5):660–665

Martínez MA, Ceballos M, Suris M et al (2013) Áfidos y sus parasitoides en sistemas urbanos de producción de hortalizas en Cuba. Rev Colomb Entomol 39(1):13–17

Martínez MA, Duarte L, Baños HL et al (2014) Míridos depredadores (Hemiptera: Heteroptera: Miridae) en cultivos de tomate y tabaco en Cuba. Rev Protección Veg 29(3):204–207

Mattiacci L, Dicke M (1995) The parasitoid *Cotesia glomerata* (Hymenoptera: Braconidae) discriminates between first and fifth larval instars of its host *Pieris brassicae*, on the basis of contact cues from frass, silk, and hervibore-damaged leaf tissue. J Insect Behav 8(4):485–498

Meira AL, Pratissoli D, Souza LP et al (2011) Seleção de espécies de *Trichogramma* sp. em ovos da traça-das-crucíferas, *Plutella xylostella*. Caatinga 24(2):1–8

Monnerat RG, Bordat D, Branco MC et al (2000) Efeito de *Bacillus thuringiensis* Berliner e inseticidas químicos sobre a traça-das-crucíferas, *Plutella xylostella* (L.) (Lepidoptera: Yponomeutidae) e seus parasitóides. An Soc Entomol Bras 29(4):723–730

Montoya AL, Pérez SP, Wolff M (2012) The diversity of flower flies (Diptera: Syrphidae) in Colombia and their Neotropical distribution. Neotrop Entomol 41(1):46–56

Morales MN, Köhler A (2008) Comunidade de Syrphidae (Diptera): diversidade e preferências florais no Cinturão Verde (Santa Cruz do Sul, RS, Brasil). Rev Bras Entomol 52(1):41–49

Moreno LLV (2002) Avances en el control biológico de *Bemisia tabaci* en la región neotropical. Manejo Integr Plagas Agroecol 66:82–95

Moreno LLV, Figueroa SC, Pérez AC et al (2010) Diagnóstico de la utilización de entomófagos y entomopatógenos para el control biológico de insectos por los agricultores en Cuba. Fitosanidad 14(3):159–169

Onody HC (2009) Estudo da fauna de himenópteros parasitoides associados a hortas orgânicas e da utilização de extratos vegetais no controle de *Plutella xylostella* (Lepidoptera, Plutellidae). Universidade Federal de São Carlos, Tese

Onody HC, Melo IF, Penteado-Dias AM (2012) Abundância, riqueza e diversidade de espécies de *Eiphosoma* Cresson, 1865 (Hymenoptera, Ichneumonidae, Cremastinae) associadas a hortas orgânicas. Idesia 30(1):115–120

Pineda S, Henry TJ, Corrales-Madrid JL et al (2017) First records of the Dicyphine plant bug *Nesidiocoris tenuis* (Hemiptera: Heteroptera: Miridae: Bryocorinae) in Mexico. Proc Entomol Soc Wash 119(2):290–295

Ramos T de O (2015) Couve consorciada com sorgo e feijão-guandu na ocorrência de pulgões e insetos predadores. Tese, Universidade Estadual Paulista "Júlio de Mesquita Filho"

Resende AL, Silva EE, Silva VB et al (2006) Primeiro registro de *Lipaphis pseudobrassicae* Davis (Hemiptera: Aphididae) e sua associação com insetos predadores, parasitoides e formigas em couve (Cruciferae) no Brasil. Neotrop Entomol 35(4):551–555

Resende ALS, Lixa AT, Santos CMA et al (2011) Comunidade de joaninhas (Coleoptera: Coccinellidae) em consórcio de couve (*Brassica oleraceae* var. acephala) com coentro (*Coriandrum sativum*) sob manejo orgânico. Rev Bras Agroecol 6(1):81–89

Rodrigues SMM, Bueno VHP, Sampaio MV et al (2004) Influência da temperatura no desenvolvimento e parasitismo de *Lysiphlebus testaceipes* (Cresson) (Hymenoptera: Braconidae, Aphidiinae) em *Aphis gossypii* Glover (Hemiptera: Aphididae). Neotrop Entomol 33(3):341–346

Rukmowati-Brotodjojo RR, Walter GH (2006) Oviposition and reproductive performance of a generalist parasitoid (*Trichogramma pretiosum*) exposed to host species that differ in their physical characteristics. Biol Control 39(3):300–312

Sacramento FZ, Tamashiro LAG, Shultz H et al (2013) Uso do coentro (*Coriandrum sativum*) como planta atrativa para inimigos naturais em cultivos orgânicos de hortaliças. In: Anais do 13° Simpósio de Controle Biológico, Embrapa, Brasília, 15–18 set 2013

Santos-Cividanes TM, Cividanes FJ, Ribeiro AA et al (2010) Diversidade de Coccinellidae na cultura do quiabeiro em Ribeirão Preto, São Paulo. Pesqui Tecnol 7(2):1–5

Sengonca C, Kranz J, Blaeser P (2002) Attractiveness of three weed species to polyphagous predators and their influence on aphid populations in adjacent lettuce cultivations. Anz Schadlingskd 75(6):161–165

Shah PA, Pell JK (2003) Entomopathogenic fungi as biological control agents. Appl Microbiol Biotechnol 61(5/6):413–423

Silva JBC, Giordano LB, Furumoto O et al (2006) Cultivo de tomate para industrialização, 2a edn. Embrapa Hortaliças, Brasília

Silveira LCP, Bueno VHP (2003) *Orius insidiosus* (Say, 1832) (Heteroptera, Anthocoridae): sensibilidade ao fotoperíodo e diapausa reprodutiva? Rev Bras Entomol 47(4):631–635
Silveira LCP, Bueno VHP, Pierre LSR et al (2003) Plantas cultivadas e invasoras como habitat para predadores do gênero *Orius* (Wolff) (Heteroptera: Anthocoridae). Bragantia 62(2):261–265
Silveira LCP, Berti Filho E, Pierre LSR et al (2009) Marigold (*Tagetes erecta* L.) as an attractive crop to natural enemies in onion fields. Sci Agric 66(6):780–787
Souza IL, Marucci RC, Silveira LCP et al (2018) Effects of marigold on the behavior, survival and nutrient reserves of *Aphidius platensis*. BioControl 63(4):543–553
Takahashi KM (2003) Aspectos bioecológicos e potencial de parasitismo de *Encarsia formosa* (Gahan) (Hymenoptera, Aphelinidae) sobre *Bemisia tabaci* biótipo B (Gennadius) (Hemiptera, Aleyrodidae) em couve, tomate e soja. Tese, Escola Superior de Agricultura "Luiz de Queiroz"
Tinjacá (2016) Innovación sostenible en control biológico en sistemas de manejo integrado de plagas. In: Anales 43° Congreso Sociedad Colombiana de Entomología, Simposio 2, Sociedad Colombiana de Entomología, Manizales, 27–29 Jul 2016
Togni PHB, Cavalcante KR, Langer LF et al (2010) Conservação de inimigos naturais (Insecta) em tomateiro orgânico. Arq Inst Biol 77(4):669–676
Van Lenteren JC, Bolckmans K, Köhl J et al (2018) Biological control using invertebrates and microorganisms: plenty of new opportunities. BioControl 63(1):39–59
Van Steenis MJ (1993) Intrinsic rate of increase of *Aphidius colemani* Vier. (Hym., Braconidae), a parasitoid of *Aphis gossypii* Glov. (Hom., Aphididae), at different temperatures. J Appl Entomol 116(1/5):192–198
Vaz LAL, Tavares MT, Lomônaco C (2004) Diversidade e tamanho de himenópteros parasitoides de *Brevicoryne brassicae* L. e *Aphis nerii* Boyer de Fonscolombe (Hemiptera: Aphididae). Neotrop Entomol 33(2):225–230
Vázquez LL, Fernández E, Lauzardo J (2005) Manejo Agroecológico de Plagas en Fincas de la Agricultura Urbana (MAPFAU). Cidisav, La Habana
Velásquez C, Gerding M (2006) Evaluación de diferentes especies de *Trichogramma* spp. para el control de *Helicoverpa zea* (Boddie) (Lepidoptera: Noctuidae). Agric Téc 66(4):411–415

Chapter 29
Ornamental Plants

Lívia Mendes Carvalho, Brígida Souza, and Ana Luiza Viana de Sousa

29.1 Introduction

The production of flowers and ornamental plants generates a wide variety of products with a worldwide value estimated at US$ 55 billion in 2016 (Van Rijswick 2016). In Latin America, Colombia, Ecuador, Costa Rica, Mexico, the Dominican Republic, and Guatemala stand out in the global trade of flowers and ornamental plants, particularly when it comes to the colors and shapes of their products. Colombia is the leading exporter in Latin America and holds second place in the world ranking of exporters of cut flowers, right after the Netherlands (Buainain and Batalha 2007; Worldatlas 2017). In Brazil, the professionalization and commercial dynamism of floriculture is rather recent; however, in 2016, this industry grew by 6% specifically in the domestic market, when compared to the previous year (Junqueira and Peetz 2016; Instituto Brasileiro de Floricultura 2017).

Flowers and ornamental plants are valued for their visual aspect; therefore, all efforts are concentrated on the obtainment of quality products and the avoidance of damages throughout the entire production chain, from production to commercialization. One of the main problems associated with quality loss during cultivation is related to insect pests. Several pests may occur in ornamental plants, depending on the species and on the type and place of cultivation. The main groups of pests asso-

L. M. Carvalho (✉)
Empresa de Pesquisa Agropecuária de Minas Gerais (EPAMIG), Campo Experimental Risoleta Neves, São João del Rei, MG, Brazil
e-mail: livia@epamig.br

B. Souza
Department of Entomology, Federal University of Lavras (UFLA), Lavras, MG, Brazil

A. L. V. de Sousa
Rutgers University, New Brunswick, NJ, USA

B. Souza et al. (eds.), *Natural Enemies of Insect Pests in Neotropical Agroecosystems*, https://doi.org/10.1007/978-3-030-24733-1_29

ciated with these crops are mites (Acari), thrips (Thysanoptera), aphids (Aphididae), whiteflies (Aleyrodidae), mealybugs (Pseudococcidae), lepidopteran larvae, fungus gnats (Sciaridae), and coleopterans, which may increase to population densities large enough to cause damages to the products to be commercialized, thus demanding the use of control measures.

Pest control in floriculture aims at obtaining an acceptable product that is free from symptoms, from visible injuries, and from any organisms. In general, this is hard to achieve; therefore, pesticides are used in most cases. However, several factors, such as pest resistance to active principles and the need to comply with environmental requisites, have motivated minimum use of these products and the adoption of sustainable methods to protect the crops against pests (Marsh and Gallardo 2009; Bueno et al. 2016; Mengistie et al. 2017; Toumi et al. 2017).

The development of technologies that allow the production of quality flowers with minimum or no pesticide use is a key factor to improve the flower industry in Latin America. The preservation of biodiversity and the biological control of pests are fundamental practices in the sustainable production of ornamental plants, which have been adopted worldwide, especially in Latin-American countries. The biological control in floriculture has been regarded as a need to reduce dependence on pesticides and the problems that arise from using them, in addition to preserving the quality of the products (Parrella and Nicholls 2006; Marsh and Gallardo 2009; Bueno et al. 2016). This strategy may reduce production costs and raise prices by trading products free from chemical residues (Nicholls et al. 1998; Parrella and Nicholls 2006; Yong et al. 2014).

Some flower companies have adopted agricultural practices that aim at obtaining compliance certificates, thus affecting competition in market. This strategy has been adopted by producers all over the world, aiming at having better access to flower exportation markets (Rikken 2010; Smit and Snijders 2014; Mengistie et al. 2017). Certified flowers are more accessible to consumers in flower shops and specific websites. According to Rikken (2010), the new consumption standards, the pressure of the media, and the campaigns of non-governmental organizations (NGOs) have made consumers interested in knowing the conditions in which flowers and ornamental plants are produced.

Research related to the sustainable production of flowers and ornamental plants is still incipient, and many farmers have used the same techniques that are applied to the production of fruits and vegetables (Yong and Levya 2010; Reis et al. 2015). In this chapter, we gather some information about the use of more sustainable practices and about the use of biological control agents against mites and insect pests in flower and ornamental plant crops, focusing on the results obtained in Latin-American countries.

29.2 Pest Management

The integrated pest management (IPM) in ornamental plants has been more and more used. In Colombia, for example, since 1996, the flower industry has adopted the program *Finca Florverde*. This program promotes sustainable production by means of agricultural practices that comply with the international standards of quality and responsibility (both social and environmental) from cultivation to post-harvest. Through this program, producers have become certified and received the *Florverde®* certificate, in compliance with standards from GlobalGAP and Rainforest Alliance™ (Smit and Snijders 2014; Colmenárez et al. 2016; Florverde Sustainable Flowers 2017). In this program, pest management is based on three principles: (1) the use of reliable monitoring systems that provide guidance and support for decision-making processes, (2) the priority to the use of control strategies that do not involve chemicals, and (3) the rational and safe use of pesticides, that is, only when strictly necessary and only in the required amounts (Florverde Sustainable Flowers 2017). The international certification of Colombian flowers brought improvements for workers and for the environment, cutting the use of pesticides in half, making water reuse possible, eradicating child labor, etc. (Valcárcel-Calderón 2013). Currently Florverde® Sustainable Flowers produces approximately 1.3 billion cut flowers per year in more than 2000 certified hectares (Florverde Sustainable Flowers 2017).

In Ecuador, the flower industry has also adopted the fundamental principle of safe and sustainable production, focused on certifications such as Flor Ecuador Certified®, Rainforest Alliance™, FairTrade® Certified, GlobalGAP, Veriflora® Certified, and Organic Certified. Certified companies dedicated to the cultivation and the export of equatorial flowers follow social and environmental requisites, such as the safe and efficient use of pesticides. In this context, producers have sought for more ecological alternatives to decrease pests in their cultures (Pro Ecuador 2015).

In Brazil, IPM and biological control have been adopted in the production of roses (*Rosa* spp.), lilies (*Lilium* spp.), spaths (*Spathiphyllum* spp.), begonias (*Begonia* spp.), and orchids (Orchidaceae), with a significant reduction in the use of pesticides (Wit et al. 2009; Bueno et al. 2016). In the past few years, rose producers in Brazil have been interested in certification systems that focus on social and environmental responsibilities, that is, certificates that ensure quality and sustainability in the production process (Carvalho et al. 2019), which is an important factor that adds value to their brands and makes their products strong competitors in the market.

In the lily crops of Holambra, SP, for example, the integrated pest and disease management was implemented by an entirely new design of a production system, which is implemented over 13.500 m^2. The main phytosanitary problems were aphids, fungus gnats, leafminer, thrips and larvae, and *Botrytis*, *Rhizoctonia*, *Phytophthora*, *Pythium*, and *Fusarium*. More than 30 commercial brands of pesticides were used to control such pests. These products were gradually eliminated from the production system by the implementation of tactics that included the introduction of control agents based on the colonization of the substrate with *Trichoderma*, *Metarhizium*, *Beauveria*, and other agents present in biofertilizers. Furthermore, pulverizations with *Trichoderma*, *Clonostachys*, *Metarhizium*, *Beauveria*, and *Bacillus thuringiensis* var. *israelensis* were made. No pesticides were used, except for flowers destined to the external market, due to phytosanitary regulations. The chemical control in the past would cost R$10.00/m^2/year, and now, after the use of integrated pest and disease management, it costs R$3.00/m^2/year (Wit et al. 2009).

A similar system was also adopted for the *Spathiphyllum* sp., of which the main pest is fungus gnats and the main disease is caused by *Cylindrocladium spathiphylli*, in addition to *Pythium* and *Phytophthora*. The fungicides available in the market are not registered for use, nor are they effective as expected. Thus, all pesticides used to control pests and diseases in the culture were replaced with alternative techniques. The cultivation substrate was enriched with biofertilizers and with *Trichoderma*. The plants were pulverized with other biocontrol agents (*Trichoderma* spp., *Metarhizium anisopliae*, *Clonostachys rosea*, *Beauveria* sp., *B. thuringiensis* var. *israelensis*, and *Bacillus subtilis*) and a fish extract. The sanitization of plants and the use of adhesive traps became a routine (Wit et al. 2009).

In Chile, producers have used methods for pest management applied to organic productions. These methods are based on the principles of ecological soil management, of ecological pest management, of disease and spontaneous plant management, of local biodiversity management, as well as on the increase and establishment of a natural biological control. The Flower-Growing Management Plan describes the main problems and, for each of them, specifies natural enemies, preventive measures, alternative measures, and inputs. For example, for the control of pests and diseases, they recommend, as a prevention, the use of plants as repellent barriers and the use of flowers or aromatic plants as barriers inside and outside greenhouses or in open fields, by planting species such as calendula (*Calendula* spp.), basil (*Ocimum basilicum* L.), thyme (*Thymus vulgaris* L.), and marigold (*Tagetes* spp.), among others (Muturana and Rozas 2009).

In Colombia, the tropical flowers and foliages are also produced for exportation as bouquets or decorative arrangements. Likewise, these products are subject to pest attacks. Aristizábal et al. (2013) carried out surveys on pest management practices associated with the commercial production of tropical foliages (*Cordyline*, *Dracaena*, *Anthericum*, *Pittosporum*, and *Phormium*, among others) and heliconias (*Heliconia* spp.). The biological control methods reported by farmers were the application of entomopathogenic fungi *Isaria* (=*Paecilomyces*) *fumosorosea*, *Beauveria bassiana*, and *M. anisopliae*, in addition to the release of lacewings (Chrysopidae) and parasitoid wasps belonging to genus *Trichogramma* (Trichogrammatidae). About 70% of the farmers use yellow adhesive traps, especially to monitor populations of thrips, which is a pest that restricts exportation. In the current study, natural enemies occurred naturally and helped control pests, among which the most abundant were predators belonging to families Coccinellidae,

Chrysopidae, Reduviidae, Lycidae, and Formicidae and other parasitoids from families Ichneumonidae, Braconiae, and Tachinidae.

Reducing the use of pesticides also helps in the natural control of pest populations, especially by increasing the number of predators and parasitoids in the cultivated area. Thus, it is necessary to choose selective pesticides that are specific to beneficial organisms and to observe compatibility of agricultural practices so as to ensure the effectiveness of biological agents. Parrella and Nicholls (2006) reported that some Colombian floriculturists have developed pest management programs in which natural enemies are released in areas surrounding the greenhouses, thus helping reduce the migration of pests into the greenhouses.

The adoption of proper IPM practices and the rational use of pesticides, among other measures, help reduce pest infestation and increase the beneficial fauna, thus being characterized as efficient and sustainable methods for the production of quality flowers (Carvalho et al. 2012; Almeida et al. 2014).

29.3 Augmentative Biological Control

Several flower producers in Latin America have adopted the release of natural enemies as part of crop management. These control strategies consist of inoculating/inundating beneficial agents, which have been effective in programs that are correctly implemented. These introductions have sought to virtually eliminate pest populations right at the beginning of the production cycle and to keep them at reduced levels until harvest. Fixed numbers of specimens are released per unit area, as soon as pests are seen in the plants (Nicholls et al. 1998; Parrella and Nicholls 2006). Some biological agents are used for pest control in ornamental plants in Colombia, such as *Phytoseiulus persimilis* (Athias-Henriot) (Phytoseiidae) to control spider mites; *Encarsia formosa* Gahan (Aphelinidae) and *Verticillium lecanii* (Zimmermann) for whiteflies; *Aphidius matricariae* Haliday (Braconidae), *Aphidoletes aphidimyza* (Rondani) (Cecidomyiidae), and *Hippodamia convergens* Guérin Menevillc (Coccinellidae) for aphids; *Diglyphus begini* (Ashm.) (Eulophidae) for leafminer; and *Amblyseius cucumeris* (Oudemans) (Phytoseiidae), *Orius* spp., and *Beauveria bassiana* Vuill. to control thrips (Parrella and Nicholls 2006). The biological products commercialized for pest control in ornamental plants in Brazil are described in Table 29.1.

The use of predatory mites has been evident in many countries, including in Latin-American countries, as a viable tool to control phytophagous mites in many of ornamental plants. In Colombia, for example, many producers use *Amblyseius* spp. and *Phytoseiulus* spp. against the spider mite *Tetranychus urticae* (Koch) (Tetranychidae), obtained from rearing of predatory mites in their property or obtained from local distributors (Valcárcel-Calderón 2013).

In anthurium (*Anthurium* spp.), chrysanthemum (*Chrysanthemum* spp.), gerbera (*Gerbera* spp.), roses, and orchids in Brazil, phytophagous mites have been controlled mostly by the predatory mites *Neoseiulus californicus* (McGregor) (Phytoseiidae) and *Phytoseiulus macropilis* (Banks) (Phytoseiidae) (Bueno and

Table 29.1 Biological products commercialized in Brazil for pest control in ornamental plants

Biological agent	Species	Target pest (scientific name)	Commercial name
Fungus	*Beauveria bassiana*	Spider mite (*Tetranychus urticae*)	Atrevido® Ballvéria® BeauveControl® Beauveria Oligos WG® Beauvetec® BioBVB® Bioveria WP® Boveria-Turbo® Boveril WP PL63® Boveryd® Excellence Mig-66® Exteminator Bio® Hovex® Interceptor®
Fungus	*Beauveria bassiana*	Whitefly (*Bemisia tabaci* race B)	Atrevido® BeauveControl® BeauveriaOligos WG® Beauvetec® BioBVB® Bioveria WP® Boveria-Turbo® Boveril WP PL63® Boveryd® Boveryd FR25® Excellence Mig-66® Exteminator Bio® Hovex® Interceptor®
Predator mite	*Neoseiulus californicus*	Spider mite (*Tetranychus urticae*)	Califorce.® Neomip® Neomip Max® Spical®
Predator mite	*Phytoseiulus macropilis*	Spider mite (*Tetranychus urticae*)	Celta® Macromip Max®
Predator mite	*Stratiolaelaps scimitus*	Fungus gnats (*Bradysia matogrossensis*)	Entomite® Stratiomip®
Predator	*Orius insidiosus*	Thrips (*Frankliniella occidentalis*)	OriusIBI®

Source: Data from the public access website, Brazil (2019)

Poletti 2009; Barbosa et al. 2017). *P. macropilis* is recommended in cases of high infestation of *T. urticae*, which demands a rapid reduction of its population density. *N. californicus* is recommended either to maintain pests at low rates or to be released right at the beginning of the infestation, in order to avoid its spread (Barbosa et al. 2017). Both predators may be acquired from companies that produce biological control agents; however, some predatory mites have been multiplied by the flower producers themselves in mini-factories installed in their properties. In rose bush farms in the cities of Itapeva-MG, Andradas-MG, São Benedito-CE, and Ubajara-CE, Brazil, for instance, the producers stopped the application of chemical acaricides against *T. urticae*. This species ceased to be the main problem in cultivations and, when they occur, the producers use predatory mites obtained from rearing in their property.

In azalea (*Rhododendron* spp.), anthurium, and other ornamental plants cultivated in Brazilian greenhouses, the predatory mite *Stratiolaelaps scimitus* (Womersley) (Laelapidae) has been used to control the fungus gnats *Bradysia matogrossensis* (Lane) (Sciaridae). Fungus gnats, also known as "sciara," are pests of ornamental plants, because their larvae feed on root tissues. In sprout cultivations, the larvae attack the root system and made plants more susceptible to diseases. The inundation release of this predatory mite has been recommended right after the sprouts are planted, which is when the infestation by fungus gnats is still low. This practice has ensured a successful biological control in these crops (Bueno and Poletti 2009; Barbosa et al. 2017).

The control of *Aphis gossypii* Glover (Aphididae) in chrysanthemum was obtained with two releases of the parasitoid *Lysiphlebus testaceipes* (Cresson) (Braconidae) (0.15 and 0.24 female/m^2, respectively); for the control of the thrips *Frankliniella occidentalis* Pergande (Thripidae), the predator *Orius insidiosus* (Say) (Anthocoridae) was released at the rate of 1.5 *Orius*/m^2; and in gerbera, the control of *F. occidentalis* was obtained after the use of 1.2 *O. insidiosus* per plant (Silveira et al. 2004; Rodrigues et al. 2005). In gypsophila (*Gypsophila paniculata* L.), in Colombia, the control of the leafminer *Liriomyza huidobrensis* (Blanchard) (Agromyzidae) was obtained by introducing and preserving the parasitoid *Diglyphus begini* (Ashmead) (Eulophidae) (Cure and Cantor 2003).

Research conducted at the Department of Entomology of the Federal University of Lavras (UFLA), Lavras-MG, Brazil, has been directed toward the use of predators to control the main pests found in rose bushes, inside protected environments. *Chrysoperla externa* (Hagen) (Chrysopidae), *Hippodamia convergens* Guérin-Méneville, *Cycloneda sanguinea* (Linnaeus), and *Eriopis connexa* (Germar) (Coccinellidae) have been studied for the control of the aphids *Rhodobium porosum* (Sanderson), *Macrosiphum rosae* (L.), and *Macrosiphum euphorbiae* (Thomas) (Aphididae) (Salamanca et al. 2015; Gamboa et al. 2016). *C. externa* and *Cryptolaemus montrouzieri* Mulsant (Coccinellidae) have been studied for the control of citrus mealybug (*Planococcus citri*) (Risso) (Pseudococcidae). The effects of combination of *O. insidiosus* and *N. californicus* about predation of *T. urticae* and their interaction with other pests (e.g., the thrips *F. occidentalis* and the whitefly *Bemisia tabaci* (Gennadius) (Aleyrodidae)) are also the purpose of studies on the biological control of rose bushes cultivated in protected environments.

Part of the research done at UFLA includes studies on mass production and on methods of release and transportation of eggs and larvae of *C. externa* for the control of rose bush pests (Amaral et al. 2013; Sousa et al. 2016; Bezerra et al. 2017). According to Bueno et al. (2016), 100,000 to millions of this species of lacewings are commercialized every week in Latin America.

29.4 Vegetation Diversification

Pest management programs must be associated with cultural practices and habitat modifications that aim at a more stable biological control system in the long run and that are not based only on massive introductions of beneficial agents (Parrella and Nicholss 2006). Abundance and diversity of natural enemies in the cultivation area depend on the diversity of the associated vegetation (Altieri and Nicholls 2010; Nicholls et al. 2015); therefore, applying plant diversification strategies in areas of ornamental plant cultivations may reduce pest attack through the attraction and conservation of natural enemies.

Although most studies on the relationship between cultivation diversity and pest incidence are focused on vegetables and fruit trees, more recent research has sought to evaluate the effects of diversification on flower crops. However, there is little information on the association of ornamental plant species with natural enemies, on how to manage them properly, and on the best species that provide pollen, nectar, habitat, etc. To choose the best plant or group of plants, one should consider the pest to be controlled, the reality of farmers, and the practices they use.

The flowering plants have been studied for use in conservative biological control programs, because they release organic volatile compounds that may either repel pests and reduce their populations directly or attract natural enemies. In Brazil, coriander (*Coriandrum sativum* L.), when associated with rose bush crops, attracted adults of *C. externa* and helped increase the number of eggs 1.7× more than if the roses were kept alone (Salamanca et al. 2015). Flowers of dill (*Anethum graveolens* L.), coriander (*C. sativum*), and fennel (*Foeniculum vulgare* Mill.) allowed the survival and reproduction of *C. externa* up to the third generation under laboratory conditions, pointing out the value of these plants for conservative biological control programs (Resende et al. 2017). Basil flowers (*O. basilicum*), for example, attracted the predator *Ceraeochrysa cubana* (Hagen) (Chrysopidae) and enabled a higher rate of survival of larvae and adults, when compared to *Mentha piperita* L., *Melissa officinalis* L., and *Cordia verbenacea* DC (Batista et al. 2017). The presence of marigold flowers (*Tagetes erecta* L.) helped the establishment of *O. insidiosus* and the control of thrips in rose bushes (Bueno et al. 2009).

In Cuba, flower production increased considerably in the last decade due the movement of urban agriculture. In these areas, the farmers have been cultivating several species of flowers associated with the cultivation of vegetables, fruits, and trees. By applying agronomical practices, producers have managed to prevent or suppress pests in crops. These areas are provided with diversified associations, including repellent plants, barrier plants, and reservoirs of bioregulating organisms.

Species of Asteraceae, such as marigolds and sunflowers (*Helianthus annuus* L.), attract beneficial insects, due to the characteristics of their flowers. Sunflowers are also used as living barriers and bring certain advantages, like acting as repellents to some pests and as shelters and food to natural enemies (Yong et al. 2014). In Cuban urban areas, marigold (*Tagetes* sp.), basil (*O. basilicum*), Cuban oregano (*Plectranthus amboinicus* (Lour.) Spreng.), mint (*Mentha* spp.), rosemary (*Rosmarinus officinalis* L.), thyme (*T. vulgaris*), calendula (*Calendula officinalis* L.), and sesame (*Sesamum indicum* L.) are used in the diversification and explored as commercial products. According to these producers, diversification enables, in addition to other advantages, the increase and preservation of natural enemies, which are responsible for establishment the insect communities in the entire system (Vázquez and Fernández 2007; Yong et al. 2007, 2014; Yong and Levya 2010).

In Peru, a system typical to the Mesoamerican traditional agriculture, which consists of cultivation flowers such as African lily (*Agapanthus africanus* (L.) Hoffmanns) and chrysanthemum (*Chrysanthemum indicum* L.) in urban greenhouses managed with plant diversification, still persists because of its multifunctional role. These flowers are cultivated on the edges of annual crops, such as corn, beans, pumpkin, and fava bean, besides several types of fruit trees. Live fences made of tree species, such as sweet acacia (*Acacia farnesiana* L.), eucalyptus (*Eucalyptus rostrata* Schltdl.), pepper tree (*Schinus molle* L.), pine tree (*Pinus teocote* Schltdl.), cedar (*Cedrela fissilis* Vell.), and holm oak (*Quercus ilex* L.), are also used. These practices help promote biodiversity and consist of a means to manage the use and the preservation of local biological resources (Flores-Sánchez et al. 2012).

In Colombia, heliconia crops, for example, provide an ideal habitat for the biological cycle of certain natural enemies belonging to orders Diptera, Coleoptera, Hymenoptera, and Hemiptera, due to the different phenological, structural, and environmental conditions they offer. There were a larger number of insects associated with species *Heliconia rostrata* (Ruiz & Pav.) and *Heliconia wagneriana* Petersen, especially because their bracts create a favorable microenvironment for the biological cycle of many insect groups and offer shelter and food (Henao and Ospina 2008).

In Brazil, research was done on the effects of plant diversification with species that are adequate for green manure and with plants that attract natural enemies associated with rose crops, along with good agricultural practices (Table 29.2). The green manure calopogonium (*Calopogonium mucunoides* L.) and marigold (*Tagetes erecta* L.) contributed to the increase in populations of natural enemies and to the reduction of mites and aphids. These studies demonstrated that it is possible to conduct a sustainable rose crop, with correct fertilization, without the excessive use of fertilizers and pesticides (Carvalho et al. 2012, 2013; Almeida et al. 2014; Reis et al. 2015). The method of conservative biological control by using plant diversification may be important for the control of several pests; however, more research on ornamental plants crop systems is needed.

Table 29.2 Agricultural practices used in rose crop systems and their purpose

Agricultural practices	Purpose
Construction of flower beds perpendicularly to the inclination	Soil conservation
Fertilization by adding magnesium thermophosphate, cattle manure, and wood ash	Supply of phosphorus and other nutrients with products allowed by the organic crop legislations
Management fertilization with organic compounds, biofertilizers, earthworm humus, bokashi, cattle manure, and chicken manure	Supply of nutrients for plants with products allowed by organic crop legislations and that may be produced by the farmer, which avoids dependence on external inputs
Interspaced cultivation of peanuts (*Arachis pintoi* cv. Amarillo), radishes (*Raphanus sativus* L.), and jack bean (*Canavalia ensiformis* cv. Comum)	Increase biodiversity in the system; green manure (periodic cutting and placing onto flower beds); protection, preservation of humidity, and increase of organic fertilization in the soil
Cultivation of pigeon pea (*Cajanus cajan* L.) and gliricidia (*Gliricidia sepium* (Jacq.) Steud.) around the crops	Increase biodiversity in the system; windbreaker; green manure (cutting branches every three months and depositing them onto the beds)
Cultivation of coriander (*Coriandrum sativum* L.) and marigold (*Tagetes* spp.) in between rose bush beds	Increase shelter and food availability for natural enemies by diversification species in the environment
Cultivation of sesame (*Sesamum indicum* L. cv. BRS Seda) and peppermint (*Mentha piperita* L.) around the crops	Increase biodiversity in the system; preserve natural enemies; repel ants
Weekly monitoring	Application of the integrated pest and disease management to take measures about the preventive or curative alternative control
Pulverization with raw milk (20%), Bordeaux mixture, sodium bicarbonate solution (0.1%), and horsetail extract (*Equisetum* spp.)	Disease control
Pulverization with neem oil (0.1%), cow's urine, ash mixture, *Metarhizium anisopliae*, *Beauveria bassiana*, plant extracts such as castor bean (*Ricinus communis* L.), citronella grass (*Cymbopogon winterianus* Jowitt), and bougainvillea (*Bougainvillea* spp.)	Pest control

Source: Reis et al. (2015)

29.5 Challenges at the Sustainable Production of Flowers and Ornamental Plants

Despite the biological resources and the vast Latin-American biodiversity, the sustainable pest management in ornamental plant crops is still scarce. So that good agricultural practices may be more and more applied to floriculture and ornamental plants in Latin-American countries, it is important to promote their integrated use. According to Bueno et al. (2016), the simultaneous integration of management methods for more than one pest or pathogen increases chance of success and contributes to reducing production costs.

The growing access to information and the increasing awareness of the population about the adverse effects of pesticides used in pest control have led to a larger demand for biological products (Colmenárez et al. 2016) and brought more growth opportunities for this market in Latin America. According to Agropages (2017), global companies of biological products have increased the availability of their products in Latin America due to the restrictions against pesticides. However, there are still challenges when it comes to their use in floriculture and, according to Valcárcel-Calderón (2013), the main ones are:

- *Commercial barriers:* Flowers, as opposed to vegetables (some of which are examples of successful biological control), are cosmetic products with high demands on appearance; there is no tolerance to evident damages. Besides, many products are sold with stems and leaves, which should also be free from damage.
- *Phytosanitary barriers:* The phytosanitary regulations of buying countries, some of which have recently reduced tolerance to the presence of any living individual in the product to be marketed, in addition to impeding quarantine pests (the presence of a biological agent (predator/parasitoid) may be a reason for rejection of an entire flower lot during trade).
- *Cultural barriers:* The flower industry is still highly dependent on pesticides. Field monitoring processes are more general than detailed. Biological control has been regarded as a replacement for chemical control, which creates false expectations as to speed of action.
- *Barriers to the supply of biological control agents:* There are still few biological control agents available. Companies of biological agents are relatively new and cannot count on a wide range of products. This situation is more favorable for entomopathogens; however, it is still necessary to ensure the quality and efficiency of these biocontrol agents.

Furthermore, in the flower industry, the variety of species and cultivars is far wider than in other agricultural industries, which include herbaceous and wood plants and annual and perennial plants, which have different needs in terms of irrigation, fertilization, and cultivation environment. The use of more sustainable practices that ensure the production of quality flowers is different for each species, and to apply them, more scientific research and perseverance are needed in this area (Reis et al. 2015).

29.6 Final Considerations

In the past decades, the international market has had a higher demand for clean flowers, limiting the excessive use of pesticides, such as insecticides and fungicides. Flowers have been increasingly produced within rigorous quality control standards, from planting to harvest and commercialization. The proper pest control is a fundamental factor to reach such quality and should not be treated as an isolated practice. On the contrary, for a successful management of these organisms, one should approach the entire system holistically, comprehending and managing the ecological interactions that ensure the maintenance of pest populations at low densities and that keep their natural enemies around, which demands proper control and prevention strategies. However, a general change in the Latin-American flower industry, regarding the adoption of biodiversity management and the biological control of pest, will only be possible with more intense actions by the industries. Investments in research and technology are necessary in places where more sustainable pest management strategies are essential tools for flower and ornamental plant producers.

References

Agropages (2017) Adopción de métodos de control biológicos para protección de cultivos en Latino América. http://news.agropages.com/News/NewsDetail%2D%2D-23925.htm. Accessed 2 Nov 2017

Almeida EFA, Carvalho LM, Reis SN et al (2014) Tecnologias para produção sustentável de flores. InfAgropecu 35:99–106

Altieri MA, Nicholls C (2010) Diseños agroecológicos para incrementar la biodiversidad de entomofauna benéfica en agrosistemas. SOCLA, Medellín

Amaral BB, Souza B, Bezerra CES et al (2013) Storing eggs of *Chrysoperla externa* (Hagen, 1861) (Neuroptera: Chrysopidae) for management of large-scale rearing. Açoreana 9:103–109

Aristizábal LF, Ospina KA, Vallejo UA et al (2013) Entomofauna associated with *Heliconia* spp. (Zingiberales: Heliconiaceae) grown in the central area of Colombia. Fla Entomol 96(1):112–119

Barbosa MFCB, Demite PR, Moraes GJ et al (2017) Controle biológico com ácaros predadores e seu papel no manejo integrado de pragas. Promip, São Paulo

Batista MC, Fonseca MCM, Teodoro AV et al (2017) Basil (*Ocimum basilicum* L.) attracts and benefits the green lacewing *Ceraeochrysa cubana* Hagen. Biol Control 110:98–106

Bezerra CES, Amaral BB, Souza B (2017) Rearing *Chrysoperla externa* larvae on artificial diets. Neotrop Entomol 46(1):93–99

Brasil. Ministério da Agricultura, Pecuária e Abastecimento (2019) AGROFIT: sistemas de agrotóxicos fitossanitários. Ministério da Agricultura, Pecuária e Abastecimento, Brasília

Buainain AM, Batalha MO (2007) Cadeia produtiva de flores e mel. IICA, Brasília

Bueno VHP, Poletti M (2009) Progress with biological control and IPM strategies in protected cultivation in Brazil. In: Proceedings of the 49° IOBC/WPRS working group "integrated control in protected crops",Cabi, Crete, 6–11 Sept 2009

Bueno VHP, Silva AC, Carvalho LM et al (2009) Control of thrips with *Orius insidiosus* in greenhouse cut roses: use of a banker plant improves the performance of the predator. In:

Proceedings of the 49° of the IOBC/WPRS working group "integrated control in protected crops," Cabi, Crete, 6–11 Sept 2009
Bueno VHP, Van Lenteren JC, Bettiol W et al (2016) Controle biológico em cultivo protegido. In: Halfeld-Vieira BA, Marinho-Prado JS, Nechet KL et al (eds) Defensivos agrícolas naturais: uso e perspectivas. Embrapa Meio Ambiente, Brasília, pp 457–504
Carvalho LM, Almeida K, Taques TC et al (2012) Manejo de pragas em cultivo de roseira de sistema de produção integrada e sistema convencional. Biosci J 28(6):938–944
Carvalho LM, Almeida K, Taques TC et al (2013) Integrated production of roses: influence of the soil management in the occurrence of pests and natural enemies. Acta Hortic (970):361–366
Carvalho LM, Almeida EFA, Reis SN et al (2019) Rosas. In: Paula Junior T, Venzon M (eds) 101 culturas: manual de tecnologias agrícolas. Epamig, Belo Horizonte, pp 675–690
Colmenárez Y, Vásquez C, Corniani N et al (2016) Implementation and adoption of integrated pest management approaches in Latin America: challenges and potential. In: Gill HK, Goyal G (eds) Integrated Pest Management (IPM): environmentally sound pest management. Intech, Rijecka, pp 1–18
Cure JR, Cantor F (2003) Atividade predadora e parasítica de *Diglyphus begini* (Ashm.) (Hymenoptera: Eulophidae) sobre *Liriomyza huidobrensis* (Blanch.) (Diptera: Agromyziidae) em cultivos de *Gypsophila paniculata* L. Neotrop Entomol 32(1):85–89
Flores-Sánches D, Navarro-Garza H, Carballo-Carballo A et al (2012) Sistemas de cultivo y biodiversidad periurbana: estudio de caso en la Cuenca del Río Texcoco. Agric Soc Desarro 9(2):209–223
Florverde® SustainableFlowers (2017) Estándar Florverde para la producción sostenible de flores y ornamentales: versión 7.0. Asociación Colombiana de Exportadores de Flores, Colombia
Gamboa S, Souza B, Morales R (2016) Actividad depredadora de *Chrysoperla externa* (Neuroptera: Chrysopidae) sobre *Macrosiphum euphorbiae* (Hemiptera: Aphididae) en cultivo de *Rosa* sp. Rev Colomb Entomol 42(1):54–58
Henao ER, Ospina KA (2008) Insectos benéficos asociados a cultivos de heliconias en el eje cafetero colombiano. Bol Cient Mus Hist Nat 12:157–166
Instituto Brasileiro de Floricultura (2017) Mercado de flores. http://www.ibraflor.com/site/2017/11/04/mercado-de-flores-vera-longuini/. Accessed 2 Nov 2017
Junqueira AH, Peetz MS (2016) Agronegócio da floricultura na economia brasileira. In: Agrianual 2017. Anuário da Agricultura Brasileira. IEG/FNP, São Paulo, pp 267–268
Marsh TL, Gallardo K (2009) Adopting biological control for ornamental crops in greenhouses. CAB Reviews: perspectives in agriculture, veterinary science, nutrition and natural resources 4(22):1–92
Mengistie BT, Mol APJ, Oosterveer P (2017) Governance of agro-pesticide through private environmental and social standards in the global cut flower chain from Ethiopia. Ambio 46(7):797–811
Muturana E, Rozas ME (2009) Alternatives in Chile. In: Gonzalez FB (ed) Alternatives to endosulfan in Latin America. IPEN, Texcoco, pp 13–14
Nicholls C, Parella M, Alteri M (1998) Advances and perspectives in the biological control of greenhouse pest with special reference to Colombia. Integr Pest Manag Rev 3(2):99–109
Nicholls C, Altieri MA, Vázquez LL (2015) Agroecología: princípios para la conversión y el rediseño de sistemas agrícolas. Agroecología 10(1):61–72
Parella M, Nicholls C (2006) El Control biologico de plagas de invernadero en Colombia: avances y perspectivas.org/wp-content/uploads/2010/10/CapituloXIclara1. Pdf. Accessed 15 Nov 2017
Pro Ecuador (2015) Analísis sectorial de flores de verano. www.proecuador.gob.ec. Accessed 10 Nov 2017
Reis SN, Almeida EFA, Carvalho LM et al (2015) Produção agroecológica e orgânica de flores. InfAgropecu 36:65–73
Resende ALS, Souza B, Ferreira RB et al (2017) Flowers of Apiaceous species as sources of pollen for adults of *Chrysoperla externa* (Hagen) (Neuroptera). Biol Control 106:40–44

Rickken M (2010) The European market for fair and sustainable flowers and plants. Trade for Development Centre, Brussels

Rodrigues SMM, Bueno VHP, Sampaio MV (2005) Efeito da liberação inoculativa sazonal de *Lysiphlebus testaceipes* (Hym.: Aphidiidae) na população de *Aphis gossypii* (Hem.: Aphididae) em cultivo de crisântemo em casa de vegetação comercial. Bol San Veg Plagas 31:199–207

Salamanca J, Pareja M, Rodriguez-Saona C et al (2015) Behavioral responses of adult lacewings, *Chrysoperla externa*, to a rose–aphid–coriander complex. Biol Control 80:103–112

Silveira LCP, Bueno VHP, Van Lenteren JC (2004) *Orius insidiosus* as biological control agent of thrips in greenhouse chrysanthemums in the tropics. Bull Insectol 57(2):103–109

Smit E, Snijders R (2014) Green opportunities with an orange touch in Colombia: business leads in the horticultural sector. Netherlands Enterprise Agency, Netherlands

Sousa ALV, Souza B, Bezerra CES et al (2016) Facilitated harvesting of eggs from laboratory-reared *Chrysoperla externa* (Neuroptera: Chrysopidae). Rev Colomb Entomol 42(2):133–136

Toumi K, Joly L, Vleminckx C et al (2017) Risk assessment of florists exposed to pesticide residues through handling of flowers and preparing bouquets. Int J Environ Res Public Health 14(5):526–544

Valcárcel-Calderón F (2013) El control biológico de plagas en la floricultura colombiana. In Memorias do 40° congreso colombiano de entomología, Sociedad Colombiana de Entomología, Bogotá, 10–12 Jul 2013

Van Rijswick C (2016) World floriculture map 2016: Equator countries gathering speed. RaboResearch Food & Agribusiness. https://research.rabobank.com/far/en/sectors/regional-food-agri/world_floriculture_map_2016.html. Accessed 2 Nov. 2017

Vázquez LL, Fernández E (2007) Manejo agroecológico de plagas y enfermedades en la agricultura urbana. Estudio de caso Ciudad de La Habana, Cuba. Agroecología 2:21–31

Wit JPW, Kievitsbosh RA, Bettiol W (2009) Integração de métodos físicos e biológicos para o controle de doenças e pragasemlírio e espatifilo. In: Bettiol W, Morandi MAB (eds) Biocontrole de doenças de plantas: uso e perspectivas. EmbrapaMeioAmbiente, Jaguariúna, pp 331–336

Worldatlas (2017) Global leaders in cut flower exports. https://www.worldatlas.com/ articles/global-leaders-in-cut-flower-exports.html. Accessed 7 Nov 2017

Yong A, Leyva A (2010) The floristic biodiversity in agricultural systems. Cul Trop 31:5–11

Yong A, Calves E, Ponce M et al (2007) Las escuelas de agricultores como estrategia de capacitación para pequeños productores. Cul Trop 28(4):5–8

Yong AC, Calves ES, Vidal ZT et al (2014) Producción de flores en la agricultura urbana de San José de las Lajas, Cuba. Cul Trop 35(3):94–100

Chapter 30
Pasture

Alexander Machado Auad and Sandra Elisa Barbosa da Silva

30.1 Introduction

Owing to its excellent conditions for agricultural and livestock activities, Latin America is a privileged subcontinent. The abundance of forage resources, highly characteristic of the region, has made the livestock industry one of its main production activities (Holmann et al. 2006). In fact, Brazil, Argentina, and Mexico are the biggest beef and milk producers worldwide. Together with Paraguay and Uruguay, they are also the biggest beef exporters (United States Department of Agriculture 2017). The main feature of livestock in the above-mentioned countries and in other Latin American countries is that high percentage of their herds feed on pasture.

In spite of the abundance in forage resources, Latin American livestock has a low productivity index due to degraded pastures, mainly caused by the great diversity of associated pest insects (Padilla et al. 2009). This factor is related to pasture formation, featuring extensive areas composed by a small number of grass species and cultivars, especially of the genus *Brachiaria*, which favor the emergence of pest outbreaks.

Researchers have tried to develop technologies for pest control that would take into account ecological, economic, and social principles. Biological control is the employment of products already present on the market or discussed in the literature that are potential natural enemies (Table 30.1).

Biological control agents found in nature or on the market should be better employed in the pasture system. It is a true challenge for researchers prior to intro-

A. M. Auad (✉)
Embrapa Gado de Leite/Entomologia Department, Juiz de Fora, MG, Brazil
e-mail: alexander.auad@embrapa.br

S. E. B. da Silva
Federal University of Lavras (UFLA), Department of Entomology, Lavras, MG, Brazil

B. Souza et al. (eds.), *Natural Enemies of Insect Pests in Neotropical Agroecosystems*, https://doi.org/10.1007/978-3-030-24733-1_30

Table 30.1 Main pest insects of pasture and natural enemy groups reported in the literature as a product already on the market or with potential for use. (Figure: authors)

Groups of pest insects	Groups of natural enemies				
	Entomopathogens			Entomophagous insects	
	Fungi	Bacteria	Nematodes	Parasitoids	Predators
Spittlebugs	■		▧	▧	▧
Bugs	■		▧	▧	▧
Aphids				▧	▧
Caterpillars		■		▧	▧
Mealybugs				▧	▧
Ants	■				
Termites	■				

■ Commercialized product

▧ Entries in the literature as a potential natural enemy

ducing a management program with the agents. The current chapter lists the main entomophagous and entomopathogenic agents and their potential against pasture pest insects.

30.2 Pasture Pest Insects

Entomofauna associated with pasture makes a long list and several insects may reach high infestation rates that may cause severe economic liability.

30.2.1 Spittlebug (Cercopidae)

Spittlebugs are the most important pest insects on Latin American pastures. They are widely distributed from southern USA to Argentina, where native genera are associated with common forage grass such as *Brachiaria* sp., *Pennisetum purpureum* Schumacher (elephant grass), and sugarcane.

Although nymphs may harm forage grass, adults cause the greatest liabilities due to the toxins injected, which interfere in the photosynthesis activity of the plant and cause the yellowing and death of leaves and, frequently, of the plants as well (Byers and Wells 1966; Cosenza 1982; Valério and Nakano 1998). According to Holmann and Peck (2002), milk and beef production may be drastically reduced (up to 54%), coupled to a 30% rise in cost, when herds feed on pastures with high spittlebug infestation levels (50 adults/m^2).

Aeneolamia, *Deois*, *Mahanarva*, *Zulia*, *Notozulia*, and *Prosapia* are the main spittlebug genera associated with pastures. *Aeneolamia* species have a wider geographic distribution and may occur from Mexico to Argentina. *Deois* and *Mahanarva* occur in eastern Bolivia, Peru, Ecuador, and Venezuela. They are predominant in Brazil. Whereas the genus *Zulia* features also in Colombia, the genus *Prosapia*, although occurring in several South American countries, is particularly important in Central America (Sotelo and Cardona 2001). On the other hand, *Notozulia* occurs with great frequency in Brazil, Paraguay, and Argentina (Paladini 2012).

30.2.2 Aphids (Aphididae)

Aphids or sap-sucking insects may be often found in pastures at high population level rates, causing decrease in forage quantity and quality (Auad et al. 2009; Oliveira et al. 2010). As the insects suck the sap of plants, toxins present in their saliva alter the metabolism of plants, causing deformations. Aphids form colonies of females that reproduce by thelytoky parthenogenesis in countries with hot climate. *Sipha flava* (Forbes) or the yellow sugarcane aphid is a species with high distribution rates in Latin America and a wide range of host species. Aphids cause great liabilities in grasses, particularly elephant grass and *Cynodon* sp., whereas *Rhopalosiphum padi* (L.), *Rhopalosiphum maidis* (Fitch), and *Hysteroneura setariae* (Thomas) destroy signal grass (Embrapa Gado de Leite 2009). Furthermore, liabilities may also be caused by attacks involving *Acyrthosiphon pisum* (Harris), *Acyrthosiphon kondoi* Shinji, *Therioaphis trifolii* (Monell), and *Aphis frangulae* Kaltenbach in alfalfa crops.

30.2.3 Bugs

30.2.3.1 *Blissus* sp. (Lygaeidae)

Nymphs and adults are found in the soil moving around the grass on which they feed (Pereira and Silva 1988). When they suck the sap, the bugs cause lesions in the roots and stem bases, causing the withering and yellowing of leaves, dried leaves, dwarfism, and, in certain cases, the death of the plant (Giraldo et al. 2011).

Significantly common in Cuba, Colombia, and Brazil (Valério 2000), the bugs preferentially attack cultivars of the genus *Digitaria* such as Pangola, Swazi, and Survenola. They also devastate pastures with *Cynodon* sp. and *Brachiaria* sp. (León 2014).

30.2.3.2 *Collaria* spp. (Miridae)

Species of the genus *Collaria* are associated with several types of grasses in several Latin American countries. *Collaria oleosa* (Distant) has the widest geographic distribution, occurring in several countries of Central and South America (Carvalho and Fontes 1981).When feeding, nymphs and adults harm the leaves of the host plant and cause chlorotic spots that decrease the area for photosynthesis (Estrada 2002; Giraldo et al. 2011; Silva et al. 2013).

30.2.3.3 *Scaptocoris* spp. (Cydnidae)

The brown burrower bugs *Scaptocoris castanea* (Perty) and *Scaptocoris carvalhoi* Becker can be found throughout Brazil and cause severe damage to several forage grasses (Sousa 2002; Valério 2006). Featuring underground habits, nymphs and adults of the insects harm the plants since they suck the sap of roots (Oliveira et al. 2000). Dead spots in dry plants start emerging and may reach several hectares of pasture (Sousa 2002; Valério 2006). They cause the death of grass tufts in areas with high populations. The areas are replaced by invading plants, characterizing an initial stage of pasture degradation.

30.2.4 Caterpillars

30.2.4.1 *Mocis latipes* (Gueneé) (Noctuidae)

Mocis latipes (striped grass looper), the most relevant caterpillar in pastures, is reported to de-leaf grasses in several Latin American countries, such as Brazil, Cuba, Argentina, Venezuela, Colombia, and others. The attacks of the insect are sporadic in dead spots with grasses totally without leaves around the central midrib (Centro Internacional de Agricultura 1982; Giraldo et al. 2011). Loss of leaf area jeopardizes photosynthesis rates and the productivity of the pasture (León 2014).

30.2.4.2 *Spodoptera frugiperda* (Smith) (Noctuidae)

The fall armyworm, *S. frugiperda*, is native to tropical and subtropical countries of the American continent, with wide distribution in Central and South America. It feeds on tender and axillary buds of forage grasses. The armyworm may totally consume the entire plant during severe infestations (Giraldo et al. 2011).

30.2.5 Cochineal Bugs

The cochineal bug *Antonina graminis* Maskell (Pseudococcidae) feeds almost exclusively on grasses. Dispersions occur only during the first instar when they start moving. In other stages, they fix themselves on the feeding spot and lose their legs. The insects attack the grass sprouts, starting from the stem from where they suck the sap. The drying and the death of the plants occur generally on pasture dead spots.

30.2.6 Ants (Formicidae)

Leaf-cutting ants despoil plants of their leaves at all growth stages. The leaves become the substrate for the development of fungus parasites used as food for the entire colony. A silvopastoral survey conducted during 35 months showed that only 10% of ant species under analysis are de-leafing fauna (Auad et al. 2010). However, they cause terrible losses in pastures. Liabilities caused by leaf-cutting ants are correlated with the size of the ant hill (Loeck et al. 2001). According to Valério (2005), the leaf-cutting ants *Atta bisphaerica* Forel (pasture killer ant) and *Atta capiguara* Gonçalves (brown leaf cutter) and leafcutters of the genus *Acromyrmex* are relevant pests in pastures in several Brazilian states.

30.2.7 Termites (Termitidae)

The most common termite genera of pastures are *Cornitermes* and *Syntermes*. They reduce the useful area of pastures, impede the movement of agriculture machines, and depreciate plantations. *Cornitermes* species build their nests on the soil surface, whereas *Syntermes* nests are mainly underground, more widely spread, deeper, and less compact than those built by *Cornitermes*. Termites mainly feed on leaves, branches, seeds, roots, and vegetal wastes deposited on the soil surface. They also feed on live vegetal tissues, which is probably the case of *Cornitermes cumulans* (Kollar). In Brazil, *Syntermes* species forage on green and dry leaves.

30.3 Biological Control Applied to Pasture Pests

30.3.1 Entomopathogens

30.3.1.1 Fungi

The fungus *Metarhizium anisopliae* (Metchnikoff) Sorokin has been particularly focused for the control of the spittlebug, the main biotic problem in pastures. The fungus has been pinpointed as a viable control alternative. In fact, *M. anisopliae* hits both nymphs and adults, while dead insects are the primary inoculum of spore dissemination of the microbial agent due to infection (León 2014). The above-mentioned control method with *M. anisopliae* has been greatly employed in Guatemala, Costa Rica, Panama, Venezuela, Ecuador, and Brazil for the control of spittlebug populations (Garcia et al. 2012). The microorganism has been recently introduced and applied in Mexico (Hernández-Domínguez et al. 2016). In Cuba, it has been produced and applied for the control of *M. latipes* (Vázquez et al. 2010).

Since the fungus does not survive in sufficient amounts for the control of the spittlebug in pasture systems, it has to be multiplied in vitro and then released on the pastures. The bio-insecticide is sold in ready-to-use formulations. Regardless of its formulation, the product must have a good origin, with guaranteed viability and without any sort of contamination. The product must reach the entire plant, especially its basal sections, precisely where the nymphs develop. Applications must be undertaken on the second and third nymph generations according to population levels.

Efficiency percentages in spittlebug control by *M. anisopliae* lie between 10% and 60%. The quality of the fungus applied per unit area, the application method, the employed isolate, and climatic conditions during applications may cause variations in efficiency. Temperature at 26–27 °C, high relative air humidity, absence of heavy rainfall, and low solar radiation are favorable conditions for greater efficiency. The application of the product in the early morning and in the evening to avoid the withering of the spores and in doses determined by the manufacturer is recommended to increase the efficacy of the fungus (Giraldo et al. 2011). The main control limitations for *M. anisopliae* are related to climatic conditions and to the variability in the virulence of the microorganism isolate. Such control strategy is slow but efficient when employed at the right moment and with the specific product.

Besides *M. anisopliae*, fungus *Beauveria bassiana* (Bals.-Criv.) Vuill. can also be employed for pest control in pastures. For instance, the entomopathogen is commercialized in Mexico to control *Collaria* sp. and *Aeneolamia* sp. (Mascarin and Jaronski 2016).

30.3.1.2 Bacteria

Bacillus thuringiensis Cohn (Bt) is the main bacterium used for pest control. It is currently the most commercialized biological insecticide. Bt exists naturally and may be employed as a biological insecticide strain against most caterpillars and other insects, such as coleopterans and dipterans (Carmona 2002). The bacterium is effective in pastures against *S. frugiperda* and *M. latipes.* Bioinsecticide-based products should be applied on caterpillars at the first growth stages since the insecticide spores and crystals should be ingested so that they could be toxic to the pest.

Bacillus thuringiensis is produced in several Latin American countries (Capalbo et al. 2008). Its application to pastures is more frequent in Brazil and Cuba, aiming against *M. latipes* (Vázquez et al. 2010; Polanczyk et al. 2012). Although the bioinsecticide is highly efficient against *S. frugiperda* in pastures, its application is concentrated in corn and cotton crops in several countries. Bt can be used concomitantly with other biological control agents such as other entomopathogens. In fact, Polanczyk and Alves (2005) had excellent results with the simultaneous use of Bt and *Heterorhabditis* sp. nematodes against *S. frugiperda* in the laboratory.

30.3.2 Entomophagous Insects

30.3.2.1 Parasitoids

Parasitoids, particularly of the genus *Trichogramma*, have been used in several Latin American countries. They parasitize the eggs of insects from different orders and may be multiplied easily and economically in the laboratory with alternative hosts (Cruz et al. 1999; Cruz and Monteiro 2004). They control *S. frugiperda* and *M. latipes* in pastures. *Trichogramma* sp. is produced and released for the control of *M. latipes* in Cuba (Vázquez et al. 2010).

The parasitoids are released by acquiring capsules with the parasitized eggs of alternative hosts. They are strategically placed in contaminated pasture areas for the later emergence of adults that parasitize the eggs of the pests.

30.4 Potential Agents for Pest Control in Pastures

Several natural enemies have been reported in Brazilian pastures (Fig. 30.1) and many studies have been devoted for the possible biological control of associated pests.

Fig. 30.1 Some natural enemies associated with pasture studied in Brazil. Spittlebug nymphs infected by fungus *M. anisopliae* (**a**), nematode *Steinernema riobravis* (**b**), predators *Salpingogaster nigra* (**c**), species of Asilini (**d**), preying of aphids by *Harmonia* sp. (**e**), *Diomus seminulus* (**f**), larva of *Chrysoperla externa* (**g**), preyed upon by *Aphidius colemani* (**h**) and by fungus *Lecanicillium* sp. (**i**). (Photos: Laboratório de Entomologia, Empresa Brasileira de Pesquisa Agropecuária – Embrapa Gado de Leite)

30.4.1 Entomopathogens

30.4.1.1 Fungi

Although spittlebug control by *M. anisopliae* is a well-known method, other scantily studied fungi, which naturally infect the insects, also exist. Species of the genus *Entomophthorales* have been detected at epizootic levels in sugarcane spittlebugs (Guagliumi 1972; Valério and Koller 1982). Mora et al. (2004) registered fungus *Batkoa* sp. as causing up to 90% mortality rate in *Mahanarva andigena* (Jacobi) (Cercopidae) adults on sugarcane plantations in Ecuador. In field experiments, the fungus *Furia* sp. was found at epizootic levels causing high mortality rates of *Deois schach* (Fabricius) (Cercopidae) and infecting *Deois flavopicta* (Stål) (Cercopidae). These examples reveal their bioinsecticide capacity (Leite et al. 2002).

Several wild fungi may be more virulent than commercial strains of *M. anisopliae*. Campagnani et al. (2017) isolated fungi naturally occurring in spittlebugs in

Brazil and tested their virulence against the eggs and nymphs of insects. It has been shown that a *Fusarium* species and a *Metarhizium* species were more effective in laboratory studies and greenhouses than commercially acquired *M. anisopliae*. The authors registered that spittlebugs were not extant in the habitat of the fungi. Consequently, the use of these strains may be viable for the biological control of spittlebugs.

Besides spittlebug control, *M. anisopliae* may also be employed against other pests in pastures. For instance, in Brazil, Samuels et al. (2002) reported high virulence rates of the fungus in *Blissus antillus* Leonard (Lygaeidae) eggs in the laboratory. Alves et al. (1995) also demonstrated high efficiency of *M. anisopliae* in the control of *C. cumulans*. In field studies in Colombia, *B. bassiana* proved to be highly effective against *Collaria scenica* (Stål) (Miridae). In fact, it is a possible agent for the population control of this insect in silvopastoral conditions (Bautista et al. 2014). In vitro studies by Barboza et al. (2011) showed high mortality rates (80%) when they applied *B. bassiana* on *C. scenica* at a concentration of 10^9 conidia.

30.4.1.2 Nematodes

The employment of entomopathogenic nematodes (EPNs) (Nematoda) is a highly promising method in spittlebug control since they are extant within the same milieu as the nymphs of insects. They are also efficient against other pests that inhabit the soil and live in cryptal environments. The main EPN species belong to the families Steinernematidae and Heterorhabditidae (Rhabditida), each of which has a symbiotic relationship with specific bacteria. Several reports deal with EPNs infecting spittlebug species. In Venezuela, Poinar and Linares (1985) reported an approximate 50% infection of *Aeneolamia varia* (Fabricius) (Cercopidae) adults and nymphs by *Hexamermis dactylocercus* Poinar Jr. & Linares (Mermithidae). Ferrer et al. (2004) suggested the use of EPNs for the control of *A. varia* in sugarcane plantations in Venezuela when they verified that *Heterorhabditis bacteriophora* Poinar (Rhabditida) caused 71.3–75.4% mortality rate after applying doses ranging between 50 and 100 million of nematodes per hectare under cultivation.

Isolates *Steinernema riobravis* Cabanillas, Poinar, & Raulston and *Steinernema feltiae* (Filipjev) had 92% efficiency on nymph mortality of *Mahanarva spectabilis* (Distant) (Cercopidae). However, isolates *Heterorhabditis amazonensis* RSC1 (Andaló, Nguyen, & Moino Jr.), *Steinernema carpocapsae* (Weiser), *Steinernema anomali* (Ryazan), and *Heterorhabditis* sp. JPM3 demonstrated 68–80% efficiency on the mortality of spittlebugs in the laboratory. *Steinernema riobravis* Cabanillas, Poinar, & Raulston, *S. feltiae*, and *H. amazonensis* RSC1 isolates in greenhouses showed greater virulence against *M. spectabilis* nymphs (Batista et al. 2014). Naranjo et al. (2012) detected pathogenicity of *Steinernema* sp. and *Heterorhabditis* sp. on *C. scenica* nymphs and adults and demonstrated that the two nematode species are able to infect and kill the bugs. In greenhouses, the nematodes also cause high mortality rates of the Miridae bug in foragers, with population control (Naranjo et al. 2013).

30.4.1.3 Parasitoids

There are few studies on parasitism in spittlebugs. *Anagrus* (Mymaridae) species, which are widespread in tropical countries, are reported to parasite hosts of the suborder *Auchenorrhyncha* (Triapitsyn 2002). In Brazil, Valério and Oliveira (2005) have detected the parasitism of *Notozulia entreriana* (Berg) (Cercopidae) eggs by *Anagrus urichi* Pickles. They detected 13% parasitism rate after the exposure of 100 spittlebug eggs, retrieved from the laboratory, to eight specimens of the parasitoid. According to the authors, results may indicate the multiplication of this natural enemy for the control of cercopids.

Giraldo et al. (2011) reported parasitoids *Chelonus antillarum* Marsh (Braconidae) and *Lespesia* sp. (Tachinidae) against *S. frugiperda* with high mortality rates of caterpillars. Furthermore, several studies have shown the efficaciousness of *Telenomus podisi* Ashmead (Scelionidae) in the control of the noctuid caterpillar.

30.4.1.4 Predators

Spittlebug nymphs are preyed by *Salpingogaster nigra* Schiner (Syrphidae) larvae. This type of fly occurs in Brazil, Colombia, Trinidad and Tobago, Mexico, Belize, Venezuela, and Costa Rica (Rojo et al. 2003). It is the natural enemy of all known species of sugarcane spittlebugs and other types of spittlebugs (Mendonça et al. 2005). Parra et al. (2012) reported that in vitro production of *S. nigra* adult requires approximately 40 spittlebug nymphs. The preying capacity plus its specificity, reproduction and high fertility rates, and short life cycle turn the fly into an important control agent of spittlebugs (Guagliumi 1970). Studies on *S. nigra* are rare and no information is available on mass production. Veríssimo (2018) recently investigated the biological aspects of the species and verified its possibility in in vitro development. The specificity of the predator and its low larval viability still impair mass reproduction. The fly *Porasilus barbiellini* Curran (Asilidae) is an important predator of spittlebug adults (Bueno 1987). Although the natural occurrence of predator flies that decrease spittlebug populations is a well-known fact, research on these natural enemies is still scarce, restricting its application in biological control.

Ants of the genera *Pachycondyla*, *Camponotus*, *Ectatomma*, *Pheidole*, *Labidus*, and *Mycocepurus* (Formicidae) are also control agents of pasture pests. In fact, they are predators in the systems and have been registered carrying spittlebug nymphs (Sujii et al. 2002). For instance, *Pachycondyla obscuricornis* Emery is a voracious predator species of *D. flavopicta* in pastures (Sujii et al. 2004).

Ladybirds (Coccinellidae) and chrysopids (Chrysopidae) are highly efficient predators for aphid control in pastures. Coccinellids *Diomus seminulus* (Musant) and *Harmonia axyridis* Pallas and chrysopid larvae of *Chrysoperla externa* (Hagen) have been reported to feed on *S. flava* in forager grasses cultivated in greenhouses. Specimens of *Cycloneda sanguinea* (L.) (Coccinellidae) were seen feeding on other forager aphids on the field (Embrapa Gado de Leite 2009), suggesting their efficaciousness in the control of these insects.

Keeping in mind natural and sequential control (as, for instance, earwig *Dorus* sp. of order Dermaptera, chrysopids, and predator bugs (Anthocoridae such as *Orius* spp., Reduviidae, and Nabidae)), it is important that natural enemies are kept in the cultivation milieu so that de-leafing caterpillars could be controlled.

30.5 Final Remarks

The enormous extension of pasture land in Latin America coupled to the reduced number of forager species and cultivars employed in the formation of pastures has caused pest outbreaks in pastoral systems. The issue is a conscience-raising stance for the support of research with new strategies and methods that involve biological control to lessen the effects of insect pests.

Plant resistance is currently the main tactic for the control of the spittlebug, the most relevant pest in pastures. However, the technique must be integrated to other control tactics so that the phytosanitary management of the insect could be warranted, with low impact on the environment and on human and animal health. Natural or commercialized suppression agents should be better exploited. Even though full basic knowledge, such as the levels of liabilities of pasture pests, is still unavailable, a management program should be implemented in order to include the use of biological agents.

The employment of biological control in pastures basically depends on research results that complement already known information. Research on further control strategies, albeit scantily employed for the population decrease of forager insect pests, is a true challenge for researchers.

References

Alves SB, Almeida JEM, Oino AJ et al (1995) Uso de *Metarhizium anisopliae* e *Beauveria bassiana* no controle de *Cornitermes cumulans* (Kollar, 1832) em pastagens. Ecossistema 20:50–57

Auad AM, Alves SO, Carvalho CA et al (2009) The impact of temperature on biological aspects and life table of *Rhopalosiphum padi* (Hemiptera: Aphididae) fed with signal grass. Fla Entomol 92(4):569–577

Auad AM, Resende TT, Machado VS et al (2010) Levantamento de formigas (Hymenoptera) em ambientes silvipastoril. In: Anais do 10° congresso de pós-graduação da UFLA. Universidade de Lavras, Lavras

Barboza MR, Silva DN, Lustosa SBC et al (2011) Patogenicidade do fungo entomopatogênico *Beauveria bassiana* sobre o percevejo *Collaria scenica* (Hemiptera: Miridae). Ambiência 7(3):473–480

Batista ESP, Auad AM, Andaló V et al (2014) Virulence of entomopathogenic nematodes (Rhabditida: Steinernematidae, Heterorhabditidae) to spittlebug *Mahanarva spectabilis* (Hemiptera: Cercopidae). Arq Inst Biol 81(2):145–149

Bautista LG, Cardona JA, Soto A et al (2014) Entomopathogenic activity of three fungi on *Hortensia similis* (Hemiptera: Cicadellidae) and *Collaria scenica* (Hemiptera: Miridae) in silvopastoral systems. Bol Cient Mus His Nat Univ Caldas 18(1):188–196

Bueno VHP (1987) Aspectos biológicos e ritmo diário das atividades de *Porasilus barbiellinii* predador da cigarrinha-das-pastagens. PesqAgropec Bras 22(9/10):903–915

Byers RA, Wells HD (1966) Phytotoxemia of Coastal bermudagrass caused by the two-lined spittlebug, *Prosapia bicincta* (Homoptera: Cercopidae). Ann Entomol Soc Am 59(6):1067–1071

Campagnani MO, Campos WG, Amorim SS et al (2017) Hemiptera: Cercopidae) in an Amazon silvopastoral system Fla Entomol 100:426–432

Capalbo D, Moraes I, Arantes O et al (2008) Produção de bactérias entomopatogências na América Latina. In: Alves SB, Lopes RB (eds) Controle microbiano de pragas na América Latina- avanços e desafios. Fealq, Piracicaba, pp 239–256

Carmona A (2002) Aislamiento y caracterización parcial de una cepa de *Bacillus thuringiensis* tóxica a *Spodoptera frugiperda* (Lepidoptera: Noctuidae). Bioagro 14(1):3–10

Carvalho JCM, Fontes AV (1981) Mirídeos neotropicais, CCXXV: revisão do gênero *Collaria* Provancher no Continente Americano (Hemiptera). Experientiae 27(2):1–46

Centro Internacional de Agricultura (1982) Tropical pastures program anual report 1981. Cali, Colômbia

Cosenza GW (1982) Resistência de gramíneas forrageiras a cigarrinha-das-pastagens *Deois flavopicta* (Stål 1854), 2a ed. Embrapa, Planaltina

Cruz I, Monteiro RMA (2004) Controle Biológico da lagarta do cartucho do milho *Spodoptera frugiperda* utilizando o parasitoide de ovos *Trichogramma*. Embrapa, SeteLagoas

Cruz I, Figueiredo M, Matoso MJ (1999) Controle biológico de *Spodoptera frugiperda* utilizando o parasitoide de ovos *Trichogramma*. Embrapa, Sete Lagoas

Embrapa Gado de Leite (2009) Ocorrência de afídeos (Hemiptera: Aphididae) e seus inimigos naturais em forrageiras cultivadas em casa-de-vegetação. Embrapa, Juiz de Fora

Estrada J (2002) Pastos y forrajes para el trópico colombiano. Editorial Universidad de Caldas, Manizales

Ferrer F, Arias M, Trelles A et al (2004) Posibilidades del uso de nematodos entomopatógenos para control de *Aeneolamia varia* en caña de azúcar. Manejo Integr Plagas Agroecol 72:39–43

Garcia A, Castro U, Arenas Y (2012) Selection of entomopathogenic fungi to control spittlebugs (Hemiptera: Cercopidae) of sugarcane in Colombia. Rev Colomb Entomol 38(2):252–259

Giraldo C, Reyes LK, Molina J (2011) Manejo integrado de artrópodos y parásitos en sistemas silvopastoriles intensivos. Proyecto Ganadería Colombiana Sostenible. Manual 2. GEF, Bogotá

Guagliumi P (1970) As cigarrinhas dos canaviais (Hom. Cercopidae) no Brasil. (VI. Contribuição) - A nova nomenclatura das espécies mais importantes. Bras Açucar 76(1):75–90

Guagliumi P (1972) Pragas da cana-de-açúcar no nordeste do Brasil. Instituto do Açúcar e do Álcool, Rio de Janeiro

Hernández-Domínguez C, Guzmán-Franco AW, Carrillo-Benítez MG et al (2016) Specific diversity of *Metarhizium* isolates infecting *Aeneolamia* spp. (Hemiptera: Cercopidae) in sugarcane plantations. Neotrop Entomol 45(1):80–87

Holmann F, Peck DC (2002) Economic damage caused by spittlebugs (Homoptera: Cercopidae) in Colombia: a first approximation of impact on animal production in *Brachiaria decumbens* pastures. Neotrop Entomol 31(2):275–284

Holmann FJ, Rios LR, Carulla JE et al (2006) A producción de leche y su relación con los mercados: caso Colombiano. In: Anales do 10° seminário de pastos y forrajes. Centro Internacional de Agricultura Tropical, Cali, 21–22 Apr 2006

Leite LG, Alves SB, Takada HM et al (2002) Occurrence of Entomophthorales on spittlebugs pests of pasture in eastern São Paulo state, Brazil. Arq Inst Biol 69(3):63–68

León L (2014) Control biológico de plagas en pastos y forrajes tropicales. In: Stagnaro CG, Bury NM, Belloso ES (eds) Logros y desafíos de la ganadería doble propósito. Fundación Girarz, Venezuela, pp 294–301

Loeck AE, Grutzmacher DD, Storch G (2001) Distribuição geográfica de *Atta sexdens piriventris* Santschi, 1919, nas principais regiões agropecuárias do estado do Rio Grande do Sul. Rev Bras Agrociênc 7(1):54–57

Mascarin GM, Jaronski ST (2016) The production and uses of *Beauveria bassiana* as a microbial insecticide. World J Microbiol Biotechnol 32(11):1–26

Mendonça AF, Flores S, Saenz C (2005) Cigarrinhas da cana-de-açúcar na América Latina e Caribe. Insecta, Maceió

Mora JM, Moreira KM, Alvarado DG (2004) El salivazo de la caña de azúcar, *Mahanarva andigena* (Homoptera, Cercopidae). Publ Téc 4:1–7

Naranjo N, Montero DAV, Saenz A (2012) Primer reporte de patogenicidad por nemátodos entomopatógenos sobre la chinche de los pastos *Collaria scenica* Stål (Hemiptera: Miridae). Entomotropica 26(3):117–125

Naranjo N, Montero DAV, Sáenz A (2013) Control de la chinche de los pastos *Collaria scenica* (Hemiptera: Miridae) con nematodos entomopatógenos en invernadero. Rev Bras Ciênc Agràr 8(1):90–94

Oliveira LJ, Malaguido AB, Nunes Júnior J et al (2000) Percevejo castanho da raiz em sistemas de produção de soja. Embrapa-Soja, Londrina

Oliveira AS, Souza B, Auad AM et al (2010) Respostas biológicas de *Sipha flava* (Forbes, 1884) (Hemiptera: Aphididae) alimentados em diferentes genótipos de capim-elefante. Rev Ciênc Agron 41(1):107–112

Padilla C, Crespo G, Sardiñas Y (2009) Degradación y recuperación de pastizales. Rev Cuba Ciênc Agríc 43(4):351–354

Paladini A (2012) Filogenia dos Cercopídeos Neotropicais (Hemiptera, Cercopidae, Tomaspidinae). Tese, Universidade Federal do Paraná

Parra YXG, Castro VU, Bustillo AE et al (2012) La fecundidad y longevidad de hembras de *Salpingogaster nigra* Schiner (Diptera: Syrphidae) alimentadas con flores de *Parthenium hysterophorus y Emilia sonchifolia*. In: 8° Congreso Atalac-Tecnicaña. Asociación de Técnicos Azucareros de Latinoamérica y el Caribe, Atalac, p 2012

Pereira RP, Silva SC (1988) Ocorrência de *Blissus leucopterus* em pastagem do Estado do Rio de Janeiro. Pesq Agropec Bras 23(5):551–553

Poinar GOJ, Linares B (1985) Hexamermis dactylocerus sp.n. (Mermithidae: Nematoda) a parasite of *Aeneolamia varia* (Cercopidae:Homoptera) in Venezuela. Rev Nematol 8:109–111

Polanczyk RA, Alves SB (2005) Interação entre *Bacillus thuringiensis* e outros entomopatógenos no controle de *Spodoptera frugiperda*. Manejo Integr Plagas Agroecol 74:24–33

Polanczyk RA, Bortoli SA, Bortoli CP (2012) *Bacillus thuringiensis*-based biopesticides against agricultural pests in Latin America. In: Larramendy ML, Soloneski S (eds) Integrated pest management and pest control-current and future tactics. In Tech, Croatia, pp 445–462

Rojo SG, Gilbert FMA, Nieto MGJM et al (2003) A world review of predatory hoverflies (Diptera, Syrphidae: Syrphinae) and their prey. CIBIO, Alicante

Samuels RI, Coracini DLA, Santos CM et al (2002) Infection of *Blissus antillus* (Hemiptera: Lygaeidae) eggs by the entomopathogenic fungi *Metarhizium* anisopliae and *Beauveria bassiana*. Biol Control 23(3):269–273

Silva DM, Moraes JC, Auad AM et al (2013) Selection of signal grass genotypes for resistance to *Collaria oleosa* (DISTANT, 1883) (Hemiptera: Miridae). J Agric Biol Sci 8(5):385–390

Sotelo G, Cardona C (2001) Manejo integrado del salivazo de los pastos con énfasis en resistencia varietal. In: Herrero M, Ramírez A, Joaquín N (eds) Manejo y evaluación de pasturas tropicales. Centro Internacional de Agricultura Tropical, Cali, pp 117–125

Sousa CR (2002) Composição populacional e mobilidade no solo do percevejo castanho *Atarsocoris brachiariae* (Hemiptera:Cydnidae). Dissertação,Universidade Federal de Goiás

Sujii ER, Garcia MA, Fontes EMG et al (2002) Predation as a mortality factor in populations of the spittlebug, *Deois flavopicta* Stål (Homoptera: Cercopidae). Neotrop Entomol 31(4):581–588

Sujii ER, Garcia MA, Fontes EMG et al (2004) *Pachycondyla obscuricornis* as natural enemy of the spittlebug *Deois flavopicta*. Pesq Agropec Bras 39(6):607–609

Triapitsyn S (2002) Descriptive notes on a new and other little known species of *Anagrus* Haliday, 1833 (Hymenoptera: Mymaridae) from the New World tropics and subtropics. Entomotropica17:213–223

United States Department of Agriculture (2017) Foreign agricultural service. https://apps.fas.usda.gov/psdonline/app/index.html#/app/downloads/. Accessed 5 March 2017

Valério JR (2000) Percevejo-das-gramíneas: *Blissus leucopterus* ou *Blissus antillus*? Embrapa-CNPGC, Campo Grande

Valério JR (2005) Insetos-praga em pastagens tropicais. Inf Agropecu 26(226):98–110

Valério JR (2006) Considerações sobre a morte de pastagens de *Brachiaria brizantha* cv. Marandu em alguns Estados do Centro e Norte do Brasil: enfoque entomológico. Embrapa Gado de Corte, Campo Grande

Valério JR, Koller WW (1982) Cigarrinhas das pastagens: inimigos naturais encontrados na região de Campo Grande, MS. Embrapa-CNPAC, Campo Grande

Valério JR, Nakano O (1998) Danos causados pelo adulto da cigarrinha *Zulia entreriana* na produção e qualidade de *Brachiaria decumbens*. Pesq Agropec Bras 23(5):447–453

Valério JR, Oliveira MCM (2005) Parasitismo de ovos de cigarrinhas-das-pastagens (Homoptera: Cercopidae) pelo microhimenóptero *Anagrus urichi* Pickles (Hymenoptera: Mymaridae) na região de Campo Grande, MS. Neotrop Entomol 34(1):137–138

Vázquez LL, Moreno S, Carr A et al (2010) Diagnóstico de la utilización de entomófagos y entomopatógenos para el control biológico de insectos por los agricultores en Cuba. Fitosanidad 14(3):159–169

Veríssimo BA (2018) Avaliação do potencial de Syrphidae e Asilidae como agentes de controle biológico das cigarrinhas-das-pastagens. Dissertação, Universidade Federal de Juiz de Fora

Chapter 31
Rice

Bruno Zachrisson

31.1 Introduction

The conventional agricultural production model has encouraged research and the development programs to implement innovative insect pest management strategies that promote the sustainability of agricultural crops. These strategies were specifically targeted at the cultivation of rice (*Oryza sativa* L.), which is considered one of the main agricultural products in tropical agroecosystems. Rice contributes to food security of half of the world population (Food and Agriculture Organization of the United Nations 2014). In Latin America, mainly in Brazil, Colombia, Guyana, Ecuador, and Panama, the importance of rice cultivation is reflected according to the production proportional to the cultivated area, providing 27% of the daily calories consumed by the population of these countries (Food and Agriculture Organization of the United Nations 2014).

A reduction of 12% in the performance of these crops is attributed to the incidence of insect pests (Vivas and Notz 2011). For this reason, the use of biotic resources—in this case the beneficial entomofauna—for efficient and rational insect pest management in the rice agroecosystem is a priority. The use of sustainable alternatives based on economic and environmental parameters is a choice against the excessive use of insecticides in agricultural ecosystems. In this way, natural biological control is one of the viable management variables for reducing insect pest populations in these crops.

The information collected allowed identifying and knowing the bioecology of oophagous parasitoids in different rice fields. The role of parasitoids, predators, and

B. Zachrisson (✉)
Biological Control Laboratory, Institute of Agricultural Research of Panama (IDIAP), Chepo, Panama
e-mail: bruno.zachrissons@idiap.gob.pa

B. Souza et al. (eds.), *Natural Enemies of Insect Pests in Neotropical Agroecosystems*, https://doi.org/10.1007/978-3-030-24733-1_31

entomopathogenic fungi in the agroecosystems of annual crops lies in the reduction of insect pest populations during the different stages of insect pest, not only egg parasitoids, prior to the damage that immature stages and adults can cause. The efficiency and profitability of this control measure stand out because it is consistent with integrated pest management and integrated crop management. Therefore, this chapter establishes the interactions between insect pests and natural enemies in different phenological stages of the rice crop.

The knowledge of the bioecology of oophagous parasitoids reported for the rice agroecosystem, related to climatic or abiotic factors, such as temperature, relative humidity, and solar radiation, among others, is specific to each production area. In this way, the biology, ecology, and behavior of the parasitoid species reported provided valuable information for the implementation of multiplication programs for mass breeding of egg parasitoids of the key insects in these crops. The specificity of the insect-parasitoid relationship is a determining factor in the success of biological control programs for insect pests. In addition, it is essential to consider the development stage of the insects (target of control) that offers comparative advantages in terms of management, such as technical input for the natural biological control of insect pests in this agricultural area. However, predators and entomopathogenic fungi have also been used for the management of the complex of the insect pests in this crop.

The results of the studies conducted in the last decade have been focused on identifying the species and assessing the important bioecological parameters for the colonization, establishment, and adaptation stages of insect pests, considering the direct relationship with the increase in the population of egg parasitoids. The reduction of insect pest populations below the level of economic damage is fundamental to the profitability of the production.

Integrated insect pest management in rice fields has been mainly targeted at *Tagosodes orizicolus* (Muir) (Delphacidae) and *Oebalus insularis* (Stål) (Pentatomidae), which are considered key insect pests in Panama and Central America (Table 31.1) (Zachrisson 2010a; Zachrisson et al. 2014a). However, there are other species that can cause damage to the different structures of the plant at different phenological stages, among them stand out *Lissorhoptrus* sp. (Curculionidae) (rice water weevil), *Hydrellia* sp. (Ephydridae) (leaf miner), *Diatraea tabernella* (Dyar) (Crambidae) (sugarcane borer), *Diatraea saccharalis* (Fabricius) (Crambidae) (sugarcane borer), *Rupela albinella* (Cramer) (Crambidae) (white stem borer), *Spodoptera frugiperda* (J. E. Smith) (Noctuidae) (fall armyworm), *Panoquina* sp. (Hesperiidae) (long-winged skipper), *O. insularis* (Pentatomidae) (rice stink bug), *Oebalus poecilus* Dallas (Pentatomidae) (rice stink bug), and *Oebalus ypsilongriseus* (De Geer) (Pentatomidae) (rice stink bug). The insect pests cited are considered key pests in Brazil (Krinski 2014), Ecuador (Pérez-Iglesias et al. 2017), Colombia (Pantoja 1997), and Panama (Zachrisson 2010b).

The dynamic population of these insect pests depends on the abiotic conditions and the regulation of the biocontrol agents. Some species of the family Pentatomidae, such as *Tibraca limbativentris* Stål (Pentatomidae) and *Euschistus nicaraguensis* (Rolston) (Pentatomidae), have exhibited a recent population increase, which has

Table 31.1 Degree of importance of the main insect pests that affect rice cultivation (*Oryza sativa* L.) in the production areas of Latin America

Order or suborder	Family	Species	Phenological stage[a]	Degree of importance[b]
Coleoptera	Curculionidae	*Lissorhoptrus* sp.	Establishment	2
Diptera	Ephydridae	*Hydrellia* sp.	Vegetative	2
Heteroptera	Pentatomidae	*Tibraca limbativentris*	Vegetative-Reproductive	2
Heteroptera	Pentatomidae	*Glyphepomis spinosa*	Vegetative-Reproductive	2
Heteroptera	Pentatomidae	*Hypatropis inermis*	Vegetative-Reproductive	2
Auchenorrhyncha	Delphacidae	*Tagosodes oryzicolus*	Vegetative-Reproductive	1
Lepidoptera	Crambidae	*Diatraea tabernella*	Vegetative-Reproductive	2
Lepidoptera	Crambidae	*Diatraea saccharalis*	Vegetative-Reproductive	3
Lepidoptera	Crambidae	*Rupela albinella*	Vegetative-Reproductive	2
Lepidoptera	Noctuidae	*Spodoptera frugiperda*	Vegetative-Reproductive	3
Lepidoptera	Hesperiidae	*Panoquina* sp.	Vegetative	3
Heteroptera	Pentatomidae	*Oebalus insularis*	Reproductive-Maturation	1
Heteroptera	Pentatomidae	*Oebalus poecilus*	Reproductive-Maturation	1
Heteroptera	Pentatomidae	*Oebalus ypsilongriseus*	Reproductive-Maturation	1
Heteroptera	Pentatomidae	*Euschistus nicaraguensis*	Reproductive-Maturation	2

[a]Phenological stage
[b]1, important; 2, moderately important; 3, sporadically important

coincided with high temperatures resulting from climate change in some commercial plots in Panama. Other species such as *Glyphepomis spinosa* Campos & Grazia (Pentatomidae) (Alves et al. 2012) and *Hypatropis inermis* Stål (Pentatomidae) (Krinski et al. 2015) were reported in Brazil in the last years, and the nymphs and adults feed on the stems by sucking on the plant sap.

However, the interaction of these insects with natural control agents mainly depends on the phenological stage susceptible to the attack and the reproductive strategy of the insects. Due to the reproductive capacity of parasitoids, such as *Telenomus rowani* (Gahan) (Scelionidae), *Trichogramma pretiosum* (Riley) (Trichogrammatidae), and *Telenomus podisi* (Ashmead) (Platygastridae), they are considered efficient control agents for *R. albinella*, *S. frugiperda*, and *O. insularis*, respectively (Table 31.2). Despite having moderate importance, the management of

Table 31.2 Parasitoids reported in rice cultivation (*Oryza sativa*) in Latin America

Family	Insect pest	Family	Natural enemy	Location (distribution)	Natural incidence of parasitoids[a]
Crambidae	*Diatraea saccharalis*	Trichogrammatidae	*Trichogramma pretiosum*	Panama, Colombia, Ecuador	H
Crambidae	*Rupela albinella*	Scelionidae	*Telenomus rowani*	Panama	H
Crambidae	*Spodoptera frugiperda*	Trichogrammatidae	*Trichogramma pretiosum*	Panama, Brazil, Colombia, Ecuador	M
Delphacidae	*Tagosodes orizicolus*	Encyrtidae Dryinidae	*Anagrus* sp. *Haplogonatopus hernandezae*	Panama, Costa Rica	L
Pentatomidae	*Oebalus insularis*	Platygastridae	*Telenomus podisi*	Panama, Colombia, Brazil	H
Pentatomidae	*Tibraca limbativentris*	Platygastridae Scelionidae	*Telenomus podisi* *Trissolcus urichi*	Panama, Brazil	M
Pentatomidae	*Euschistus nicaraguensis*	Platygastridae	*Telenomus podisi*	Panama, Honduras	M

[a]High (H, 80–100%), moderate (M, 60–79%), low (below 50%)

the mentioned insect pest species largely depends on the rate of natural parasitism in the production areas.

Currently, in the production areas of Latin America, *T. orizicolus* and *O. insularis* are responsible for the significant reduction in the profitability of rice cultivation (Pantoja 1997; Mora-Kepfer and Espinoza 2009; Zachrisson 2010a; Pérez-Iglesias et al. 2017). The high rate of natural parasitism of *T. rowani* and *T. podisi*, associated with *R. albinella* and *O. insularis*, indicates a high degree of control in the assessed areas (Zachrisson 2010b). In addition, other species of the family Pentatomidae, such as *T. limbativentris* (Zachrisson et al. 2014b) and *E. nicaraguensis*, were recently reported in extensive areas of rice cultivation in different locations, such as Panama (Zachrisson et al. 2018).

31.2 Insect-Plant Interaction

The natural control of insect pests in rice by means of rational management of this agroecosystem stands out due to the reduction of these arthropod populations. The relationship between egg parasitoids as biological agents for insect pest control should be taken into consideration when choosing alternative methods for the rational management of the agroecosystem. In addition, egg parasitoids have also been reported as predator species in irrigated rice in southern Brazil, grouped in the families Chrysopidae (Neuroptera), Coccinellidae (Coleoptera), Forficulidae (Dermaptera), and Carabidae (Coleoptera), among others (Gonzalez et al. 2014; Acosta et al. 2017). The use of *Metarhizium anisopliae* (Clavicipitaceae) and *Beauveria bassiana* (Clavicipitaceae) is applied to the management of different species of insect pests in several crops (Alves et al. 2008). However, there are no reports that confirm the application of these entomopathogenic fungi for the management of insect pests in rice crop.

The agroecosystem, which is the basic unit for insect pest management, integrates the abiotic and biotic factors that influence the trophic interactions. For this reason, the success of biological control programs—among them the natural biological control—largely depends on the knowledge of the parasitoid-insect-plant relationship. This way, it is necessary to know the biological and reproductive aspects of the insect pests of economic importance and their natural enemies.

Some authors have stressed the influence of the selected variety, the sowing density, and the crop phenology, which are considered the variables that may affect the efficiency in insect parasitism (Hassan et al. 1988; Botelho 1997). Some studies have suggested that the architecture of the plant, depending on the variety and sowing density, influences the parasitism rate of species of the families Platygastridae and Trichogrammatidae, which include *T. podisi* and *T. pretiosum*. In addition, another aspect to be taken into consideration is the effect of secondary metabolites, which depending on the variety of the crop and the plant phenological stage might also influence the parasitism rate of these species.

The insect pest bioecology, which includes biological and reproductive parameters, is related to the different phenological stages of the crops, depending on the development of the seedling, and the vegetative, reproductive, and maturation stages. In this way, the regulation of insect pests during the various phenological stages by means of egg parasitoids will determine the extent of the damage caused in the plants and the effect on the development of the crops. In addition, the synchronization between natural enemies and insects (target of control) determined by abiotic conditions that promote balance in the agroecosystem should be taken into consideration (Hassan et al. 1988).

31.3 Biology and Behavior of the Key and Emerging Insect Pests in Rice Cultivation

31.3.1 Tagosodes orizicolus *(Muir) (Delphacidae)*

Tagosodes orizicolus is considered a key insect pest in rice cultivation due to the mechanical damage that it causes in the plant by the injection of toxins and the transmission of phytopathogens, such as "rice hoja blanca virus" (RHBV). The recommended management for the populations of this vector insect is aimed at the mechanical damage and the transmission of viruses, varying in accordance with the economic damage levels and action thresholds. The described situations determine the management strategies considering the use of RHBV-resistant varieties (Galvis et al. 1982). The populations that transmit the viral diseases of this vector insect can have potential for vertical transmission, also known as transovarian transmission (Galvis et al. 1982). The natural control of eggs by *Anagrus* sp. Halliday (Mymaridae) and *Haplogonatus hernandezae* Olmi (Dryinidae) is sporadic and reduced in areas cultivated with the variety "Costa Rica-1113" (CR-1113) in the eastern region of Panama and Costa Rica (Mora-Kepfer and Espinoza 2009).

The recommended management for the populations of the vector insect that causes mechanical damage and transmits the virus is mainly aimed at the use of RHBV-resistant varieties and selective insecticides for sucking insects. The initial damage features perforations or incisions in the mesophyll of the leaves for feeding or oviposition. The eggs of *T. orizicolus* are oval shaped and white in color. The number of eggs can vary and 200 eggs of *T. orizicolus* per leaf that oviposit during a period of 3 days have already been reported (Galvis et al. 1982). The nymph stage of the insect, which usually consists of five instars, can vary depending on the temperature (Galvis et al. 1982). The nymphs of *T. orizicolus* exhibit brown dorsal longitudinal stripes, which persist in the adult stage (Galvis et al. 1982). Adult specimens have membranous wings, which are yellow in female specimens and always clearer than those of male specimens. The size of male specimens is two to three millimeters, always smaller than the size of female specimens. The longevity of adult specimens varies from 14 to 24 days for males and 24 to 36 days for females. This is a reproductive behavior that favors the increase of the oviposition rate.

31.3.2 Oebalus insularis *(Pentatomidae)*

Oebalus insularis, known as stink bug, is one of the most important insect pests in rice fields of Panama and Central America (Rodriguez et al. 2006; Cherry and Nuesly 2010; Vivas and Notz 2010). The inoculation of toxins and phytopathogens occurs during the milky stage of grain development through the insertion of the stylus of *O. insularis* nymphs and adults when they feed on plants (Rodriguez et al. 2006). This phenomenon, known as "grain discoloration," is directly related to *O. insularis* infestation, since more than 0.7 insects per panicle affect qualitatively and quantitatively crop yields (Vivas and Notz 2011).

Currently, there are no reports of varietal resistance to this insect. Therefore, it is difficult to control it through genetic improvement. This fact promotes the viability of biological control during the egg stage of *O. insularis* by means of *T. podisi*, which is considered the most viable proposal for the management of this insect (Rodriguez et al. 2006). This way, a protocol for the mass rearing of this egg parasitoid has been developed and implemented, with the purpose of obtaining *O. insularis* oviposition for the production of parasitized eggs and their subsequent release in rice fields (Zachrisson 2014).

The nutritional adaptation of *O. insularis* to various weed species as an alternative source of food can influence the biological and reproductive performance of this insect (Naresh and Smith 1983; Zachrisson et al. 2014a). In Latin America, *Echinochloa colona* (L.) Link is the main alternative host of *O. insularis* and responsible for the early migration of this insect in rice fields (Naresh and Smith 1983; Rashid et al. 2005). In addition to *E. colona*, other weed species, such as *Echinochloa crus-galli* (L.) Link, *Paspalum conjugatum* (P. J. Bergius), and *Paspalum virgatum* (L.), are associated with the feeding and reproduction of other *Oebalus* (Stål) species (Rashid et al. 2005) (Table 31.3, Fig. 31.1).

Table 31.3 Parasitism rates of eggs that are parasitoids of *Oebalus insularis* (Pentatomidae) in host weeds in rice production plots in Panama (2015–2016)

Weed species	Family	Parasitism rate (%)	Parasitism rate/*Telenomus podisi* (%)	Parasitism rate/*Trissolcus basalis* (%)
Cyperus iria	Cyperaceae	14.2 e[a]	89.4 b	10.6 c
Cyperus rotundus	Cyperaceae	9.8 f	81.2 c	18.8 b
Echinochloa colona	Poaceae	92.6 a	96.4 a	3.6 e
Echinochloa crus-pavonis	Poaceae	78.3 b	82.4 c	17.6 b
Eleusine indica	Poaceae	23.1 d	78.3 d	21.7 a
Ischaemum rugosum	Poaceae	21.9 d	82.6 c	17.4 b
Paspalum virgatum	Poaceae	38.9 c	86.1 b	13.9 d

[a]Means followed by the same letter in each column present no significant statistical difference ($P < 0.05$)

(a) (b)

Fig. 31.1 Nymphs (**a**) and adults (**b**) of *Oebalus insularis* (Pentatomidae) feeding in *Echinochloa colona*. (Photo: Bruno Zachrisson (2017))

31.3.3 Rupela albinella *(Crambidae)*

This insect is found in crops between 35 and 40 days after germination during the vegetative stage. The damage is caused by first instar larvae that penetrate the stem and feed on the vessels that conduct nutrients and water (xylem and phloem), causing the yellowing of the leaves in the top region of the plant. Eggs of *R. albinella* are oviposited in the upper face of leaves and they appear grouped and overlapping, covered by a cottony mass and parasitized with *T. rowani*. The parasitism rates of eggs of this insect found in the different areas of rice fields—such as the eastern region of Panama—ranged from 79% to 88%. However, the indiscriminate use of broad-spectrum insecticides has reduced the population of natural enemies, such as *T. rowani*, considered to be the main control agent for this insect.

31.3.4 Tibraca limbativentris *(Pentatomidae)*

The rice stalk stem bug *T. limbativentris* is considered a potential insect present in the main rice fields in Panama, with a relative abundance of 2.7% (Zachrisson et al. 2014b). This insect inserts its stylus into the stem and injects toxins that cause the sterility of the stem in the formation stage. The population increased between 2009 and 2012, reaching higher levels than nine insects per sweep net. This phenomenon gave rise to the use of insecticides in production fields located in the eastern region of Panama (Zachrisson et al. 2014b) (Table 31.4).

The intensive use of chemical control unilaterally aimed at the reduction of the populations of *T. limbativentris* poses the need of other control alternatives, such as biological control during the egg stage. However, the reduced information about the

Table 31.4 Percentage of *Tibraca limbativentris* (Pentatomidae) eggs parasitized with *Telenomus podisi* (Platygastridae) collected in Panama

Location	Variety	Number of eggs collected	Number of emerging parasitoids	Parasitism rate (%)
Paso Blanco Chichebre	IDIAP-L7 IDIAP-38	186 342	158 325	84.9 95.0
Total	–	**528**	**483**	**91.5**

association between parasitoids and *T. limbativentris* was restricted to the eggs of species of family Pentatomidae (Margaría et al. 2009). The presence of *Ooencyrtus submetallicus* (Howard), *T. podisi*, and *Trissolcus urichi* (Crawford) (Scelionidae) was reported for the first time in the state of Maranhão, Brazil (Maciel et al. 2007). In addition, the parasitoids *T. podisi* and *T. urichi* were reported as potential biological control agents in the state of Santa Catarina, Brazil (Riffel et al. 2010). The association between *T. limbativentris* and *T. urichi*, among other species, emphasizes the need of understanding the parasitoid-pest-crop interaction for the implementation of biological control programs (Riffel et al. 2010).

In Panama, the parasitism rates reported were higher than 80%. *Telenomus podisi* was the only species reported in the experimental areas in which insecticides were not being used (Table 31.4). This report confirmed the high natural parasitism potential of *T. podisi*, considered a promising biological control agent for *T. limbativentris*, corroborating the results obtained by different researchers (Maciel et al. 2007; Riffel et al. 2010; Zachrisson et al. 2014b).

The adaptation of *T. podisi* to the abiotic conditions of rice fields in the east of Panama confirmed the high parasitism rate of this parasitoid associated with the egg mass of *T. limbativentris*. It is worth noting that the association between *T. podisi* and *T. limbativentris* was the first report of this interaction in Panama, which also contributed to the strengthening of biological control programs for the species of the family Pentatomidae in rice (Zachrisson et al. 2014b). The incorporation of management practices favorable to the conservation of *T. podisi* strengthens the holistic approach of integrated pest management programs in this agricultural sector.

31.3.5 Euschistus nicaraguensis *(Pentatomidae)*

The sub-family Pentatominae, which includes *E. nicaraguensis*, has 116 genera and approximately more than 667 species reported in the Neotropical region (Rolston 1974). The geographic distribution of this species includes several countries in Central America between Panama and Honduras (Maes 1994). It feeds on different weed species of this area as well as rice (Maes 1994; Arismendi 2002). Despite reports of the occurrence of *E. nicaraguensis* in rice fields in Panama (Maes 1994, Arismendi 2002), this species of Pentatomidae is sporadically found and its population is small.

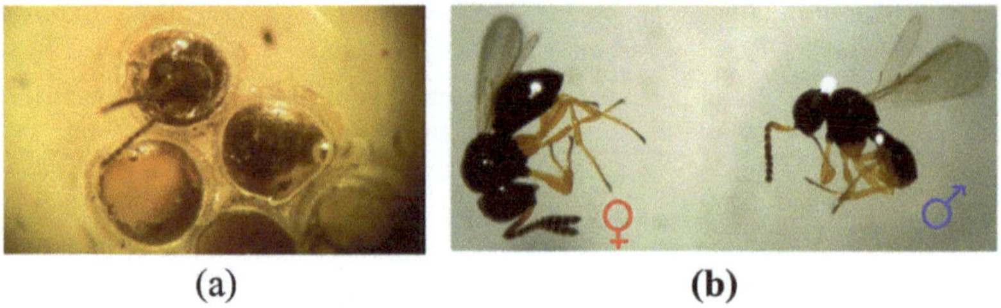

Fig. 31.2 Adult of *Telenomus podisi* (Platygastridae) (**a**) emerging from eggs of *Euschistus nicaraguensis* (Pentatomidae) and males (♂) and females (♀) (**b**) of *Telenomus podisi* (Platygastridae). (Photo: Bruno Zachrisson (2014))

Table 31.5 Parasitism rates of *Euschistus nicaraguensis* (Pentatomidae) eggs caused by *Telenomus podisi* (Platygastridae) in the rice agroecosystem of Panama

Species	Family	Total number of eggs collected/mass	Total number of parasitized eggs	Parasitism rate (%)
Oryza sativa	Poaceae	76 (3)[a]	58	82.9
Cyperus rotundus	Cyperaceae	49 (2)	22	44.9
Echinochloa colona	Poaceae	68 (3)	52	76.5

[a]Number of egg masses

However, the colonization, adaptation, and biological development of *E. nicaraguensis* in weed species associated with the rice agroecosystem have been insufficiently studied. Similarly, the interaction between the natural control of this insect species and the population of biological control agents—specifically egg parasitoids—is considered a new association between *E. nicaraguensis* and *T. podisi*, defined as a key component in the integrated management of rice agroecosystems (Fig. 31.2).

The rate of *T. podisi* parasitism (85.5%) confirmed the regulation of this insect population found in an experimental rice field planted with the variety IDIAP-38 (Table 8). The parasitism of *E. nicaraguensis* eggs reported in weed species surrounding this area varied from 42.0% to 65.0% (Table 31.5) (Zachrisson et al. 2018). This fact indicated the reduction of the insect population in areas surrounding rice fields, which could be considered a reservoir of *T. podisi* during the sowing time (Zachrisson et al. 2018).

The incidence of *E. nicaraguensis* in weed species of the families Asteraceae (*Eclipta prostrata*), Cyperaceae (*Cyperus rotundus* L., *Fimbristylis littoralis* L.), and Poaceae (*Echinochloa colona* (L.) Link) in rice fields in Panama requires a rethinking of integrated pest management programs. The trophic interaction observed between *E. nicaraguensis*, as potential pest, and *T. podisi* provides relevant information for the implementation of natural biological control.

31.4 Final Considerations

The natural biological control of insect pests in the rice agroecosystem by means of oophagous parasitoids is a successful management measure, which can be strengthened through the rational use of selective insecticides for the species of natural enemies. The results of the studies conducted in rice fields of tropical agroecosystems suggest the sustainable and rational management of this sector.

The conservation of natural reservoirs composed by weed species will promote the natural parasitism of species such as *T. rowani* and *T. podisi*, which regulate the population of *R. albinella* and *O. insularis*, respectively. The high potential of *T. rowani* and *T. podisi* in tropical agroecosystems associated with the specific features of the crops—such as the duration of the cycle of the recommended varieties and secondary metabolites—will define the plant-insect-parasitoid interaction. This aspect deserves to be further studied taking into consideration trophic interactions present in the rice agroecosystem.

Trophic interactions are relevant dynamic processes for the implementation of natural biological control programs, highlighting the relevance of new insect-parasitoid association. However, the increase in temperature during the last decade in the main rice fields in the tropical region confirms the need of performing periodic sampling of insects and natural enemies indicating their population density at different phenological stages of the crops. Thus, it will be possible to select potential control agents, specifically egg parasitoids with ideal biological and reproductive characteristics adapted to the rice agroecosystem.

The synergy of egg parasitoids associated with the predation activity and the effect of the applications of entomopathogenic fungi guarantee the reduction of the population of insect pests in rice, promoting the sustainability of the crop.

Acknowledgments The authors are thankful to the "Sistema Nacional de Investigaci6n" (SNI) and the "Secretaría Nacional de Ciencia, Tecnología e Innovaci6n" (SENACYT), Panama, Panama, for the financial support.

References

Acosta LG, Janhke SM, Redaelli LR et al (2017) Insect diversity in organic rice fields under two management system of levees vegetation. Braz J Biol 77(4):731–744

Alves SB, Lopes RB, Vieira SB et al (2008) Fungos entomopatogênicos usados no controle de pragas na América Latina. In: Alves SB, Lopes RB (eds) Controle microbiano de pragas na América Latina. FEALQ, Piracicaba, pp 69–104

Alves TM, Barrigossi JAF, Quintela ED (2012) Life cycle of *Glyphepomis spinosa* Campos & Grazia (Hemiptera: Pentatomidae): a new pest of rice in Brazil. Neotrop Entomol 41(6):437–441

Arismendi N (2002) Pentatomidae en Honduras: Listado, distribución y biología de especies [thesis]. Escuela Agrícola Panamericana, El Zamorano

Botelho PSM (1997) Eficiência de *Trichogramma* em campo. In: JRP P, Zucchi RA (eds) Trichogramma e o controle biológico aplicado. FEALQ, Piracicaba, pp 303–318

Cherry R, Nuesly G (2010) Establishment of a new stink bug pest, *Oebalus insularis* (Hemiptera: Pentatomidae), in Florida rice. Fla Entomol 93(2):291–293

Food and Agriculture Organization of the United Nations (2014) A regional rice strategy for sustainable food security in Asia and the Pacific. FAO, Bangkok

Galvis Y, Gonzalez J, Reyes J (1982) Descripción y daño de los insectos que atacan al arroz, en América Latina. CIAT, Costa Rica

Gonzalez ML, Jahnke SM, Morais RM et al (2014) Diversidad de insectos depredadores en área orizícola orgánica y de conservación, en Viamão, RS, Brasil. Rev Colomb Entomol 40(1):120–128

Hassan S, Kolher E, Rost WM (1988) Mass production and utilization of *Trichogramma*: 10. Codling moth *Cydia pomonella* and the summer fruit tortrix moth *Adoxophyes orana* (Lepidoptera: Tortricidae). BioControl 33(4):413–420

Krinski D (2014) Artrópodes asociados ao arroz de terras altas, Oryza sativa, em Novo Progresso, estado do Pará, níveis de danos e estratégias para manejo [tese]. Universidade Federal do Paraná, Curitiba, p 2014

Krinski D, Foerster LA, Grazia J (2015) *Hypatropis inermis* (Hemiptera: Pentatomidae); first record on rice crops. Rev Bras Entomol 59(1):12–13

Maciel AA, Lemos RNS, Souza JR et al (2007) Parasitismo de ovos de *Tibraca limbativentris* Stål (Hemiptera: Pentatomidae) na cultura de arroz no Maranhão. Neotrop Entomol 36(4):616–618

Maes JM (1994) Catálogo de los pentatomoidea (Heteroptera) de Nicaragua. Rev Nicar Entomol 28:1–19

Margaría CB, Loiácono MS, Lanteri AA (2009) New geographic and host records for scelionid wasps (Hymenoptera) parasitoids of insects pests in South America. Zootaxa 2314:41–49

Mora-Kepfer F, Espinoza AM (2009) Parasitism and predation of the planthopper *Tagosodes orizicolus* (Homoptera: Delphacidae) by dryinid parasitoid in Costa Rica. Rev Biol Trop 57(1):203–211

Naresh JS, Smith CM (1983) Development and survival of rice stink bugs (Hemiptera: Pentatomidae) reared on different hosts plants at four temperatures. Environ Entomol 12(5):1496–1499

Pantoja A (1997) Artrópodos relacionados con el arroz en América Latina. In: Pantoja A, Fischer A, Correa-Victoria F et al (eds) MIP en arroz: manejo integrado de plagas (Artrópodos, Enfermedades y Malezas). CIAT, Cali, pp 59–98

Pérez-Iglesias H, Rodríguez-Delgado I, Cun-Carrión J (2017) Insectos-plagas y enfermedades en el cultivo del arroz. In: Pérez-Iglesias H, Rodríguez-Delgado I (eds) Cultivos tropicales de importancia económica en Ecuador (Arroz, Yuca, Caña de azúcar y Maíz). Utmach, Quito, pp 121–133

Rashid T, Johnson DT, Bernhardt JL (2005) Feeding preference, fecundity and egg hatch of rice stink bug on artificial diet, rice and alternative host grasses. Southwest Entomol 30(4):257–262

Riffel CT, Prando HF, Boff MI (2010) Primeiro relato de ocorrência de *Telenomus podisi* (Ashmead) e *Trissolcus urichi* (Crawford) (Hymenoptera: Scelionidae) como parasitoides de ovos de percevejos-do-colmo-do-arroz, *Tibraca limbativentris* (Stål) (Hemiptera: Pentatomidae), em Santa Catarina. Neotrop Entomol 39(3):447–448

Rodriguez PG, Navas D, Medianero E et al (2006) Cuantificación del daño ocasionado por *Oebalus insularis* (Heteroptera: Pentatomidae) en el cultivo del arroz (Oryzica-1) en Panamá. Rev Colomb Entomol 32(2):131–135

Rolston LH (1974) Revision of the genus *Euschistus* in Middle America (Hemiptera: Pentatomidae, Pentatominae). Entomol Am 48(1):23–25

Vivas LE, Notz A (2010) Determinación del umbral y nivel de daño económico del chinche vaneador del arroz, sobre la variedad cimarrónen Calabozo estado de Guárico, Venezuela. Agron Trop 60(3):271–281

Vivas LE, Notz A (2011) Distribución espacial en poblaciones de *Oebalus insularis* Stal (Hemiptera: Pentatomidae) en el cultivo de arroz, estado Guárico, Venezuela. Revista UDO Agrícola 11(1):109–125

Zachrisson B (2010a) Bioecología, daños y muestreos de plagas, en el cultivo del arroz. IDIAP, Panamá

Zachrisson B (2010b) Avances en el control biológico de plagas de arroz (*Oryza sativa*), por medio de parasitoides oofagos, en Panamá. Agricultural Research Institute of Panamá, Panamá

Zachrisson B (2014) Protocolo de multiplicación de *Oebalus insularis* Stal (Heteroptera: Pentatomidae), en condiciones abióticas controladas: subsidio para la cría masiva de *Telenomus podisi* Ashmead (Hymenoptera: Platygastridae). IDIAP, Panama

Zachrisson B, Costa V, Bernal J (2014a) Incidencia natural de parasitoides de huevos de *Oebalus insularis* Stal (Heteroptera: Pentatomidae) en Panamá. Idesia 32(2):119–121

Zachrisson B, Margaría CB, Loiácono M et al (2014b) Parasitismo de huevos de *Tibraca limbativentris* (Hemiptera: Pentatomidae), en arroz (*Oryza sativa* L) en Panamá. Rev Colomb Entomol 40(2):189–190

Zachrisson, B; Polanco, P. Natural parasitism of Oebalus insularis Stal (Heteroptera: Pentatomidae) eggs in host weeds associated with rice cultivation in Panama. IOSR Journal of Agriculture and Veterinary Science (OISR-JAVS), 10(4):1–4, 2017. AQCJ https://doi.org/10.9790/2380-1004010104

Zachrisson B, Grazia J, Polanco P et al (2018) New reports of host plants of *Euschistus nicaraguensis* Rolston, 1972 (Heteroptera: Pentatomidae) and natural parasitism of *Telenomus podisi* Ashmead, 1893 (Hymenoptera: Platygastridae) in the rice agricultural ecosystem in Panama. Braz J Biol 78(3):593–594

Chapter 32
Soybean

Alexandre de Sene Pinto and Regiane Cristina Oliveira de Freitas Bueno

32.1 Introduction

Soybean is the main Brazilian culture and one of the leaders in Argentina and Paraguay, making it the most important group in Latin America (USDA 2019). In Brazil, soybean expanded from south to the middle-west due to the advances in the production system in the last years, as well as the development of new cultivars, which guarantee higher productive potential and suitable characteristics to plant in different periods of the year and regions. However, the attack of insect pests is one of the main issues faced in this culture, since they occur in all of the plant development cycles, causing injuries and damages to different plant structures, ending in the soybean production loss.

The attack of a complex of caterpillars and typical bugs is considered the most important in soybean. The lepidopterans reduce the photosynthetic rate when they consume the leaves and, if they feed on the pods, decrease the grain production. The typical bugs are more important in the reproductive period of soybean and, by directly feeding on the grains, reduce the seeds' quality (Panizzi et al. 2012; Corrêa-Ferreira and Azevedo 2002).

Among the lepidopterans that cause damages, the soybean looper caterpillar *Chrysodeixis includens* (Walker) (Noctuidae) is named as the key pest of the culture. However, this pest was considered secondary before the occurrence of the soybean rust, *Phakopsora pachyrhizi* (Sydow & P. Sydow), when it was naturally controlled by the fungus *Nomuraea rileyi* (Farlow) Sansom, which decreased the

A. de Sene Pinto (✉)
Centro Universitário Moura Lacerda, Ribeirão Preto, SP, Brazil
e-mail: alexandre@occasio.com.br

R. C. O. de Freitas Bueno
Faculdade de Ciências Agrárias, UNESP, Botucatu, SP, Brazil
e-mail: regiane.bueno@unesp.br

B. Souza et al. (eds.), *Natural Enemies of Insect Pests in Neotropical Agroecosystems*, https://doi.org/10.1007/978-3-030-24733-1_32

incidence, possibly due to the frequent use of fungicides in the culture (Moscardi et al. 2012).

Other lepidopteran species are also considered important, such as *Spodoptera cosmioides* Walker (Noctuidae), *Spodoptera eridania* Cramer (Noctuidae), *Helicoverpa armigera* (Hübner) (Noctuidae), and *Anticarsia gemmatalis* (Hübner) (Erebidae) (Bortolotto et al. 2015). *Spodoptera* and *H. armigera* cause serious damages in the reproductive phase, destroying pods and grains (Bueno et al. 2007). Other caterpillar species occur more sporadically, such as *Rachiplusia nu* (Guenée), *Trichoplusia ni* (Hübner), and *Chloridea virescens* (F.) (Noctuidae), *Omiodes indicata* (F.) (Pyralidae) (Moscardi et al. 2012), *Urbanus* spp. (Hesperiidae), and some species from Geometridae family (Pinto et al. 2008).

Nezara viridula (L.), *Piezodorus guildinii* (Westwood), and *Euschistus heros* (F.) (Pentatomidae) are the main typical bug pests in the soybean (Panizzi et al. 2012; Ávila and Grigolli 2014), but the species *Dichelops furcatus* (F.), *D. melacanthus* (Dallas), *Chinavia* spp., *Edessa meditabunda* (F.), *Acrosternum hilare* (Say), and *Thyanta perditor* (F.) (Pentatomidae) and *Neomegalotomus parvus* (Westwood) (Alydidae) also commonly occur in this culture (Pinto et al. 2008; Panizzi et al. 2012).

Secondary pests can also cause sporadic or regional damages to the culture, such as the mites *Tetranychus urticae* (Koch) (Tetranychidae) and *Polyphagotarsonemus latus* (Banks) (Tarsonemidae); the soil caterpillars and larvae *Elasmopalpus lignosellus* (Zeller) (Pyralidae) and *Agrotis ipsilon* (Hufnagel) and *Spodoptera frugiperda* (JE Smith) (Noctuidae); *Crocidosema aporema* (Walsingham) (Tortricidae); *Diabrotica speciosa* (Germar), *Cerotoma arcuata* Olivier, and *Colaspis* spp. (Chrysomelidae); *Phyllophaga cuyabana* (Moser) (Scarabaeidae); *Sternechus subsignatus* (Boheman) and *Aracanthus mourei* (Rosado Neto) (Curculionidae); *Myochrous armatus* (Baly) (Chrysomelidae); *Scaptocoris castanea* (Perty) and *S. carvalhoi* Becker (Cydnidae); and *Bemisia tabaci* (Gennadius) (Aleyrodidae) (Pinto et al. 2008; Hoffmann-Campo et al. 2012; Ávila and Grigolli 2014). The white fly, *B. tabaci*, has been causing significant damages to growers in the last years. There are many biotypes, but the B is the more aggressive one (Brown et al. 1995).

The soybean pest management is mainly made with chemical insecticides and *Bt* transgenic plants in Latin America. The use of insecticides has been causing many problems to the environment, and the selection of insect pest resistance to the chemically active principles also happens many times (Sosa-Gómez and Omoto 2012). The integrated pest management (IPM) affirms the use of biological control, conservative and/or applied, as one of at least three tactics of integrated control (Kogan 1998), but soybean is planted in more than 30 million hectares in Brazil (USDA 2019), which makes the IPM impracticable as it was stated in the 1970s (Corrêa-Ferreira et al. 2000; Gazzoni 2012).

With the advances in insect identification, occurrence prediction, pest sampling, and the biological control technology, especially the parasitoid release, the IPM has been reinvented and not been considered as an "illusion" as it was before (Ehler and Bottrell 2000). With the increase of biological control in Brazil, which raised the availability of the products, this tactic becomes an important strategy to the modern soybean (Bueno et al. 2012a, b; Parra 2014; Bortolotto et al. 2015).

Currently, the biological control takes more and more importance in the IPM programs, mainly when it emphasized the desire for a sustainable soybean culture. In this context, besides the preservation of the natural biological control, the development and use of applied biological control programs based on predators, parasitoids, and/or entomopathogens are essential. This chapter presents the main parasitoids and predators of the most important pests, briefly discusses about the habitat management (conservative biological control), and also presents the advances in the biological management (applied biological control) of the soybean pests.

32.2 Beneficial Entomofauna in the Soybean and Its Conservation

32.2.1 Predators

In the soybean culture, even though the composition of species within the group of predators can be variable according to the availability of hosts and the environmental conditions, the most common predators are spiders, followed by *Geocoris* spp. (Lygaeidae), *Nabis* (*Tropiconabis*) *capsiformis* (Germar) (Nabidae), *Callida* spp. (Carabidae), and *Orius* spp. (Anthocoridae) (Corso 1989; Cividanes and Barbosa 2001; Bueno et al. 2012a, b), among others that will be discussed.

The spiders are quite frequent and abundant (Prado et al. 1982; Moraes et al. 1991) in the soybean culture. They are generalist (Prado et al. 1982), but the identification of species has not been properly valorized. In Argentina, Liljesthröm et al. (2002) registered the presence of 28 different species grouped in Araneidae (10 species), Salticidae (7 species), Thomisidae (3 species), Anyphaenidae (2 species), Corinnidae (2 species), Lycosidae (1 species), Philodromidae (1 species), Oxyopidae (1 species), and Theridiidae (1 species), with the highest abundance in the family Thomisidae, 269 species, which represented 47.2% from the total collected individuals.

Danieli (2010) comments that the spiders must be good bioindicators of imbalances in the agroecosystem, since they are the most affected after insecticide applications.

The small typical bugs *Geocoris* are very common in the culture (Corso 1989; Cividanes and Barbosa 2001). They feed on *B. tabaci*, small caterpillars, mites, and insects' eggs (Tamaki and Weeks 1972; Waddill and Shepard 1974; Joseph and Braman 2009), even consuming until nine eggs of *A. gemmatalis* per day (Corrêa-Ferreira and Moscardi 1985).

Another small typical bug and important predator in the soybean belongs to the genus *Orius*. They consume small insects, such as newly hatched caterpillars, thrips

(*O. insidiosus* (Say) prefers them (Butler and O'Neil 2007, 2008)), aphids, white fly nymphs, and leafhoppers, besides eggs from other insects and even mites (Isenhour and Marston 1981; Isenhour and Yeargan 1982; Mccaffrey and Horsburgh 1986; Coll and Ridgway 1995; Rutledge et al. 2004). In the absence of their preys, they feed on pollen (Armer et al. 1998; Andrews and Kuhar 2010). In Brazil, *O. insidiosus* has been found also in *Bidens pilosa* (L.) (Asteraceae) and *Amaranthus* sp. (Amaranthaceae) (Silveira et al. 2003).

Easily confused with phytophagous typical bugs, *Podisus nigrispinus* (Dallas) (Pentatomidae) are common in soybean plantations in all Neotropical regions (Leite and Lara 1985; Medeiros et al. 1998a; Thomas 1992; de Clercq 2000), especially at the end of the culture cycle (Leite and Lara 1985). They are quite polyphagous (Evangelista Junior et al. 2004; Zanuncio et al. 1997). In the soybean, it is a predator of the caterpillars from the species *A. gemmatalis* (Saini et al. 1997), *R. nu* (Saini et al. 1997; Moraes et al. 1991), *C. includens* (Moraes et al. 1991), and *Spodoptera* spp. (Saini 1994) and of the eggs and nymphs of phytophagous typical bugs (Saini 1994) such as *Diabrotica speciosa* (Germar) (Chrysomelidae) (Gassen 1986), besides many other species.

Another important group is the typical bugs from the family Nabidae, *Nabis* (*Tropiconabis*) *capsiformis* (Germar) and *Nabis paranaensis* Latreille, frequently found in the soybean culture in Brazil (Moraes et al. 1991) and in many other countries in Latin America (Kerzhner 1983). They prey on small caterpillars of *C. includens* and *H. zea* (Elvin 1983), consuming daily 21 eggs or 3 third instar caterpillars of *A. gemmatalis* (Corrêa-Ferreira and Moscardi 1985), and *N. capsiformis* can prey on up to 12 eggs of *H. zea* (Parajulee et al. 2006). In Chile, it has been related to the predation of *R. nu* caterpillars by predators from the family Nabidae (Araya et al. 1997).

Other species of predatory typical bugs occur in the soybean culture, such as *Alcaeorrhynchus grandis* (Dallas) (Pentatomidae), an important soybean pest predator in Brazil, Colombia, and Mexico, consuming caterpillars (Ribeiro et al. 2010). *Tynacantha marginata* Dallas (Pentatomidae) is quite generalist, feeding on nymphs of *P. guildinii* (Panizzi and Smith 1976) or *N. viridula* (Sanchez et al. 2001), among many other preys.

Among the predatory beetles, the *Callida* spp. from the family Carabidae is common in the soybean plantations in America. It is very voracious and generalist, feeding on caterpillars, coleopterans, dipterans, and aphids (Kromp 1999). In Brazil, *Callida scutellaris* Chaudoir is one of the main predators in the soybean from Parana, consuming more than 60 caterpillars of *A. gemmatalis* in the second instar, in the larval stage, and more than 40 small caterpillars in the adult stage, per day (Corrêa-Ferreira and Pollato 1989). Another species, *Calleida decora* (F.), consumes, in 24 hours, six *C. includens* caterpillars in the first instar (Richman et al. 1980).

The Carabidae beetles more visible in the soybean plantation are from the genus *Calosoma*, based on the size and frequency of occurrence. *Calosoma granulatum* (Perty) occurs in Brazil and it is quite voracious. It consumes lepidopterans (Pasini

1995) and one adult preys on 91 small and 22 big caterpillars of *A. gemmatalis* per day (Pegoraro and Foerster 1985).

Lebia concinna (Brullé) (Carabidae) is very frequent in the soybean culture. It is smaller compared to other Carabidae beetles cited in this chapter. They feed on small insects and eggs, consuming about five *A. gemmatalis* caterpillars in the third instar in 24 hours (Corrêa-Ferreira and Moscardi 1985; Hoffmann-Campo et al. 2000).

32.2.2 Parasitoids

There are many parasitoid species associated with pest species in the soybean plantations. Those associated with the lepidopterans are well-known, since they have demonstrated good results in the control of the main pests of soybean, especially the parasitoids of eggs (Parra et al. 1987). The species of the genus *Trichogramma* (Trichogrammatidae) stand out among the other species from different groups for being easily reared (Parra 1997; Haji et al. 1998), besides being aggressive in the parasitism of eggs (Botelho 1997). In the Brazilian soybean culture, the most commonly found species is *T. pretiosum* Riley (Hohmann et al. 1989; Zachrisson 1997), parasitizing eggs from different lepidopteran pests. It is also found in Argentina and Paraguay (Querino and Zucchi 2012). In the soybean plantations, its hosts are as follows: *A. gemmatalis* (Hohmann et al. 1989; Frías et al. 1993; Foerster and Avanci 1999), *Chloridea* spp. (Martin et al. 1976), *C. virescens* (Andrade et al. 2011), *C. includens* (Martin et al. 1976; Bueno et al. 2012a, b), *E. lignosellus* (Andrade 2013), *H. armigera* (Carvalho et al. 2017), *R. nu* (Frías et al. 1993), *Spodoptera* species complex (Siqueira et al. 2012), and *T. ni* (Martin et al. 1976; Milanez et al. 2009).

In eggs of the genus *Spodoptera*, it is common to find the egg parasitoid *Telenomus remus* Nixon (Platygastridae). In Brazil, this parasitoid is not deeply studied (Pomari-Fernandes et al. 2014), but in some countries of Latin America, it is well known. In Venezuela, it has been used in biological control programs, with a parasitism reaching 90% of *Spodoptera frugiperda* (JE Smith) (Noctuidae) eggs (Hernández et al. 1989; Ferrer 2001).

Copidosoma floridanum (Ashmead) (Encyrtidae) is an endoparasitoid of caterpillars from the subfamily Plusiinae (Noctuidae) and an important regulator of the *C. includens* population in soybean plantations in all America. Also known as *Litomastix truncatellus* (Dalman) or *Copidosoma truncatellum* (Dalman) (Moraes et al. 1991; Maruya et al. 2001), Noyes (1988) stated that the correct one is *C. floridanum* to the Neotropical region. This parasitoid is able to generate more than 2000 identical parasitoids with only one egg (Strand 1989), showing its high potential of *C. includens* control, even though the parasitized caterpillars increase their consumption of leaves to about 35% until they die of parasitism consequences (Hunter and Stoner 1975).

Other caterpillar parasitoids, such as *Microcharops bimaculata* (Ashmead) and *M. anticarsiae* Gupta (Ichneumonidae), present in the USA and Central and South

America (Patel and Habib 1998; Marques et al. 1979); *Euplectrus* spp. (Eulophidae) (Wall and Berberet 1974; Marques et al. 1979), especially *E. chapadae* Ashmead which parasitizes *A. gemmatalis* (Puttler et al. 1980) and *S. frugiperda* (Murua and Virla 2004) in Brazil; and *E. plathypenae* (Howard) and *E. furnius* (Walker) in Argentina (Dequech 2002), are also important in the populational balance of lepidopteran pests.

Among the parasitoids of typical bug eggs, *Telenomus podisi* Ashmead and *Trissolcus basalis* (Wollaston) (Platygastridae) stand out in Brazil, among the 23 (Corrêa-Ferreira 1986; Foerster and Queiróz 1990; Kishino and Alves 1994; Medeiros et al. 1997, 1998b; Corrêa-Ferreira 2002), with *T. podisi* being the most commonly associated with the main species of typical bugs of the culture, *E. heros* (Foerster and Queiróz 1990; Medeiros et al. 1998b). They have high capacity of searching for hosts (Orr 1988). The females of *T. basalis* and *T. podisi* present high reproductive potential, with mean fertility of 250 eggs per female, in *N. viridula* eggs, and 211, in *E.heros* eggs, respectively, being laid in the first week of the female life (Pacheco and Corrêa-Ferreira 2000).

Besides these species, *Trissolcus urichi* (Crawford) was registered for the first time in Brazil in 1995, parasitizing eggs of *Edessa meditabunda* (F.), *E. heros*, and *Thyanta perditor* (F.) (Pentatomidae) (Corrêa-Ferreira and Moscardi 1995), and this species is quite aggressive when parasitizing the eggs (Sujii et al. 2002).

There is also the species of parasitoids that attack adults of typical bugs, represented by the dipterans of the family Tachinidae and microhymenopterans. Although they are not specific, they present such preference to some species, such as *Trichopoda giacomelli* (Blanchard) (Tachinidae) to *N. viridula* and *Hexacladia smithii* Ashmead (Encyrtidae) to *E. heros* adults.

H. smithii parasitize about six adults of *E. heros* (Nunes and Corrêa-Ferreira 2002a, b) and also parasitize *E. meditabunda* (Panizzi and Corrêa-Ferreira 1997) and *D. melacanthus* (Corrêa-Ferreira 2002). Species of *Hexacladia* are also reported in Argentina, Mexico, and Peru (Burks 1972; Rasplus et al. 1990). In Brazil, it is also cited that *H. blanchardi* De Santis parasitizes adults of *E. meditabunda* (Medeiros et al. 1998b).

Trichopoda giacomelli (Blanchard) (Tachinidae) preferably attacks adults of *N. viridula*, but parasitizes species of the family Pentatomidae (Corrêa-Ferreira 1984; La Porta 1987; Panizzi and Corrêa-Ferreira 1997). The females lay more than 275 eggs (Coombs 1997) during their lives. The biological control with this species, in Brazil and Argentina (Liljesthröm 1980, 1981), and with *T. pennipes* (F.), in the USA (Buschman and Whitcomb 1980; Harris and Todd 1982), is quite discussed due to the superparasitism that normally happens in typical bugs.

32.2.3 Natural Enemies' Conservation in the Soybean Culture

The Latin American grower which produces soybean does not tolerate the minimum defoliation on the plants, applying insecticides much before the culture reaches the control level. However, the infestations of caterpillars that cause some defoliation in soybean plantation are beneficial to the grower, since defoliation permits higher penetration of sunlight in the canopy and, thus, guarantees the development of the pods in the basal and medium part of the plant, promoting a better production of grains (Gassen 2001). The increase of the grower's tolerance to the defoliation guarantees a better maintenance of natural enemies in the crop, which needs a minimum populational rate of the pests.

Different from the sugarcane, another important culture in Latin America, the soybean is planted in quite large areas, with few wild woods in their borders, which could contribute to the maintenance of the natural enemies inside the culture. Keeping intersperse cultures or borders of the crop is a tactic not accepted by the growers as they are hard to be installed in the current system of the culture conduction.

On the other hand, the planting system of soybean must interfere in the occurrence of natural enemies during the harvest. The occurrence of spiders is observed during all soybean cycles, improving from the vegetative phase to the maximum density that occurs normally in the maintenance stage of the plants (Prado et al. 1982; Leite and Lara 1985). However, the occurrence of the highest number of spiders can be influenced by the different systems of planting. In the southeast of Brazil, Cividanes (2002) observed a higher number of spiders from the species *Lycosa erythrognatha* (Lucas) (Lycosidae), *Oxyopes salticus* Hentz (Oxyopidae), *Ailluticus* sp. (Salticidae), *Apopyllus silvestrii* (Simon) and *Camillina pulcher* (Keyserling) (Gnaphosidae), and *Goeldia* sp. (Titanoecidae) in direct planting systems.

Chemical insecticide use affects the populations of non-target organisms in the soybean culture, including many predators and parasitoids. Danieli (2010), evaluating the application of different insecticides in the soybean, verified that *Geocoris* spp., *Podisus* sp., and *Nabis* spp. from the family Reduviidae, *Lebia* sp., *Cycloneda sanguinea* (L.) (Coccinellidae), and species of Hymenoptera are negatively affected by many insecticides and the recovery of their populations is slower with permethrin, methomyl, lambda-cyhalothrin, and methamidophos. Zucoloto (2017) verified that chlorantraniliprole does not affect the populations of Carabidae, but it is harmful to other coleopterans, ants, dipterans, and spiders. On the other hand, the carabideans are negatively affected by acephate, as well as ants and dipterans.

Danieli (2010) did not verify the impact of the evaluated insecticides over the *T. podisi* parasitoid released in the field, but Zucoloto (2017) observed that acephate and chlorantraniliprole decreased the parasitism of *E. heros* eggs in the field, until at least 10 days after the application.

32.3 Applied Biological Control of Soybean Pests with Macroorganisms

In the soybean culture, the biological control of pests with parasitoids has been increasing in Brazil. The two parasitoids that are started to be used are *T. pretiosum*, for the lepidopteran pests' eggs, and *T. podisi*, in the control of typical bug pests' eggs.

The success of the egg parasitoid releases in the soybean basically depends on knowing the bioecological characteristics of the parasitoid and its interaction with the target host, such as the number of parasitoids to be released, the period and the number of releases, the releasing method, the relation with the phenology of the plant, the density of other natural enemies present in the agroecosystem, and the local climatic conditions (Bourchier and Smith 1996), and many of these items need to be better studied in the soybean agroecosystem.

Furthermore, the applied biological control depends on the perfect integration with the insecticide pulverization that needs to be selective to the parasitoids which will be released.

Trichogramma presents a high potential to be used in the soybean IPM of almost all the lepidopteran pest species, such as the soybean caterpillar, *A. gemmatalis*, and of other species that recently became important, such as the soybean looper, *C. includens*, the species complex from the genus *Spodoptera*, and other species of less occurrence, such as *O. indicata*, *E. lignosellus*, *Etiella zinckenella* (Treitschke) (Pyralidae), and *C. aporema*, among others. The species of *Trichogramma*, usually generalist, are, therefore, biological control agents with potential to control all these species of lepidopteran pests.

The parasitoid *T. pretiosum* is started to be used in order to control the lepidopteran pests in the soybean, in Brazil, in 2012, with the entrance of the quarantine species *H. armigera* in the country (Czepak et al. 2013). Until then, this parasitoid was used only in organic areas, in about 8000 hectares (Vasconcelos 2012), reaching the top in 2014, with almost 800,000 treated hectares. With the register of new molecules to control *H. armigera* in Brazil, the area treated with *Trichogramma* currently decreased to less than 200,000 hectares.

The moment for this parasitoid's release is determined by the beginning of the males' capture in pheromone traps for the main lepidopteran pests. It is made through two or three releases of 50,000–100,000 parasitoids per hectare in the following weeks, during the vegetative period of the culture, and the other two, again in the following weeks, at the beginning of flowering, but in doses of 100,000–200,000 parasitoids per hectare, since the lepidopteran pest pressure is higher at this time and the presence of *C. includens* requires higher doses of parasitoids (Bueno et al. 2012a, b).

The growers that use the IPM release the parasitoids only in the vegetative period, and during the reproductive period, they use insecticides, because *T. pretiosum* fails when controlling *C. includens*, since it prefers eggs of *A. gemmatalis*, *S. cosmioides*, or *H. armigera* (Oliva 2016; Baggio 2017). Other options to the chemical insecticides are the application of *Bacillus thuringiensis* (Berliner) or entomopathogen viruses, both to small caterpillars.

Trichogramma pretiosum is released on the pupal stage, without capsules for protection, by mechanically spreading the pupae (treated or not with repellent to predators) by airplanes or drones.

In the 1990s, it was developed the control of the typical bug *N. viridula* eggs by releasing *T. basalis* in the south of the country (Corrêa-Ferreira 1993; Corrêa-Ferreira and Moscardi 1996; Corrêa-Ferreira 2002). This program was limited to the Parana state, Brazil, focusing on those typical bug pests. It reached more than 18,000 hectares of soybean and reduced from 2.8 to 1.3 insecticide applications during all harvests (Corrêa-Ferreira et al. 2000; Corrêa-Ferreira 2002).

With the soybean culture expansion to other hotter regions in the middle and north of the country, the more important typical bug pest was *E. heros*, as it is more adapted to higher temperatures (Frugeri et al. 2012a, b). The mentioned biological control program was developed for *T. basalis*, but *T. podisi* is more common in *E. heros* eggs (Pacheco et al. 1999) and more indicated to the new productive regions of soybean, which are hotter (Rocha 2016).

The multiplication of *T. podisi* in industrial scale was very hard, because it depended on the rearing of the typical bugs and the soybean plants or others (Corrêa-Ferreira 1985), and, for this reason, such biological control program did not advance. However, in 2012, an artificial diet to *E. heros* was developed (Mendoza et al. 2016), and it finally gets simpler to multiply *T. podisi*, being produced by the company Bug Biological Agents from 2014 onward.

The release of this parasitoid is made just after the emergence of the first adults in the field, generally one week before the beginning of the soybean flowering, in the borders of the plantation. A new release is then made between 15 and 20 days after the first, and another release is required in areas with high pest pressure or on those that will be harvested at last. The released dose of *T. basalis* was made of 5000 parasitoids per hectare (Corrêa-Ferreira 2002), the same which was recently demonstrated to *T. podisi* (Gomes 2016; Tibaldi 2017).

As for the release of *T. basalis* in the 1990s, *N. viridula* eggs with the parasitoids' pupae were placed on cards, which were enclosed into small packages with holes to the exit and entrance of the adults. These packages were manually distributed in the field, hanging them on the plants located in the borders of the plantation (Corrêa-Ferreira 2002; Polanczyk et al. 2006). Nowadays, *T. podisi* is mechanically released in small capsules with holes to the exit and entrance of the adults, by airplanes or drones. The releases still happen in demonstration fields, because commercial registration for the parasitoid is pending in Brazil. It is still not used in Argentina and Paraguay.

The biological management of soybean pests using *T. pretiosum* and *T. basalis* or *T. podisi* has showed good results, being as efficient as the chemical insecticides (Figueiredo 2009; Agostinho 2010; Corrêa-Ferreira et al. 2010) or even more (Mendes Neto 2014), depending, probably, on the climatic conditions during the harvest.

The releasing cost of *T. pretiosum* is US$30.00–40.00 (product, transport, and application), and for *T. podisi* it will be US$35.00–50.00 when applied in Brazil.

32.4 Final Remarks

The habitat management, favoring the natural biological control, is still very incipient in Latin American soybean areas. Since the 1970s, the knowledge about the parasitoids and predators of soybean pests advanced a lot, indicating that the management of agroecosystems in their favor should evidence the benefits of pest management. The spiders are the main predators in the soybean environment, but they are not much studied and preserved, since many insecticides rapidly decrease their populations. In the last years, growers have been searching for information about the selectivity of chemical products which they use on their plantations and the damages that they can cause to the natural enemies, indicating a change in their position.

On the other hand, the applied biological control has been increasing to more than 70% per year in this culture in Brazil, especially about the microorganisms used to control the pests and diseases and as plant growth promoters. The use of *T. pretiosum* to control the eggs of lepidopteran pests has increased in the Brazilian middle-west, the most important region to the soybean culture, especially against *A. gemmatalis*, *C. includens*, *H. armigera*, and *Spodoptera* spp. in the vegetative period. In the reproductive period, a little time before the flowering, the grower has prioritized the use of chemical insecticides, because the pressure of such caterpillars is much higher. However, the use of viruses and *B. thuringiensis* could be an option to replace the insecticides and to help *T. pretiosum* in this control. The parasitoid *T. podisi* will rapidly be applied in Brazil in the coming years, against the phytophagous typical bugs.

The use of drones to release these parasitoids favored the grower's acceptance to the biological control, since it quite facilitated the field application. New parasitoids might be part of the biological options, such as *T. remus* to control the eggs of *Spodoptera* spp., which certainly are taking the soybean in Latin America to the status of culture that practices a truly integrated pest management.

References

Agostinho R (2010) Comparação entre manejo biológico e convencional de lagartas e percevejos na cultura da soja em Ribeirão Preto, SP. Monografia, Centro Universitário Moura Lacerda, Ribeirão Preto, 56p

Andrade RS (2013) Criação de *Elasmopalpus lignosellus* (Zeller, 1848) (Lepidoptera: Pyralidae) em dieta artificial em diferentes temperaturas e seleção de linhagens de *Trichogramma pretiosum* Riley, 1879 (Hymenoptera: Trichogrammatidae) para seu controle em arroz. Dissertação de Mestrado, ESALQ/USP, Piracicaba, 72p

Andrade GS, Pratissoli D, Dalvi LP, Desneux N, dos Santos Junior HJG (2011) Performance of four *Trichogramma* species (Hymenoptera: Trichogrammatidae) as biocontrol agents of *Heliothis virescens* (Lepidoptera: Noctuidae) under various temperature regimes. J Pest Sci 84:313–320

Andrews H, Kuhar T (2010) The minute pirate bug (*Orius*). Virginia Polytechnic Institute and State University. Disponível em https://pubs.ext.vt.edu/3002/3002-1437/3002-1437.html

Araya JE, Arretz P, Lamborot L, Guerreiro MA (1997) Natural control by predators of *Rachiplusia nu* (Guenée) (Lepidoptera: Noctuidae) on *Phaseolus beans* in Central Chile. Z Pflanzenkr Pflanzenschutz 104:147–152

Armer CA, Wiedenmann RN, Bush DR (1998) Plant feeding site selection on soybean by the facultatively phytophagous predator *Orius insidiosus*. Entomol Exp Appl 86:109–118

Ávila CJ, Grigolli JFJ (2014) Pragas da soja e seu controle. In: Lourenção ALF, Grigolli JFJ, Melotto AM, Pitol C, Gitti DC, Roscoe R (Orgs) Tecnologia e Produção Soja 2013/2014, 1st edn. Midiograf, Curitiba, p 109–168, 247p

Baggio RC (2017) Preferência hospedeira de *Trichogramma pretiosum* por ovos de lepidópteros em diferentes temperaturas e umidades relativas do ar em laboratório. Monografia, Centro Universitário Moura Lacerda, Ribeirão Preto, 34p

Bortolotto OC, Pomari-Fernandes A, Bueno RCOF, Bueno AF, Kruz YKS, Queiroz AP, Sanzovo A, Ferreira RB (2015) The use of soybean integrated pest management in Brazil: a review. Agron Sci Biotechnol 1:25–32

Botelho PSM (1997) Eficiência de *Trichogramma* em campo. In: Parra JRP, Zucchi RA (eds) *Trichogramma* e o controle biológico aplicado. FEALQ, Piracicaba, p 303–318, 324p

Bourchier RS, Smith SM (1996) Influence of environmental conditions and parasitoid quality on field performance of *Trichogramma minutum*. Entomol Exp Appl 80:461–468

Brown JK, Frohlich DR, Rossel RC (1995) The sweetpotato or silverleaf whiteflies: biotypes of *Bemisia tabaci* ou a species complex? Annu Rev Entomol 40:511–534

Bueno RCOF et al (2007) Sem barreira. Rev Cultiv 93:12–15

Bueno AF, Sosa-Gómez DR, Corrêa-Ferreira BS, Moscardi F, Bueno RCOF (2012a) Inimigos naturais das pragas da soja. In: Hoffmann-Campo CB, Corrêa-Ferreira BS, Moscardi F (eds) Soja: manejo integrado de insetos e outros artrópodes-praga. Embrapa, Brasília, p 493–629, 858p

Bueno RCOF, Parra JRP, Bueno AF (2012b) *Trichogramma pretiosum* parasitism and dispersal capacity: a basis for developing biological control programs for soybean caterpillars. Bull Entomol Res 102:1–8

Burks BD (1972) The genus *Hexacladia* Ashmead (Hymenoptera: Encyrtidae). Proc Entomol Soc Wash 74:363–371

Buschman LL, Whitcomb WH (1980) Parasites of *Nezara viridula* (Hemiptera: Pentatomidae) and other Hemiptera in Florida. Fla Entomol 63:154–162

Butler CD, O'Neil RJ (2007) Life history characteristics of *Orius insidiosus* (Say) fed diets of soybean aphid, *Aphis glycines* Matsumura, and soybean thrips, *Neohydatothrips variabilis* (Beach). Biol Control 40:339–346

Butler CD, O'Neil RJ (2008) Voracity and prey preference of insidious flower bug (Hemiptera: Anthocoridae) for immature stages of soybean aphid (Hemiptera: Aphididae) and soybean thrips (Thysanoptera: Thripidae). Environ Entomol 37:964–972

Carvalho GS, Silva LB, Veras MS, dos Santos ECB, Almeida MS, Reis SS (2017) Biological parameters of three *Trichogramma pretiosum* strains (Riley, 1879) (Hymenoptera: Trichogrammatidae) on eggs *Helicoverpa armigera* (Hübner, 1805) (Lepidoptera: Noctuidae). Acta Sci Biol Sci 39:349–355

Cividanes FJ (2002) Efeitos do sistema de plantio e da consorciação soja-milho sobre artrópodes capturados no solo. Pesq Agrop Brasileira 37:15–23

Cividanes FJ, Barbosa JC (2001) Effects of no-tillage and of soybean-corn intercropping on natural enemies and pests. Pesq Agrop Brasileira 36:235–241

Coll M, Ridgway RL (1995) Functional and numerical responses of Orius insidiosus (Heteroptera: Anthocoridae) to its prey in different vegetable crops. Ann Entomol Soc Am 88:732–738

Coombs MT (1997) Influence of adult food deprivation and body size on fecundity and longevity of *Trichopoda giacomelli*: a South American parasitoid of *Nezara viridula*. Biol Control 8:119–123

Corrêa-Ferreira BS (1984) Incidência do parasitoide *Eutrichopodopsis nitens* Blanchard, 1966 em populações do percevejo verde *Nezara viridula* (Linnaeus, 1758). An Soc Entomol Bras 13:321–330

Corrêa-Ferreira BS (1985) Criacao massal do percevejo verde, *Nezara viridula* (L). Londrina, EMBRAPA-CNPSo, 16p. (EMBRAPA-CNPSo, Documentos, 11)

Corrêa-Ferreira BS (1986) Ocorrência natural do complexo de parasitoides de ovos de percevejos da soja no Paraná. An Soc Entomol Bras 15:189–199

Corrêa-Ferreira BS (1993) Utilização do parasitóide de ovos *Trissolcus basalis* (Wollaston) no controle de percevejos da soja. Londrina, Embrapa-CNPSo, 40p. (Circular Técnica, 11)

Corrêa-Ferreira BS (2002) *Trissolcus basalis* para o controle de percevejos da soja. In: Parra JRP, Botelho PSM, Corrêa-Ferreira BS, Bento JMS (eds) Controle biológico no Brasil: parasitoides e predadores. Manole, São Paulo, p 449–476, 609p

Corrêa-Ferreira BS, Azevedo J (2002) Soybean seed damage by different species of stink bugs. Agric For Entomol 4:145–150

Corrêa-Ferreira BS, Moscardi F (1985) Potencial de consumo dos principais insetos predadores ocorrentes na cultura da soja. Resultados de Pesquisa de Soja 1984/85, p 79. (Documentos 15)

Corrêa-Ferreira BS, Moscardi F (1995) Técnicas de armazenamento de ovos do percevejo-verde visando à multiplicação do parasitoide *Trissolcus basalis* (Wollaston). Pesq Agrop Brasileira 28:1247–1253

Corrêa-Ferreira BS, Moscardi F (1996) Biological control of soybean stink bugs by inoculative releases of *Trissolcus basalis*. Entomol Exp Appl 79:1–7

Corrêa-Ferreira BS, Pollato SLB (1989) Biologia e consumo do predador *Callida* sp. (Coleoptera: Carabidae) criado em *Anticarsia gemmatalis* Hübner, 1818. Pesq Agrop Brasileira 24:923–927

Corrêa-Ferreira BS, Domit LA, Morales L, Guimarães RC (2000) Integrated soybean pest management in micro river basins in Brazil. Int Pest Manag Rev 5:75–80

Corrêa-Ferreira BS, Alexandre TM, Pelizzaro EC, Moscardi F, Bueno AF (2010) Práticas de manejo de pragas utilizadas na soja e seu impacto sobre a cultura. Londrina, Embrapa Soja, 16p. (Embrapa Soja, Circular técnica, 78)

Corso IC (1989) Efeito de inseticidas sobre inmigos narturais de insetos pragas da soja. Resultados de Pesquisa de Soja 1988/89. p 32–36

Czepak C, Albernaz KC, Vivan LM, Guimarães HO, Carvalhais T (2013) Primeiro registro de ocorrência de *Helicoverpa armigera* (Hübner) (Lepidoptera: Noctuidae) no Brasil. Pesqui Agropecu Trop 43:110–113

Danieli T (2010) Eficácia de inseticidas utilizados no controle de lagartas desfolhadoras e impacto na artropodofauna da soja RR. Monografia, Centro Universitário Moura Lacerda, Ribeirão Preto, 101p

De Clercq P (2000) Predaceous stinkbugs (Pentatomidae: Asopinae). In: Schaefer CW, Panizzi AR (eds) Heteroptera of economic importance. CRC Press, Boca Raton, p 737–789, 850p

Dequech STB (2002) *Campoletis flavicincta* (Hym. Ichneumonidae): ocorrência, criação e interação com *Spodoptera frugiperda* (Lep.: Noctuidae) e *Bacillus thuringiensis aizawai*. Tese de Doutorado, Universidade Federal do Rio Grande do Sul, Porto Alegre, 79p

Ehler LE, Bottrell DG (2000) The illusion of integrated pest management. Issues Sci Technol 16:61–64

Elvin MK (1983) Quantitative estimation of rates of arthropod predation on velvetbean caterpillar (Anticarsia gemmatalis Hübner) eggs and larvae in soybeans. Ph.D. Dissertation, University of Florida, 249p

Evangelista Junior WS, Gondim Junior MGC, Torres JB, Marques EJ (2004) Fitofagiade *Podisus nigrispinus* em algodoeiro e plantas daninhas. Pesq Agrop Brasileira 39:413–420

Ferrer F (2001) Biological control of agricultural insect pests in Venezuela; advances, achievements, and future perspectives. Biocontrol News Inf 22:67–74

Figueiredo FSGD (2009) Comparação entre o manejo biológico e o convencional de lagartas e percevejos na cultura da soja em Ribeirão Preto, SP. Monografia, Centro Universitário Moura Lacerda, Ribeirão Preto, 67p

Foerster FA, Avanci MRF (1999) Parasitóides de ovos de *Anticarsia gemmatalis* Hübner (Lepidoptera: Noctuidae) na cultura da soja. An Soc Entomol Bras 28:545–548

Foerster LA, Queiróz JM (1990) Incidência natural de parasitismo em ovos de pentatomídeos da soja no Centro-Sul do Paraná. An Soc Entomol Bras 19:221–232

Frías EA, Ovruski SM, Popich SB (1993) Parasitoides de huevos de lepidópteros noctuidos encontrados en cultivos de soja y su evaluación como agentes de control. CIRPON. Rev Invest 9:29–35

Frugeri AP, Barbosa GC, Bortolotto OC, Silva GV, Siqueira F, Bueno AF (2012a) O aquecimento global e o *Euschistus heros* (Fabricius) (Hemiptera: Pentatomidae). In: Congresso Brasileiro de Entomologia, 24, 2012, Curitiba. Anais... Curitiba, SEB

Frugeri AP, Barbosa GC, Bortolotto OC, Silva GV, Siqueira F, Bueno AFD (2012b) Influência da temperatura sobre os aspectos biológicos de *Euschistus heros* (Fabricius) (Hemiptera: Pentatomidae). In: Jornada Acadêmica da Embrapa Soja, 7, 2012, Londrina. Resumos expandidos.... Embrapa Soja, Londrina, pp 25–29. (Embrapa Soja. Documentos, 333)

Gassen DN (1986) Parasitos, patógenos e predadores de insetos associados à cultura do trigo. Embrapa-CNPT, Passo Fundo, 86p

Gassen DN (2001) O desfolhamento e a planta de soja. Rev Plant Direto 1:26

Gazzoni DL (2012) Perspectivas do manejo de pragas. In: Hoffmann-Campo CB, Corrêa-Ferreira BS, Moscardi F (eds) Soja: manejo integrado de insetos e outros artrópodes-praga. Embrapa, Brasília, pp 789, 858p–829

Gomes MG (2016) Quantidade liberada do parasitoide *Telenomus podisi* no controle de ovos do percevejo-marrom, *Euschistus heros*, em soja. Monografia, Centro Universitário Moura Lacerda, Ribeirão Preto, 34p

Haji FND, Jimenez Velasquez J, Bleicher E, Alencar JA, Haji AT, Diniz RS (1998) Tecnologia de produção massal de *Trichogramma* spp. Embrapa-CPATSA, Petrolina, 24p

Harris VE, Todd JW (1982) Longevity and reproduction of the southern green stink bug, *Nezara viridula*, as affected by parasitization by *Trichopoda pennipes*. Entomol Exp Appl 31:409–412

Hernández D, Ferrer F, Linares B (1989) Introducion de *Telenomus remus* Nixon (Hym.: Scelionidae) para controlar Spodoptera frugiperda (Lep.: Noctuidae) en Yaritagua-Venezuela. Agron Trop 39:199–205

Hoffmann-Campo CB, Moscardi F, Corrêa-Ferreira BS, Oliveira LJ, Sosa-Gómez DR, Panizzi AR, Corso I, Gazzoni DL, de Oliveira EB (2000) Pragas da soja no Brasil e seu manejo integrado. Embrapa Soja, Londrina, 70p. (Embrapa Soja, Circular Técnica, 30)

Hoffmann-Campo CB, Corrêa-Ferreira BS, Moscardi F (2012) Soja: manejo integrado de insetos e outros artrópodes-praga. Embrapa, Brasília, 858p

Hohmann CL, Silva SMT, Santos WJ (1989) Lista preliminar de Trichogrammatidae encontrados no Paraná. An Soc Entomol Bras 18:203–206

Hunter K, Stoner A (1975) *Copidosoma truncatellum*: effect of parasitization on food consumption of larval *Trichoplusia ni*. Environ Entomol 4:381–382

Isenhour DJ, Marston NL (1981) Seasonal cycles of Orius insidious (Hemiptera: Anthocoridae) in Missouri soybean. J Kansas Entomol Soc 54:129–142

Isenhour D, Yeargan KV (1982) Oviposition sites of *Orius insidious* (Say) and *Nabis* spp. in soybean (Hemiptera: Anthocoridae and Nabidae). J Kansas Entomol Soc 55:65–72

Joseph SV, Braman SK (2009) Predatory potencial of *Geocoris* spp. and *Orius insidious* on fall armyworm in resistant and susceptible turf. J Econ Entomol 102:1151–1156

Kerzhner IM (1983) Airbone *Nabis capsiformis* (Heteroptera: Nabidae) from the Atlantic, Indian, and Pacific oceans. Int J Entomol 25:273–275

Kishino K, Alves RT (1994) Utilização de inimigos naturais no controle de insetos-pragas na região dos Cerrados. In: Embrapa – Centro de Pesquisa Agropecuária dos Cerrados. Relatório Técnico do Projeto Nipo-Brasileiro de Cooperação de Pesquisa Agrícola nos Cerrados 1987-1992. Brasília, Embrapa-CPAC/JICA p 127–155, 516p

Kogan M (1998) Integrated pest management: historical perspectives and contemporary development. Annu Rev Entomol 43:243–270

Kromp B (1999) Carabid beetles in sustainable agriculture: a review on pest control efficacy, cultivation impacts and enhancement. Agric Ecosyst Environ 74:187–228

La Porta NC (1987) Aspectos biologicos de *Trichopoda giacomelli* (Blanchard) Guimarães, 1971 (Diptera: Tachinidae), parasitoide de *Nezara viridula* (L.) 1758 (Hemiptera: Pentatomidae). Rev Soc Entomol Argent 44:433–439

Leite LG, Lara FM (1985) Flutuação populacional de insetos e inimigos naturais associados à cultura da soja em Jaboticabal, SP. An Soc Entomol Bras 14:45–57

Liljesthröm G (1980) Contribución al estúdio sistemático y ecológico de dípteros taquinideos neotropicales. Dissertação de Mestrado, Universidad Nacional de La Plata, Buenos Aires

Liljesthröm G (1981) Algunas consideraciones sobre la dinámica poblacional de *Nezara viridula* (L.) (Hemiptera: Pentatomidae) e interaciones con *Trichopoda giacomelli* (Blanchard, 1966) (Diptera: Tachinidae). Neotropica 27:11–16

Liljesthröm G, Minervino E, Castro D, González A (2002) La comunidad de arañas del cultivo de soja en la provincia de Buenos Aires, Argentina. Neotrop Entomol 31:197–209

Marques GL, Ben JR, Corso IC (1979) Parasitas de *Anticarsia gemmatalis* Hubner, 1818 em Passo Fundo, RS. In: Seminario Nacional de Pesquisa de Soja, 1, 1978, Londrina. Anais…, vol 2. EMBRAPA-CNPSo, Londrina, pp 107–111

Martin PB, Lingren PD, Greene GL, Ridgway RL (1976) Parasitization of two species of Plusiinae and *Heliothis* spp. after releases of *Trichogramma pretiosum* in seven crops. Environ Entomol 5:991–995

Maruya WI, Pinto AS, Gravena S (2001) Parasitoides e *Nomuraea rileyi* (Farlow) Samson em lagartas desfolhadoras (Lepidoptera) na cultura da soja. Bol San Veg Plagas 27:561–567

Mccaffrey JP, Horsburgh RL (1986) Biology of *Orius insidiosus* (Hemiptera: Anthocoridae): a predador in Virginia apple orchards. Envion Entomol 15:984–988

Medeiros MA, Schimidt FVG, Loiácono MS, Carvalho VF, Borges M (1997) Parasitismo e predação em ovos de *Euschistus heros* (Fab.) (Heteroptera: Pentatomidae) no Distrito Federal, Brasil. An Soc Entomol Bras 26:397–401

Medeiros RS, Lemos WP, Ramalho FS (1998a) Efeitos da temperatura no desenvolvimento de *Podisus nigrispinus* (Dallas) (Heteroptera, Pentatomidae), predador do curuquerê-do-algodoeiro (Lepidoptera, Noctuidae). Rev Bras Entomol 42:121–130

Medeiros MA, Loiácono MS, Borges M, Schimidt FVG (1998b) Incidência natural de parasitoides em ovos de percevejos (Hemíptera: Pentatomidae) encontrados na soja no Distrito Federal. Pesq Agrop Brasileira 33:1431–1435

Mendes Neto W (2014) Comparação entre manejo biológico e convencional de pragas na cultura da soja. Monografia, Centro Universitário Moura Lacerda, Ribeirão Preto, 35p

Mendoza AC, da Rocha ACP, Parra JRP (2016) Lyophilized artificial diet for rearing the Neotropical *Euschistus heros* (Hemiptera: Pentatomidae). J Insect Sci 16:1–9

Milanez AM, Pratissoli D, Polanczyk RA, Bueno AF, Tufik CBA (2009) Avaliação de *Trichogramma* spp. para o controle de *Trichoplusia ni*. Pesq Agrop Brasileira 44:1219–1224

Moraes RR, Loeck AE, Belarmino LC (1991) Inimigos naturais de *Rachiplusia nu* (Guenée, 1852) e de *Chrysodeixis includens* (Walker, 1857) (Lepidoptera: Noctuidae) em soja no Rio Grande Do Sul. Pesq Agrop Brasileira 26:57–64

Moscardi F, Bueno AF, Sosa-Gómez DR, Roggia S, Hoffmann-Campo CB, Pomari AF, Corso IC, Yano SAC (2012) Artrópodes que atacam as folhas da soja. In Hoffmann-Campo CB, Corrêa-Ferreira BS, Moscardi F (Eds) Soja: manejo integrado de insetos e outros artrópodes-praga. Brasília, Embrapa, p 213–334, 858p

Murua G, Virla EG (2004) Contribution to the biological knowledge of *Euplectrus platyhypenae* (Hymenoptera: Eulophidae) a parasitoid of *Spodoptera frugiperda* (Lepidoptera: Noctuidae) in Argentina. Folia Entomol Mex 43:171–180

Noyes JS (1988) *Copidosoma truncatellum* (Dalman) and *C. floridanum* (Ashmead) (Hymenoptera, Encyrtidae), two frequently misidentified polyembryonic parasitoids of caterpillars (Lepidoptera). Syst Entomol 13:197–204

Nunes MC, Corrêa-Ferreira BS (2002a) Danos causados à soja por adultos de *Euschistus heros* (Fabricius) (Hemíptera: Pentatomidae), sadios e parasitados por *Hexacladia smithii* Ashmead (Hymenoptera: Encyrtidae). Neotrop Entomol 31:109–113

Nunes MC, Corrêa-Ferreira BS (2002b) Desempenho alimentar e sobrevivência de *Euschistus heros* parasitado por *Hexacladia smithii* em sementes de soja. Pesq Agrop Brasileira 37:1219–1223

Oliva MB (2016) Preferência hospedeira e quantidade liberada de *Trichogramma pretiosum* para ovos de múltiplas espécies de lepidópteros em algodoeiro. Monografia, Centro Universitário Moura Lacerda, Ribeirão Preto, 55p

Orr DB (1988) Scelionid wasps (Wollaston) as biological control agents: a review. Fla Entomol 71:506–528

Pacheco DJP, Corrêa-Ferreira BS (2000) Parasitismo de *Telenomus podisi* Ashmead (Hymenoptera: Scelionidae) m populações de percevejos pragas da soja. An Soc Entomol Bras 29:295–302

Pacheco DJP, Corrêa-Ferreira BS, de Oliveira MCN (1999) Flutuação populacional dos percevejos-pragas da soja e seus parasitóides de ovos em relação à fenologia da soja. Arq Inst Biol 66:1–8

Panizzi AR, Corrêa-Ferreira BS (1997) Dynamics in the insect fauna adaptation to soybean in the tropics. Trends Entomol 1:71–88

Panizzi AR, Smith JG (1976) Observações sobre inimigos naturais de *Piezodorus guildinii* (Westwood, 1837) (Hemíptera: Pentatomidae) em soja. An Soc Entomol Bras 5:11–17

Panizzi AR, Bueno AF, da Silva FAC (2012) Insetos que atacam vagens e grãos. In: Hoffmann-Campo CB, Corrêa-Ferreira BS, Moscardi F (eds) Soja: manejo integrado de insetos e outros artrópodes-praga. Embrapa, Brasília, p 335–420, 858p

Parajulee MN, Shrestha RB, Leser JF, Wester DB, Blanco CA (2006) Evaluation of the functional response of selected arthropod predators on bollworm eggs in the laboratory and effect of temperature on their predation efficiency. Environ Entomol 35:379–386

Parra JRP (1997) Técnicas de criação de Anagasta kuehniella, hospedeiro alternativo para produção de Trichogramma. In: Parra JRP, Zucchi RA (eds) Trichogramma e o controle biológico aplicado. FEALQ, Piracicaba, p 121–150, 324p

Parra JRP (2014) Biological control in Brazil: an overview. Sci Agric 71:345–355

Parra JRP, Zucchi RA, Silveira Neto S (1987) Biological control of pests through egg parasitoids of the genera *Trichogramma* and/or *Trichogrammatoidea*. Mem Inst Oswaldo Cruz 82:153–160

Pasini A (1995) Biologia e técnica de criação do predador *Calosoma granulatum* Perty, 1830 (Coleoptera: Carabidae), em *Anticarsia gemmatalis* Hüebner, 1818 (Lepidoptera: Noctuidae), lagarta da soja. Tese de Doutorado, Piracicaba, ESALQ/USP, 67p

Patel PN, Habib MEM (1998) *Microcharops anticarsiae* Gupta (Hym., Ichneumonidae), an endoparasitoid of the velvetbean caterpillar, *Anticarsia gemmatalis* (Lepid., Noctuidae) and morphology of its immature stages. J Appl Entomol 122:369–375

Pegoraro RA, Foerster LA (1985) Observações sobre o ciclo evolutivo e hábitos alimentares de *Calosoma granulatum* Perty, 1830 (Coleoptera: Carabidae) em laboratório. An Soc Entomol Bras 14:269–275

Pinto AS, Parra JRP, de Oliveira HN (2008) Guia de campo de pragas e insetos benéficos da soja. CP 2, Piracicaba, 64p

Polanczyk RA, Pinto AS, Rossi MM, Esquesário LC (2006) Controle biológico de pragas da soja. In: Pinto AS, Nava DE, Rossi MM, Malerbo-Souza DT (Orgs) Controle biológico de pragas: na prática. CP 2, Piracicaba, p 83–96, 287p

Pomari-Fernandes A, Bueno AF, De Bortoli SA, Menezes Junior AO (2014) *Telenomus remus* no controle do gênero *Spodoptera*: do laboratório ao campo. Cad Agroecol 9:1–5

Prado PCN, Cunha HF, Silva AL (1982) Ocorrência dos principais insetos-pragas da soja e seus inimigos naturais em Santa Helena de Goiás. Pesqui Agropecu Trop 12:31–44

Puttler B, Gordh G, Long SH (1980) Bionomics of *Euplectrus plutteri*, new species, an introduced parasite of the velvetbean caterpillar, *Anticarsia gemmatalis*, from South America. Ann Entomol Soc Am 73:28–35

Querino RB, Zucchi RA (2012) Guia de identificação de *Trichogramma* para o Brasil, 2nd edn. Embrapa, Brasília, 103p

Rasplus JY, Pluot-Sigwalt D, Llosa JF, Couturier G (1990) *Hexacladia linci*, n.sp. (Hymenoptera: Encyrtidae) endoparasite de *Lincus malevolus* Rolston (Heteroptera: Pentatomidae). Ann Entomol Soc Fr 26:255–263

Ribeiro RC, Lemos WP, Bernardino AS, Buecke J, Müller AA (2010) Primeira ocorrência de *Alcaeorrhynchus grandis* (Dallas) (Hemíptera: Pentatomidae) predando lagartas desfolhadoras do dendezeiro no estado do Pará. Neotrop Entomol 39:131–132

Richman DB, Hemenway RC Jr, Whitcomb WH (1980) Field cage evaluation of predators of the soybean looper, *Chrysodeixis includens* (Lepidoptera: Noctuidae). Envion Entomol 9:315–317

Rocha ACPD (2016) Dinâmica de interações tróficas de *Euschistus heros* (Hemiptera: Pentatomidae) com *Telenomus podisi* e *Trissolcus basalis* (Hymeniptera: Platygastridae). Dissertação de Mestrado, ESALQ/USP, Piracicaba, 105p

Rutledge CE, O'Neil RJ, Fox TB, Landis DA (2004) Soybean aphid predators and their use in integrated pest management. Ann Entomol Soc Am 97:240–248

Saini E (1994) Aspectos morfologicos y biologicos de *Podisus connexivus* Bergroth (Heteroptera: Pentatomidae). Rev Soc Entomol Argent 53:35–42

Saini E, Quintana G, Rios M (1997) Interacción entre el depredator *Podisus connexivus* (Heteroptera: Pentatomidae) y el virus de la poliedrosis nuclear de *Rachiplusia nu* (Lepidoptera: Noctuidae) en condiciones de laboratorio. Rev Soc Entomol Argent 56:105–108

Sanchez MC, Borges B, Grazia J (2001) *Tynacantha marginata* Dallas (Heteroptera: Pentatomidae), predator of green stink bug (*Nezara viridula* L.) in Venezuela. Entomotropica 16:213–214

Silveira LCP, Bueno VHP, Pierre LSR (2003) Plantas cultivadas e invasoras como habitat para predadores do gênero *Orius* (Wolff) (Heteroptera: Anthocoridae). Bragantia 62:261–265

Siqueira JR, Bueno RCOF, Bueno AF, Vieira SS (2012) Preferência hospedeira do parasitoide de ovos *Trichogramma pretiosum*. Ciênc Rural 42:1–5

Sosa-Gómez DR, Omoto C (2012) Resistência a inseticidas e outros agentes de controle em artrópodes associados à cultura da soja. In: Hoffmann-Campo CB, Corrêa-Fereira BS, Moscardi F (Orgs) Soja: Manejo integrado de insetos e outros artrópodes-praga, 1st edn. Embrapa, Londrina, p 673–723, 859p

Strand MR (1989) Development of polyembryonic parasitoid *Copidosoma floridanum* in *Trichoplusia ni*. Entomol Exp Appl 50:37–46

Sujii ER, Costa MLM, Pires CSS, Colazza S, Borges M (2002) Inter- and intra-guild interactions in egg parasitoid species of the soybean stink bug complex. Pesq Agrop Brasileira 37:1541–1549

Tamaki G, Weeks RE (1972) Efficiency of three predators, *Geocoris bullatus*, *Nabis americoferus* and *Cocinella transversogutata*, used alone or in combination against three insect prey species, *Myzus persicae*, *Ceramica picta*, and *Mamestra configurata*, in a greenhouse study. Environ Entomol 1:258–263

Thomas DB (1992) Taxonomic synopsis of the Asopinae Pentatomidae (Heteroptera) of the western hemisphere. Entomological Society of America, Lanham, 156p

Tibaldi GR (2017) Controle de ovos em diferentes infestações de Euschistus heros com liberações de diferentes quantidades de Telenomus podisi na soja. Monografia, Centro Universitário Moura Lacerda, Ribeirão Preto, 29p

USDA – United States Department of Agriculture (2019) World agriculture production. Circular series 3, 32p

Vasconcelos Y (2012) Inseto contra inseto. Rev Pesqui FAPESP 195:68–73

Waddill VV, Shepard M (1974) Potential of *Geocoris punctipes* [Hemiptera: Lygaeidae] and *Nabis* spp. (Hemiptera: Nabidae) as predators of *Epilachna varivestis* [Coleoptera: Coccinellidae]. Entomophaga 9:421–426

Wall R, Berberet RC (1974) The life cycle of *Euplectrus platyhypenae*, a gregarious external parasitoid of Peanut Foliage Feeders in Oklahoma. Environ Entomol 3:744–746

Zachrisson BSA (1997) Bioecologia de *Trichogramma pretiosum* Riley, 1879, para o controle de *Anticarsia gemmatalis* Hübner, 1818, na cultura da soja. Tese de Doutorado, Piracicaba, ESALQ/USP, 106p

Zanuncio JC, Torres JB, Bernardo DL, De Clercq P (1997) Effects of prey switching on nymphal development of four species of stinkbugs. Meded Fac Landbouwwet Univ Gent 62:483–490

Zucoloto GC (2017) Impacto de acefato e clorantraniliprole sobre organismos de superfície de solo e *Telenomus podisi* em soja. Monografia, Centro Universitário Moura Lacerda, Ribeirão Preto, 60p

Chapter 33
Sugarcane

Alexandre de Sene Pinto and Sóstenes Eduardo Leal Trujillo

33.1 Introduction

Brazil is the biggest sugarcane producer in South America. It has produced 633.26 million tons of harvest in 2017/2018 from its 10.24 million hectares of planted areas (8.73 million of harvested hectares) (Conab 2018).

An arthropodofauna associated to the sugarcane culture is quite complex. Due to the extensive and continual cultivated areas in several regions, especially in São Paulo state, this agroecosystem becomes convenient to the attack of different pests (Guagliumi 1972/1973; Mendonça 1996b). Some of them stand out because of the frequency in which they occur and also for the damages they cause, as in the sugarcane borer *Diatraea saccharalis* (F.) (Crambidae), considered a key pest to the culture, and the rootshopper, *Mahanarva fimbriolata* (Stål) (Cercopidae), a problem in many Brazilian states (Mendonça et al. 2005; Pinto et al. 2006a, b, 2009; Dinardo-Miranda 2008).

Diatraea saccharalis is not the only species registered in Brazil, since *Diatraea flavipennella* (Box) also occurs in the northeast region (Mendonça et al. 1996a). However, more recently, Solis and Metz (2016) reported the occurrence of 15 species from this genus in several plant species in Brazil and found that *D. flavipennella* was a junior synonym of *Diatraea impersonatella* (Walker) (Francischini et al. 2017). Besides the taxonomic issues related to the genus *Diatraea*, it was discovered that *M. fimbriolata* is actually formed by a complex of species, prevailing *M. spectabilis* (Distant) (and not *M. fimbriolata*) also occurring *M. liturata* (Le Peletier & Serville) (Alves and Carvalho 2014) and other unidentified species.

A. de Sene Pinto (✉)
Centro Universitário Moura Lacerda, Ribeirão Preto, SP, Brazil
e-mail: alexandre@occasio.com.br

S. E. Leal Trujillo
Department of Entomology, Lavras, MG, Brazil

B. Souza et al. (eds.), *Natural Enemies of Insect Pests in Neotropical Agroecosystems*, https://doi.org/10.1007/978-3-030-24733-1_33

Other pests affect the sugarcane production, as the lepidopterans *Telchin licus licus* (Drury) (Castniidae), predominant in the north and northeast of the country (Mendonça et al. 1996b). Individuals of *T. licus licus* were found in 1983, in Minas Gerais (Planalsucar 1983), and in 2007, in sugarcane plantation around the city Limeira, SP (Almeida et al. 2007; Pinto 2008; Pinto et al. 2009), besides *Elasmopalpus lignosellus* (Zeller) (Pyralidae) (Viana 2004) and *Hyponeuma taltula* (Schaus) (Noctuidae) (Guagliumi 1973; Guagliumi and Mendes 1974; Pinto et al. 2009). It is also observed the occurrence of the coleopterans *Migdolus fryanus* (Westwood) (Cerambycidae) (Bento et al. 1995), *Sphenophorus levis* Vaurie (Precetti and Arrigoni 1990) (which is certainly a complex of species), *Metamasius hemipterus* (L.) (Curculionidae) (Zorzenon et al. 2000), scarabeids, and *Conoderus* (Elateridae) (Mendonça 1996b; Pinto et al. 2009; Moraes and Ávila 2014).

Among the hemipterans, it was registered *Scaptocoris* spp. (Cydnidae) (Fernandes et al. 2004), aphids, and mealybugs (Pinto et al. 2009). Mites can also infest sugarcane plantations (Moraes and Flechtmann 2008), as *Oligonychus grypus* (Almeida et al. 2016).

In Argentina, the species of pests are similar from the ones in Brazil (Salvatore et al. 2009). In Guatemala, there are other leafhoppers that attack the sugarcane culture, as *Aeneolamia postica* (Walker) and *Prosapia simulans* (Walker) (Cercopidae), and, besides *D. saccharalis*, other borer species cause damages, as *Diatraea crambidoides* (Grote), which surpass *D. saccharalis* in occurrence, and *Diatraea magnifactella* (Dyar) and *Eoreuma loftini* (Dyar) (Crambidae) (del Bosque et al. 2012), both also common in the south of Mexico. There is *Saccharosydne saccharivora* (Westwood) (Delphacidae) as well, which attacks the aboveground parts of the plant, besides a complex of soil pests, as scarabeids [(*Phyllophaga* spp., *Podischnus agenor* (Olivier), *Euetheola bidentata* (Burmeister) (Scarabaeidae)], brown stink bug [*Scaptocoris* spp. (Cydnidae)], larvae from *Dipropus* spp., *Horistonotus* spp., *Agrypnus* spp., *Dilobitarsus* spp., and others (CENGICAÑA 2017).

This chapter will discuss the biological control of the main sugarcane pests, highlighting the natural enemies' preservation, habitat management, and applied biological control, especially in Brazil.

33.2 Sugarcane Beneficial Entomofauna and Habitat Management

A wide diversity of natural enemies of *Diatraea* spp. has been reported. In Colombia, Bustillo (2011) mentions the dipteran parasitoids *Billaea claripalpis* (van der Wulp) and *Lydella minense* (Townsend) (Tachinidae); the hymenopterans *Trichogramma exiguum* (Pinto and Platner) (Trichogrammatidae), *Telenomus alecto* (Crawford) (Platygastridae), *Apanteles diatraeae* (Muesebeck), *Cotesia flavipes* (Cameron), and *Agathis stigmateus* (Cresson) (Braconidae); the wasps *Polistes* sp. and *Pepsis* sp. (Vespidae); and the beetle from the family Coccinellidae *Coleomegilla maculata* (De Geer), also common in Brazil (Prezoto et al. 1994; Giannotti et al. 1995; Mendonça 1996c; Locher 2012; Silva 2012a, b). Vargas et al. (2011) reported the parasitism of the larvae and pupae of *Diatraea* sp. by *Tetrastichus howardii* (Olliff) (Eulophidae) in Brazilian plantations.

Several studies reported a vast diversity of natural enemies associated to sugarcane leafhoppers. Bustillo (2011) highlights the nymph predator *Salpingogaster nigra* (Schiner) (Syrphidae) and the eggs predator ants *Pheidole* sp., *Solenopsis* sp., *Paratrechina* sp., and *Camponotus blandus* (Smith) (Formicidae). Two genera of Mymaridae, mentioned to Brazil, include cercopid egg parasitoid species, as *Anagrus* sp. (Reis et al. 1983; Pires et al. 1993) and *Acmopolynema hervali* (Gomes) (Gomes 1948; Marques and Vilas Boas 1985). Triapitsyn (2002) referred to *Anagrus urichi* Pickles (Mymaridae) as an important Neotropical species to natural biological control of many cercopid species and redescribed the species based on individuals obtained from eggs of the leafhopper *Deois flavopicta* (Stål) coming from Brazil and Guiana.

The vegetation corridors attract, conserve, and distribute natural enemies to the productive agroecosystem (Gómez Mendez 2017; Peres et al. 2009). In the nearest areas to the biological corridors, there is higher diversity of natural enemies, which reduces followed by the increase of the distance between environmental refuges (Gómez Mendez 2007, 2017; Peres et al. 2009; Haro et al. 2018). The families Diapriidae, Scelionidae, Ceraphronidae, Platygastridae, Braconidae, Figitidae, Encyrtidae, Mymaridae, Ichneumonidae, and Chalcididae (Hymenoptera) were the most frequently found near the biological corridors, many of them related to the biological control of some important sugarcane pests (Gómez Mendez 2017; Márquez et al. 2017). It is possible the parasitoids movement among the sugarcane plantation and the biological corridors composed by trees, since the majority of the collected individuals from both systems was allocated to the same morphospecific groups, with a decreasing distribution pattern regard to the distance of the diversification area (Gómez Mendez 2017). However, the spatial and temporal distribution of the beneficial entomofauna associated with the sugarcane plantation has the Diptera order as the most abundant, followed by the Hymenoptera (Márquez et al. 2017).

The maintenance of the adjacent cultures is an important conservation strategy of natural enemy species (Altieri et al. 2003). The sugarcane plantation supports the maintenance of some of these species, but disfavors others. The wasps nesting (Vespidae) inside the plantation, as well as their prey foraging, indicate the establishment of these natural enemies in the culture. *Polistes simillimus* Zikán and *Polistes lanio lanio* (F.) (Vespidae) were observed foraging (Prezoto et al. 1994; Giannotti et al. 1995; Locher 2012; Silva 2012a, b) and nesting (Silva 2012a, b) in these environments. In Juiz de Fora city, MG, Brazil, 20 wasp species (nine predators) built their nests in sugarcane plantations, which *Agelaia vicina* (Saussure) was the most abundant and *Mischocyttarus drewseni* Saussure and *P. simillimus* qualified as constant (Silva 2012a, b).

On the other hand, some investigations suggest that the wasps only use the sugarcane plantations to forage, while the near woods are used as places to their nest building, evidencing the importance of maintaining these areas as refuge. Locher (2012) registered 11 species of common wasps in the plantations and near woods, mentioning *Polybia ignobilis* (Haliday), *Agelaia pallipes* (Olivier), *A. vicina*, *Polybia* (*gr. occidentalis*) sp., and *M. drewseni* as constant in both environments; however, the nesting was observed only in the woods, whereas the sugarcane plantation is used as foraging area. Tanaka Junior and Noll (2011) reported 25 wasp species in wood areas close to the plantations, considering that *Brachygastra moebiana* (Saussure), *Metapolybia docilis* Richards, *Mischocyttarus ignotus* Zikán, *Mischocyttarus paulistanus* Zikán, and *Mischocyttarus consimilis* Zikán have not been reported in these environments yet.

Predator ants are considered the most important natural enemies and biological indicators of environmental imbalances (Fittkau and Klinge 1973; Silva and Brandão 1999) in sugarcane plantations. Although the burnings of these plantations are still used to clean the area in order to facilitate the harvesting, this tactic has been rapidly decreased in Brazil since 1990. The fire negatively affects the ant's community, which generally recovers quite fast (Araújo et al. 2005; Santos et al. 2012; Benazzi et al. 2013). Other tactics, as herbicides, fertilizers (Fernandes et al. 2000; Santos et al. 2012), and insecticides applications (Masson et al. 2012), also negatively affect some ant populations found in these plantations, and the recovery could not be so fast (Masson et al. 2012). On the other hand, the straw kept in the sugarcane plantation soil increases the diversity and support the occurrence of soil organisms, especially the predators (Benazzi et al. 2013).

The diversity of ant species in sugarcane plantations is inferior to those observed in near woods (Santos et al. 2012). In Pernambuco, from 47 ant species found in these plantations (and 111 species in adjacent Atlantic rainforest), *Pheidole radoszkowskii* Mayr was the most frequent, abundant, and constant; *Paratrechina* sp. was constant; and *Crematogaster victima* Mayr, *Labidus* sp., and *Crematogaster crinosa* Mayr (Formicidae) were quite abundant. With regard to the epigeic species, even though the number of ants had been six times higher in the sugarcane plantations than in the woods, the culture and its tactics negatively influenced the composition, diversity, and number of individuals per species (Santos et al. 2012).

Based on this information, it is possible to conclude that the natural enemies conservation in sugarcane plantation areas is a strategy which must be considered (Vargas et al. 2006; Altieri and Nicholls 2009; Haro et al. 2018; Souza et al. 2018), managing the habitat in order to preserve, attract, and even increase their populations. The association of some plant species with the sugarcane can be positive to the pest management in this culture. In Colombia, for example, it is known that floral nectaries of *Bidens pilosa* L., *Acmella oppositifolia* (Lamarck) (Asteraceae), *Mangifera indica* L. (Anacardiaceae), and *Phylanodiflora* (=*Lippianodiflora*) L. Michx. (Verbenaceae) are used as food source to *Genea jaynesi* (Aldr.) (Tachinidae), a parasitoid of *Diatraea* spp. (Vargas et al. 2006), and the maintenance or even planting of such species close to the sugarcane plantations must favor the control of this pest, since the agricultural tactics negatively affect the diversity of plant species (Alba et al. 2015). *Crotalaria juncea* L., *Crotalaria spectabilis* Roth., *Canavalia ensiformis* (L.) (Fabaceae), *Zea mays* L. (Poaceae), *Tagetes erecta* L. (Asteraceae), and *Cucurbita moschata* Duch. (Cucurbitaceae) also benefit the *Diatraea* spp. control when intercalated with sugarcane (Márquez 2017).

The unproductive and the conventional areas of sugarcane culture are not considered biodiversity factors but actually banks of weeds (Vargas et al. 2006; Gómez Mendez 2017). It is important, in the plantation, to favor the presence of arboreal plants that are not competitors to the sugarcane, which flower and produce nectar to support beneficial insects (Bustillo Pardey 2009; Jamont et al. 2014). However, the survival of such plants can be affected by drought periods and by intense management of sugarcane, as the application of herbicides (Altieri and Nicholls 2009; Bustillo Pardey 2009).

Even though the biodiversity management in sugarcane plantations is an important element in the populational regulation of pests, it is important to consider that the associated vegetation can become a host for the pests, and, therefore, the control of these areas must be carefully monitored. For instance, populations of leafhoppers *Aeneolamia varia* (F.) (Cercopidae) can establish themselves on weeds as *Brachiaria* sp., *Cynodon nlemfuensis* Vanderyst, and *Rottboellia* sp. (Poaceae) (Bustillo Pardey 2013; Vargas and Gutiérrez 2017).

33.3 Applied Biological Control of Sugarcane Pests in Brazil

The biological control of sugarcane pests is accomplished in more than 50% of cultivating areas from Brazil (Parra 2014). The sugarcane borer, *D. saccharalis*, is mainly managed using the parasitoid *C. flavipes*, in more than 3 million hectares (Parra et al. 2010; Cherubin 2018), and *Trichogramma galloi* Zucchi (Trichogrammatidae), in about 2 million hectares all around the country (Cherubin 2018). The rootshoppers of the genus *Mahanarva* were controlled with the fungus *Metarhizium anisopliae* (Metsch.) Sorok. in 2 million hectares until 2006, when the insecticide Ethiprole appeared, with high residual power, which resulted in a decrease of areas with biological control application. With the improvement of the fungus application technology, in 2016, the grower has returned to the use of biological insecticide in about 3 million hectares (Cherubin 2018).

The biological control of the sugarcane borer in Brazil started in the 1950s, with the importation of the parasitoid fly *Lixophaga diatraeae* (Townsend) (Tachinidae) from Cuba. These insects were reared in laboratory conditions and then released, but they did not achieve the expected success, and only adapted to the Amapá state (Mendonça 1996a). Afterwards, the native flies *L. minense* and *B. claripalpis* started to be reared and released in sugarcane plantations, even though the parasitism rates accomplished were low in almost all country, except in São Paulo state (Botelho and Macedo 2002). Furthermore, the rearing system was quite onerous, but, even so, such flies were multiplied in laboratory for many years and released in sugarcane plantations. Later, these flies were replaced by *C. flavipes*, imported from

Trinidad-Tobago, India, and Pakistan (Botelho et al. 1980; Mendonça 1996a; Botelho and Macedo 2002) in the 1970s. Until the last years, this parasitoid has been the most used in the sugarcane borer control in Brazil (Parra 2014; Cherubin 2018).

Cotesia flavipes is reared in laboratory almost in the same way as it was reared 50 years ago, with few advances (Cano et al. 2006). With regard to the rearing methodology, scarce investigations were concluded in the last years, as the substitution of formol by a combination of less dangerous substances (thiocyanato-bis-methylene and 2-thiocyanomethyl-thiobenzothiazole 10%) to the sugarcane borer rearing (Silva 2012a, b) and a definition of the best diet to rear *D. saccharalis* and *C. flavipes* (Lima 2011). As the increasing of researches with microorganisms that affect the rearing of insects in laboratory, it was discovered that *Nosema* (Nosematidae), regularly found in *D. saccharalis* rearing all around Brazil, negatively affects the biology and the caterpillar searching capacity of the parasitoids (Simões et al. 2012). This results in the reduction of *C. flavipes* efficiency in the control of *D. saccharalis* in different regions of Brazil, causing a dissatisfaction in the growers. Recently, some recommendations were proposed in order to disinfect the sugarcane borer creations, which must be considered by companies and mills that produce this braconid.

At the same way, the release technology has not considerably changed since it was implanted in the country in 2010 (Botelho et al. 1980; Botelho and Macedo 2002). Despite the use of *C. flavipes* being established in Latin America, such technology barely follows the precepts defined by Pinto and Parra (2002), which guarantee the success of a biological control program.

In the last years, the release technology of *C. flavipes* changed a lot. The decision of the release moment depends on the sampling methods of the pest that needs to be precise and simple to guarantee the success of the control. Until the 2010s, *D. saccharalis* caterpillars sampling was made in two points of 10 linear meters (5 m in two parallel lines) per hectare. Carvalho (2011) determined that two parallel lines of the sampling were necessary to a reliable evaluation, regardless the size of such lines, and Dinardo-Miranda et al. (2011) defined a new methodology based on that information. Thus, Cerci et al. (2013) concluded that population samplings in two points of 2 linear meters (1 meter in two parallel lines) per hectare are enough to the periodic survey.

The release of *C. flavipes* in the sugarcane plantations, which was manually made in four points per hectare, changed to eight points with 750 parasitoids in each recipient of release (Ivan 2012). The number of parasitoids released changed from 6000 individuals to 9000, when low caterpillar infestations (about 1000 per hectare) were registered (Ivan et al. 2013). More recently, *C. flavipes* adults retained in capsules are mechanically released by drones in 25 m lines. Another method developed in order to release *C. flavipes* is based on the liberation of a group of pupae (mass) via drone, without the capsule's protection. The pupae are treated with repellents (Mateus Junior 2017) avoiding the predation (Rodrigues et al. 2014), without any effect on the emergency of the parasitoid (Vinha et al. 2014).

When it is opted by the control of sugarcane borer only by *C. flavipes*, it is necessary at least two releases of parasitoid during the harvest. As the use of such insect generally does not decrease the infestation less than 3% (intensity rate of infestation), there is a current tendency of using *C. flavipes* only in sugarcane plantation borders, especially those well developed and the ones which are waiting for the harvest.

The slow reduction of the pest population by *C. flavipes* facilitated the entrance of another parasitoid to control the eggs of *D. saccharalis*. Studied since the 1980s in Brazil, *T. galloi* is preferred due to promoting the populational reduction of the pest still in the egg stage (Botelho 1985), which reduces the damage caused by the entrance of the caterpillar in the plant stem (Pinto et al. 2013). With the increase of *T. galloi* use, the monitoring of *D. saccharalis* changed to traps of virgin female pheromone to the attraction of males and then the definition of the ideal moment of the parasitoid release (Conti 2012). This methodology is so efficient and inexpensive that became common in a short period in about half of the sugarcane plantation area (about 5 million hectares), defining not only the moment of *T. galloi* and *C. flavipes* release but also the chemical insecticide application (Pinto et al. 2006a, b, 2018).

Since the beginning of the 2010s, *T. galloi* has been mechanically and aerially released in Brazil, initially by airplanes, but since 2017 drones have been used. The release is predominantly made with unprotected pupae and spread in the agricultural area, a technique formerly defined by Mills et al. (2000). The predation of such minute pupae is minimum, since only five of them are released per square meter (50,000 per hectare). The efficiency is the same as found in the manual release (Arroyo et al. 2014). Guatemala and Colombia have also been using this technology since 2017. To the sugarcane borer control, some areas require only a series of three *T. galloi* releases; the majority requires two series and irrigated areas, with vinasse application, with susceptible varieties, etc., reaching three series of release.

With regard to the control of other sugarcane pests, there are few advances. In the rootshopper's management, *Mahanarva* spp., since the 1970s it has been used the fungus *M. anisopliae* (Marques et al. 1981; Alves et al. 2008). When the option is to perform the control of *Mahanarva* spp. only with the fungus, at least two applications are necessary. Currently, it has been used a chemical insecticide and half dose of fungus in the same application to control this pest. This subject is discussed in Chap. 20 of this book.

The control of *S. levis* is predominantly performed by using chemicals, but in 2018 the fungus *M. anisopliae* started to be used in larvae and *Beauveria bassiana* (Bals.) Vuill. to the pupae and adults (Delfanti 2012; Salvador Neto 2012; Litholdo et al. 2013), applied with implements that cut the socks of sugarcane and inject the product 15 cm depth in the soil (or apply 30% on the soil and 70% in it). The fungus *B. bassiana* has been applied on the soil in the mating period and when the adults get exposed. *B. bassiana* is quite efficient in the sugarcane giant borer control, *T. licus licus* (Mendonça et al. 1996b), especially when applied via liquid just after the cut, with pulverizer attached to the harvester (Pinto et al. 2013).

The biological control of pests is more inexpensive than the chemical in the most cited cases, besides not causing environmental impacts, not contaminating the food, and not being related to resistance. Nowadays, the sugarcane borer control using *C. flavipes* sums about US$7.00 and *T. galloi* about US$17.00 per application (one of the first and three of the second), including the price of the product, transport, and application. The control of *Mahanarva* spp. exclusively with *M. anisopliae* sums about US$23.00, but if it is applied with a chemical insecticide, this value decreases to half of the price.

In sugarcane plantations, biological control has been used more and more as the first option in the control of the most important pests, reserving the chemical insecticides only to the areas where the control was not properly made or where the right moment of a bioproduct application was lost.

33.4 Final Remarks

The habitat management of sugarcane plantations and the biodiversity as part of the pest management have been barely researched to Central and South America regions. This is due to the studies being based in conventional management, since the sugarcane is planted as monoculture. In the pest management, the natural enemies' conservation must be considered as a strategy, encouraging the establishment and the proliferation of them in the agroecosystems.

On the other hand, the growers from South America have been using isolated biological control tactics or associated to chemical insecticides in order to control the main pests. In Brazil, in more than half of the sugarcane plantation area, the biological control is being used with the systematic releases of *C. flavipes* and *T. galloi* parasitoids, and the fungus *B. bassiana* and *M. anisopliae* to other pests, with costs mostly inferior to the chemical control.

References

Alba R, Jiménez R, Castellanos L (2015) Relación de los biorreguladores de plagas con la diversidad vegetal y las prácticas agroecológicas. Rev Agroecosistemas 3:435–445

Almeida LCD, Dias Filho MM, Arrigoni EDB (2007) Primeira ocorrência de *Telchin licus* (Drury, 1773), a broca gigante da cana-de-açúcar, no Estado de São Paulo. Rev Agric 82:223–226

Almeida LCD, Moraes GJD, Chabregas SM, Arrigoni EDB (2016) Primeiro relato da ocorrência do ácaro *Oligonychus grypus* em cana-de-açúcar no Estado de São Paulo. Rev Agric 84:173–174

Altieri M, Nicholls C (2009) Biodiversidad y manejo de plagas en agroecosistemas. Icaria Editorial, Barcelona, 247p

Altieri MA, Silva NE, Nicholls CI (2003) O papel da biodiversidade no manejo de pragas. Ribeirão Preto, Holos, 226p

Alves RT, Carvalho GS (2014) Primeiro registro das espécies de cigarrinhas-da-raiz da cana-de-açúcar *Mahanarva spectabilis* (Distant) e *Mahanarva liturata* (Le Peletier & Serville) atacando canaviais na região de Goianésia (GO), Brasil. Arq Inst Biol 81:83–85

Alves SB, Lopes RB, Vieira SA, Tamai MA (2008) Fungos entomopatogênicos usados no controle de pragas na América Latina. In: Alves SB, Lopes RB (eds) Controle microbiano de pragas na América Latina. FEALQ, Piracicaba, pp 69–110, 414p

Araújo RA, Araújo MS, Goring AHR, Guedes RNC (2005) Impacto da queima controlada da palhada da cana-de-açúcar sobre a comunidade de insetos locais. Neotrop Entomol 34:649–658

Arroyo BM et al (2014) Liberação aérea de *Trichogramma galloi*, pela técnica de espalhamento de pupas desprotegidas, no controle de ovos de *Diatraea saccharalis* em cana-de-açúcar. In: Congresso Brasileiro de Entomologia, 25, Goiânia. Resumos... Goiânia: Embrapa Arroz e Feijão/UFG

Benazzi ES, Bianchi MO, Correira MEF, Lima E, Zonta E (2013) Impactos dos métodos de colheita da cana-de-açúcar sobre a macrofauna do solo em área de produção no Espírito Santo – Brasil. Semin Ciênc Agrár 34:3425–3442

Bento JMS, Vilela EF, Della Lúcia TMC, Leal WS, Novaretti WRT (1995) *Migdolus*: biologia, comportamento e controle. [s.n.], Salvador, 58p

Botelho PSM (1985) Tabela de vida ecológica e simulação da fase larval de *Diatraea saccharalis* (Fabricius 1794) (Lepidoptera: Pyralidae). Tese (Doutorado em Entomologia), Esalq/USP, Piracicaba, 110f

Botelho PSM, Macedo N (2002) *Cotesia flavipes* para o controle de *Diatraea saccharalis*. In: Parra JRP, Botelho PSM, Corrêa-Ferreira BS, Bento JMS (eds) Controle biológico no Brasil: parasitoides e predadores. Manole, São Paulo, pp 409–425, 609p

Botelho PSM, Macedo N, Mendes AC, Silveira Neto S (1980) Aspects of the population dynamics of *Apanteles flavipes* (Cameron) and support capacity of its host *Diatraea saccharalis* (Fabr.). In: Congress of the International Society of Sugar Cane Technologists, 18, Manila, Proceedings, Manila

Bustillo Pardey A (2009) Acciones para reducir las poblaciones de *Diatraea*. Cenicaña 31:10–16

Bustillo Pardey A (2011) Parasitoides, Predadores y Entomopatogenos que afecta la caña de azúcar en Colombia. Carta Trimestral, Cenicaña 3-4, 11p.

Bustillo Pardey A (2013) Insectos plaga y organismos benéficos del cultivo de la caña de azúcar en Colombia. Cenicaña, Cali, 164p

Cano MAV, Santos EM, Pinto ADS (2006) Produção de *Cotesia flavipes* para o controle da broca-da-cana. In: Pinto ADS (Org) Controle de pragas da cana-de-açúcar. Biocontrol, Sertãozinho, pp 21–24, 64p

Carvalho JDS (2011) Distribuição espacial e amostragem sequencial de *Diatraea saccharalis* (Fabricius, 1794) (Lepidoptera: Crambidae), na cultura de cana-de-açúcar. Tese (Doutorado em Agronomia), FCAV/Unesp, Jaboticabal, 61f

CENGICAÑA (2017) Guía de Buenas Prácticas Agrícolas en Caña de Azúcar. 84p. Disponível em: https://cengicana.org/. Acesso em: 10 out 2017

Cerci JVF et al (2013) Comparação entre métodos de amostragem de *Diatraea saccharalis* em cana-de-açúcar. In: Congresso Brasileiro de Fitossanidade, 2, Jaboticabal, Resumo expandido, Jaboticabal

Cherubin N (2018) Cana-de-açúcar: referência quando o assunto é controle biológico. RPA News 202:6–14

Conab – Companhia Nacional de Abastecimento (2018) Acompanhamento da safra brasileira de cana-de-açúcar (vol 5), Brasília: Conab, 71p

Conti BFD (2012) Exigências térmicas e monitoramento de *Diatraea saccharalis* (Fabricius, 1794) (Lepidoptera: Crambidae), como suporte ao seu controle biológico. Tese (Doutorado em Entomologia), ESALQ/USP, Piracicaba, 120f

del Bosque LAR, Pérez RL, Vázquez HM, Valdez JA (2012) Manejo integrado de barrenadores en caña de azúcar en el sur de Tamaulipas. Instituto Nacional de Investigaciones Forestales, Agrícolas y Pecuarias, México, 18p

Delfanti LADA (2012) Eficácia do fungo *Metarhizium anisopliae* no controle de *Sphenophorus levis* e efeito em outras pragas de solo em cana-de-açúcar. Monografia (Trabalho de Conclusão de Curso de Agronomia), Centro Universitário Moura Lacerda, Ribeirão Preto, 30f

Dinardo-Miranda LL (2008) Pragas. In Dinardo-Miranda LL, de Vasconcelos ACM, Landell MGDA (eds) Cana-de-açúcar. IAC, Campinas, pp 349–422, 882p

Dinardo-Miranda LL, Fracasso JV, Perecin D (2011) Variabilidade espacial de populações de *Diatraea saccharalis* em canaviais e sugestão de método de amostragem. Bragantia 70:577–585

Fernandes WD, Cruz MA, Faccenda O, Valente TO (2000) Impacto de herbicidas em uma guilda de formigas predadoras. Rev Bras Herbic 1:225–231

Fernandes PM, Oliveira LJ, Sousa CR, Czepak C, Barros RG (2004) Percevejos castanhos. In: Salvadori JR, Ávila CJ, Silva MTB (eds) Pragas de solo no Brasil. Embrapa Trigo/Embrapa Agropecuária Oeste/Fundacep Fecotrigo, Passo Fundo/Dourados/Cruz Alta, pp 477–493, 541p

Fittkau EJ, Klinge H (1973) On biomass and trophic structure of Central Amazonian rainforest ecosystems. Biotropica 5:2–14

Francischini FJB, de Campos JB, Alves-Pereira A, Gomes Viana JP, Grinter CC, Clough SJ, Zucchi MI (2017) Morphological and molecular characterization of Brazilian populations of *Diatraea saccharalis* (Fabricius, 1794) (Lepidoptera: Crambidae) and the evolutionary relationship among species of *Diatraea* Guilding. PLoS One 12:e0186266. https://doi.org/10.1371/journal.pone.0186266

Giannotti E, Prezoto F, Machado VLL (1995) Foraging activity of *Polistes lanio lanio* (Fabr.) (Hymenoptera, Vespidae). An Soc Entomol Bras 24:455–463

Gomes JG (1948) *Acmopolynema hervali* n. sp., parasito de ovos de *Tomaspis liturata* (Chalcidoidea, Mymaridae). Rev Bras Biol 8:417–420

Gómez Mendez HA (2007) Influência de corredor de vegetação na riqueza e abundância de Scarabaeinae (Insecta: Coleoptera) e de parasitóides (Insecta: Hymenoptera) em um agroecossistema de cafeeiro. Dissertação (Mestrado em Entomologia), Universidade Federal de Lavras, Lavras, 37p

Gómez Mendez HA (2017) Vegetação de entorno e sua influência sobre insetos praga e parasitoides em cultivos de cana e milho na Guatemala. Tese (Doutorado em Entomologia), Universidade Federal de Lavras, Lavras, 59p

Guagliumi P (1972/1973) Pragas da cana-de-açúcar. Nordeste do Brasil, MIC/IAA, 622p. (Coleção Canavieira 10)

Guagliumi P (1973) Situação das cigarrinhas e das brocas nos canaviais do estado de Santa Catarina, e descobrimento de uma nova praga da cana. Bras Açucar 81:10–13

Guagliumi P, Mendes AC (1974) Três novas pragas da cana-de-açúcar no Estado de São Paulo. Bras Açucar 82:80–81

Haro MM, Silveira LCP, Wilby A (2018) Stability lies in flowers: plant diversification mediating shifts in arthropod food webs. PLoS One 13:e0193045

Ivan EAF (2012) Uso de *Cotesia flavipes* (Cam.) (Hymenoptera: Braconidae) no controle de *Diatraea saccharalis* (F.) (Lepidoptera: Crambidae) em cana-de-açúcar. Monografia (Trabalho de Conclusão de Curso de Agronomia), Centro Universitário Moura Lacerda, Ribeirão Preto, 56f

Ivan EAF, Dias TAM, Vasconcelos GDR, Pinto ADS (2013) Quantidade liberada de *Cotesia flavipes* para o controle de baixa infestação de *Diatraea saccharalis* em cana-de-açúcar. In: Simpósio de Controle Biológico, 13, Bonito, Resumo, Dourados

Jamont M, Dubois-Pot C, Jaloux B (2014) Nectar provisioning close to host patches increases parasitoid recruitment, retention and host parasitism. Basic Appl Ecol 15:151–160

Lima AAE (2011) Comparação de dietas artificiais para criação de *Diatraea saccharalis* (Lepidoptera: Crambidae) e avaliação da qualidade de *Cotesia flavipes* (Hymenoptera: Braconidae) em criações massais. Dissertação (Mestrado em Sanidade Vegetal, Segurança Alimentar e o Ambiente), Instituto Biológico, Apta, São Paulo, 95p

Litholdo MG et al (2013) Doses do fungo *Beauveria bassiana* GR no controle de *Sphenophorus levis* em cana-de-açúcar. In: Reunião Brasileira sobre Pragas de Solo, 13, Rondonópolis, Resumo expandido, Rondonópolis

Locher GDA (2012) Estudo comparativo da diversidade de vespas sociais (Hymenoptera, Vespidae) em mata ciliar e cultura de cana-de-açúcar na região de Ipeúna, SP. Dissertação (Mestrado em Zoologia), Instituto de Biociências de Rio Claro, Unesp, Rio Claro, 73p

Marques EJ, Vilas Boas AM (1985) Recuperação do parasito *Acmopolynema hervali* de ovos de *Mahanarva posticata* em Carpina-PE. Pesq Agrop Brasileira 20:271–271
Marques EJ, Villas Boas AM, Pereira CEF (1981) Orientações técnicas para a produção do fungo entomógeno *Metarhizium anisopliae* (Metsch.) Sorok. em laboratórios setoriais. Boletim Técnico Planalsucar 3:5–23
Márquez JM (2017) Efecto de cultivos intercalados sobre el comportamiento del daño por Barrenador del tallo *Diatraea* spp. en caña de azúcar. Cengicaña 16:66–67
Márquez JM et al (2017) Entomofauna benéfica (Hymenoptera: Parasítica) en caña de azúcar, asociada a las áreas con infestación del barrenador del tallo. Cengicaña 16:223–241
Masson MB, Pinto ADS, Fernandes EM, Magro Neto J, Pereira LA, Lopes VL, dos Santos AC (2012) Impacto de fipronil e tiametoxam sobre organismos não-alvo de superfície de solo em cana-de-açúcar. In: Workshop de Agroenergia, 6, Ribeirão Preto
Mateus Junior RDR (2017) Uso de misturas de produtos em pupas de *Cotesia flavipes* como repelentes de predadores para a liberação aérea em cana-de-açúcar. Monografia (Trabalho de Conclusão de Curso de Agronomia), Centro Universitário Moura Lacerda, Ribeirão Preto, 45f
Mendonça AF (1996a) Distribuição de *Diatraea* spp. (Lep.: Pyralidae) e de seus principais parasitoides larvais no Continente Americano. In: Mendonça AF (ed) Pragas da cana-de-açúcar. Insetos & Cia, Maceió, pp 83–121, 239p
Mendonça AF (1996b) Guia das principais pragas da cana-de-açúcar na América Latina e Caribe. In: Mendonça AF (ed) Pragas da cana-de-açúcar. Insetos & Cia, Maceió, pp 3–48, 239p
Mendonça AF (1996c) Pragas da cana-de-açúcar. Insetos & Cia, Maceió, 239p
Mendonça AF, Moreno JA, Risco SH, Rocha ICB (1996a) As brocas da cana-de-açúcar, *Diatraea* spp. (Lep., Pyralidae). In: Mendonça AF (ed) Pragas da cana-de-açúcar. Insetos & Cia, Maceió, pp 51–82, 239p
Mendonça AF, Viveiros AJA, Sampaio Filho F (1996b) A broca-gigante da cana-de-açúcar, *Castnia licus* Drury, 1770 (Lep.: Castniidae). In: Mendonça AF (ed) Pragas da cana-de-açúcar. Insetos & Cia, Maceió, pp 133–167, 239p
Mendonça AF, Flores S, Sáenz CE (2005) Cigarrinhas da cana-de-açúcar na América Latina e Caribe. In: Mendonça AF (ed) Cigarrinhas da cana-de-açúcar: controle biológico. Insecta, Maceió, pp 3–48, 317p
Mills N et al (2000) Mass releases of *Trichogramma* wasps can reduce damage from codling moth. Calif Agric 54:22–25
Moraes GC, Ávila CJ (2014) Insetos-praga associados ao solo na cultura da cana-de-açúcar, no Estado de Mato Grosso do Sul. Dourados, Embrapa Agropecuária Oeste, 37p. (Boletim de Pesquisa e Desenvolvimento, Embrapa Agropecuária Oeste, 66)
Moraes GJ, Flechtmann CHW (2008) Manual de Acarologia: acarologia básica e ácaros de plantas cultivadas no Brasil. Holos, Ribeirão Preto, 308p
Parra JRP (2014) Biological Control in Brazil: an overview. Sci Agric 71:420–429
Parra JRP, Botelho PSM, Pinto ADS (2010) Controle biológico de pragas como um componente-chave para a produção sustentável da cana-de-açúcar. In: Cortez LAB (Org) Bioetanol de cana-de-açúcar: P&D para produtividade e sustentabilidade. Blucher, São Paulo, pp 441–450, 992p
Peres FSC, Fernandes OA, Silveira LCP, da Silva CSB (2009) Cravo-de-defunto como planta atrativa para tripes em cultivo protegido de melão orgânico. Bragantia 68:953–960
Pinto ADS (2008) Manejo racional de pragas na expansão do canavial. Idea News 8:40–50
Pinto ADS, Parra JRP (2002) Liberação de inimigos naturais. In: Parra JRP, Botelho PSM, Corrêa-Ferreira BS, Bento JMS (eds) Controle biológico no Brasil: parasitoides e predadores. Manole, São Paulo, pp 325–342, 609p
Pinto ADS, Garcia JF, Botelho PSM (2006a) Controle biológico de pragas da cana-de-açúcar. In: Pinto ADS, Nava DE, Rossi MM, Malerbo-Souza DT (Orgs) Controle biológico de pragas: na prática. CP 2, Piracicaba, pp 65–74, 287p
Pinto ADS, Garcia JF, de Oliveira HN (2006b) Manejo das principais pragas da cana-de-açúcar. In: Segato SV, Pinto ADS, Jendiroba E, de Nóbrega JCM (Orgs) Atualização em produção de cana-de-açúcar. CP 2, Piracicaba, pp 257–280, 415p

Pinto ADS, Botelho PSM, de Oliveira HN (2009) Guia ilustrado de pragas e insetos benéficos da cana-de-açúcar. CP 2, Piracicaba, 160p
Pinto ADS, Lopes VL, de Lima AA (2013) Manejo de pragas da cana-de-açúcar. In: Santos F, Borém A (Orgs) Cana-de-açúcar: do plantio à colheita. Universidade Federal de Viçosa, Viçosa, pp 73–87, 257p
Pinto ADS, Rodrigues LR, Oliva MB (2018) Uso de armadilhas de feromônio para a broca-da-cana, *Diatraea* spp., em canaviais. Piracicaba, Occasio, 24p
Pires CSS, Fontes EMG, Sujii ER, Fernandes HMC, Gomes DF (1993) Ocorrência de *Anagrus* sp. (Hymenoptera: Mymaridae) parasitando ovos de *Deois flavopicta* (Stal) (Homoptera: Cercopidae) em pastagens do Brasil Central. An Soc Entomol Bras 22:411–413
Planalsucar (1983) Relatório anual 1983. Planalsucar-IAA, Piracicaba, 164p
Precetti AACM, Arrigoni EB (1990) Aspectos biológicos e controle do besouro *Sphenophorus levis* Vaurie, 1978 (Coleoptera: Curculionidae) em cana-de-açúcar. São Paulo, Copersucar, 15p. (Boletim Técnico Copersucar, especial)
Prezoto F, Giannotti E, Machado VLL (1994) Atividade forrageadora e material coletado pela vespa social *Polistes simillimus* Zikán, 1951 (Hymenoptera, Vespidae). Insecta 3:11–19
Reis PR, Botelho W, Mendes WBA (1983) Recomendações para o controle das cigarrinhas-das-pastagens. Belo Horizonte, Epamig, 15p. (Boletim Técnico, 1)
Rodrigues LR, Vinha FB, Pinto ADS, Rossi MM (2014) Avaliação do ácido tricloro-isocianúrico na repelência de predadores de pupas de *Cotesia flavipes* (Cam.) (Hymenoptera: Braconidae) espalhadas pelo solo na liberação aérea. In: Congresso Brasileiro de Entomologia, 25, Goiânia, Anais... Goiânia
Salvador Neto N (2012) Doses dos fungos *Metarhizium anisopliae* e *Beauveria bassiana* GR no controle de *Sphenophorus levis* e corós em cana-de-açúcar. Monografia (Trabalho de Conclusão de Curso de Agronomia), Centro Universitário Moura Lacerda, Ribeirão Preto, 25f
Salvatore AR, López G, Willink E (2009) Plagas en el cultivo de caña de azúcar. In: Romero ER, Digonzelli PA, Scandaliaris J (eds) Manual del cañero. Estación Experimental Agroindustrial Obispo Colombres, Las Talitas, pp 143–150, 232p
Santos MPCJD, Carrano-Moreira AF, Torres JB (2012) Diversidade de formigas epigeicas (Hymenoptera: Formicidae) em floresta ombrófila densa e em cultivo de cana-de-açúcar, no município de Igarassu, PE. Rev Bras Ciênc Agrár 7:648–656
Silva MGMFD (2012a) Substituição do formaldeído na dieta artificial para criação de *Diatraea saccharalis* (Lepidoptera: Crambidae): efeito na biologia do hospedeiro e de seu parasitoide *Cotesia flavipes* (Hymenoptera: Braconidae). Dissertação (Mestrado em Sanidade Vegetal, Segurança Alimentar e o Ambiente), Instituto Biológico, Apta, São Paulo, 81p
Silva NJDJ (2012b) Diversidade de vespas sociais em cultivo de cana-de-açúcar. Dissertação (Mestrado em Ecologia Aplicada ao Manejo e Conservação de Recursos), Universidade Federal de Juiz de Fora, 53p
Silva RR, Brandão CRF (1999) Formigas (Hymenoptera: Formicidae) como indicadores da qualidade ambiental e da biodiversidade de outros invertebrados terrestres. Biotemas 12:55–73
Simões RA, Reis LG, Bento JMS, Solter L, Delalibera I (2012) Biological and behavioral parameters of the parasitoid *Cotesia flavipes* (Hymenoptera: Braconidae) are altered by the pathogen *Nosema* sp. (Microsporidia: Nosematidae). Biol Control 63:164–171
Solis MA, Metz MA (2016) An illustrated guide to the identification of the known species of *Diatraea* Guilding (Lepidoptera: Crambidae: Crambinae) based on genitalia. ZooKeys 565:73–121
Souza IL, Marucci RC, Silveira LCP, de Paulo NCP, Lee JC (2018) Effects of marigold on the behavior, survival and nutrient reserves of *Aphidius platensis*. BioControl 63:543–553
Tanaka Junior GM, Noll FB (2011) Diversity of social wasps on semideciduous seasonal forest fragments with different surrounding matrix in Brazil. Psyche 2011:1–8
Triapitsyn SV (2002) Descriptive notes on a new and other little-known species of *Anagrus* Haliday, 1833 (Hymenoptera: Mymaridae) from the New World tropics and subtropics. Entomotropica 17:213–223

Vargas G, Gutiérrez Y (2017) Manejo de las poblaciones del salivazo *Aeneolamia varia*. Carta Inf Cenicaña 5:18–19

Vargas G, Obando V, Gómez L (2006) *Jaynesleskia jaynesi*: otra alternativa para el manejo de *Diatraea* spp. Carta Trimest Cenicaña 28:3–5

Vargas EL, Pereira FF, Tavares MT, Pastori PL (2011) Record of *Tetrastichus howardii* (Hymenoptera: Eulophidae) parasitizing *Diatraea* sp. (Lepidoptera: Crambidae) in sugarcane crop in Brazil. Entomotropica 26:143–146

Viana PA (2004) Lagarta-elasmo. In: Salvadori JR, Ávila CJ, Silva MTB (eds) Pragas de solo no Brasil. Embrapa Trigo/Embrapa Agropecuária Oeste/Fundacep Fecotrigo, Passo Fundo/Dourados/Cruz Alta, pp 379, 541p–408

Vinha FB, Rodrigues LR, Pinto ADS, Rossi MM (2014) Aspectos biológicos de *Cotesia flavipes* (Cam.) (Hymenoptera: Braconidae) após a aplicação do ácido tricloro-isocianúrico para a repelência de predadores para a liberação aérea. In: Congresso Brasileiro de Entomologia, 25, Goiânia, Anais... Goiânia

Zorzenon FJ, Bergmann EC, Bicudo JEA (2000) Primeira ocorrência de *Metamasius hemipterus* (Linnaeus, 1758) e *Metamasius ensirotris* (Germar, 1824) (Coleoptera, Curculionidae) em palmiteiros dos gêneros *Euterpe* e *Bactris* (Arecaceae) no Brasil. Arq Inst Biol 67:265–268

Chapter 34
Microbial Control of Sugarcane Pests

José Eduardo Marcondes de Almeida

34.1 Introduction

Sugarcane has important microbial control programs in Latin America, mainly using entomopathogenic fungi. However, other microbial agents are being studied and introduced to control pests in this crop. Virtually, only Argentina, Bolivia, Chile, Guyana, Paraguay, Suriname, and Haiti do not apply microbial control despite having sugarcane cultivation areas.

Among the countries that use microbial pest control, Brazil has the largest sugarcane cultivation area with approximately 9 million hectares (Conab 2018). Cuba has an area of 1.8 million hectares (EcuRed 2015). Mexico, a major sugar exporter accounting for 4% of the world production, has an area of 700,000 ha (Secretaria de Agricultura Ganaderia, Desarrollo Rural, Pesca y Alimentacion 2015). Colombia accounts for 1.1% of the world sugar market with more than 200,000 ha (Asocaña 2014). Venezuela has 1 million hectares of sugarcane cultivation with great use of microbial control (Alves and Lopes 2008). The sugarcane cultivation area corresponds to 130,000 ha in Ecuador (Centro de Investigación la Cana de Azúcar del Ecuador 2013). The sugarcane culture in Peru has been highlighted by its high productivity, over 100 tons/hectare in a planted area of 81,000 ha (Ministry of Agriculture Peru 2013). Guatemala is the third largest sugar producer in Central America, followed by Mexico and Cuba in planted area with over 200,000 ha (Melgar 2014). The Central America countries and the Caribbean have conditions to develop sugarcane cultivation; thus, Nicaragua, El Salvador, Honduras, Puerto Rico, and Costa Rica also have significant sugarcane area with pest management based on microbial control, mainly entomopathogenic fungi (Chacón 2012).

J. E. M. de Almeida (✉)
Instituto Biológico, Agência Paulista de Tecnologia dos Agronegócios (APTA), Campinas, SP, Brazil
e-mail: jemalmeida@biologico.sp.gov.br

B. Souza et al. (eds.), *Natural Enemies of Insect Pests in Neotropical Agroecosystems*, https://doi.org/10.1007/978-3-030-24733-1_34

The objective of this chapter is to make a survey of information on the research and application of microorganisms in the biological control of pests of sugarcane culture in order to help researchers in new studies and scientific improvement.

34.2 Sugarcane Pests in Latin America

Sugarcane pests may be divided into three ecological categories in Latin America: stalk borers, sapsuckers and leaf feeders, and soil pests (Table 34.1).

34.3 Microbial Control of the Main Sugarcane Insect Pests in Latin America

34.3.1 Stalk Borers

34.3.1.1 Lepidoptera

The microbial control of *Diatraea saccharalis* (Fabricius) (Crambidae) stalk borer was carried out in the 1980s using DsGV granulovirus, and it achieved some success in small areas. However, the program developed by the University of Campinas (UNICAMP), Brazil, did not achieve the expected success due to the large sugarcane area and the need of high virus volume, which increased production costs (Alves 1986).

The application of *Bacillus thuringiensis* (Berliner) as bait to control *D. saccharalis* has also been studied and showed 76.7% decrease in the attack on cane internodes in greenhouse. It indicates that this bacterium may also be an alternative to control stalk borers (Alves et al. 1981).

The application of entomopathogenic fungi to control *D. saccharalis* has been showing high use potential for over 30 years. Alves et al. (1985) found that the CL_{50} of *Beauveria bassiana* (Bals.-Criv.) Vuill. for pre-pupa was 1.3×10^5 conidia/mL, and, under field conditions, sugarcane stalks were infested with third instar caterpillars previously treated with suspensions of the fungus at the concentrations of 3.7×10^7 and 3.7×10^8 conidia/mL. It was found that caterpillar has mortality of 47.5% and 56%, respectively, with harm reduction in the stalks.

The selection of isolates is one of the most important stages in the microbial control program, along with the need to work with multiple fungal isolates, featuring them according to their morphology, isoenzymes, plague virulence, and DNA. Soza-Gómez et al. (1994) conducted a featuring study on *Beauveria* spp. isolates to control *D. saccharalis* and found biological differences among nine *B. bassiana* pathotypes and one *Bactris brongniartii* Mart. pathotype. The LT_{50} values ranged from 5.9 days in the most virulent isolate to 8 days or more in the less virulent ones. They also found that isolates from the same region showed high similarity levels, thus suggesting variations among geographical populations.

Table 34.1 Major sugarcane pests in Latin America under the use of microbial control agents

Type	Order	Species	Occurrence	Reference
Stalk borers	Lepidoptera	*Diatraea saccharalis* *Diatraea flavipennella*	Latin America	Mendonça et al. (1996a)
		Telchin licus licus	Brazil	Mendonça et al. (1996b), Dinardo-Miranda (2008)
Sapsuckers and leaf feeders	Hemiptera	*Mahanarva posticata*	Brazil	Alves and Lopes (2008), Pinto et al. (2009)
		Mahanarva fimbriolata	Brazil	Almeida and Batista Filho (2006), Mendonça (1996)
		Aeneolamia varia	Colombia, Peru, Ecuador, Venezuela	Mendonça (2005)
		Aeneolamia postica	Central America	Mendonça (2005)
		Mahanarva bipars	Venezuela	Mendonça (2005)
		Perkinsiella saccharicida	Ecuador	Badilla et al. (2004)
Root feeders	Coleoptera	*Sphenophorus levis*	Brazil	Gallo et al. (2002)
		Migdolus fryanus	Brazil	Pinto et al. (2009)
	Isoptera	*Heterotermes tenuis*	Brazil	Almeida and Batista Filho (2008)

The use of entomopathogenic fungi to control stalk borers was studied 30 years ago; however, there was no continuity due to the success of the *Cotesia flavipes* (Cameron) (Braconidae) parasitoid. However, new studies on the application of entomopathogenic fungi have become necessary due to the large expansion of sugarcane areas, especially in warmer areas, and the growing demand for the parasitoid in the culture. Importantly, microbial control agents should be used in *D. saccharalis* or other *Diatraea* species when the caterpillars are still in the first stage of growth, while they have not penetrated the sugarcane stalk.

Beauveria bassiana and *Metarhizium anisopliae* (Metchnikoff) Sorokin isolates have been most recently selected due to the increased sugarcane area in Brazil. LC_{50} of 5×10^8 *M. anisopliae* conidia/mL and 1×10^8 *B. bassiana* conidia/mL were found in second instar caterpillars. The isolates were tested for production in rice and for field-testing efficiency, and the dosage of 6×10^{12} conidia/ha was found in *B. bassiana* and that of 1×10^{13} conidia/ha was found in *M. anisopliae*. *D. saccharalis* microbial control is still not applied on large scale because of the fungus dosage cost, which is higher than that of the juvenoid pesticide registered for the pest. However, their average efficiency is the same, 60% control (Zappelini 2009; Zappelini et al. 2010).

The control of *D. saccharalis* using *B. bassiana* was studied in Cuba as a control alternative under field conditions. According to Estrada et al. (1997), the application of 10^{12} conidia/ha on homogeneous pest populations resulted in yield of 10 tons/ha.

According to Estrada et al. (2004), the presence of *B. bassiana* fungus is common in *D. saccharalis* caterpillars in Cuban cane fields, a fact that proves that this fungus is an important biological control agent that should not be neglected.

In Argentina, two *B. bassiana* isolates were selected for *D. saccharalis* with TL_{50} of 2 days and 90% efficiency (Alves and Lecuona 1996). A wettable powder biopesticide was also developed based on *B. bassiana* with two isolates selected for *D. saccharalis* at the concentration of 1.5×10^{10} conidia/g with 82.5% laboratory efficiency (Arcas et al. 1999).

34.3.2 Sapsuckers and Leaf Feeders

34.3.2.1 Hemiptera

The microbial control of *Mahanarva posticata* (Stål) (Cercopidae), a leafhopper, in Northeastern Brazil had its first positive results in 1969, and, since 1975, *M. anisopliae* production plants have been installed in the states of Sergipe, Alagoas, Pernambuco, and Paraíba by sugar and alcohol plants and by the association of sugarcane growers, Associação dos Plantadores de Cana de Alagoas (ASPLANA), and they produce the locally sourced PL43 isolate. From 1977 to 1991, 670,000 ha of sugarcane were sprayed, producing more than 40 tons of *M. anisopliae* conidia (Marques 1992).

The PL43 isolate is still produced by solid fermentation in rice, in factories in northeastern Brazil. However, the applied area reaches 250,000 ha in the northeastern sugarcane region, using on average 5×10^{12} conidia/ha (Marques et al. 2010).

Mahanarva fimbriolata (Stål) (Cercopidae), a root hopper, has become a relevant sugarcane pest since 1997, mainly in São Paulo State, Brazil, and it spread out to all Brazilian cane fields within 15 years. The Biological Institute (SAA/APTA) in partnership with the College of Agriculture of “Luiz de Queiroz” (ESALQ/USP) and the Federal University of São Carlos (UFSCar) initially conducted a research for selecting *M. anisopliae* isolates to control this pest. The IBCB 348, IBCB 425, and ESALQ 1037 isolates were selected as the most virulent with good conidia production in rice, with average of 1×10^9 conidia/g of rice (Loureiro et al. 2005).

The application of the *M. anisopliae* fungus has been held in São Paulo State in approximately 350,000 ha (2013/2014), with 60% average control efficiency, thus allowing sugarcane producers to coexist with the plague. This area reaches 1.5 million ha in Brazil. IBCB 425 is the most produced isolate to control *M. fimbriolata* in the country, and it is used by 37 companies (Almeida 2014).

The fungus application is recommended at 5×10^{12} conidia/ha. It must be washed, and the suspension must be sprayed close to the cane roots with flow of 300 L/ha or by air flow of 30 L/ha. It is also possible to perform the aerial rice + fungus application at 7–10 kg/ha (Almeida et al. 2004; Almeida and Batista Filho 2006; Alves et al. 2010).

Metarhizium anisopliae fungus was researched mainly to control *Aeneolamia varia* (Fabricius) (Cercopidae) in Colombia because, since 2007, it has been causing damage to sugarcane in the country (Cuarán et al. 2012). The microbial control program for this pest began with studies on *M. anisopliae* and *Paecilomyces lilacinus* (Thom) Samson selection in *A. varia* (Garcia et al. 2012). The *M. anisopliae* CCMa01 isolate, which was selected after some selection researches, is the most productive isolate, and it effectively controls *A. varia* (Obando et al. 2013). The *M. anisopliae* dosage used to control *A. varia* reaches 5×10^{13} conidia/ha (Solarte et al. 2012). *M. anisopliae* fungus is also applied to *A. varia* in Peru and Venezuela (Alves and Lopes 2008).

Ecuador also uses *M. anisopliae* and *B. bassiana* in *Perkinsiella saccharicida* Kirkaldy (Delphacidae) leafhopper due to a study on the selection of isolates from these fungi, resulting in commercial isolates produced in rice (Badilla et al. 2004).

In Central America, Mexico also has a microbial control program for *Aeneolamia postica* (Walker) (Cercopidae) since the 1970s. This program uses selected isolates to control the most important leafhopper in this region (Torres et al. 2013). Likewise, Guatemala also applies *M. anisopliae* to control *A. postica*, although its efficiency is not as high as in Mexico, thus requiring the selection of isolates (Melgar 2014). Costa Rica, El Salvador, Nicaragua, and Puerto Rico apply *M. anisopliae* to control *Aeneolamia* spp. since the 1980s, and they achieve important results in the control of this pest, which affects this sugarcane region in Latin America (Badilla 2002; Alves and Lopes 2008).

34.3.3 Root Feeders

34.3.3.1 Coleoptera

The microbial control of *Sphenophorus levis* Vaurie (Curculionidae) – a sugarcane root borer – began by selecting *M. anisopliae* (ESALQ 259) and *B. bassiana* (ESALQ 447) isolates at ESALQ/USP. The herein used strategy consisted of applying 500 g conidia/ha in baits made of sugarcane stems, which were lengthwise cut and impregnated with the fungi conidia. Two hundred (200) baits were distributed per hectare. The adult insects were attracted by the baits, became infected with the applied fungus, and died in the bait or next to it, spreading the disease within the area due to the high density of baits (Badilla and Alves 1991).

The application of entomopathogenic nematodes has also been investigated and produced satisfactory results in *S. levis* control. It may be applied in combination with fungi. The Biological Institute, in partnership with the Center for Sugarcane Technology (CTC – Centro de Tecnologia Canavieira) and UFSCar, conducted researches using *Heterorhabditis indica* Poinar, Karunakar, and David and *Steinernema* spp. nematodes as well as *M. anisopliae* and *B. bassiana* entomopathogenic fungi. *Steinernema* spp. caused 70% mortality in adults at the dose of 1 × 10^8 infective juveniles/ha under laboratory conditions. The application of nematodes produced sugarcane gains of 15 tons/ha under field conditions (Leite et al. 2006).

Entomopathogenic nematodes isolate selection researches were conducted, and *Steinernema braziliense* (Nguyen et al.) IBCBn 06 was the most virulent one. Likewise, the association with thiamethoxam 250 WG chemical insecticide also produced sugarcane weight gains of up to 28 ton/ha in places where the nematode was applied with underdose of 500 g thiamethoxam 250 WG/ha (Leite et al. 2012).

The *S. braziliense* IBCBn 06 isolate is produced by the sponge method in in vitro production through a partnership of the Biological Institute with the company Bio Controle Métodos de Controle de Pragas Ltda in Indaiatuba, São Paulo, Brazil. This isolate possesses registration in the Ministry of Agriculture and Livestock, MAPA de Bio NEP®, and a dosage of 1 kg/ha in 400 L of water (ground application) is recommended (Brasil 2015).

Simi (2014) selected *M. anisopliae* IBCB 383 and *B. bassiana* IBCB 170 isolates to control *S. levis*. The author found that when the *S. braziliense* nematode was applied along with *M. anisopliae,* the effect was antagonistic under field condition, and when it was combined with *B. bassiana*, there was synergistic effect on the control of *S. levis* in sugarcane.

Migdolus fryanus (Westwood) (Cerambycidae) microbial control is still experimental, but some research findings demonstrate the potential use of entomopathogenic nematodes to control this sugarcane pest. Pizano et al. (1985) and Arrigoni et al. (1986) found *Steinernema glaseri* (Nguyen and Smart) and *Steinernema carpocapsae* (Weiser) parasitism on *M. fryanus* larvae under field conditions; however, the results were not satisfactory. More recently, Machado et al. (2005) found *S. glaseri* and *H. indica* parasitism on *M. fryanus* eggs and larvae under laboratory condition, reaching 76% parasitism in dosages of 600 infective juveniles/larva in each nematode species.

34.3.3.2 Blattodea

The microbial control of termites in sugarcane was developed from studies using baits or traps for subterranean termites, especially the *Heterotermes tenuis* (Hagen) (Rhinotermitidae) species.

The Termitrap® bait was developed under laboratory and field conditions at ESALQ/USP, Piracicaba, SP, Brazil, from tests performed on materials that are attractive to *H. tenuis*. The corrugated cardboard was the most attractive one, and it was then used for making baits for survey studies on termites in cane, population fluctuation, and control. This bait allowed conducting a survey on termite species in Piracicaba region for 3 years, and *H. tenuis* was the most attracted species; however, it was also possible to collect the following Termitidae species: *Cornitermes cumulans* (Kollar), *Procornitermes* spp., *Nasutitermes* spp., *Anoplotermes* spp., *Constrictotermes* spp., *Ruptitermes* spp., *Syntermes molestus* (Burmeister), *Cylindrotermes* sp., *Neocapritermes* spp., and the Rhinotermitidae specie *Coptotermes havilandi* (Wasmann). The damages to sugarcane roots were associated with *H. tenuis* and *Procornitermes* sp. (Alves et al. 1985; Almeida and Alves 1995, 2009; Almeida et al. 1999).

Studies for selecting *B. bassiana* isolates that are pathogenic to *H. tenuis* were performed in the laboratory using over 100 isolates. It was found that ESALQ 634 isolate was the most pathogenic, causing over 90% mortality in the population of *H. tenuis* soldiers and workers (Almeida et al. 1997).

Several experiments were performed under field conditions in order to investigate *H. tenuis* population decline. Imidacloprid and fipronil insecticides as well as triflumuron growth regulator were studied in association with *B. bassiana* fungus in small concentrations in baits. Some experiments showed decline in *H. tenuis* population by applying monitoring baits. It demonstrates the potential application of these active principles to control this pest in baits; however, it was observed that higher concentrations resulted in repellency. Thus, it is important to conduct more repellency tests under laboratory and field conditions, as well as to study other active principles. Fungi were not repellent to termites. However, the experiments showed that the termite population reduction was slower, thus requiring further research to evaluate the control with marked population and longer evaluation time of at least 6 months (Almeida and Alves 1996, 1999; Almeida et al. 2000, 2003).

34.4 Final Considerations

The microbial control in Latin America has needs that are favorable to soil and climate. Studies on the biodiversity of entomopathogenic organisms are still needed in order to learn new species or even greater variety of isolates for researches with selection of virulent isolates to sugarcane pests in Brazil. Maintaining entomopathogen agent collections is extremely necessary to the development of biopesticides, which represents another challenge to this process since it involves an important legislative factor (Alves and Lopes 2008).

Most bio-factories in Latin America are handmade, since they directly depend on hand labor, which is not always qualified, a fact that increases the cost of the process. In addition, they do not use technologies involved in Good Manufacturing Practices, which makes them less competitive. Qualitative and quantitative analyses related to the concentration, viability, purity, and virulence or potency in entomopathogens have not been standardized. There might be some progress related to entomopathogenic bacteria. As for the other agents, there are no defined concepts regarding these parameters.

Legislation is a challenge for the development of biopesticides, as it is not specific to biological products, both in the register and in the distribution of resources through the use of genetic patrimony, although some Latin American countries, such as Brazil, have specific legislation on the use of genetic heritage in the country, according to the Nagoya Protocol (Escobar 2015).

The interest in the production of biopesticides based on fungi, bacteria, nematodes, and viruses has also increased, leading the research sector to invest in new production and formulation techniques, as well as in laboratory and field efficiency studies. The companies interested in this market follow this trend and create new job opportunities, thus generating income and protecting the environment, since it is considered a low-risk technology.

References

Almeida JEM (2014) Cigarrinha-da-raiz da cana: como controlar? Rev Cultiv G C 177:30–31

Almeida JEM, Alves SB (1995) Seleção de armadilhas para a captura de *Heterotermes tenuis* (Hagen). An Soc Entomol Bras 24(3):619–624

Almeida JEM, Alves SB (1996) Mortalidade de *Heterotermes tenuis* (Hagen) atraídos por armadilhas com *Beauveria bassiana* (Bals.) Vuill. e imidacloprid. An Soc Entomol Bras 25:507–512

Almeida JEM, Alves SB (1999) Controle de *Heterotermes tenuis* (Hagen, 1858) (Isoptera; Rhinotermitidae) em cana-de-açúcar com iscas Termitrap® associadas ao fungo entomopatogênico *Beauveria bassiana* (Bals.) Vuill. e/ou a inseticidas em época seca. Arq Inst Biol 66(2):85–90

Almeida JEM, Alves SB (2009) Atividade de forrageamento de *Heterotermes tenuis* (Hagen) (Isoptera: Rhinotermitidae) em canade-açúcar. Arq Inst Biol 76(4):613618

Almeida JEM, Batista Filho A (2006) Controle biológico da cigarrinha-da-raiz da cana-de-açúcar com o fungo *Metarhizium anisopliae*. Bol Téc/Inst Biol 16:1–19

Almeida JEM, Batista Filho A (2008) Cupins: pragas em Áreas Agrícolas. Bol Téc/Inst Biol 22:1–20

Almeida JEM, Alves SB, Pereira RM (1997) Selection of *Beauveria* spp. isolates of the termite *Heterotermes tenuis* (Hagen, 1858). J Appl Entomol 121(5):539–543

Almeida JEM, Alves SB, Walder JMM (1999) Tamanho da área de forrageamento do cupim subterrâneo *Heterotermes tenuis* (Isoptera; Rhinotermitidae) em cana-de-açúcar. Sci Agric 56(2):313–316

Almeida JEM, Alves SB, Almeida LC (2000) Controle de *Heterotermes tenuis* (Hagen) (Isoptera; Rhinotermitidae) e *Cornitermes cumulans* (Kollar) (Isoptera; Termitidae) com inseticida fipronil associado ao fungo entomopatogênico *Beauveria bassiana* (Bals.) Vuill. em isca atrativa na cultura da cana-de-açúcar (*Saccharum officinarum* L.). Arq Inst Biol 67(2):235–241

Almeida JEM, Batista Filho A, Alves SB et al (2003) Avaliação de inseticidas e fungos entomopatogênicos para o controle de cupins subterrâneos da cana-de-açúcar. Arq Inst Biol 70:347–353

Almeida JEM, Batista Filho A, Santos AS (2004) Controle da cigarrinha da raiz da cana-de-açúcar, *Mahanarva fimbriolata*, com o fungo entomopatogênico *Metarhizium anisopliae*. STAB 22(4):42–45

Alves SB (1986) Controle microbiano de insetos. Manole, São Paulo

Alves SB, Lecuona RE (1996) Utilización de hongos entomopatógenos. In: Lecuona RE (ed) Microorganismos patógenos empleados en el control microbiano de insectos plagas. Talleres Gráficos Mariano, Buenos Aires, pp 241–254

Alves SB, Lopes RB (2008) Controle microbiano de pragas na América Latina. FEALQ, Piracicaba

Alves SB, Melo LAS, Pereira RM (1981) Controle da broca-da-cana, *Diatraea saccharalis* (Fabr., 1994) (Lep., Pyralidae) com *Bacillus thuringiensis* Berliner, na forma de isca. Ecossistema 6:105–112

Alves SB, Pádua LEM, Azevedo EMWM et al (1985) Controle da broca da cana-de-açúcar pelo uso de *Beauveria bassiana*. Pesqui Agropecu Bras 20(4):403–406

Alves LFA, Neves PMJO, Faria MR (eds) (2010) Recomendações para utilização de fungos entomopatogênicos no controle de pragas. CP 2: Rede Entomofungo, Piracicaba

Arcas JÁ, Díaz BM, Lecuona RE (1999) Bioinsecticidal activity of conidial and dry mycelium preparations of two isolates of *Beauveria bassiana* against the sugarcane borer *Diatraea saccharalis*. J Biotechnol 67(2/3):151–158

Arrigoni EB, Dinardo LL, Conde AJ et al (1986) Aplicação de *Neoplectana carpocapsea* (Weiser, 1955) em condições de campo para o controle de *Migdolus* spp. (Coleoptera: Cerambycidae). Nematol Bras 8:181–189

Asocaña (2014) Informe anual 2013–2014. Asocana, Cali. http://www.asocana.org/modules/documentos/10572.aspx. Accessed 15 May 2015

Badilla FF (2002) Un programa exitoso de control biolgico de insectos plaga de la caña de azúcar en Costa Rica. Manejo Integr Plagas Agroecol 64:77–87

Badilla FF, Alves SB (1991) Controle do gorgulho da cana-de-açúcar *Sphenophorus levis* Vaurie, 1978 (Coleoptera: Curculionidae) com *Beauveria* spp. em condições de laboratório e campo. An Soc Entomol Bras 20:251–253

Badilla FF, Jara W, Gordillo W (2004) Control del saltahojas de la caña de azúcar *Perkinsiella saccharicida* con los hongos *Metarhizium anisopliae* y *Beauveria bassiana* en el Ingenio San Carlos, en Ecuador. Manejo Integr Plagas Agroecol 73:29–34

Brasil. Ministério da Agricultura, Pecuária e Abastecimento (2015) Agrofit: sistema de agrotóxicos fitossanitários. Agrofit, Brasília. http://extranet.agricultura.gov.br/agrofit_cons/principal_agrofit_cons. Accessed 7 May 2015

Centro de Investigación la Caña de Azúcar del Ecuador (2013) Caña de Azúcar: cultivo para sostenibilidade. Cincae, El Triunfo. http://cincae.org/cana-de-azucar-cultivo-para-la-sostenibilidad/. Accessed 15 May 2015

Chacón HNS (2012) Dinámica económica de la industria azucarera centroamericana: el caso de Guatemala, Honduras, El Salvador y Nicaragua. http://bdigital.zamorano.edu/bitstream/11036/945/1/T3369.pdf. Accessed 20 May 2015

Conab (2018) Informações agropecuárias – safras. https://www.conab.gov.br/info-agro/safras/serie-historica-das-safras/item/7736-cana-de-acucar-area-total. Accessed 14 Aug 2018

Cuarán VL, Valderrama UC, Pardey AEB (2012) Metodo para evaluar el daño de los salivazos (Hemiptera: Cercopidae) sobre la caña de azúcar. Rev Colomb Entomol 38(2):171–176

Dinardo-Miranda LL (2008) Pragas. In: Dinardo-Miranda LL, Vasconcelos ACM, Landell MGA (eds) Cana-de-açúcar. Instituto Agronômico, Campinas, pp 349–422

EcuRed (2015) Cultivo de la caña de azucar. http://www.ecured.cu/index.php/Cultivo_de_la_ca%C3%B1a_de_az%C3%BAcar. Accessed 15 May 2015

Escobar H (2015) Brazil cut red tape stifling biodiversity studies. Science 348(6238):952–953

Estrada ME, Romero M, Snowball M (1997) Aplicación de *Beauveria bassiana* en la lucha biológica contra *Diatraea saccharalis*. Cana Azucar 15(1):39–43

Estrada ME, Romero M, Rivero MJ et al (2004) Presencia natural de *Beauveria bassiana* (Bals.) Vuill. em el cultivo de la caña de azúcar (*Saccharum* sp. híbrido) en Cuba. Rev Iberoam Micol 21:42–43

Gallo D, Nakano O, Silveira Neto S (2002) Entomologia agrícola. FEALQ, Piracicaba

Garcia AD, Bustillo AEP, Castro UV et al (2012) Selección de hongos entomopatógenos para controlar salivazos (Hemiptera: Cercopidae) en la caña de azúcar em Colombia. Rev Colomb Entomol 38(2):252–259

Leite LG, Tavares FM, Batista Filho A (2006) Alternativa de controle. Rev Cultiv G C 83:30–33

Leite LG, Tavares FM, Batista Filho A (2012) Eficiência de nematoides entomopatogênicos e inseticidas químicos contra *Sphenophorus levis* e *Leucothyreus* sp. em cana-de-açúcar. Pesqui Agropecu Trop 42(1):40–48

Loureiro ES, Batista Filho A, Almeida JEM (2005) Seleção de isolados de *Metarhizium anisopliae* (Metsch.) Sork. contra a cigarrinha-da-raiz da cana-de-açúcar, *Mahanarva fimbriolata* (Stal) (Hemiptera: Cercopidae) em laboratório. Neotrop Entomol 34(5):791–798

Machado LA, Habib M, Leite LG (2005) Patogenicidade de nematoides entomopatogênicos a ovos e larvas de *Migdolus fryanus* (Westwood, 1863) (Coleoptera; Vesperidae). Arq Inst Biol

72:221–226

Marques EJ (1992) Controle microbiano de cigarrinha-das-folhas (Hemiptera: Cercopidae) com *Metarhizium anisopliae* (Metsch.) Sorok.: eficiência e limitações. In: Resumos do 3° Simpósio de *Controle* Biológico, Sociedade Entomológica do Brasil, Águas de Lindóia, 12–16 out 1992

Marques EJ, Faria MR, Almeida JEM et al (2010) Recomendações para o uso de micoinseticidas destinados ao controle da cigarrinha-da-folha da cana-de-açúcar. In: Alves LFA, Neves PMOJ, Faria MR (eds) Recomendações para utilização de fungos entomopatogênicos no controle de pragas. CP 2: Rede Entomofungo, Piracicaba, pp 29–32

Melgar M (2014) Desarrollo tecnológico de la agroindústria azucarera y perspectivas. In: Melgar M, Meneses A, Orozco H et al (eds) El cultivo de la caña de azucar em Guatemala. Artemis, Guatemala City, pp 1–32

Mendonça AF (1996) Pragas da cana-de-açúcar. Insetos & Cia, Maceió

Mendonça AF (2005) Cigarrinhas da cana-de-açúcar. Insetos & Cia, Maceió

Mendonça AF, Moreno JÁ, Risco SH et al (1996a) As brocas da cana-de-açúcar, *Diatraea* spp. (Lep., Pyralidae). In: Mendonça AF (ed) Pragas da cana-de-açúcar. Insetos & Cia, Maceió, pp 51–82

Mendonça AF, Viveiros AJ, Sampaio Filho F (1996b) A broca gigante da cana-de-açúcar, *Castnia licus* Drury, 1770 (Lep., Castiniidae). In: Mendonça AF (ed) Pragas da cana-de-açúcar. Insetos & Cia, Maceió, pp 131–168

Ministério da Agricultura do Peru (2013) Caña de Azúcar: principales aspectos de la cadena agroproductiva. Dirección de Información Agraria, Peru

Obando JAB, Bustillo AEP, Castro UV et al (2013) Selección de cepas de *Metarhizium anisopliae* para el control de *Aeneolamia varia* (Hemiptera: Cercopiae). Rev Colomb Entomol 39(1):26–33

Pinto AS, Botelho PSM, Oliveira HN (eds) (2009) Guia ilustrado de pragas e insetos benéficos da cana-de-açúcar. CP 2: Rede Entomofungo, Piracicaba

Pizano MA, Aguillera MM, Monteiro AR et al (1985) Incidence of *Neoplectana glaseri* Steiner, 1929 (Nematoda: Steinernematidae) parasitizing *Migdolus fryanus* (Westwood, 1863) (Col.: Cerambycidae). Entomol News 17:9–10

Secretaria de Agricultura Ganaderia, Desarrollo Rural, Peca y Alimentacion (2015) La Caña de Azúcar em México. http://www.siap.sagarpa.gob.mx/siaprendes/contenidos/3/03-cana-azucar/contexto-3.html. Accessed 15 May 2015

Simi LD (2014) Controle de *Sphenophorus levis* e *Conotrachelus humeropictus* pelo uso combinado de nematoides e fungos entomopatogênicos. Tese, Universidade Estadual de São Paulo

Solarte JAM, Bustillo AEP, Castro UV et al (2012) Eficacia de *Metarhizium anisopliae* para controlar *Aeneolamia varia* (Hemiptera: Cercopiae), en caña de azúcar. Rev Colomb Entomol 38(2):177–181

Soza-Gómez DR, Alves SB, Milani MT (1994) Characterization and phonetic analysis of geographical isolates of *Beauveria* spp. Pesqui Agropecu Bras 29(3):401–409

Torres MC, Cortez HMM, Ortiz CFG (2013) Caracterización de aislamientos nativos de *Metarhizium anisopliae* y su patogenicidad hacia *Aeneolamia postica*, en Tabasco, México. Rev Colomb Entomol 39(1):40–46

Zappelini LO (2009) Seleção de isolados dos fungos entomopatogênicos *Beauveria bassiana* e *Metarhizium anisopliae* visando ao controle da broca da cana-de-açúcar, *Diatraea saccharalis* (Lepidoptera: Crambidae). Dissertação, Instituto Brasileiro para Segurança do Paciente de São Paulo

Zappelini LO, Almeida JEM, Batista Filho A et al (2010) Seleção de isolados do fungo entomopatogênico *Metarhizium anisopliae* (Metsch.) Sorok. visando ao controle da broca da cana-de-açúcar *Diatraea saccharalis* (Fabr. 1794). Arq Inst Biol 77(1):75–82

Chapter 35
Weeds

Kátia de Lima Nechet, Marcelo Diniz Vitorino, Bruno Sérgio Vieira, and Bernardo de Almeida Halfeld-Vieira

Worldwide, weeds represent about 0.1% of the flora presenting ecological attributes that can negatively interfere in agroecosystems and in native ecosystems (Cousens and Mortimer 2000). Considering the global agricultural productivity, weeds represent one of the major limiting factors involving multiple deleterious effects on crops, such as competition for water, light, mineral nutrients from the soil, production of allelopathic compounds, interference with harvesting, contamination of the product harvested with seeds and other plant parts, and the increase of the moisture content on the product harvested. In addition, weeds can serve as alternative hosts for pests and diseases of cultivated crops (Auld 1998; Duke 2015). In native ecosystems, the introduction of a plant can cause changes in nutrient recycling, hydrology, and energy budgets, which can greatly reduce the abundance or survival of native species (Mack et al. 2000). In this sense, the biotic invasions are recognized as one of the main causes of loss in biodiversity worldwide (Randall 1996). Furthermore, people mobility, global market, and the lack of trade regulations for ornamental plants have promoted alien invasions.

Although herbicides configure the dominant resources used for weed control, they do not provide a complete solution to the complex challenge presented by weeds (Walsh et al. 2013). Moreover, environmental impacts, such as reduction of the biodiversity, groundwater contamination, and the widespread evolution of resistance to herbicides in weed species have forced the development of alternative, nonchemical control strategies (Owen 2016; Shaw 2016). Important practices for sustainable weed control in cropping systems have been studied, and the integrated

K. de Lima Nechet (✉) · B. de Almeida Halfeld-Vieira
Embrapa Meio Ambiente, Jaguariúna, SP, Brazil
e-mail: katia.nechet@embrapa.br

M. D. Vitorino
Fundação Universidade Regional de Blumenau, Blumenau, SC, Brazil

B. S. Vieira
Universidade Federal de Uberlândia, Monte Carmelo, MG, Brazil

B. Souza et al. (eds.), *Natural Enemies of Insect Pests in Neotropical Agroecosystems*, https://doi.org/10.1007/978-3-030-24733-1_35

management using cultural, mechanical, biological, and chemical control tactics has been proposed to build a holistic approach to provide the main crop benefits over weeds (Harker and O'Donovan 2013). However, these practices are not widely adopted (Young et al. 2017).

In this context, the biological control of weeds involving the use of natural enemies to reduce the population of target plant below threshold (Watson 1991) has been frequently investigated and applied as an isolated control strategy to solve intractable weed problems (Charudattan and Dinoor 2000). Macro- and microorganisms are used as biological control agents in two broad approaches: the classical biological control and the inundative biological control, also known as the bioherbicide approach (Hallet 2005).

The classical biological control involves the discovery of an effective and highly host-specific agent in its native environment, which is export to a region where the target weed had been previously introduced (McFadyen 1998). Generally, this approach has been almost exclusively used for the control of extensive infestations in relatively undisturbed environments, such as aquatic systems, forests, and rangelands (DiTomaso 2000; Charudattan 2005). Although there are limitations and still challenges, many examples of successful classical biological control programs using insects and fungi are registered in the literature (Barreto 2009; Paynter et al. 2015). It is important to point out that spatial distribution and abundance of the host have influenced the establishment and maintenance of the biological control agent (Cousens and Croft 1995). Thus, a minimum density of the target weed overtime is necessary for a successful integrated weed management.

Latin America is the center of origin of many invasive plant species that are considered target weeds for biological control programs, including aquatic and terrestrial weeds (Cronk and Fuller 1995; Julien and Griffiths 1998). The first successful program of biological control of weed, initiated in 1902, was carried out by the Hawaii Department of Agriculture through the introduction of insects from Mexico to control *Lantana camara* in Hawaii (Waterhouse and Norris 1987). In 1925, the moth *Cactoblastis cactorum* from Argentina was introduced in Australia to control prickly pear (*Opuntia stricta*) (McFadyen and Wilson 1997). The positive impacts of these introductions in the control of those weeds are considered an example for studies in this area. The research activities and collaboration carried out between institutions from countries with alien weeds and countries of origin of those weeds reflect the importance of Latin American biodiversity as a source of biological control agents (Fowler et al. 2013; Macedo et al. 2016).

The bioherbicide approach is the application of a plant pathogen through appropriate formulations to control target weeds similarly to herbicides (TeBeest 1996). The literature has reported a number of pathogens, mainly fungi, under investigation for their potential as bioherbicide. However, there are few currently registered bioherbicides, and efforts to overcome this situation as well as opportunities for the exploitation of microorganisms in weed management have been discussed (Hallet 2005; Marrone 2014). A detailed explanation of the process involved in a bioherbicide development is reported in Barreto (2009).

In Latin America, the studies of potential bioherbicide focusing on fungi as control agents are restricted to Brazilian institutions. The studies aim at controlling weeds considered of increasing importance for Brazilian agricultural crops due to the low efficacy of herbicides. The first program for developing a mycoherbicide in Brazil was performed by Embrapa Soja in the 1980s and targeted wild poinsettia (*Euphorbia heterophylla*). Although the results indicated good potential of the fungus *Bipolaris euphorbiae* to control wild poinsettia, problems such as plant natural resistance, lack of commercial interest, and availability of new herbicides led to the failure of the program. Since then, many projects have demonstrated that there is a big diversity of phytopathogenic fungi in Brazil with potential to control weeds. The status of each program varies from prospecting fungi to the attempt to register a product as mycoherbicide (Borges Neto and Pitelli 2004; Vieira and Barreto 2005, 2010; Guatimosim et al. 2015; Nechet et al. 2006, 2016).

Besides the scientific development and achievements, governments have devised international documents that can affect weed control. The United Nations Conference on Environment and Development (UNCED), in its Agenda 21, recognized integrated pest management as the preferred strategy to reduce the dependence on chemical pesticides, including herbicides. Since the Convention on Biological Diversity, the biological control has been pointed out as priority component of pest management (Glowka et al. 1994).

In the next topic, some ongoing projects considering the biological control of weeds in Brazil are presented. The classical approach focused on the South American contributions to control alien weeds through the introduction of insects as biological control agents. The bioherbicide approach presented studies with *Bidens pilosa* and *Ipomoea* spp., which are important weeds for the agricultural production in Brazil.

35.1 Classical Approach

As previously mentioned, South America, remarkably Brazil, is considered the center of origin of many weed species with worldwide potential to become widespread alien invasors, such as water hyacinth (*Eichornnia crassipes*-Pontederiaceae), lantana (*Lantana camara*-Verbenaceae), water lettuce (*Pistia stratiotes*-Araceae), salvinia (*Salvinia molesta*-Salviniaceae), Barbados gooseberry (*Pereskia aculeata*-Cactaceae), Brazilian peppertree (*Schinus terebinthifolia*-Anacardiaceae), strawberry guava (*Psidium cattleianum*-Myrtaceae), tropical soda apple (*Solanum viarum*-Solanaceae), bugweed (*Solanum mauritianum*-Solanaceae), miconia (*Miconia calvescens*-Melastomataceae), and flaxleaf fleabane (*Conyza bonariensis*-Asteraceae). Some of these species have practices of biological control strategies already adopted, however with different levels of effectiveness. The most successful examples of biological control are observed on plants whose natural enemies have found conditions to establish their populations, e.g., water hyacinth, salvinia, and tropical soda apple. Other plants, such as lantana, Barbados gooseberry, and strawberry guava had control agents introduced and established, although with unsatisfactory control rate.

For these species, we can point out some studies aiming at prospection, selection, and potential biocontrol agents of weeds performed by South American researchers in cooperation with international institutes, resulting in classical biological control programs with different levels of success, such as:

1. *Gratiana boliviana* (Coleoptera: Cassidinae) introduced from Argentina and Paraguay to control tropical soda apple in Florida, USA
2. *Neochetina eichhorniae* and *Neochetina bruchi* (Coleoptera: Curculionidae) from Argentina to control water hyacinth on the coast of the Gulf of Mexico, USA, African continent, and Australia. Moreover, the introduction of *Eccritotarsus catarinensis* (Hemiptera: Miridae) from Brazil and Peru in South Africa, Benin, and China
3. *Lema basicostata* (Coleoptera: Chrysomelidae) from Brazil for the control of wandering Jew (*Tradescantia fluminensis*-Commelinaceae) in New Zealand
4. *Heteropsylla spinulosa* (Hemiptera: Psyllidae) from Brazil for the control of giant sensitive plant (*Mimosa diplotricha*-Fabaceae) in Australia, Guam, Micronesia, and Papua New Guinea
5. *Penthobruchus germaini* (Coleoptera: Chrysomelidae) from Argentina for the control of Mexican palo-verde (*Parkinsonia aculeata*-Fabaceae) in Australia
6. *Lysathia* sp. (Coleoptera: Chrysomelidae) from Brazil for the control of parrot feather (*Myriophyllum aquaticum*-Haloragaceae) in Australia
7. *Cyrtobagous salviniae* (Coleoptera: Erirhinidae) from Brazil for the control of other macrophyte known as Kariba weed (*Salvinia molesta*-Salviniaceae) in Australia, Botswana, USA, Fiji, Ghana, India, Kenya, Malaysia, Mauritania, Namibia, Papua New Guinea, Senegal, Sri Lanka, Zambia, and Zimbabwe
8. *Uroplata girardi* (Coleoptera: Chrysomelidae) from Brazil and Argentina for the control of the ornamental species lantana in Australia, South Africa, Guam, Hawaii, Mauritius, Solomon Islands, New Caledonia, and Samoa (Winston et al. 2014).

Recently, new biocontrol agents have been selected focusing on the control of South American plant species. Some introductions are the leaf-gall insect of *Tectococcus ovatus* (Hemiptera: Eriococcidae) from Brazil to control the strawberry guava (Fig. 35.1a), an alien weed introduced in Hawaii, USA (Johnson 2005; Wessels et al. 2007); *Phenrica guerini* (Coleoptera: Chrysomelidae) (Zimmermann et al. 2009) and *Catorhintha schaffneri* (Hemiptera: Coreidae) (Paterson et al. 2014a, b) from Brazil to control Barbados gooseberries, an alien invasive species in South Africa (Fig. 35.1b) first recorded by the Cape Town Botanical Gardens (Paterson et al. 2011); as well as the request and releasing approval of *Calophya latiforceps* (Hemiptera: Psyllidae) and *Pseudophilothrips ichini* (Thysanoptera: Phlaeotripidae) (Wheeler et al. 2016) to control Brazilian peppertree, in Florida, USA.

Fig. 35.1 (**a**) Leaf-gall insect *Tectococcus ovatus* (Hemiptera: Eriococcidae) on strawberry guava (*Psidium cattleianum*) in Hawaii. (**b**) Barbados gooseberries (*Pereskia aculeata*) invasion in South Africa. (**c**) Anthracnose lesions on beggartick (*Bidens pilosa*) plants caused by *Colletotrichum truncatum* isolate UFU 280 in natural conditions. (**d**) Dead plant of beggartick after inoculation with *C. truncatum* isolate UFU 280 conidia. (**e**) Leaf spot on moningglory plant (*Ipomoea nil*) caused by *Cercospora* sp. in natural conditions. (**f**) Plant defoliation induced by the inoculation with *Cercospora* sp. (Photos: (**a** and **b**) Marcelo Diniz Vitorino; (**c** and **d**) Bruno Sérgio Vieira; (**e** and **f**) Kátia de Lima Nechet)

The most important alien invaders in Brazil are classified based on their aggressivity, geographical distribution, and options of classical biological control agents in other ecosystems, such as:

1. Yellow bells (*Tecoma stans*-Bignoniaceae), originally from Mexico and southern USA, introduced as ornamental plant which has become invasive in southern Brazil. This species is also considered invasive in South Africa, where ladybug

Mada polluta (Coleoptera: Coccinellidae) (Madire 2013) originally from Mexico was introduced as a biological control agent.

2. Japanese raisin tree (*Hovenia dulcis*-Rhamnaceae), originally from China, Korea, and Japan. Among other uses, this plant is commonly used in Brazil as a windbreak plant in poultry and pig farms. It is considered one of the most important alien invaders in Brazil, especially in the south region.
3. South African lovegrass (*Eragrostis plana*-Poaceae) from South Africa, which was introduced as potential pasture plant, is an invader to native fields in southern Brazil, with estimates of over 3 million hectares invaded in Rio Grande do Sul State.
4. White ginger (*Hedychium coronarium*-Zingiberaceae), native from tropical Asia, believed to have been introduced in times of slavery when African slaves brought these species of plants to use as mattresses (Reaser et al. 2005). This is an important invasive plant of hydromorphic areas in Brazil, remarkably in coastal areas (Kissmann and Groth 1991), occupying 12 phytophysiognomies (Zenni and Ziller 2011).
5. Rubber vine (*Cryptostegia grandiflora*-Apocynaceae), native from Madagascar in Africa, was introduced as ornamental plant due to the beauty of its flowers and climbing habit. It is an invasive species of the Cerrado or open forests with endemic native occurrence of carnauba palms (*Copernicia prunifera*) in northeastern Brazil. This palm has been seriously endangered by the invasion of rubber vine, which uses the trunk of carnauba palms as growth support.
6. Mexican sunflower or tithonia (*Thytonia diversifolia*-Asteraceae), originally from Mexico, is found in tropical and subtropical areas (Pérez et al. 2009; Medina et al. 2009) and was introduced with ornamental purposes in Brazil. It is considered an invasive species of Santa Catarina Restinga ecosystem which performs an important role in the local marine life (CONSEMA 2012). Nowadays, this species is spreading to the Itajaí valley. In South Africa, this plant has *Physonota maculiventris* (Coleoptera: Chrysomelidae) as a biocontrol agent (Mphephu et al. 2017).

These examples illustrate how Brazilian biodiversity acts as a source of biological control agents of weeds worldwide. On the other hand, unfortunately, until now, no introductions of any foreign insects to control weeds through classic control strategies have become a reality in Brazil.

35.2 Inundative Approach

Target Weed: Beggartick (*Bidens pilosa* L.)

Biological Agent: *Colletotrichum truncatum.*

Beggartick is a pantropical weed of great economic importance for the Brazilian agriculture. It is present in almost all the national territory, mainly in the agricultural areas of the Center-South region (Holm 1991; Kissmann and Groth 1991). The weed is very prolific and has a short cycle, capable of producing up to three generations a year usually forming dense infestations. In addition to causing productivity losses due to the interference on cultivated plants, the presence of beggartick is aggravated since there have been several cases of resistance to the herbicides acetolactate synthase (ALS) in Brazil, and the confirmed cases of cross-resistant biotypes of ALS inhibitors from sulfonylureas and imidazolinones groups (Gelmini et al. 2002).

During a survey of phytopathogenic fungi associated with *B. pilosa* in the region of Alto Paranaíba, municipality of Iraí de Minas (Minas Gerais, Brazil), beggartick plants showing deep anthracnose lesions, some dead, were observed (Fig. 35.1c). These plants were collected and sent to LAMIF (Laboratory of Microbiology and Phytopathology) of the Universidade Federal de Uberlândia-Campus Monte Carmelo. The fungus was morphologically and molecularly identified as *Colletotrichum truncatum* isolate UFU 280, and its pathogenicity was confirmed. The potential of this fungus was verified as a mycoherbicide to control *B. pilosa*.

Preliminary observations have shown that *C. truncatum* isolate UFU 280 has natural characteristic to produce conidia in liquid culture media, and the fermentation conditions are known to trigger its conidia production (Jackson et al. 1996). A basic set of conditions for the mass production of conidia in submerged culture, in 6–9 days, are growth in liquid culture medium malt extract (using suspension of conidia as inoculum), pH 9.0, agitation in orbital shaker at 150 rpm, temperatures ranging from 20–25 °C (Vieira et al. 2018). However, additional tests at an industrial scale on larger capacity bioreactors under precisely controlled conditions for the various parameters are still necessary to fully confirm the commercial viability of the mass production of conidia of *C. truncatum* isolate UFU 280.

Regarding the efficiency of *C. truncatum* isolate UFU 280 as mycoherbicide for the beggartick, greenhouse experiments were delineated considering the effect of the inoculum concentration, the phenological stage of plants, and the leaf wetness period on disease severity. The parameters were conidial suspensions at concentrations of 0, 10^3, 10^4, 10^5, 10^6, and 10^7 conidia mL^{-1}, with four different phenological stages (2 pairs of leaves, 4 pairs of leaves, inflorescence at flowering and plants with fruits), and 0, 6, 12, 24, and 48 h of leaf wetness period. In order to verify the host range of this isolate, inoculations were carried out on species chosen according to the centrifugal-phylogenetic method (Wapshere 1974), with emphasis on species of the Asteraceae family, along with economically important species, such as Asteraceae-*Conyza sumatrensis* (horseweed), *B. pilosa* (beggartick), *Bidens subalternans* (beggartick), *Lactuca sativa* (lettuce), Poacea-*Zea mays* (corn), Fabaceae-*Phaseolus vulgaris* (bean), *Glycine max* (soybean), Amaranthaceae-*Beta vulgaris* (beet), Malvaceae-*Gossypium herbaceum* (cotton), Rubiaceae-*Coffea arabica* (coffee), Euphorbiaceae-*Euphorbia heterophylla*, Poaceae-*Digitaria insularis*, Solanaceae-*Solanum lycopersicum* (tomato), and Rutaceae-*Citrus sinensis* (orange).

The results demonstrated that all inoculum concentrations tested had 100% control (dead plants), showing a wide concentration range of conidia for a total control of the target species of weed (Fig. 35.1d). Control levels of 100% were obtained when *C. truncatum* isolate UFU 280 conidia suspensions were applied in the phenological stages of 2 and 4 pairs of leaves, resulting in the death of the beggartick plants 7 days after the applications. In the two other tested phenological stages (flowering inflorescence and fruiting plants), the applications of conidia suspensions of *C. truncatum* isolate UFU 280 resulted in unsatisfactory levels of control. In these cases, the beggartick plants showed only yellowing and small lesions on leaves and stems. These results were already expected since the recommended phenological stage for the control of most weed species is 2–6 leaves or 2–5 tillers for chemical herbicides (Cabral et al. 2013). The period of leaf wetting showing the highest efficiency in the control of beggartick plants was 48 h, obtaining 100% of dead plants. The other periods tested were not enough to cause lesions in the beggartick-inoculated plants. This result can be explained by the germination of conidia of the genus *Colletotrichum*, occurring only in the presence of high relative humidity during a long period. This result represents a limitation to the development of a commercial mycoherbicide based on *C. truncatum* isolate UFU 280 conidia. This current limitation can be overcome with suitable liquid formulations to be recommended at the time of field application, such as the addition of some wetting adjuvants. The isolate of *C. truncatum* isolate UFU 280 was inoculated on plants belonging to 14 species of 10 different plant families and was only pathogenic to *B. pilosa* and *B. subalternans*.

Other aspects of great relevance to allow a proper evaluation of the commercial viability of a *C. truncatum* isolate UFU 280 mycoherbicide are the selection of a process for preserving the propagules, which allows an adequate shelf life, the development of an adequate formulation, and the demonstration of the efficacy of that isolate in field conditions. These parameters are under investigation, aiming at developing a future mycoherbicide to control beggartick plants.

Target Weed: Morningglories (*Ipomoea* spp.)

Biological Agent: *Bipolaris* sp.; *Colletotrichum* sp.; *Cercospora* sp.

The species *Ipomoea hederifolia*, *I. quamoclit*, and *I. grandifolia*, popularly known as morningglories, are considered weeds of increasing importance, especially in non-burning sugarcane farming. These weeds are aggressive vines that interfere with culture productivity and mechanical harvesting, affecting the efficiency of machines and crop quality. The chemical control using herbicides in these conditions has low efficacy (Silva and Monquero 2013). In this context, these species are considered weed targets for alternative control.

In Brazil, few fungi associated with *Ipomoea* spp. have been reported (Mendes et al. 1998). In previous survey performed by Embrapa Meio Ambiente, São Paulo, Brazil, 17 fungi associated with *Ipomoea* spp. were identified and were restricted to the genera *Bipolaris*, *Colletotrichum*, *Alternaria*, and *Cercospora*. The associations observed were *Bipolaris* × *I. nil* and *I. quamoclit*; *Colletotrichum* × *I. cairica*; *Alternaria* × *I. cairica* and *I. quamoclit*; and *Cercospora* × *I. nil*, *I. triloba*, *I. hederifolia*, *I. indivisa*, and *I. grandifolia*. The most frequent association was *Cercospora* × *I. nil*. The symptoms caused by the fungi *Bipolaris*, *Colletotrichum*, and

Cercospora were leaf spots and by *Alternaria* were the leaf yellowing and leaf blight. In nature, the genus *Cercospora* occurs more frequently in species of *Ipomoea* (Fig. 35.1e). Furthermore, the species *I. quamoclit*, *I. hederifolia*, and *I. grandifolia*, which are highlighted as important weeds in sugarcane crop, are hosts of the fungi *Bipolaris*, *Cercospora*, and *Alternaria*. In this survey, we have indicated the diversity of fungi associated to the target weed, which can be investigated as potential biological control agents.

All the isolates were inoculated on the species *I. nil*, *I. grandifolia*, *I. hederifolia*, and *I. quamoclit* for pathogenicity testing and host specificity. The strains of *Bipolaris* sp. were not pathogenic to the four species of *Ipomoea*. The strain of *Colletotrichum* sp. was specific to *I. nil*. One strain of *Alternaria* sp. was pathogenic to *I. nil*, *I. hederifolia* and *I. quamoclit*, whereas the other was pathogenic only to *I. nil*. The six strains of *Cercospora* sp. had different results: two of them were specific to *I. nil*; one was pathogenic to *I. nil* and *I. hederifolia*; and three strains were pathogenic to *I. nil*, *I. hederifolia*, and *I. grandifolia.* Considering the observed results, the following strains were selected for detailed studies of biological control: one strain of *Alternaria* sp. once it was the only pathogenic to *I. quamoclit*, and one strain of *Cercospora* sp. which caused irregular, pale brown spots on leaves surrounded by a yellow halo on *I. nil*, *I. hederifolia*, and *I. grandifolia*. Later, those leaves turned yellow resulting in early abscission (Fig. 35.1f). This chapter focused on the results of the *Cercospora* sp. isolate.

Epidemiological studies based on the effects of propagule concentration, leaf wetness period, and phenological stage on the development of cercospora leaf spot on *Ipomoea nil* were conducted to determine the optimum conditions for the progress of the disease and to the induction of plant defoliation. The parameters considered for inoculum concentration were 2×10^5, 2×10^6, and 2×10^7 conidia mL^{-1}; for leaf wetness period were 0, 6, 12, 24, and 48 h after inoculation; and for the phenological stages were "plants with up to 3 leaves," "plants with 3–5 leaves," "plants with open flowers," and "plants with fruits." Additionally, host range was determined based on 15 plants species selected according to Wapshere (1974), belonging to the subclass Asteridae, with emphasis on members of Convolvulaceae, Solanaceae, and Compositae, as well as safeguards species such as sugarcane, coffee, soybean, and corn.

The results demonstrated that, among the epidemiological parameters investigated, an inoculum concentration of 2×10^7 conidia mL^{-1}, a minimum leaf wetness of 24 h and plants from phenological stages "plants with 3–5 leaves" to "plants with fruits" were the best parameters to induce plant defoliation and high disease incidence on *Ipomoea nil*. These parameters were used as protocol to study the effectiveness of the isolate of *Cercospora* sp. on other species of *Ipomoea* spp. Although the symptoms of leaf spot were observed on *I. grandifolia*, *I. hederifolia*, and *I. quamoclit*, the plant defoliation was restricted to *I. grandifolia* and *I. hederifolia.* The *Cercospora* sp. isolate was specific to *I. nil*, *I. grandifolia*, *I. hederifolia*, and *I. quamoclit* so the non-target species are not susceptible to the isolate (Nechet and Halfeld-Vieira 2019).

In this study, a fungus isolate from mycobiota of morningglories was selected for mycoherbicide purposes, and its effect on plants is dependent on a high inoculum concentration and some specific conditions of wetness period. The plant defoliation was considered the main effect that can reduce the growth of these weeds in a crop. Besides, the isolate was specific to three species of *Ipomoea* and do not pose a risk to economically important plants.

The study represents the first attempt to prospect fungi as an alternative control of *Ipomoea* spp. For the first time, the "morningglories complex" was target weeds in a program of biological control. The choice was based on the increasing importance of morningglories as weeds, mainly on non-burning sugarcane and on organic sugarcane crops, which have expanded the production area in Brazil. Regardless of the situation, there is no alternative control to chemical herbicides. The great obstacle pointed out by our results was the inoculum production, once the selected isolate belongs to the genus *Cercospora* sp., which has been known as a difficult fungus to sporulate in vitro conditions. However, future research projects focusing on nutrient sources of substrate, fermentation techniques, and adjuvants can overcome this problem.

In this sense, although the current results are not applied immediately to the producers, they represent important steps in the future development of a mycoherbicide to control *Ipomoea* spp., which involves the discovery of potential candidates as biological control agents, and epidemiological studies related to disease development on target weed.

35.3 Final Remarks

The use of natural enemies to weed control in Latin America has been investigated considering both the classical and the bioherbicide approaches. This region has a wide diversity of organisms that can be exploited as biological control agents. Although the undergoing studies are limited to few teams of research institutions, mainly in Brazil, the increased importance of biological control of weeds can stimulate other Latin American scientists to contribute to this area.

References

Auld BA (1998) On the social value of biological control of weeds. Int J Soc Econ 25(6/8):1199–1206

Barreto RW (2009) Controle biológico de plantas daninhas com fitopatógenos. In: Bettiol W, Morandi MAB (eds) Biocontrole de doenças de plantas: uso e perspectivas. Embrapa Meio Ambiente, Jaguariúna, pp 101–128

Borges Neto CL, Pitelli RA (2004) Adjuvantes e herbicidas e a infectividade de *Fusarium graminearum*, agente potencial de biocontrole de *Egeria densa* e *Egeria naja*. Planta Daninha 22:77–83

Cabral PHR, Jakelaitis A, Cardoso IS et al (2013) Interferência de plantas daninhas na cultura do sorgo cultivado em safrinha. Pesqui Agropecu Trop 43(3):308–314

Charudattan R (2005) Ecological, practical, and political inputs into selection of weed targets: what makes a good biological control target? Biol Control 35(3):183–196

Charudattan R, Dinoor A (2000) Biological control of weeds using plant pathogens: accomplishment and limitations. Crop Prot 19(8/10):691–695
CONSEMA (2012) Resolução Consema n° 08, de 14 de setembro de 2012. Reconhece a Lista Oficial de Espécies Exóticas Invasoras no Estado de Santa Catarina e dá outras providências. DO St Catarina 429:3–6
Cousens R, Croft AM (1995) Weed populations and pathogens. Weed Res 40:63–82
Cousens R, Mortimer M (2000) Dynamics of weed populations. Cambridge University Press, Cambridge
Cronk QCB, Fuller JL (1995) Plant invaders. Chapman and Hall, London
DiTomaso JM (2000) Invasive weeds in rangelands: species, impacts, and management. Weed Sci 48(2):255–265
Duke SO (2015) Proving allelopathy in crop–weed interactions. Weed Sci 63(1):121–132
Fowler SV, Barreto RW, Dodd S et al (2013) *Tradescantia fluminensis*, an exotic weed affecting native forest regeneration in New Zealand: ecological surveys, safety tests and releases of four biocontrol agents from Brazil. Biol Control 64(3):323–329
Gelmini GA, Victória Filho R, Novo MCSS (2002) Resistência de *Bidens subalternans* aos herbicidas inibidores da enzima acetolactato sintase utilizados na cultura da soja. Planta Daninha 20(2):319–325
Glowka L, Burhenne-Guilman F, Synge H et al (1994) A guide to the convention on biological diversity. IUCN, Gland and Cambridge
Guatimosim E, Pinto HJ, Pereira OL et al (2015) Pathogenic mycobiota of the weeds *Bidens pilosa* and *Bidens subalternans*. Trop Plant Pathol 40(5):64–83
Hallet SG (2005) Where are the bioherbicides. Weed Sci 53(3):404–415
Harker KN, O'Donovan JT (2013) Recent weed control, weed management, and integrated weed management. Weed Technol 27(1):1–11
Holm L (1991) The world's worst weeds-distribution and biology. Krieger Publishing Company, Malabar
Jackson MA, Schisler DA, Slininger PJ et al (1996) Fermentation strategies for improving the fitness of a bioherbicide. Weed Technol 10(3):645–650
Johnson MT (2005) Petition for field release of *Tectococcus ovatus* (Homoptera: Eriococcidae) for classical biological control of strawberry guava, *Psidium cattleianum* Sabine (Myrtaceae). Hawaii. http://www.fs.fed.us/psw/topics/ecosystemprocesses/tropical/invasive/. Accessed 2 Nov 2017
Julien MH, Griffiths MW (1998) Biological control of weeds. A world catalogue of agents and their target weeds. CABI Publishing, Wallingford
Kissmann KG, Groth D (1991) Plantas infestantes e nocivas. Basf Brasileira, São Paulo
Macedo DM, Pereira OL, Hora Júnior BT et al (2016) Mycobiota of the weed *Tradescantia fluminensis* in its native range in Brazil with particular reference to classical biological control. Australas Plant Pathol 45(1):45–56
Mack RN, Simberloff D, Lonsdale WM et al (2000) Biotic invasions: causes, epidemiology, global consequences, and control. Ecol Appl 10(3):689–710
Madire LG (2013) Biology and host range of *Mada polluta*, a potential biological control agent of *Tecoma stans* in South Africa. Biocontrol Sci Technol 23(8):944–955
Marrone PG (2014) The market and potential of biopesticides. In: Gross AD, Coats JR, Duke SO et al (eds) Biopesticides: state of the art and future opportunities. ACS Symposium Series. American Chemical Society, Washington, pp 245–258
McFadyen REC (1998) Biological control of weeds. Annu Rev Entomol 43:369–393
McFadyen REC, Wilson B (1997) A history of biological control of weeds. In: Julien M, White G (eds) Biological control of weeds: theory and practical applications. ACIAR, Canberra, pp 17–22
Medina MG, García DE, González ME et al (2009) Variables morfo-estructurales y de calidad de la biomasa de *Tithonia diversifolia* en la etapa inicial de crecimiento Morpho-structural. Zootec Trop 27(2):121–134

Mendes MAS, Silva VL, Dianese JC et al (1998) Fungos em plantas no Brasil. Embrapa-SPI/ Embrapa-Cenargen, Brasília

Mphephu TE, Olckers T, Simelane DO (2017) The tortoise beetle *Physonota maculiventris* (Chrysomelidae: Cassidinae) is suitable for release against the weedy Mexican sunflower *Tithonia diversifolia* (Asteraceae) in South Africa. Biocontrol Sci Technol 27(4):510–524

Nechet KL, Halfeld-Vieira BA (2019) Development of *Cercospora* leaf spot on *Ipomoea* weed species for biological control. BioControl 64(2):185–195

Nechet KL, Barreto RW, Mizobuti ES (2006) *Bipolaris euphorbiae* as a biological control agent for wild poinsettia (*Euphorbia heterophylla*): host-specificity and variability in pathogen and host populations. Biol Control 51(2):259–275

Nechet KL, Barreto RW, Vieira BS (2016) Uso dos fungos *Sphaceloma poinsettiae* e *Bipolaris euphorbiae* como mico-herbicidas no controle de *Euphorbia heterophylla*. In: Halfeld-Vieira BA, Marinho-Prado JS, Nechet KL et al (eds) Defensivos agrícolas naturais: uso e perspectivas. Embrapa, Brasília, pp 147–176

Owen MDK (2016) Diverse approaches to herbicide-resistant weed management. Weed Sci 64(1):570–584

Paterson ID, Hoffmann JH, Klein H et al (2011) Biological control of Cactaceae in South Africa. Afr Entomol 19(2):230–246

Paterson ID, Mdodana LA, Mpekula O et al (2014a) A promising biological control agent for the invasive alien plant, *Pereskia aculeata* Miller (Cactaceae), in South Africa. Biocontrol Sci Technol 24(10):1083–1095

Paterson ID, Vitorino MD, Cristo SC et al (2014b) Prioritisation of potential agents for the biological control of the invasive alien weed, *Pereskia aculeata* (Cactaceae), in South Africa. Biocontrol Sci Technol 24(4):407–425

Paynter Q, Fowler SV, Hayes L et al (2015) Factors affecting the cost of weed biocontrol programs in New Zealand. Biol Control 80:119–127

Pérez A, Montejo I, Iglesias JM et al (2009) *Tithonia diversifolia* (Hemsl.) A. Gray. Pastos Forrajes 32:1–15

Randall JM (1996) Weed control for the preservation of biological diversity. Weed Technol 10:370–383

Reaser JK, Galindo-Leal C, Ziller SR (2005) Visitas indesejadas: a invasão de espécies exóticas. In: Galindo-Leal C, Câmara IG (eds) Mata Atlântica: biodiversidade, ameaças e perspectivas. Fundação SOS Mata Atlântica. Conservação Internacional, Belo Horizonte, São Paulo, pp 390–403

Shaw DR (2016) The "wicked" nature of the herbicide resistance problem. Weed Sci 64(1):552–558

Silva PV, Monquero PA (2013) Influência da palha no controle químico de plantas daninhas no sistema de cana crua. Rev Bras Herb 12(1):94–103

TeBeest DO (1996) Biological control of weeds with plant pathogens and microbial pesticides. Adv Agron 56:115–137

Vieira BS, Barreto RW (2005) *Lewia chlamidosporiformans* sp. nov. from *Euphorbia heterophylla*. Micotaxon 94:245–248

Vieira BS, Barreto RW (2010) Liquid culture production of chlamydospores of *Lewia chlamidosporiformans* (Ascomycota: Pleosporales), a mycoherbicide candidate for wild poinsettia. Australas Plant Pathol 39(2):154–160

Vieira BS, Dias LVSA, Langoni VD et al (2018) Liquid fermentation of *Colletotrichum truncatum* UFU 280, a potential mycoherbicide for beggartick. Australas Plant Pathol 47(3):277–283

Walsh M, Newman P, Powles S (2013) Targeting weed seeds in-crop: a new weed control paradigm for global agriculture. Weed Technol 27(3):431–436

Wapshere AJ (1974) A strategy for evaluating the safety of organisms for biological control. Ann Appl Biol 77(2):201–211

Waterhouse DFW, Norris KR (1987) Biological control Pacific prospects. ACIAR, Canberra

Watson AK (1991) The classical approach with plant pathogen. In: TeBeest DO (ed) Microbial control of weeds. Chapman and Hall, New York, pp 3–23

Wessels FJ, Cuda JP, Johnson MT et al (2007) Host specificity of *Tectococcus ovatus* (Hemiptera: Eriococcidae), a potential biological control agent of the invasive strawberry guava, *Psidium cattleianum* (Myrtales: Myrtaceae), in Florida. Biol Control 52(4):439–449

Wheeler GS, McKay F, Vitorino MD et al (2016) Biological control of the Invasive Weed *Schinus terebinthifolia* (Brazilian Peppertree): a review of the project with an update on the proposed agents. Southwest Nat 15(8):15–34

Winston RL, Schwarzländer M, Hinz HL et al (2014) Biological control of weeds: a world catalogue of agents and their target weeds, 5th edn. USDA Forest Service, Morgantown

Young SL, Pitla SK, Van Evert FK et al (2017) Moving integrated weed management from low level to a truly integrated and highly specific weed management system using advanced technologies. Weed Res 57(1):1–5

Zenni RD, Ziller SR (2011) An overview of invasive plants in Brazil. Braz J Bot 34(3):431–446

Zimmermann HG, Moran VC, Hoffmann JH (2009) Invasive cactus species (Cactaceae). In: Muniappan R, Reddy GVP, Raman A (eds) Biological control of tropical weeds using arthropods. Cambridge University Press, Cambridge, pp 108–129

Chapter 36
Plant Diseases

Flávio Henrique Vasconcelos de Medeiros and Júlio Carlos Pereira da Silva

36.1 Introduction

Although frequently encountered in nature, epidemic plant diseases are rare in plant-diversified environments and frequent in monocultures, which is predominant for most crops (Agrios 2005). Furthermore, the plant disease only occurs when a virulent pathogen encounters a susceptible plant host at a specific timing, i.e., under favorable environmental conditions (Campbell and Madden 1990). Therefore, preventive-based disease management strategies should be the prioritized over curative-based ones, especially for highly epidemic diseases, such as rusts, by the release of one or more microbial species involved in the biocontrol (Van Lenteren et al. 2018) or a synergistic action of buildup of pathogen-antagonistic species and breaking-down the pathogen life cycle reached through crop rotation (Gil et al. 2008).

In the middle age in Europe, wheat was the most important staple food, and it was sown year after year in the same field, without fungicide seed or plant treatment. The monoculture and the lack of technology for disease management resulted in increasing epidemic outbreak of take-all disease (*Gaeumannomyces graminis* var. *tritici*). Religion played an important role in the daily explanations of the phenomena, such as a miracle, and the diseased plant patches would gradually reduce all of a sudden, increasing the crop yield. Later, in the twentieth century, the phenomenon was scientifically deciphered (Weller et al. 1988). In such suppressive soils, there is a selective buildup in the population of *Pseudomonas fluorescens* complex that produce the broad-spectrum antibiotic 2,4-DAPG (*phlD*$^{+}$) during the monoculture of wheat or barley. It is so-called soil suppressiveness and has been implicated in the "sudden" reduction in the disease progress over the years (Baker and Cook 1974). Such soil suppressiveness is so-called specific and can be eliminated by pasteurization (60 °C, 30 min) and the fumigation of the soil or crop.

F. H. V. de Medeiros (✉) · J. C. P. da Silva
Department of Plant Pathology, Universidade Federal de Lavras, Lavras, MG, Brazil
e-mail: flaviomedeiros@dfp.ufla.br

B. Souza et al. (eds.), *Natural Enemies of Insect Pests in Neotropical Agroecosystems*, https://doi.org/10.1007/978-3-030-24733-1_36

On the other hand, crop rotation also leads to the buildup in the community of antagonistic microorganisms (*Trichoderma*, *Gliocladium*, and actinomycetes) and a reduction in the community of soilborne pathogens (*Fusarium solani* and *Sclerotinia minor*) (Gil et al. 2008), therefore leading to a general suppressiveness (Cook and Rovira 1976).

Regardless of the considered suppressiveness (general or specific), the conducive soil can be restored by hampering the plant disease by amending a small portion of the suppressive to a conducive one (Mendes et al. 2011). In other ways, the microbial community could be transferred, and this was one of the bases of the biological control of plant diseases, particularly the soilborne ones, for which few control alternatives are available.

Therefore, biological control of plant diseases is defined as the use of a microorganism or its derivatives to control another microorganism or quenching the disease onset, i.e., by either using biological agent, their metabolites, or its components or even adopting crop management practices that boost up the population of native antagonists (Baker and Cook 1974).

In Latin America, the biological control of pests has improved in research and adoption in the field for the management plant diseases, particularly for nematodes (gall and lesion) and white mold (*Sclerotinia sclerotiorum*), with approximately five million hectares treated with biocontrol agents of plant disease (Van Lenteren et al. 2018). The growth in the biofungicide market in Latin America in 2017 was estimated to be 15.2% when compared to 2012 (Bettiol et al. 2014). In Brazil, the first biological product to control plant disease was only registered in 2008, and, presently, there are still few products registered with the Brazilian government (15 in total) (Agrofit 2017). Currently, the use of biocontrol agents is growing in Latin America. One of the reasons for this, besides sociocultural aspects, is related to the small amount of companies and products available to growers. The mass production of disease biocontrol agents, making it possible to supply a large demand of antagonists, is important for the evolution of such disease control strategy. In Latin America, there are several biocontrol companies with efficient, good-quality, and an affordable price product, just seeking government regularization (Stefanova 2007; Bettiol and Ghini 2009; Bettiol et al. 2014). Therefore, changes in the research and history of biological control of plant disease in Brazil are ongoing and promising.

The knowledge of possible new biocontrol agents to be used has increased in recent years, thanks to advances in molecular biology research and field trails that confirm their efficacy in the disease management (Chenthamara and Druzhinina 2016). Most studies still take into account only few groups of microorganisms as plant disease control antagonists. This is due to the success over the years of using some genera, such as *Trichoderma* and *Bacillus*, against several plant pathogens of important crops all around the world. Many Latin American countries have these two genera as the active mechanism in most of their commercial biological products to control plant diseases (Stefanova 2007; Bettiol et al. 2014; Yanet-Suárez and Cabrales 2016; Pérez-Torres et al. 2018). However, nowadays, other fungi and bacteria have been performing similar activity and are also considered in the portfolio of microbial biocontrol agents, for instance, *Cladosporium cladosporioides*, for the management of coffee postharvest diseases (Chalfoun et al. 2013).

In our chapter, we have gone through the main biocontrol agents commercially used for the management of plant diseases and their modes of action, with emphasis on its use in Latin America. Currently, the highest diversity of microbial biocontrol agents are based on fungi due to its peculiar mode of action, and, therefore, this group of microorganism has been more dissected than bacteria or viruses.

36.2 Microorganisms Involved in Biological Control of Plant Diseases

The antagonistic microorganisms used as biocontrol agents are usually classified by their sites of action as suppressors of soil diseases (belowground), phylloplane protectors (aboveground), postharvest protectors (fruits, vegetables, and flowers), and spermosphere protectors (seeds and seedlings). On the other hand, a biocontrol agent can be classified or applied by its mechanism of action through the production of toxic metabolites (antibiosis), capacity of competition for space or nutrient, growth promotion, parasitic activity, or resistance induction (Baker and Cook 1974). However, for the selection of biocontrol agents, it should prioritize the disease control in vivo or high-throughput screening approaches that would firstly exclude the non-desirable microorganisms rather than the mechanism of action or the type of interaction involved in the process (Köhl et al. 2011).

Nevertheless, some features are important to be taken into account for obtaining a viable antagonist microorganism aiming at plant disease control practices. Thus, they must be genetically stable and not lose its mode of action, grow in cheap and easy-to-produce artificial medium, be efficient at low concentrations, reach several pathogens in various crops, be compatible with other field practices, such as chemical and cultural control, and be harmless to humans. Therefore, efficient microorganisms may (a) induce plant resistance; (b) compete with pathogens for nutrients, sites, or oxygen; (c) produce secondary metabolites with an effect antagonistic to the pathogen; and (d) directly attack the pathogen through parasitism or predation (Druzhinina et al. 2011). Furthermore, the use of nonantagonistic organisms that are contributors to the plant health can be considered in the control of phytopathogens; and the use of attenuated virus strains by preimmunization may be used in inoculating plants to prevent the virus from causing disease.

The most successful history in biological control of a virus disease was the eradication of the *Citrus tristeza virus* (CTV). CTV was eradicated from the main commercial citrus in Brazil by preimmunization or bud donor plants, and all grafted plantlets were immune to the virulent strain of the virus with apparent no fitness cost to the plant over the years and at no extra cost to the grower.

Another benefit that is commonly triggered by the biocontrol treatment is growth promotion. This is a common feature of root-inhabiting microbes and is a result of the production of hormones, improving the nutrient and water uptake (Hu et al. 2017). The effect of the plant growth promoter does not directly target the pathogen, but rather the host, in an induction of tolerance-like disease control, i.e., the pathogen is able to infect the plant, but with the plant growth promotion mediated by the symbiont microbe, the plant compensates such loss and assures sustainable yields.

In addition to the effectiveness in the control, the biocontrol agent must follow some criteria to be successfully formulated as a biological product. Aboveground biocontrol agents should be able to withstand critical conditions such as high solar incidence, lack of humidity, and fluctuations in temperatures (high and low ones). Postharvest organisms should not grow at 37 °C (human internal body temperature) and do not produce residual compounds such as toxins (Köhl et al. 2011). There are several organisms with potential use for plant diseases control, but compared to the vast diversity of microorganisms already known, only a small part has been studied by their antagonist activity against plant pathogens.

The most widely applied microorganisms for the biological control of plant diseases are yeast-like fungi, myceliogenic fungi, bacteria (eubacteria and actinomycetes), and attenuated pathogenic viruses, which are further detailed below.

36.2.1 Fungi

Fungi are organisms that produce mycelium and conidia, which are widespread in nature, playing multiple roles in the ecosystem. Fungi are classified into three different kingdoms (*Chromista*, *Protozoa*, and *Fungi*), which means they have different lifestyles and structure yet share similarities regarding nutrient acquisition and reproduction and, specially, in the association with plants. While agronomist devotes more time to study the plant pathogenic ones, most of the encountered fungi are not pathogenic and even play an important role in nutrient acquisition to the plant either as symbionts (such as mycorrhizae) or plant litter decomposers (such as *Penicillium* spp. and *Chaetomium* spp.). Biological control agents can be found in any kingdom or phylum of fungi. However, most of fungal-based biocontrol products used against plant pathogens have an Ascomycota as an agent (Fig. 36.1). In the soil, the competition for nutrient and space is a constant challenge for the establishment of such fungi, and frequently they also deploy special features such as chitinase production, which is crucial for the parasitism of other fungi, insect, or nematode egg, all of which have chitin as constituent of its cell wall/outer layer. Therefore, the soil harbors biocontrol agents, and one of the strategies of biocontrol is the specific enrichment in such populations by delivering them into the soil. Hence, *Trichoderma* spp. (Fig. 36.1b) is the most widely deployed fungus that acts in the parasitism of microorganisms in the soil and also promotes the plant growth when in association with roots. Brazil is the largest consumer of *Trichoderma*-based products in the world (Lorito et al. 2010), with over two million hectares treated at least once with the biocontrol agent (Van Lenteren et al. 2018). It is mainly recommended for white mold management (*S. sclerotiorum*), targeting the sclerotia (resting structure), therefore reducing the apotecia formation and ascospore release.

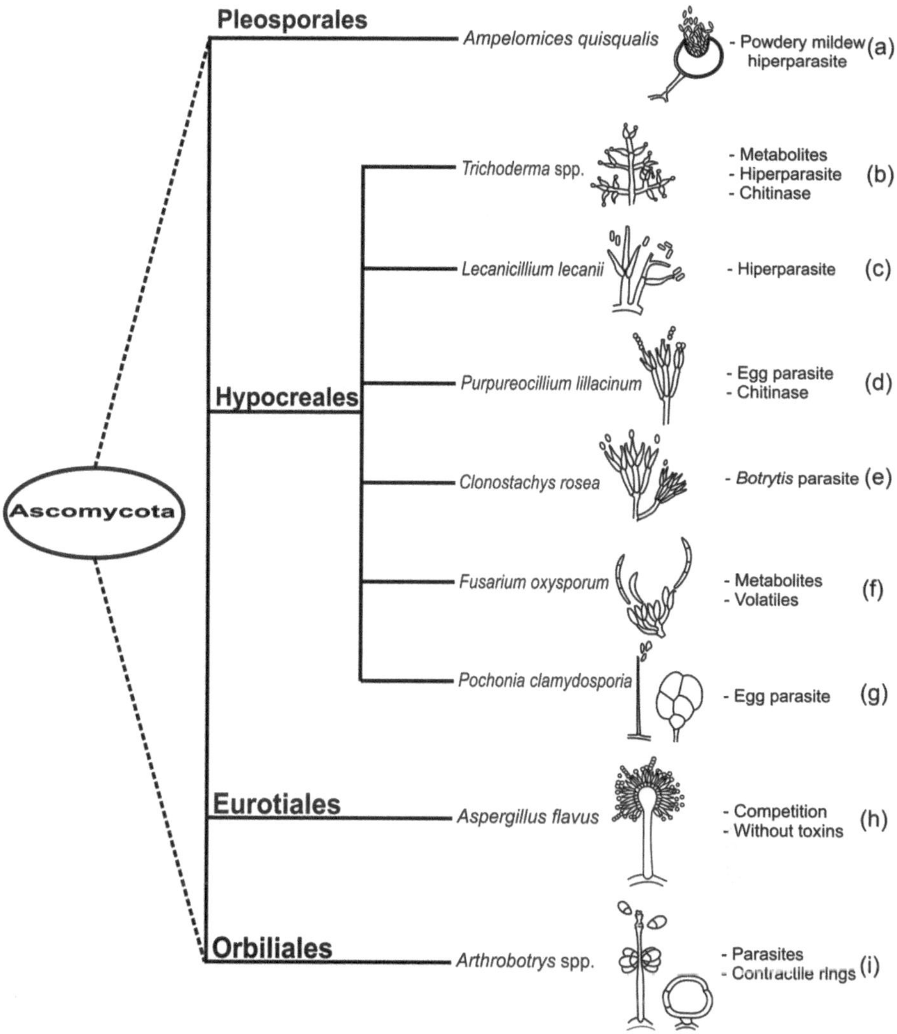

Fig. 36.1 Some of the most used *Ascomycota* fungi as control agents in bio-based treatments against plant pathogens. (Authors: Flávio H.V. Medeiros, Julio Carlos P. Silva)

Another soil inhabitant also implicated in chitinase production is *Purpureocillium lilacinum* (Fig. 36.1d), and, although it can parasite other fungi and insects, it is recommended for nematode egg parasitism, reducing the initial inoculum of the pathogen applied as soil drench upon sowing. Although *P. lilacinum*is is found in high populations in the soil, it is not the only egg parasite, and many other microorganisms have already been isolated from the egg masses and implicated in the reduction of the initial inoculum (Costa et al. 2015). One of them, based on *Pochonia chlamydosporia* (Fig. 36.1g), has been used to develop a commercial product and is currently used worldwide (Dallemole-Giaretta et al. 2012).

Pochonia chlamydosporia belongs to the *Verticillium*-like group, it sporulates poorly but produces chlamydospores, which are the propagules used as the biocontrol delivered product. Similarly to *P. lilacinum*, it is intended that a soil amendment reduces the initial inoculum. Interestingly, it has been reported to be intrinsically associated to the roots, promoting plant growth (Larriba et al. 2015) and even infecting the sedentary *Meloidogyne* sp. female (Monteiro et al. 2017).

Another strategy used in the management of soilborne nematodes is predatory fungus, such as *Arthrobotrys dactyloides* (Fig. 36.1i). Such group of fungi deploys of contractile rings that capture juvenile and adult nematodes when passing through it. Once captured, the fungus penetrates the pseudoceloma of the worm and sporulates over it, restarting a new cycle. Although frequently isolated from soils, the developed commercial products based on such group of fungi never gained much attention from the market compared to the abovementioned egg parasites, due to its variable control efficacy among different fields. The explanation for such variability was postulated by Barron (2003). Most of the nematode trapping fungi presents cellulosic and lignocellulosic composition and play an important role in wood decay and organic matter cycling. Such substrate is carbon-rich and therefore nitrogen-poor. For protein synthesis, the fungus needs extra supply of nitrogen, which is found in the nematode they entrap, an evolutionary trait similar to the carnivorous plant. Eventually, their biocontrol performance is enhanced when the fungus is applied in carbon-rich soils, such as the stubble-rich no-till fields.

Another group of fungi frequently isolated from the soil is *Cladosporium cladosporioides*. It is a highly sporulating fungus with strong saprophytic activity that may compete for nutrient with plant pathogens, such as apple scab (*Venturia inaequalis*), that overwinter in the stubble and reduce its inoculum, which is tackled in its ascospore release when leaves are treated with the fungus. The fungus can be sprayed over plants and protects buds and leaves from the pathogen infection and can replace the fungicide spray using the same fungicide-based disease forecast system (Köhl et al. 2015). Through a similar competition mechanism, *C. cladosporioides* has been implicated in the postharvest management of mycotoxin in coffee berries, when the berries are sprayed in the field (Chalfoun et al. 2013).

The competition is a common mode of action through which other fungi control plant disease, especially the postharvest ones. Several yeasts are used in the biological control (*Sporobolomyces roseus*, *Aureobasidium pullulans*, and *Cryptococcus laurentii*) because of their fast growing, highly sporulating nature, and ability to survive under drought stress. That is considered an advantage for formulation and assures its performance in the management of disease of the phylloplane and fructoplane, where the fluctuations in humidity and temperature are frequent. Furthermore, yeasts rarely produce the same antibiotic used in medicine (Berdy 2005), and, therefore, the end consumers of the yeast-treated fruits will unlikely exert an antibiotic selection pressure on the human pathogenic bacteria and impact on the gut bacterial community.

Myceliogenic fungi are also implicated in the postharvest disease control, but its mode of action is rather different from yeasts, for example, the fungi *Clonostachys* spp. (Fig. 36.1e). They are frequently encountered in nature, particularly on falling leaves and fruits, such as white or yellowish mycelium and conidia. The fungus colonizes the apoplast of plant tissues and hamper the development of plant pathogens, such as *Botrytis cinerea*. It is even able to parasite the pathogen mycelium and to reduce it, when it is at a low inoculum level. Therefore, its performance is related to the timing of delivery when it is sprayed after the pathogen establishment, i.e., at curative basis, the biocontrol does not occur. On the other hand, when it is sprayed upon blooming, not only the disease is avoided, but also there are reports that the fruit has a longer shelf life likely due to its lower pathogen inoculum but also residual colonization by the biocontrol fungus (Abreu et al. 2014). Noteworthy, to assure growers will always be dealing with a preventive application of the biocontrol, it is of great importance to collect and bury all gray mold symptomatic fruit and leaves and also to control spider mites that cause wound and act as an inoculum spreader of *B. cinerea*. Such management can also be achieved with biocontrol, through the delivery of predator mites (Morandi et al. 2000).

Conversely, plant pathogens are also sought to act as biocontrol agents. *Aspergillus flavus* (Fig. 36.1h) defective in the production of aflatoxin and cyclopiazonic acid are delivered to the field upon flowering; they colonize grains or nuts, and whenever a mycotoxigenic fungus reaches such grain, it does not encounter space or free nutrient for infection (Dorner and Lamb 2006). In Africa, a total reduction of aflatoxin has been reached with a multi-sponsor initiative of reduction mycotoxicosis in maize (Atehnkeng et al. 2008).

A similar use of non-virulent pathogenic fungal species was attempted in the 1980s–1990s with the Fuseclean, a product that had a nonpathogenic strain of *Fusarium oxysporum* (Fig. 36.1f) as its active ingredient and when applied to the soil could reduce the infection by virulent strains of *F. oxysporum* (Alabouvette et al. 1993). Recent studies about the risk of gene flow in *F. oxysporum* raise concern about the risks of such introduction of biocontrol agent (Ma et al. 2010). However, the metabolites produced by the fungus can be explored. *F. oxysporum* is able to produce volatile organic compounds that have nematicide activity (Terra et al. 2017).

While some biocontrol fungi are recovered from nature, multiplied in the lab, and reintroduced in the field with an important role in the plant protection against diseases, some others are reported but not used in the field, for example, *Lecanicillium lecanii* (Fig. 36.1c) and rust hyperparasites. They are reported as white mycelial growth over the brown or orange uredospores or other sporulating state in the macrocyclic rusts (Spencer and Atkey 1981). The fluctuation in the hyperparasitism has been reported and occurs mostly in the dry season when the sporulation of the pathogen is higher. Therefore, it is assumed that such parasitism would result in the reduction of the initial inoculum of the rust of perennial crops, such as coffee, reducing or delaying the epidemic outbreaks (Martins et al. 2015). However, rust epidemic depends less on the initial inoculum than necrotrophic diseases, since it has a short life cycle and is highly sporulating, being easily wind dispersed (Maffia et al. 2009). Surprisingly, such parasitism does play a role in the rust severity reduction, but there are ecological players that should be taken into account (Vandermeer et al. 2009). The authors found a strong correlation between the *Azteca instabilis* (Smith

1862) (Hymenoptera: Formicidae) nests and the trees with lower rust (*Hemileia vastatrix*) severity. Such ants present a mutualistic relationship with the green coffee scale, *Coccus viridis* (Green 1889) (Hemiptera: Coccidae), and wherever there is an ant nest, the population of the scale tends to buildup due to its protection against predators assured from the ant. However, such bodyguards could not avoid the scale to become sick, and they were infected with *L. lecanii*, a fungus with strong chitinolytic activity. With the high population of the scale, the infection rapidly turned into an epidemic, and the fungus abundantly sporulated over the insects. In turn, next to the scale, a rust pustule was frequently found, which is also another chitin-rich substrate (Liu et al. 2011), and therefore on the menu of the fungus, such parasitism was accompanied by the gradual decrease in the disease severity in the field.

The hyperparasitism is an important mode of action of biocontrol agents and is generally a result of coevolution, and, as such, a biocontrol agent would have a narrow host range, controlling few diseases. However, such biocontrol is worthy when the disease is of great importance. In greenhouse production systems, simply avoiding overhead irrigation controls most of the diseases. However, the powdery mildew disease, which is favored by low free water of the phylloplane, is still a challenge. When such epidemy occurs, a fungus that produces black picnidia (*Ampelomyces quisqualis)* is encountered over the fluffy mycelium and conidiophores of powdery mildews (Fig. 36.1a). It can be applied over infected plants and feeds on the pathogen mycelium, resulting in gradual decrease of the disease severity. A sine qua non condition for the effective parasitism is high humidity, which is also encountered in greenhouse production but not always in the field.

Another worldwide spread disease of great importance is white mold. Although *Trichoderma* spp. is widely used to manage this disease, the fungus is able to parasite the sclerotia of the pathogen; the most efficient parasite of such resting structure is *Paraconiothyrium minitans* (Zeng et al. 2014). The biocontrol is able to perceive the oxalic acid released by the germinating sclerotia by transmembrane receptors, which trigger a signal transduction cascade and gene regulation that drive its mycelial growth toward the chemical stimulus and parasitism of the sclerotium (Whipps et al. 2008). The biocontrol is largely sprayed over *S. sclerotiorum*-infested fields but has not achieved satisfactory white mold control in tropical fields, either due to the higher or variable temperature or the competition with the native microbiome.

Although classical biocontrol is reported in entomology and weed science, it is scarce in the biocontrol of plant diseases. However, in perennial tropical crops, it has already been achieved and resulted in long-term plant protection with one or few introductions of the biocontrol agent. In coconut, tar spot (*Phyllacora torrendiella*) is a devastating disease. The fungus causes leaf spots that merge and may completely defoliate the plant. The pathogen life cycle encompasses a four-month resting stage as stroma. Upon rain and high temperatures, the mature stroma exposes its asci, which releases the ascospores and starts over a new disease cycle. During its resting stage, while not yet mature, the stroma is vulnerable to hyperparasite colonization. With little disease control alternative, Sudo (1989) isolated *Acremonium alternatum* and *Acremonium persicinum* from *P. torrendiella* stroma, multiplied it in the laboratory, and sprayed it over coconut leaves. Without any further spray, the disease gradually decreased, and the coconut yield was restored over the years.

Once the vulnerable stage within the life cycle of a given plant pathogen is known, the biocontrol strategy is ought to be implemented. In rubber tree (*Hevea brasiliensis*) production, leaf blight (*Microcyclus ulei*) is the most devastating disease that has led to the extinction of rubber production in the amazon region. Its control is based on an escape strategy, i.e., planting the tree in regions not favorable to the disease but favorable to the biocontrol is a promising strategy. Similarly to tar spot, leaf blight has its resting stage in the decaying leaves as stroma that harbors the asci, which releases ascospores when mature. Such stroma is colonized by *Dycima pulvinata* and has compromised its ascospore formation (Mello et al. 2005). When still attached to the plant, the leaves are sprayed with the biocontrol fungus, and when the conditions are favorable for the disease to start over, the initial inoculum is much lower and gradually reduced over years, especially when combined to other disease management practices (Melo et al. 2008).

Noteworthy, when the parasitism is the strategy attempted for the disease control, it is important to work on the expectation of the grower that the results are achieved gradually. Generally, a field trial of at least 3 years is necessary to validate the technology and/or expect return on the investment in terms of yield. Nevertheless, once the balance between pathogen and biocontrol populations is achieved, there is no further need of biocontrol introduction in the field (Freitas et al. 2009).

In this regard, it is important to bring about our previously frustrated attempt to achieve faster results on the management of tar spot of coconut by integrating the hyperparasite biocontrol approach with fungicide. The approach was directed toward the combination of the application of *Acremonium alternatum* and fluotriafol, a triazole that is applied using a shotgun coupled with a rigid hose with an umbrella handle-like ending that allows to individually treat plants. The product is systemic and protects the newly formed leaves from the ascospore infection. The experiment was set up with the grower to compare the exclusive use of the chemical fungicide, the biocontrol application alone, no application, and both the chemical and the biocontrol application. In the evaluation, approximately 3 months after the plant sprays, it was already visible that the fungus was colonizing the stroma. However, the newly formed leaves in the fungicide treatment were completely clean from the disease. Although the approach intended to work on the expectation of the grower and deliver a fast result with plant protection, it also intended sustainability with the establishment of the biocontrol and a gradual reduction of the dependency of the chemical control and even the biocontrol. However, the grower instead aborted the research initiative and kept on with the monthly basis fungicide application.

A similar IPM approach was attempted and well succeeded in the management of witch's broom (*Moniliophthora perniciosa*). The biocontrol fungus *Trichoderma stromaticum* is able to colonize the broom and reduce in up to 99% of the basidiocarp formation (Samuels et al. 2000). However, the reduction of the initial inoculum was not enough to hamper the disease onset and progress. This complementary action was achieved by spraying plants with the biocontrol and 15 days afterward with a copper-based fungicide that protected the young plant tissues from the basidiospore infection. The combination resulted in increased yield since the first year of implementation of the IPM approach compared to either the biocontrol or the copper application on the broom pruning in Brazil (Medeiros et al. 2010). With the increasing labor costs in the cocoa production, the hyperparasite application has since gradually replaced the broom pruning in cocoa production in South America, the so-called biological pruning.

36.2.2 Bacteria

The most abundant and dynamic microorganisms are bacteria. They are initial colonizers of the plant tissue due to its fast-growing nature and the higher efficiency in nutrient use compared to eukaryote cells. Although many bacterial genera are reported as biocontrol agents, playing a role in nature, most of the commercial products available to growers are based on *Bacillus* spp. due to its easiness to mass production and formulation (Biocomes 2017).

Bacillus spp. are gram-positive, anaerobe-facultative, and endospore-forming bacteria found in the different ecosystems that play a role in the disease suppression. In a survey of soils for the suppressiveness to *Pectobacterium carotovorum* subsp. *carotovorum*, the most suppressive and most conducive soils were similar for most of the chemical and physical properties regarding the microbial composition. There was a significant and high positive correlation between the *Bacillus* spp. community and the suppressiveness measured as the relative extinction rate, while there was a negative significant correlation between the conducive and the *Bacillus* spp. community, i.e., the more *Bacillus* spp. the soil contained, the more suppressive it was to the pathogen. Therefore, the *Bacillus* spp. played a role in the disease control, and its application may result in the disease control. When seeds are treated with *Bacillus*-based products, it could colonize the roots and protect it against soilborne infection, for instance, nematodes, as mentioned earlier in this chapter.

Other bacteria have been reported as playing a role in plant protection, and its mode of action is mostly indirect, i.e., they trigger induced systemic resistance, and the plant itself produces the antimicrobial compounds that tackled the pathogen invasion and colonization. Those bacteria are *Rhizobium tropici* and *Azospirillum brasilense* that are mostly intended for nitrogen fixation in Fabaceae and Poaceae, respectively.

Another important bacterial genus implicated in the biological control is *Pasteuria* spp. It has long been reported as a nematode parasite. It forms endospores that are resting structures in the soil and have electrostatic affinity for the nematode epidermis (Preston et al. 2003). Once attached, it penetrates the pseudoceloma and gradually colonizes the worm less drastic than *Arthrobotrys dactyloides*, and mostly the nematode reaches maturity, laying less or no eggs at all. In the case of *Meloidogyne* spp., the female is actually dead and filled with *Pasteuria* endospores that are later released to the soil once the crop is harvested and the soil is plowed.

There are different species of *Pasteuria*, and it is generally specific to each species of nematode, i.e., *Pasteuria penetrans* parasites *Meloidogyne* spp., while *Pasteuria nishizawae* parasites *Heterodera glycines*. One company (*Pasteuria* Bioscience acquired by Syngenta) developed a protocol for the low-cost mass production of each *Pasteuria* species, and the product is starting to be marketed all around the world. However, little scientific information is already publicly available about its performance in the field.

Noteworthy, many other bacterial genera are reported as biocontrol agents of plant disease but are either in a red list of human health governmental agencies, such as *Burkholderia* and *Serratia* (Mendes et al. 2013), difficult to formulate as a long-shelf life manner, such as *Pseudomonas*, or slow growing and not feasible for large-scale production, such as actinobacteria.

36.2.3 Virus

Virus like agents have been studied for many years to control plant pathogens and pests (Anagnostakis 1982; Lacey et al. 2001). Weak virus strains have been used for the cross-protection or preimmunization of plants. In Brazil, the control of the *Citrus tristeza virus* and the common and ringspot virus of the Cucurbitaceae has been achieved by using preimmunization (Fulton 1986; Rezende and Pacheco 1998). Furthermore, viruses can also cause diseases in plant pathogen microorganisms, such as fungi infected with mycoparasite viruses, bacteria infected with bacteriophage viruses, and nematodes. Biological control using virus as the agent depends on the ability of the viral particles to endure at high levels in the vicinity of the target microorganism. Viruses have no motion ability and should be applied directly to host cells. They are mandatory parasites and will depend on the presence of the host cell to replicate. In addition, viruses must reach and infect hosts before environmental factors, such as temperature, humidity, and sunrays, reduce the viral population below levels that are effective for biological control (Jones et al. 2012). Thus, several considerations exist regarding the improvement of virus-pathogen interactions: density and accessibility of the target microorganism; time of virus application to optimize effectiveness; the ability to infect and replicate under environmental conditions; viral density at the site of application; viral particle stability rates; and the presence of adequate moisture to promote virus diffusion. Mycoparasite viruses can control phytopathogenic fungi. The fungal isolates of *Rosellinia necatrix*, *Diaporthe* sp., and *Cryphonectria parasitica* lost virulence in pear and apple plants when they were infected by viral particles from the Partitivirus and *Mycoreovirus* groups (Kanematsu et al. 2010).

Viruses capable of infecting bacteria are also considered biological control agents. Several examples show the ability of bacteriophages to control plant diseases (Jones et al. 2012). Nematodes, such as cyst nematodes (*Heterodera glycines*), were infected by many virus particles in soybean field and under greenhouse conditions. Several of these SCN viruses were also detected in *H. trifolii* (clover cyst) and

H. schachtii (beet cyst), but not the other tested cyst root-knot or reniform nematode species (Ruark et al. 2017).

36.2.4 Other Groups Involved in the Biological Control of Plant Diseases

Great part of studies on biological control of plant diseases is related to fungi, bacteria, and even virus. However, organisms such as protozoa, nematode, and insects may be related to the control of plant pathogens.

In nature, biological control occurs spontaneously, following a balance in the trophic chains. Particularly in the soil, such naturally occurring biological control is suppressiveness against soilborne diseases, which is rated in different levels according to the strength of the microbial community to stand increasing levels of target pathogens. Suppressive soils are defined as those where disease development is suppressed even when the pathogen is introduced on the susceptible host presence (Mendes et al. 2011). In a typical agricultural soil, the trophic-chain comprises the roots of plants, plant-parasitic nematodes, other groups of nematodes, fungi, bacteria, mites, insects, amoebas, and earthworms (Barker and Koenning 1998). Bacteria can be preyed by nematodes and protozoa and fungi by nematodes, springtail, mites, centipedes, and certain mycophagous protozoa that feed on spores and hyphae. For instance, *Gephyramoeba*, *Mayorella*, *Saccamoeba*, and *Thecamoeba* are protozoa that feed on the plant pathogen fungi *G. graminis* var. *tritici* and *Cochliobolus sativus* (Chakraborty 1983). Mite species are thought to feed on some plant pathogens. Tydeid mite (*Orthotydeus lambi*) can suppress the development of grape powdery mildew (*Uncinula necator*) on wild and cultivated grapes (English-Loeb et al. 1999). Predatory nematodes feed on soil microorganisms including plant parasitic nematodes. Predation by nematodes of the orders *Mononchida*, *Diplogasterida*, *Dorylaimida*, and *Aphelenchida* has been studied as biocontrol agents of plant parasitic nematodes (Khan and Kim 2007). The predator nematode *Mononchoides fotidens* is able to feed on juveniles of *Meloidogyne arenaria* (Khan and Kim 2005).

36.3 Final Considerations

Compared to biocontrol of plant pests, biocontrol of plant diseases is still very young in science and application. However, great potential exists in increasing the adoption of the biocontrol agents currently available in the market, especially in the development of new products for new targets that are still neglected from the biocontrol industry, such as staple food and vegetables. For such minor crops, governmental subsidies should reinforce the initiative of researchers to bring such products to the growers. For example, the initiative of the which's broom biocontrol agent in cocoa that is produced by the government (TRICHOVAB) and distributed to growers at fair cost.

Furthermore, we explore much less than the available microbial and animal resources for biocontrol of plant diseases. To promote such improvement, research should focus on naturally occurring disease suppressiveness and not only point out the most promising organisms implicated in such biocontrol but also identify the conditions that support the microbial establishment, such as nutrient amendment, pH, and moisture.

Obviously, growers will not adopt the biocontrol if he is not aware of its existence. The biocontrol topic should be included in agronomy courses or at least embedded within the plant disease management courses worldwide. Furthermore, law enforcement measures to reduce the chemical use, such as the EU 1107/2009, would promote not only the biocontrol adoption but also its success in disease control since failure of the biocontrol establishment is also due to incompatibility with the presently adopted fungicide-based IPM.

Finally, biocontrol product development should be sought in a global perspective, in different conditions, considering the nuances of each region, crop, and disease occurrence. For instance, Latin America, Africa, and Asia could simultaneously develop a biocontrol product for the management of coffee rust disease (*Hemileia vastatrix*). All of them have many tropical regions, grow coffee in vast lands, and share similar problems with rust.

References

Abreu LM, Moreira GM, Ferreira D et al (2014) Diversity of Clonostachys species assessed by molecular phylogenetics and MALDI-TOF mass spectrometry. Fungal Biol 118(12):1004–1012

Agrios GN (2005) Plant pathology. Elsevier Academic Press, Burlington

Agrofit (2017). http://agrofit.agricultura.gov.br/agrofit_cons/principal_agrofit_cons. Accessed 10 Oct 2017

Alabouvette C, Lemanceau P, Steinberg C (1993) Recent advances in the biological control of Fusarium wilts. Pest Manag Sci 37:365–373

Anagnostakis SL (1982) Biological control of chestnut blight. Science 215(4532):466–471

Atehnkeng J, Ojiambo PS, Donner M et al (2008) Distribution and toxigenicity of *Aspergillus* species isolated from maize kernels from three agro-ecological zones in Nigeria. Int J Food Microbiol 122(2):74–84

Baker K, Cook RJ (1974) Biological control of plant pathogens. WH Freeman and Company, San Francisco

Barker KR, Koenning SR (1998) Developing sustainable systems for nematode management. Annu Rev Phytopathol 36:165–205

Barron GL (2003) Predatory fungi, wood decay, and the carbon cycle. Biodiversitas 4(1):3–9

Berdy J (2005) Bioactive microbial metabolites. J Antibiot 58(1):1–26

Bettiol W, Ghini R (2009) Impactos das mudanças climáticas sobre o controle biológico de doenças de plantas. In: Bettiol W, Morandi MAR (eds) Biocontrole de doenças de plantas: uso e perspectivas. Embrapa Meio Ambiente, Jaguariúna, pp 29–48

Bettiol W, Rivera MC, Mondino P et al (2014) Control biológico de enfermedades de plantas en América Latina y el Caribe. Embrapa Meio Ambiente, Jaguariúna

Biocomes (2017). www.biocomes.eu. Accessed 10 Oct 2017

Campbell CL, Madden LV (1990) Introduction to plant disease epidemiology. Wiley, Michigan

Chakraborty S (1983) Population dynamics of amoebae in soils suppressive and non-suppressive to wheat take-all. Soil Biol Biochem 15(6):661–664

Chalfoun SM, Angélico CL, Pimenta CJ et al (2013) Viabilidade de *Cladosporium cladosporioides* no produto "Cladosporin" em diferentes temperaturas. 8° anais do simpósio de pesquisa dos cafés do Brasil, Salvador, 2013. Fapemig, Salvador, pp 1–5

Chenthamara K, Druzhinina IS (2016) Ecological genomics of mycotrophic fungi. In: Druzhinina IS, Kubicek CP (eds) Mycota series volume IV: environmental and microbial relationships. Springer, Switzerland, pp 215–246

Cook RJ, Rovira A (1976) The role of bacteria in the biological control of *Gaeumannomyces graminis* by suppressive soils. Soil Biol Biochem 8(4):269–273

Costa LSAS, Campos VP, Terra WC et al (2015) Microbiota from *Meloidogyne exigua* egg masses and evidence for the effect of volatiles on infective juvenile survival. Nematology 17(6):715–724

Dallemole-Giaretta R, Freitas LG, Lopes EA et al (2012) Screening of *Pochonia chlamydosporia* Brazilian isolates as biocontrol agents of *Meloidogyne javanica*. Crop Prot 42:102–107

Dorner J, Lamb M (2006) Development and commercial use of Afla-guard®, an aflatoxin biocontrol agent. Mycotoxin Res 22(1):33–38

Druzhinina IS, Seidl-Seiboth V, Herrera-Estrella A et al (2011) *Trichoderma*: the genomics of opportunistic success. Nat Rev Microbiol 9(10):749–759

English-Loeb G, Norton AP, Gadoury DM et al (1999) Control of powdery mildew in wild and cultivated grapes by a tydeid mite. Biol Control 14(2):97–103

Freitas LG, Podesta GS, Ferraz S et al (2009) Supressividade de solo a *Meloidogyne* spp. por *Pasteuria penetrans* nos Estados do Maranhão e Santa Catarina. In: Bettiol W, Morandi MAR (eds) Biocontrole de doenças de plantas: uso e perspectivas. Embrapa Meio Ambiente, Jaguariúna, pp 147–166

Fulton RW (1986) Practices and precautions in the use of cross protection for plant virus disease control. Annu Rev Phytopathol 24:67–81

Gil SV, Haro R, Oddino C et al (2008) Crop management practices in the control of peanut diseases caused by soilborne fungi. Crop Prot 27(1):1–9

Hu J, Wei Z, Weidner S et al (2017) Probiotic *Pseudomonas* communities enhance plant growth and nutrient assimilation via diversity-mediated ecosystem functioning. Soil Biol Biochem 113:122–129

Jones JB, Vallad GE, Iriarte FB et al (2012) Considerations for using bacteriophages for plant disease control. Bacteriophage 2(4):208–214

Kanematsu S, Sasaki A, Onoue M et al (2010) Extending the fungal host range of a partitivirus and a mycoreovirus from *Rosellinia necatrix* by inoculation of protoplasts with virus particles. Phytopathology 100(9):922–930

Khan Z, Kim YH (2005) The predatory nematode, *Mononchoides fortidens* (Nematoda: Diplogasterida), suppresses the root-knot nematode, *Meloidogyne arenaria*, in potted field soil. Biol Control 35(1):78–82

Khan Z, Kim YH (2007) A review on the role of predatory soil nematodes in the biological control of plant parasitic nematodes. Appl Soil Ecol 35(2):370–379

Köhl J, Postma J, Nicot P et al (2011) Stepwise screening of microorganisms for commercial use in biological control of plant-pathogenic fungi and bacteria. Biol Control 57(1):1–12

Köhl J, Scheer C, Holb IJ et al (2015) Toward an integrated use of biological control by *Cladosporium cladosporioides* H39 in apple scab (*Venturia inaequalis*) management. Plant Dis 99(4):535–543

Lacey LA, Frutos R, Kaya H et al (2001) Insect pathogens as biological control agents: do they have a future? Biol Control 21(3):230–248

Larriba E, Jaime MD, Nislow C et al (2015) Endophytic colonization of barley (*Hordeum vulgare*) roots by the nematophagous fungus *Pochonia chlamydosporia* reveals plant growth promotion and a general defense and stress transcriptomic response. J Plant Res 128(4):665–678

Liu W, Xie Y, Xue J et al (2011) Ultrastructural and cytochemical characterization of brown soft scale Coccus hesperidum (Hemiptera: Coccidae) infected by the *Lecanicillium lecanii* (Ascomycota: Hypocreales). Micron 42(1):71–79

Lorito M, Woo SL, Harman GE et al (2010) Translational research on *Trichoderma*: from'omics to the field. Annu Rev Phytopathol 48:395–417

Ma L-J, Van Der Does HC, Borkovich KA et al (2010) Comparative genomics reveals mobile pathogenicity chromosomes in *Fusarium*. Nature 464:367–373

Maffia L, Haddad F, Mizubuti ES (2009) Controle biológico da ferrugem do cafeeiro. In: Bettiol W, Morandi MAR (eds) Biocontrole de doenças de plantas: uso e perspectivas. Embrapa Meio Ambiente, Jaguariúna, pp 267–675

Martins S, Soares A, Medeiros F et al (2015) Contribution of host and environmental factors to the hyperparasitism of coffee rust under field conditions. Australas Plant Pathol 44(6):605–610

Medeiros F, Pomella A, Souza J et al (2010) A novel, integrated method for management of witches' broom disease in Cacao in Bahia, Brazil. Crop Prot 29(7):704–711

Mello S, Frazão H, Silva J (2005) Capacidade germinativa e infectiva de isolados de *Dicyma pulvinata* antagônicos a *Microcyclus ulei* mantidos em coleção de cultura. Agrociencia 9(2):421–426

Melo DF, Mello SCM, Mattos CRR et al (2008) Compatibilidade de *Dicyma pulvinata* com defensivos agrícolas e eficiência do biocontrole do mal-das-folhas da seringueira em campo. Pesq Agropec Bras 43(2):179–185

Mendes R, Garbeva P, Raaijmakers JM (2013) The rhizosphere microbiome: significance of plant beneficial, plant pathogenic, and human pathogenic microorganisms. FEMS Microbiol Rev 37(5):634–663

Mendes R, Kruijt M, Bruijn I et al (2011) Deciphering the rhizosphere microbiome for disease-suppressive bacteria. Science 332(6033):1097–1100

Monteiro TS, Lopes EA, Evans HC et al (2017) Interactions between *Pochonia chlamydosporia* and nematodes. In: Lopez RH, Llorca LVL (eds) Perspectives in sustainable nematode management through *Pochonia chlamydosporia* applications for root and rhizosphere health. Springer, Switzerland, pp 77–96

Morandi MA, Sutton JC, Maffia LA (2000) Relationships of aphid and mite infestations to control of *Botrytis cinerea* by *Clonostachys rosea* in rose (*Rosa hybrida*) leaves. Phytoparasitica 28(1):55–64

Pérez-Torres E, Bernal-Cabrera A, Milanés-Virelles P et al (2018) Eficiencia de *Trichoderma harzianum* (cepa a-34) y sus filtrados en el control de tres enfermedades fúngicas foliares en arroz. Bioagro 30:17–26

Preston J, Dickson D, Maruniak J et al (2003) *Pasteuria* spp.: systematics and phylogeny of these bacterial parasites of phytopathogenic nematodes. J Nematol 35(2):198–207

Rezende JAM, Pacheco D (1998) Control of papaya ringspot virus-type W in zucchini squash by cross-protection in Brazil. Plant Dis 82(2):171–175

Ruark CL, Koenning SR, Davis EL et al (2017) Soybean cyst nematode culture collections and field populations from North Carolina and Missouri reveal high incidences of infection by viruses. PLoS One 12(1):e0171514

Samuels G, Pardo-Schultheiss R, Hebbar K et al (2000) *Trichoderma stromaticum* sp. nov., a parasite of the cacao witches broom pathogen. Mycol Res 104(6):760–764

Spencer D, Atkey P (1981) Parasitic effects of *Verticillium lecanii* on two rust fungi. Trans Br Mycol Soc 77(3):535–542

Stefanova M (2007) Desarrollo, alcances y retos del biocontrol de fitopatógenos en Cuba. Summa Phytopathol 33:104–160

Sudo S (1989) Biocontrole de Catacauma torrendiella e Coccostroma palmicola, agentes causadores da lixa preta do coqueiro. In: 3ª reunião brasileira sobre controle biológico de doenças de plantas, USP/Embrapa, Piracicaba, 1989

Terra WC, Campos VP, Pedroso MP et al (2017) Volatile molecules of *Fusarium oxysporum* strain 21 are retained in water and control *Meloidogyne incognita*. Biol Control 112:34–40

Van Lenteren JC, Bolckmans K, Köhl J et al (2018) Biological control using invertebrates and microorganisms: plenty of new opportunities. BioControl 63(1):39–59

Vandermeer J, Perfecto I, Liere H (2009) Evidence for hyperparasitism of coffee rust (*Hemileia vastatrix*) by the entomogenous fungus, *Lecanicillium lecanii*, through a complex ecological web. Plant Pathol 58(4):636–641

Weller DM, Howie WJ, Cook RJ (1988) Relationship between in vitro inhibition of *Gaeumannomyces graminis* var. tritici and suppression of take-all of wheat by fluorescent pseudomonads. Phytopathology 78:1094–1100

Whipps J, Sreenivasaprasad S, Muthumeenakshi S et al (2008) Use of *Coniothyrium minitans* as a biocontrol agent and some molecular aspects of sclerotial mycoparasitism. Eur J Plant Pathol 121:323–330

Yanet-Suárez L, Cabrales CP (2016) Identificación de especies de cepas nativas de Trichoderma sp. y Bacillus sp. y evaluación de su potencial antagonista in vitro frente al hongo fitopatógeno nativo Moniliophthora roreri en el departamento de Norte de Santander. Respuestas 13(1):45–56

Zeng LM, Zhang J, Han YC et al (2014) Degradation of oxalic acid by the mycoparasite *Coniothyrium minitans* plays an important role in interacting with *Sclerotinia sclerotiorum*. Environ Microbiol 16(8):2591–2610

Part V
Experiences in the Integration of Biological Control in Pest Management Programs in Latin America

Chapter 37
Physiological and Ecological Selectivity of Pesticides for Natural Enemies of Insects

Geraldo Andrade Carvalho, Anderson Dionei Grützmacher, Luis Clepf Passos, and Rodrigo Lopes de Oliveira

37.1 Introduction

Every year, agricultural production in South American countries increases, mainly due to the use of cultivars that are more productive and with different resistance levels to pests and diseases, better soil preparation, balanced fertilization, and agrochemicals use.

Despite this scenario, several factors have contributed to prevent an even greater increase in agricultural production, highlighting the presence of numerous pests and diseases, forcing the farmers to use agrochemicals intensively. However, the high number of chemicals applications and the lack of prior knowledge of their impacts on agroecosystems can lead to an increase in production costs and cause negative consequences for the environment. An example is the reduction or elimination of beneficial populations of organisms, such as natural enemies. Consequently, ecological imbalances may occur, resulting in the appearance of secondary pests, resurgence, and selection of resistant pest populations to the compounds.

Due to negative effects of chemicals on humans and in the environment, the concept of integrated pest management (IPM) was developed focusing in the continuous incorporation of new technologies, aiming the reduction of these products in crops. Within this concept, the main objective is to maximize the effects of phytosanitary products on pests, with minimal impact on the beneficial organisms present in agroecosystems. This is very important, since in Latin America there is a great occurrence of natural enemies that can be applied in IPM programs (Melo et al. 2004; Zucchi et al. 2010).

G. A. Carvalho (✉) · L. C. Passos · R. L. de Oliveira
Federal University of Lavras (UFLA), Department of Entomology, Lavras, MG, Brazil
e-mail: gacarval@ufla.br

A. D. Grützmacher
Federal University of Pelotas (UFPel), Department of Phytosanitary, Pelotas, RS, Brazil
e-mail: adgrutzm@ufpel.tche.br

B. Souza et al. (eds.), *Natural Enemies of Insect Pests in Neotropical Agroecosystems*, https://doi.org/10.1007/978-3-030-24733-1_37

Therefore, this chapter covers several aspects related to the selectivity of phytosanitary products. There will be emphasized selectivity concepts and results of experiments regarding the effects of agrochemicals to natural enemies, as well as their implications in IPM in different crops.

37.2 Selectivity of Agrochemicals Versus MIP

To ensure the success of IPM programs, monitoring of the pest population must be carried out constantly (Gallo et al. 2002). In this way, it is recommended to divide the cultivated area into smaller subareas, aiming at the execution of specific management strategies in each one. Monitoring is done by sampling insects in the plants, and this sampling can be done with the presence of an operator (direct collect, entomological net) or without its presence (traps, with or without attractiveness), depending on the insect to be sampled. For some pests, damage to some parts of plants (fruits, leaves) is sampled instead of insect presence (e.g., in the case of coffee borer, *Hypothenemus hampei* (Ferrari) (Curculionidae), or in a complementary manner to insect sampling (such as the tomato pinworm, *Tuta absoluta* (Meirick) (Gelechiidae)).

From the collected data, it is possible to estimate the population density of the insect in the area and thus compare with literature information to determine if the population of the insect reached a level that requires human intervention for its reduction (economic threshold). If this occurs, the responsible farmer or technician must decide which control methods should be applied, basing his decision on economic, social, and environmental factors (Gallo et al. 2002). In addition, for some specific situations, there is the nonaction level, which is the population density of natural enemies capable of naturally control a pest. An example is the case of the coffee leaf miner, *Leucoptera coffeella* (Guérin-Mèneville & Perrottet) (Lyonetiidae). In this case, coffee leaves presenting mines (pest damage sign) are sampled, and in addition the state of the mines is taken into account, as they may show signs of predation. If this occurs, it is an indicative that natural enemies present in the area are exerting biological control on the pest. In the specific case of the coffee leaf miner, if more than 60% of the mines sampled show signs of predation, it is not necessary to adopt other control measures (Picanço 2010).

In this way, the sampling of pests and natural enemies is one of the foundations of IPM. If an intervention with insecticides is required, it is interesting that these products cause the least possible impact to natural enemies, released or naturally present in the area. In this context, the integration between biological and chemical control tactics should be stimulated by the use of selective products to natural enemies (Parra 2014). According to Ripper et al. (1951), selectivity can be defined as the capacity of a product to control the target pest, causing the least possible impact on beneficial organisms such as predators, parasitoids, and pollinators, and can be divided into:

Physiological Selectivity Related to the chemical nature of the insecticide and its relation to insects. Differences in the processes of absorption, penetration, transport, activation, and degradation of insecticides in each insect species may cause a product to be more toxic to the pest than to its natural enemy in a situation in which both have contact with the compound or its residues.

Ecological Selectivity Intrinsic to the beneficial organism and its habitat, this occurs due to behavior differences between pests, natural enemies, and pollinators, causing the chemical to have contact only with certain species, especially with pests. Thus, to obtain this type of selectivity, it is necessary to know the pest bioecology and beneficial arthropods that are present in the different agroecosystems.

When implementing IPM programs, the compatibility or integration between control methods is recommended. The preservation of natural enemies in the agroecosystem is one of the most important strategies in IPM programs, since the use of chemical compounds can negatively impact their populations (Degrande et al. 2002). The synthetic insecticides used to control agricultural pests belong to different chemical groups, and their modes of action can act upon different groups of insects. Unfortunately, few farmers use safe ways to assess the risks that their investment may be subject to and suffer from the attack of pests on their crops. In many cases, in order to reduce pest populations, chemical compounds are sprayed without technical orientations, generating even greater instability in the agroecosystem due to the elimination of controlling agents of these undesirable organisms. Thus, successive imbalances are generated by the impact promoted by chemical interventions, often leading to a general lack of pest control (Bueno et al. 2017).

Biological control may be responsible for population of pest arthropods stability in agroecosystems. In addition to keeping their populations in equilibrium and below the economic threshold, either by maintaining existing natural enemies or releasing mass-reared insects on field, the biological method still minimizes human intervention compared to other control methods (Bueno et al. 2017).

The use of chemicals with high toxicity and broad action spectrum is recognized as the main cause of ecological imbalances in agroecosystems. Most of the used insecticides are neurotoxic, that is, they act on the nervous system of the insects (e.g., pyrethroids, neonicotinoids, and organophosphates). In general, these products have a broad spectrum of action; therefore, they tend to be less selective (Foerster 2002). To avoid this situation, the use of selective products is considered a feasible strategy, justifying studies in this research line (Degrande et al. 2002; Foerster 2002). Many biological characteristics of natural enemies may interfere with their susceptibility to phytosanitary products, such as sex, age, size, daily habit, and development stage (Foerster 2002). In this context, works are being developed by researchers taking into account some of these characteristics, as exemplified below.

37.3 Studies of the Effects of Agrochemicals on Beneficial Organisms

Most of the results were obtained according to a methodology standardized by the "International Organization for Biological and Integrated Control of Noxious Plants and Animals" (IOBC), which takes into account the mortality caused by the products and reduction in beneficial activities (predation and parasitism), as well as biological characteristics. Other selectivity tests will also be discussed, such as persistence of the effect of pesticide residues on natural enemies, behavioral changes of these organisms, and exposure of beneficial insects to plants originated from treated seeds.

Predators of the Chrysopidae Family (Green Lacewings) Although adults of Chrysopidae feed on plant sources (pollen, nectar, and honeydew), larvae are polyphagous predators. Therefore, these insects are important agents of biological control for several species of pest arthropods in several crops of economic interest.

However, in order to maintain the population of lacewings in IPM programs that include the use of phytosanitary products, it is necessary to evaluate the impact of insecticides on these organisms. These studies allow selecting products that cause low or no-negative effects on the predator. Thus, it is possible to integrate chemical and biological methods, using selective insecticides associated with the release of these natural enemies or aiming their maintenance in agroecosystems.

In this context, Godoy et al. (2010) carried out a study regarding the effect of insecticides (thiamethoxam, imidacloprid, milbemectin, pyriproxyfen, and spirodiclofen) on adults of *Chrysoperla externa* (Hagen) and *Ceraeochrysa cubana* (Hagen) (Chrysopidae). They concluded that milbemectin, pyriproxyfen, and spirodiclofen were harmless to predators and could be recommended in the IPM programs aiming the maintenance of these natural enemies in citrus crops.

Lacewings are frequently found in coffee crops, acting as regulators of pest populations, and, therefore, Vilela et al. (2010) performed a study aiming to evaluate the effects of spirodiclofen, fenpropathrin, sulfur, and abamectin on the first three larvae instars of *C. externa*. They found that fenpropathrin (in two dosages, 0.15 and 0.30 g a.i./L) and abamectin at the highest dosage (0.0225 g a.i./L) were the most harmful products to predator larvae. These products affected the survival of nymphs of the three instars, in addition to causing changes in the larval period. Spirodiclofen (0.12 g a.i./L) caused a moderate effect, while sulfur (4.0 and 8.0 g i.a./L) was harmless to *C. externa* larvae.

Haramboure et al. (2013) evaluated the effect of different concentrations of cypermethrin, via topical application, on *C. externa* larvae in Argentina. The larvae showed symptoms of insecticide intoxication, such as tremors, reduced movement, and knockdown effect (paralysis due to insecticide intoxication). The proportion of specimens that exhibited these symptoms was higher as the dosage increased. However, at all evaluated dosages, insects recovered after 96 hours, probably due to enzymatic action related to detoxification. These results indicate that *C. externa* larvae present high tolerance to this pyrethroid. Another study, also conducted in Argentina, showed that pyriproxyfen and acetamiprid may decrease survival and increase the duration of the first larval instar of *C. externa* (Rimoldi et al. 2017).

Predatory Hemiptera Selectivity tests have been performed for different species of predator bugs. These natural enemies, both nymphs and adults, are polyphagous and have predatory habit, feeding on eggs, caterpillars, mites, aphids, and thrips. Among studied species are *Orius insidiosus* (Say) (Anthocoridae), *Macrolophus basicornis* (Stål) (Miridae), and *Podisus nigrispinus* (Dallas) (Pentatomidae).

Moscardini et al. (2013) evaluated the effect of insecticides used in tomato crops on *O. insidiosus* eggs. Abamectin, cartap, spirotetramat + imidacloprid, and flubendiamide were harmful; pymetrozine was slightly harmful; pyriproxyfen and Rynaxypyr were innocuous; however, the last product affected reproductive parameters of *O. insidiosus*, demonstrating the importance of evaluation of sublethal effects in selectivity studies.

Insecticides recommended for control rosebush pests also had the lethal and sublethal effects evaluated in *O. insidiosus* eggs and nymphs. Predator's eggs were immersed in insecticide solutions for 5 seconds and then were kept in the laboratory. The insecticide deltamethrin was harmful to eggs and fifth instar nymphs, while formetanate and spinosad were slightly harmful. Formetanate did not negatively affected the reproductive characteristics of females originated from treated eggs. Spinosad reduced the number and viability of eggs placed by females from treated fifth instar nymphs (Torres et al. 2007).

To evaluate the effects of insecticides on *O. insidiosus*, topical application can be used, that is, the product to be tested is applied directly on the back of the insect. Thus, adults of this predator were submitted to the application of chemical products used in chrysanthemum crops, and chlorothalonil was found to be harmless. Mancozeb and dicofol were slightly harmful, and lufenuron showed to be moderately harmful. The other products, bifenthrin, triazophos, methomyl, and acetamiprid, were considered harmful. Due to the low toxicity of chlorothalonil, mancozeb, and dicofol, these compounds can be recommended in IPM programs aiming at the conservation of this predator in this culture (Albernaz et al. 2009).

The mirid bug *M. basicornis* is a potential biological control agent of the tomato pinworm, *T. absoluta*, but it can be negatively influenced by several phytosanitary products used to control pests in tomato. In the laboratory, Passos et al. (2017) treated tomato leaflets with insecticides recommended for the control of this lepidopteran, and then, nymphs and adults of *M. basicornis* were exposed to the leaflets. The products not classified as harmless in laboratory were also evaluated in greenhouse. Abamectin and chlorfenapyr were toxic to nymphs and adults of the predator in all bioassays. Cartap hydrochloride was slightly toxic to adults in laboratory, but toxic to nymphs in laboratory and moderately toxic in greenhouses. Chlorantraniliprole and teflubenzuron were harmless to *M. basicornis* in most of the bioassays, except for nymphs in laboratory test, where they were classified as moderately toxic and slightly toxic. Thus, these products should be preferred in IPM programs to conserve populations of this predator.

Another exposition route of predators to insecticides in agroecosystems is through contaminated extrafloral nectar when seed treatment is used, since many of these insects use these resources to complement their feeding. The effects of insecticides used as seed treatment of soybean (thiamethoxam and chlorantraniliprole) were evaluated on the zoophytophagous predator *P. nigrispinus*. Thiamethoxam caused a mortality increase, besides affecting the fecundity, the preoviposition period, and the females' survival when compared to chlorantraniliprole. However, life expectancy of *P. nigrispinus* females was prolonged by chlorantraniliprole, which also increased the intrinsic rate of population increase (r_m) and finite growth rate (l) and reduced population doubling time (TD) when compared to thiamethoxam. Other parameters such as net reproductive rate (R_0) and mean intergenerational interval (T) were not affected by any of the insecticides. In addition, the distance traveled and mean velocity of females exposed to these products were evaluated, but these parameters were not influenced. Thus, seed treatment with these insecticides presented low (chlorantraniliprole) and moderate (thiamethoxam) risk for *P. nigrispinus* (Gontijo et al. 2017).

***Doru lineare* (Dermaptera: Forficulidae)** This is one of the most common predators in the corn crop, preying mainly *Spodoptera frugiperda* (JE Smith) (Noctuidae) eggs and caterpillars. This natural enemy is present during the whole cycle of this crop, from the seedlings emergence until harvest, and should be preserved in agricultural systems by the use of selective pesticides.

In order to know the insecticides' impact on *D. lineare*, a work was carried out evaluating products used in corn on eggs and first instar nymphs of the predator. Triflumuron, deltamethrin, diflubenzuron, lambda-cyhalothrin + thiamethoxam, alpha-cypermethrin, methoxyfenozide, lambda-cyhalothrin, chlorpyrifos, lufenuron, tebufenozide, azadirachtin, carbaryl, and spinosad were evaluated. It was concluded that triflumuron, diflubenzuron, lambda-cyhalothrin + thiamethoxam, lambda-cyhalothrin, chlorpyrifos, lufenuron, and carbaryl reduced egg viability. The nymph survival from treated eggs was reduced by deltamethrin, diflubenzuron, lambda-cyhalothrin + thiamethoxam, lambda-cyhalothrin, chlorpyrifos, and carbaryl. For nymphs in the residual contact test, all insecticides were harmful within 96 hours after exposing the insects to their residues (Zotti et al. 2010).

Predatory Coccinellids Coccinellidae family insects are important biological control agents in several crops of economic importance. Larvae and adults of some species, such as *Coleomegilla maculata* (DeGeer) and *Hippodamia convergens* (Guérin-Méneville), are voracious predators of aphids, thrips, eggs, and small caterpillars. Moscardini et al. (2015) evaluated the effect of the insecticides thiamethoxam and chlorantraniliprole, used in sunflower seed treatment, on the biological and reproductive characteristics of these predators. *C. maculata* and *H convergens* larvae and adults were exposed to the products by feeding on treated plants' extrafloral nectaries. Both insecticides extended the pupal period of *C. maculata* exposed in the larval stage. For *H. convergens*, there was a reduction in egg viability and an increase in the embryonic period of exposed insects in the larval stage, as well as a change in the sex ratio caused by thiamethoxam. These insecticides also prolonged the pupae stage of insects originated from exposed adults.

The physiological selectivity of insecticides used in cotton for eggs and third instar larvae of *Cycloneda sanguinea* (Linnaeus) (Coccinellidae) was evaluated by Pedroso et al. (2012). The insecticide triflumuron was harmless to *C. sanguinea* eggs and slightly harmful to larvae. Spinosad was slightly harmful to eggs and larvae, while the other products (chlorfenapyr, clothianidin, and imidacloprid + ß-cyfluthrin) were harmful to the predator's eggs and larvae.

In Argentina, the selectivity of four insecticides used in horticulture (pyriproxyfen, teflubenzuron, acetamiprid, and cypermethrin) to *Eriopis connexa* (Germar) (Coccinellidae), an important control agent of horticultural pests in the Neotropical region, was evaluated. All compounds caused lethal and/or sublethal effect on the predator. However, neurotoxic insecticides were more toxic than growth regulators, and the pupa presented greater susceptibility to the products in relation to the adult stage (Fogel et al. 2016).

Parasitoids *Trichogramma pretiosum* Riley (Trichogrammatidae) is one of the most used species in biological control programs in the world. Insects of this genus are generalist egg parasitoids, used mainly to control lepidopteran pest, through inundative releases in agroecosystems. *T. pretiosum* has been recorded in several countries of Latin America (Zucchi et al. 2010), assuming importance in the control of several phytophagous insects. In Brazil, for example, it is registered for some crops to control defoliant lepidopterans, such as the tomato pinworm (*T. absoluta*) and the fall armyworm (*S. frugiperda*). Therefore, it is essential to carry out studies aiming the determination of selective products to this natural enemy.

The effect of insecticides used on the tomato crop (acetamiprid, lufenuron, imidacloprid, novaluron, triflumuron, and pyriproxyfen) was evaluated on immature stages of *T. pretiosum*, and acetamiprid, imidacloprid, lufenuron, and pyriproxyfen were toxic to the parasitoid (Carvalho et al. 2010b). When these insecticides were evaluated on adults, pyriproxyfen was slightly harmful, due to reduction in parasitism capacity of female of the maternal generation. Acetamiprid, imidacloprid, lufenuron, and triflumuron were harmless to *T. pretiosum* adults and could be recommended in tomato IPM programs in order to preserve this parasitoid (Carvalho et al. 2010a)

In order to investigate the effect of other products recommended for the tomato pests' control, abamectin, acetamiprid, cartap, and chlorpyrifos were tested for *T. pretiosum* at different stages of development. The parasitism of the females exposed to the products in *Ephestia kuehniella* (Zeller) (Pyralidae) eggs was evaluated. It was observed a reduction up to 98.3% in the parasitism caused by acetamiprid and 100% reduction due to the effect of the other insecticides evaluated (Moura et al. 2006).

Moura et al. (2004) reported the residual effect of acetamiprid, chlorfenapyr, imidacloprid, thiacloprid, and thiamethoxam on the first two generations of parasitoids from *T. pretiosum* females treated in laboratory. Imidacloprid and chlorfenapyr were the most detrimental; and thiacloprid reduced the rate of parasitism, although it did not affect longevity. The deleterious effects of chlorfenapyr and thiacloprid were transmitted to the subsequent generation of this parasitoid. As thiamethoxam and acetamiprid were selective under the evaluated conditions, they are suitable for use in IPM programs in tomato.

Nörnberg et al. (2011) verified the persistence (the harmful activity duration) of ten products, recommended in the integrated fruit production, on *T. pretiosum*. Adults were exposed to grapevine leaves treated with the products, and the residual effect on parasitism at 3, 10, 17, 24, and 31 days after spraying was evaluated. The active ingredients malathion and the fungicides, tetraconazole, mancozeb, and methyl thiophanate, were classified as short-lived (<5 days of harmful action); the insecticide/acaricide abamectin as slightly persistent (5–15 days of harmful action); the insecticide chlorpyrifos as moderately persistent (16–30 days of harmful action); and phosmet and carbaryl insecticides and sulfur fungicide/acaricide as persistent (>31 days of harmful action).

Aiming the use of *T. pretiosum* along with chemicals in conventional corn crops, Stefanello Júnior et al. (2008) investigated the effect of 16 commercial formulations of insecticides registered for this crop on the parasitism capacity of this species in laboratory. Alpha-cypermethrin and lufenuron were slightly deleterious; spinosad was moderately harmful; cypermethrin, deltamethrin, deltamethrin + triazophos, trichlorfon, parathion-methyl, triazophos, lambda-cyhalothrin, malathion, fenitrothione, beta-cyfluthrin, and chlorpyrifos were harmful to *T. pretiosum* adults in the initial toxicity test. The insecticides triflumuron and novaluron were innocuous; therefore, they should be preferred in maize IPM programs where this parasitoid is used.

Besides parasitoids of eggs, such as *Trichogramma* spp., there are also species that parasitize other stages of pests' development. Luna-Cruz et al. (2015) evaluated the effect of insecticides on *Tamarixia triozae* (Burks) (Eulophidae), a parasitoid of the psyllid *Bactericera cockerelli* (Šulc) (Triozidae) nymphs, common in Solanaceae crops in Central and North America. Parasitoid adults were exposed to tomato leaflets treated with insecticides recommended for the control of the psyllid (spirotetramat, spiromesifen, beta-cyfluthrin, pymetrozine, azadirachtin, imidacloprid, abamectin, and spinosad), and all of them had at least some level of toxicity to the parasitoid. Spinosad was classified as harmful; abamectin was moderately harmful, and the others were slightly harmful. Spinosad and abamectin also showed greater persistence of the lethal effect on the parasitoid (41 and 24 days, respectively), while the other insecticides did not exceed 13 days. Thus, products that caused lower mortality and had lower persistence in the environment can be considered in *B. cockerelli* management programs.

In summary, there is a great volume of works published in recent years regarding the effects of phytosanitary products on natural enemies in Latin America. However, researches must continue as new chemical molecules are commercially registered to collaborate in the IPM programs development.

37.4 Final Considerations

Researches about the effects of chemical compounds on beneficial organisms have been left aside for many years. However, the demand of society for better agricultural products and with less chemicals has been stimulating the increase of studies in this research line. Integration between chemical and biological methods for pest control is important for the development of IPM programs in different crops of economic value in the Neotropical region.

In the first studies on selectivity, only the mortality caused by the compounds on beneficial insects was emphasized. Researches that considered the sublethal effects of the products on the populations of these organisms were scarce. Over time, researchers have also become concerned about the effects of compounds on the behavior and reproduction of natural enemies and pollinators. In addition, the structure for research in Brazil and other Latin-American countries, such as Argentina and Colombia, has greatly improved. Therefore, there are better conditions for the study of pesticide selectivity to natural enemies in laboratory, semifield and field conditions.

Phytosanitary products are used in great quantity for pest control in Latin-American countries (especially in Brazil, considered the greatest consumer of chemical products in the world). Thus, it is of utmost importance to conduct researches that seek the rational use of these compounds, respecting the factors of natural mortality and the natural enemies' action in agroecosystems.

Acknowledgments The authors are grateful to the National Council for Scientific and Technological Development (CNPq), CAPES Foundation (Brazilian Ministry of Education), and the Minas Gerais State Foundation for Research Aid (FAPEMIG) for their collaboration.

References

Albernaz KC, Carvalho GA, Carvalho BF et al (2009) Toxicidade de pesticidas para adultos de *Orius insidiosus* (Say, 1832) (Hemiptera: Anthocoridae). Arq Inst Biol 76(4):589–595

Bueno AF, Carvalho GA, Santos AC et al (2017) Pesticide selectivity to natural enemies: challenges and constraints for research and field recommendation. Ciênc Rural 47(6):1–10

Carvalho GA, Godoy MS, Parreira DS et al (2010a) Selectivity of growth regulators and neonicotinoids for adults of *Trichogramma pretiosum* (Hymenoptera: Trichogrammatidae). Rev Colomb Entomol 36(2):195–201

Carvalho GA, Godoy MS, Parreira DS et al (2010b) Effect of chemical insecticides used in tomato crops on immature *Trichogramma pretiosum* (Hymenoptera: Trichogrammatidae). Rev Colomb Entomol 36(1):10–15

Degrande PE, Reis PR, Carvalho GA et al (2002) Metodologia para avaliar o impacto de pesticidas sobre inimigos naturais. In: Parra JRP, Botelho PSM, Corrêa-Ferreira BS et al (eds) Controle biológico no Brasil: parasitóides e predadores. Manole, São Paulo, pp 71–94

Foerster LA (2002) Seletividade de inseticidas a predadores a parasitóides. In: Parra JRP, Botelho PSM, Corrêa-Ferreira BS et al (eds) Controle biológico no Brasil: parasitoides e predadores. Manole, São Paulo, pp 95–114

Fogel MN, Schneider MI, Rimoldi F et al (2016) Toxicity assessment of four insecticides with different modes of action on pupae and adults of *Eriopis connexa* (Coleoptera: Coccinellidae), a relevant predator of the Neotropical Region. Environ Sci Pollut Res 23(15):14918–14926

Gallo D, Nakano O, Silveira Neto S et al (2002) Entomologia agrícola. FEALQ, Piracicaba

Godoy MS, Carvalho GA, Carvalho BF et al (2010) Seletividade fisiológica de inseticidas em duas espécies de crisopídeos. Pesqui Agropecu Bras 45(11):1253–1258

Gontijo PC, Abbade Neto DO, Oliveira RL et al (2017) Non-target impacts of soybean insecticidal seed treatments on the life history and behavior of *Podisus nigrispinus*, a predator of fall armyworm. Chemosphere 191:342–349

Haramboure M, Francesena N, Reboredo GR et al (2013) Toxicity of cypermethrin on the neotropical lacewing *Chrysoperla externa* (Neuroptera: Chrysopidae). Commun Agric Appl Biol Sci 78(2):339–344

Luna-Cruz A, Rodríguez-Leyva E, Lomeli-Flores JR et al (2015) Toxicity and residual activity of insecticides against *Tamarixia triozae* (Hymenoptera: Eulophidae), a parasitoid of *Bactericera cockerelli* (Hemiptera: Triozidae). J Econ Entomol 108(5):2289–2295

Melo MC, Dellapé PM, Carpintero DL et al (2004) Reduviidae, Miridae y Lygaeoidea (Hemiptera) recolectados en Colonia Carlos Pellegrini (Esteros de Iberá, Corrientes, Argentina). Rev Soc Entomol Argent 63(2):59–67

Moscardini VF, Gontijo PC, Carvalho GA et al (2013) Toxicity and sublethal effects of seven insecticides to egg of the flower bug *Orius insidiosus* (Say) (Hemiptera: Anthocoridae). Chemosphere 92(5):490–496

Moscardini VF, Gontijo PC, Michaud JP et al (2015) Sublethal effects of insecticide seed treatments on two Nearctic lady beetles (Coleoptera: Coccinellidae). Ecotoxicology 24(5):1152–1161

Moura AP, Carvalho GA, Rigitano RLO (2004) Efeito residual de novos inseticidas utilizados na cultura do tomateiro sobre *Trichogramma pretiosum* Riley, 1879 (Hymenoptera, Trichogrammatidae). Acta Sci Agron 26(2):231–237

Moura AP, Carvalho GA, Pereira AE et al (2006) Selectivity evaluation of insecticides used to control tomato pests to *Trichogramma pretiosum*. Biol Control 51:769–778

Nörnberg SD, Grutzmacher AD, Kovaleski A et al (2011) Persistência de agrotóxicos utilizados na produção integrada de maçã a *Trichogramma pretiosum*. Ciênc Agrotecnol 35(2):305–313

Parra JRP (2014) Biological control in Brazil: an overview. Sci Agric 71(5):420–429

Passos LC, Soares MA, Costa MA et al (2017) Physiological susceptibility of the predator *Macrolophus basicornis* (Hemiptera: Miridae) to pesticides used to control of *Tuta absoluta* (Lepidoptera: Gelechiidae). Biocontrol Sci 27(9):1082–1095

Pedroso EC, Carvalho GA, Leite MIS et al (2012) Seletividade de inseticidas utilizados na cultura algodoeira a ovos e larvas de terceiro ínstar de *Cycloneda sanguinea*. Arq Inst Biol 79(1):61–68

Picanço MC (2010) Manejo integrado de pragas. UFV, Viçosa

Rimoldi F, Fogel MA, Ronco AE et al (2017) Comparative susceptibility of two Neotropical predators, *Eriopis connexa* and *Chrysoperla externa*, to acetamiprid and pyriproxyfen: short and long-term effects after egg exposure. Environ Pollut 231(1):1042–1050

Ripper WE, Greenslade RM, Hartley GS (1951) Selective inseticides and biological control. J Econ Entomol 44(4):448–458

Stefanello Júnior GJ, Grutzmacher AD, Grutzmacher DD et al (2008) Efeito de inseticidas usados na cultura do milho sobre a capacidade de parasitismo de *Trichogramma pretiosum* Riley, 1879 (Hymenoptera: Trichogrammatidae). Arq Inst Biol 75(2):187–194

Torres FZV, Carvalho GA, Souza JR et al (2007) Seletividade de inseticidas a *Orius insidiosus*. Bragantia 66(3):433–439

Vilela M, Carvalho GA, Carvalho CF et al (2010) Seletividade de acaricidas utilizados em cafeeiro para pré-pupas e adultos de *Chrysoperla externa* (Hagen, 1861) (Neuroptera: Chrysopidae). Arq Inst Biol 57(5):505–510

Zotti MJ, Grutzmacher AD, Grutzmacher DD et al (2010) Seletividade de inseticidas usados na cultura do milho para ovos e ninfas do predador *Doru lineare* (Eschscholtz, 1822) (Dermaptera: Forficulidae). Arq Inst Biol 77(1):111–118

Zucchi RA, Querino RB, Monteiro RC (2010) Diversity and hosts of *Trichogramma* in the New World, with emphasis in South America. In: Consoli FL, Parra JRP, Zucchi RA (eds) Egg parasitoids in agroecosystems with emphasis on *Trichogramma*. Springer, New York, pp 219–236

Chapter 38
Use of Natural Chemical Products for Pest Control

Dejane Santos Alves and Jociani Ascari

38.1 Introduction

Natural products are chemical substances produced by living organisms, and, due to their biological activities, they can be employed in pest management programs. These substances may be the result of the metabolism of microorganisms, animals, and plants. Products derived from the secondary metabolism of plants have long been used for pest control; reports on the use of pyrethrum dust in Mesopotamia predate the Christian era (Pavela 2016). Botanical insecticides played a crucial role in pest control before the advent of synthetic chemical insecticides. In 1947, over 6700 (US) tons of *Derris elliptica* (Wall.) Benth roots were imported from southeastern Asia to the United States, but this number decreased to 1500 in 1963 (Wink 1993).

Nowadays, plant metabolites for arthropod control have been the subject of several studies. This is mainly due to a growing awareness of the risks caused by synthetic pesticides (Jardim et al. 2009; Lundin et al. 2015; Colmenarez et al. 2016), as well as numerous reports on the selection of arthropod populations resistant to chemical synthetics (Rais et al. 2013; Silva et al. 2016; Hidalgo et al. 2017). Even recent technologies, such as the use of genetically modified plants, have accelerated the selection of resistant populations (Monnerat et al. 2015a, b; Blanco et al. 2016), which intensifies the search for molecules that may be used to control arthropod pests.

Secondary plant metabolites can either be used as plant extracts and essential oils, mainly from organic farms, or their active principle(s) can be isolated and synthesized in laboratory, which reduces the costs of the process. Another approach is the synthesis of other substances from lead molecules of plant substances. It is

D. S. Alves (✉) · J. Ascari
Coordenação do Curso de Licenciatura em Ciências Biológicas – Universidade Tecnológica Federal do Paraná, Santa Helena, PR, Brazil
e-mail: dejanealves@utfpr.edu.br

B. Souza et al. (eds.), *Natural Enemies of Insect Pests in Neotropical Agroecosystems*, https://doi.org/10.1007/978-3-030-24733-1_38

worth mentioning that classes of synthetic chemical insecticides currently employed were originated from plant-derived substances. Carbamates were initially developed from the alkaloid physostigmine, extracted from calabar beans *Physostigma venenosum* Balf. F. (Dev and Koul 1997). Pyrethroids are synthetic analogues of pyrethrins, natural insecticidal substances obtained from the flowers of *Chrysanthemum*. Although the bioactivity of natural pyrethrins has been known for a long time, its instability in the environment prevented it from being commercialized; therefore, the creation of synthetic analogues allowed the success of its commercialization in large scale (Palmquist et al. 2012).

In recent years, the number of botanical insecticides published in Latin America has increased considerably (Fouad et al. 2014; Napal et al. 2015; Luiz et al. 2017). However, few botanical insecticides are commercialized, due to numerous factors, such as the difficulties of standardization, the competition with multinationals, and strict regulating laws. However, this scenario tends to change: by 2025, botanical insecticides are estimated to be 7% of the world biopesticide market (Isman 2015). It is also highlighted that botanical insecticides constitute an important tool to be used in integrated pest management (IPM).

This chapter will cover the natural products of plant origin and their applications in the management of arthropod pests, with emphasis on the scientific results and practical evidence of studies carried out in Latin America.

38.2 Plant-Derived Metabolites in Arthropod Pest Control

Plants and phytophagous arthropods coexist for more than 350 million years. During this period, both developed strategies to dribble each other's defensive systems. Plant defense strategies include the production of secondary metabolites. More than 50,000 substances have been identified from plant metabolism, and this number is estimated to exceed 100,000 (Yazdani et al. 2011). In this sense, the South American tropical ecosystems are the largest source of secondary metabolites in the world, since they host one-third of the botanical biodiversity of the planet (Desmarchelier 2010).

Secondary plant metabolites are commonly classified based on some criteria, such as carbon skeletal type and functional groups, plant source (e.g., piperine is obtained from plants of the genus *Piper*), and biogenesis and biosynthetic origins, among others (Shukla 2009). Alkaloids, terpenoids, and phenolic compounds are some important classes of secondary metabolites involved in plant defense against phytophagous arthropods.

Alkaloids can be defined as cyclic compounds that contain nitrogen in a negative oxidation state (Pelletier 1983); these metabolites are found in more than 150 botanical families (Bhat et al. 2005). These substances can lead to neurotoxic effects on insects. Nicotine, for example (Fig. 38.1a), present in the leaves of *Nicotiana tabacum* L., mimics the neurotransmitter acetylcholine, leading to hyperexcitation of the nervous system (El-Wakeil 2013). Despite the widespread use of this botanical

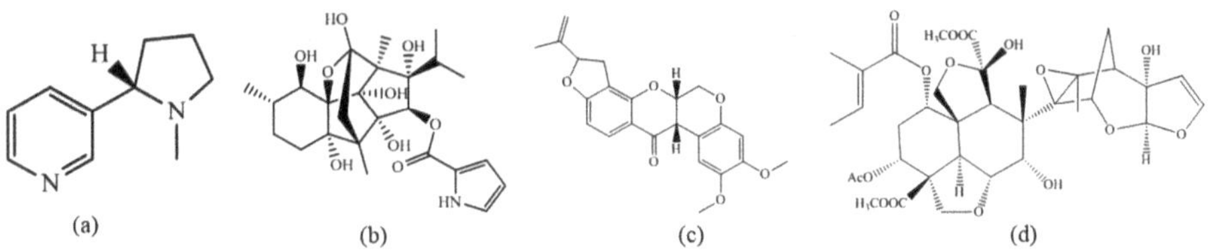

Fig. 38.1 Chemical structure of some plant-derived metabolites toxic for insects: (**a**) nicotine, (**b**) ryanodine, (**c**) rotenone, (**d**) azadirachtin A. (Figure: authors)

insecticide in the past, current laws prohibit its use in organic systems due to its high toxicity to mammals (Electronic Code of Federal Regulations 2017). Another alkaloid that deserves attention is ryanodine (Fig. 38.1b), extracted from leaves of the South American plant *Ryania speciosa* Vahl. Although efforts to develop a synthetic commercial insecticide analogous to ryanodine have been unsuccessful, because of its high toxicity to mammals, two classes of synthetic chemicals have recently emerged, resulting in commercial insecticides that target ryanodine receptors in insects (Sattelle et al. 2008).

Another important class of secondary plant metabolites are phenolic compounds, which are characterized by the presence of hydroxyl groups attached to aromatic rings. This class includes flavonoids, tannins, coumarins, anthraquinones, phenolic acids, and others (Dewick 2002). Among the phenolic compounds with insecticidal activity, rotenone deserves highlight (Fig. 38.1c). It is an isoflavonoid found in the roots of tropical Fabaceae, such as *Derris*, *Lonchocarpus* and *Tephrosia* (Isman 2006). Initially used by indigenous populations of the Americas as a fishing poison, rotenone is still used in organic farming systems. This substance acts by blocking the electron transport chain in the mitochondria, thus preventing energy production (Copping and Duke 2007). Tannins, which are known for their ability to reduce palatability to herbivores (War et al. 2012), are also noteworthy. In addition, they are able to form oxygen reactive species (Barbehenn and Constabel 2011). Efforts have been made to map tannin regulatory gene and breed genetically modified plants that overexpress this gene (Boeckler et al. 2014).

Terpenoids are one of the largest and most diverse classes of secondary metabolites (Gershenzon and Dudareva 2007). Their biosynthetic origin is derived from isoprene units, which can be classified as monoterpenes (C10), sesquiterpenes (C15), diterpenes (C20), sesterpenes (C25), triterpenes (C30), tetraterpenes (C40) and polyterpenes (C5n) (Simões et al. 2017). In particular, monoterpenes and sesquiterpenes are the components most frequently found in essential oils (Dewick 2002). However, the main barrier to the commercialization of essential oils is their high instability in the environment, due to high volatilization and thermal decomposition. To solve this problem, techniques for controlled release of essential oils through micro and nanoencapsulation have been evaluated (Khot et al. 2012; Oliveira et al. 2017). As for the mechanism of action, there is evidence that these oils interfere with neuromodulator octopamine (Kostyukovsky et al. 2002) and GABA-gated chloride channels (Priestley et al. 2003).

Currently, the most commercialized botanical insecticide in the world is neem oil, extracted from *Azadirachta indica* A. Juss. Despite being native to India, the neem tree has propagated to the Caribbean Islands, Central America, and South America (Koul et al. 1990). Neem oil is a mixture of active components, and its major compound is azadirachtin A (tetranortriterpenoid or limonoid) (Fig. 38.1d). Azadirachtin acts in brain-corpus cardiacum complex, inhibiting the release of hormones. As a result, it suppresses molting, pupation, adult formation, and fecundity (Yu 2014). In adult females, a similar mechanism of action leads to sterility (Isman 2006; Bajwa and Ahmad 2012; Ntalli and Caboni 2012).

38.3 Scientific Results and Practical Evidence in Latin American

The neotropics are estimated to host from 90,000 to 110,000 plant species (Stoll and Squeo 2012). This diversity evidences the potential of the region for the use of secondary metabolites against arthropod pests. However, there is still limited use of botanical insecticides.

Specifically in Brazil, numerous studies have been carried out in the last years aiming at the search for active metabolites against crop pests (Fouad et al. 2014; Luiz et al. 2017; Massarolli et al. 2017). In a compilation of data by Isman and Grieneisen (2014), Brazil ranked third in the list of articles on botanical insecticides published in 2012, accounting for 494 (40.9%) of 1207. Nevertheless, botanical insecticides have limited use in this country. The use of plant extracts is often limited to the use of home-prepared solutions, recommended for organic cultivation (Menezes 2005; Previero et al. 2010).

However, this scenario tends to change because biopesticide registration is increasing. So far, the Ministry of Agriculture, Livestock and Food Supply (*Ministério da Agricultura, Pecuária e Abastecimento* – MAPA) has already granted registration to 68 phytosanitary products approved to be used in organic agriculture, though only two of them are plant-derived (Brasil 2017). The botanical insecticides registered by the competent bodies to be employed in Brazil are Azamax® [UPL do Brasil Indústria e Comércio de Insumos Agropecuários S.A.], which has azaractins A and B as active ingredients, and Bioisca® [Cooperativa de Cafeicultores e Agropecuaristas – COCAPEC]. Bioisca® has the aerial part of *Tephrosia candida* D.C. and the oily extract of *Psychotria marcgravii* (A. St.-Hil.) Spreng in its composition and rotenoid saponin flavones as active ingredients. The recent registration of Bioisca® represents a great advance, given that it is the first botanical insecticide marketed in the world for the control of leaf-cutting ants.

Among the scientific studies carried out in Brazil, those using the botanical families Meliaceae (Silva et al. 2012; Giongo et al. 2015, 2016) and Piperaceae (Fazolin et al. 2005, 2016; Estrela et al. 2006; Brito et al. 2015; Volpe et al. 2016) can be highlighted. Recently, many Brazilian researchers have explored the insecticidal potential of plants belonging to the family Annonaceae (Ribeiro et al. 2016; Ansante et al. 2017; Gonçalves et al. 2017; Souza et al. 2017). Secondary metabolites, such as acetogenins (Ansante et al. 2015) and phenylpropanoids (Alves et al. 2015, 2016), have been considered responsible for toxicity in these plants. The great potential of the family Annonaceae for pest control can be exemplified by the commercialization of botanical pesticides in other countries, such as the United States and India (Isman 2006; Isman and Seffrin 2014).

In addition, research has been done in Latin America to assess safety of plant-derived metabolites for nontarget species. Some studies have also verified the selectivity of plant extracts and essential oils for important predators and parasitoids (Tavares et al. 2009; Werdin-González et al. 2013; Turchen et al. 2016; Bernardi et al. 2017; Parreira et al. 2018). These studies are of fundamental importance, because they demonstrate that botanical pesticides can be compatible with biological control.

Practical evidence shows the potential of numerous plant species abundant in the Brazilian territory and the constant use of extracts from spontaneous plants (*Momordica charantia* L.) and cultivated plants (*Annona squamosa* L., *Ocimum basilicum* L., *Manihot esculenta* Crantz., *Allium sativum* L., *Allium cepa* L. and *Ruta graveolens* L.). In addition, there is the application of oil of neem for pest control, as revealed in a study carried out with organic farmers in the State of Alagoas, northeastern Brazil (Sousa et al. 2012).

A similar study carried out in southern Brazil showed that 82% of organic farmers grow plants with insecticidal properties (*Helichrysum bracteatum* L., *Tagetes erecta* L., *Calendula officinalis* L., *Bidens pilosa* L., *Matricaria chamomilla* L., *Capsicum frutescens* L., *N. tabacum* and *Solanum lycopersicum* L.) (Dietrich et al. 2011). The unexplored potential of plants, which has been used in association with traditional knowledge, stands out. For example, in semiarid areas of Northeastern Brazil, the local population uses *Stemodia maritima* L. and *Selaginella convoluta* (Arnott) Spreng as insect repellent (Agra et al. 2007). Maybe these species also have potential to be employed for agricultural pest control.

In other Latin-American countries, the scenario is not different: many scientific works, a lot of practical evidence associated with traditional knowledge, and few commercial products. In Colombia, much research has been done with essential oils, especially for the control of insect pests in stored grains (Gallardo et al. 2012, 2014; Verbel et al. 2013). In Chile, there are various studies conducted to evaluate the toxic effect of plants of the family Rhamnaceae (Cespedes et al. 2013; Alarcón and Cespedes 2015) and Calceolariaceae (Cespedes et al. 2014, 2015) on insect pests. Regarding the genus *Calceolaria*, some species are endemic to Chile and show high resistance to insect attack in nature. For example, secondary metabolites from *Calceolaria talcana* J. Grau & C. Ehrh. are active against *Spodoptera frugiperda* (J E Smith) (Noctuidae) and fruit flies *Drosophila melanogaster* Meigen (Drosophilidae) (Muñoz et al. 2013). On the other hand, *Calceolaria integrifolia*s complex reduces feeding and growth of *Acanthoscelides obtectus* (Say) (Bruchidae) and *Epilachna varivestis* Mulsant (Coccinellidae). Its toxicity is attributed to diterpenes and naphthoquinones (Cespedes et al. 2016).

Among the studies developed in Argentina, we highlight a screening carried out with 89 native species of the central region of that country, employed on the leaf-cutting ant *Acromyrmex lundi* (Guérin) (Formicidae) and its symbiotic fungus, *Leucoagaricus gongylophorus*. The extracts from *Aristolochia argentina* Griseb and *Flourensia oolepis* Blake showed strong anti-foraging effects and inhibited fungus growth. The extract of *A. argentina* was selected for fractionation, allowing the isolation and identification of the active substance argentilactone (Napal et al. 2015). The family Annonaceae has also been the focus of research in Argentina. *Rollinia emarginata* Schltdl. showed insecticidal activity against *S. frugiperda*, and from this plant, the following compounds were isolated: vomifoliol, dehydrovomifoliol, blumenol C, loliolide, 7-epiloliolide, vanillin and dihydroactinolide (Colom et al. 2007). Likewise, we can mention the toxicity of acetogenins from the Argentine collection of *Annona cherimola* Mill. and the Bolivian collection of *Annona montana* Macfad. These metabolites caused a lethal effect on *Oncopeltus fasciatus* (Dallas) (Colom et al. 2008).

38.4 Final Remarks

The search for botanical insecticides for the control of arthropod pests is highly important. Despite the abundance of information from popular knowledge and numerous scientific studies carried out in Latin America, the commercialization of botanical insecticides is still limited. However, there is a growing interest in the search for botanical insecticides, mainly by multinational companies. Considering the increasing demand for pesticide-free products and the accelerated selection process of arthropod populations resistant to synthetic pesticides, which further intensifies the search for new molecules, the use of botanical insecticides and/or plant-derived substances for arthropods control is expected to increase.

References

Agra MF, Baracho GS, Nurit K et al (2007) Medicinal and poisonous diversity of the flora of "CaririParaibano", Brazil. J Ethnopharmacol 111(2):383–395

Alarcón J, Cespedes CL (2015) Chemical constituents and biological activities of South American Rhamnaceae. Phytochem Rev 14(3):389–401

Alves DS, Morejón RC, Machado ART et al (2015) Acaricidal activity of Annonaceae fractions against *Tetranychus tumidus* and *Tetranychus urticae* (Acari: Tetranychidae) and the metabolite profile of *Duguetia lanceolata* (Annonaceae) using GC-MS. Semina Ciênc Agrár 36(6):4119–4132

Alves DS, Machado ART, Campos VAC et al (2016) Selection of Annonaceae species for the control of *Spodoptera frugiperda* (Lepidoptera: Noctuidae) and metabolic profiling of *Duguetia lanceolata* using nuclear magnetic resonance spectroscopy. J Econ Entomol 109(2):649–659

Ansante TF, Ribeiro LP, Bicalho KU et al (2015) Secondary metabolites from Neotropical Annonaceae: screening, bioguided fractionation, and toxicity to *Spodoptera frugiperda* (J.E. Smith) (Lepidoptera: Noctuidae). Ind Crop Prod 74:969–976

Ansante TF, Ribeiro LP, Vendramim JD (2017) Acute and chronic toxicities of an annonin-based commercial bioinsecticide and a joint mixture with a limonoid-based formulation to the fall armyworm. Neotrop Entomol 46(2):216–222

Bajwa AA, Ahmad A (2012) Potential applications of neem based products as biopesticides. Health 3:116–120

Barbehenn RV, Constabel CP (2011) Tannins in plant-herbivore interactions. Phytochemistry 72(13):1551–1565

Bernardi D, Ribeiro L, Andreazza F et al (2017) Potential use of *Annona* by products to control *Drosophila suzukii* and toxicity to its parasitoid *Trichopria anastrephae*. Ind Crop Prod 110:30–35

Bhat SV, Nagasampagi BA, Sivakumar M (2005) Chemistry of natural products. Springer, Berlin

Blanco CA, Chiaravalle W, Dalla-Rizza M et al (2016) Current situation of pests targeted by Bt crops in Latin America. Curr Opin Insect Sci 15:131–138

Boeckler GA, Towns M, Unsicker SB et al (2014) Transgenic upregulation of the condensed tannin pathway in poplar leads to a dramatic shift in leaf palatability for two tree-feeding Lepidoptera. J Chem Ecol 40(2):150–158

Brasil. Ministério da Agricultura, Pecuária e Abastecimento (2017) Sistema AGROFIT. http://agrofit.agricultura.gov.br. Accessed 2 Nov 2017

Brito EF, Baldin ELL, Silva RCM et al (2015) Bioactivity of *Piper* extracts on *Tuta absoluta* (Lepidoptera: Gelechiidae) in tomato. Pesqui Agropecu Bras 50:196–202

Cespedes CL, Molina SC, Muñoz E et al (2013) The insecticidal, molting disruption and insect growth inhibitory activity of extracts from *Condalia microphylla* Cav. (Rhamnaceae). Ind Crop Prod 42:78–86

Cespedes CL, Salazar JR, Ariza-Castolo A et al (2014) Biopesticides from plants: *Calceolaria integrifolia* S.l. Environ Res 132:391–406

Cespedes CL, Aqueveque PM, Avila JG et al (2015) New advances in chemical defenses of plants: researches in Calceolariaceae. Phytochem Rev 14(3):367–380

Cespedes CL, Lina-Garcia L, Kubo I et al (2016) *Calceolaria integrifolia* s.l. complex, reduces feeding and growth of *Acanthoscelides obtectus*, and *Epilachna varivestis*: a new source of bioactive compounds against dry bean pests. Ind Crop Prod 89:257–267

Colmenarez Y, Wyckhuys K, Ciomperlik MA et al (2016) Uso do manejo integrado de pragas e controle biológico pelos agricultores na América Latina e no Caribe: desafíos e oportunidades. In: Halfeld-Vieira BA, Marinho-Prado JS, Nechet KL et al (eds) Defensivos agrícolas naturais: uso e perspectivas. Embrapa, Brasília, pp 802–853

Colom OA, Popich S, Bardon A (2007) Bioactive constituents from *Rollinia emarginata* (Annonaceae). Nat Prod Res 21(3):254–259

Colom OA, Barrachina I, Mingol IA et al (2008) Toxic effects of annonaceous acetogenins on *Oncopeltus fasciatus*. J Pest Sci 81(2):85–89

Copping LG, Duke SO (2007) Natural products that have been used commercially as crop protection agents. Pest Manag Sci 63(6):524–554

Desmarchelier C (2010) Neotropics and natural ingredients for pharmaceuticals: why isn't South American biodiversity on the crest of the wave? Phytother Res 24(6):791–799

Dev S, Koul O (1997) Insecticides of natural origin. Harwood Academic, New York

Dewick PM (2002) Medicinal natural products: a biosynthetic approach. John Wiley & Sons, Chichester

Dietrich F, Strohschoen AAG, Schultz G et al (2011) Utilização de inseticidas botânicos na agricultura orgânica de Arroio do Meio/RS. Rev Bras Agrociência 17(4):251–255

Electronic Code of Federal Regulations (e-CFR) (2017). https://www.ecfr.gov/cgi-bin/. Accessed 1 Nov 2017

El-Wakeil NE (2013) Botanical pesticides and their mode of action. Gesunde Pflanz 65(4):125–149
Estrela JLV, Fazolin M, Catani V et al (2006) Toxicity of essential oils of *Piper aduncum* and *Piper hispidinervum* against *Sitophilus zeamais*. Pesqui Agropecu Bras 41(2):217–222
Fazolin M, Estrela JLV, Valdomiro C et al (2005) Toxicity of *Piper aduncum* oil to adults of *Cerotoma tingomarianus* Bechyné (Coleoptera: Chrysomelidae). Neotrop Entomol 34(3):485–489
Fazolin M, Estrela JLV, Medeiro AFM et al (2016) Combining the essential oil of *Piper aduncum* L. with commercial insecticides. Semin Ciênc Agrár 37(6):3903–3914
Fouad HA, Faroni LRD, Tavares WDS et al (2014) Botanical extracts of plants from the Brazilian Cerrado for the integrated management of *Sitotroga cerealella* (Lepidoptera: Gelechiidae) in stored grain. J Stored Prod Res 57:6–11
Gallardo KC, Verbel JO, Stashenko EE (2012) Repellency and toxicity of essential oils from *Cymbopogon martinii*, *Cymbopogon flexuosus* and *Lippia origanoides* cultivated in Colombia against *Tribolium castaneum*. J Stored Prod Res 50:62–65
Gallardo KC, Benitez NP, Castro NP et al (2014) Plants cultivated in Choco, Colombia, as source of repellents against *Tribolium castaneum* (Herbst). J Asia Pac Entomol 17(4):753–759
Gershenzon J, Dudareva N (2007) The function of terpene natural products in the natural world. Nat Chem Biol 3(7):408–414
Giongo AMM, Vendramim JD, Freitas SDL et al (2015) Growth and nutritional physiology of *Spodoptera frugiperda* (Lepidoptera: Noctuidae) fed on Meliaceae fractions. Rev Colomb Entomol 41(1):33–40
Giongo AMM, Vendramim JD, Freitas SDL et al (2016) Toxicity of secondary metabolites from Meliaceae against *Spodoptera frugiperda* (J E Smith) (Lepidoptera: Noctuidae). Neotrop Entomol 45(6):725–733
Gonçalves GLP, Domingues VC, Ribeiro LP et al (2017) Compounds from *Duguetia lanceolata* St.- Hil. (Annonaceae) bioactive against *Zabrotes subfasciatus* (Boheman) (Coleoptera: Chrysomelidae: Bruchinae). Ind Crop Prod 97:360–367
Hidalgo RR, Otáñez XP, Carrera SG et al (2017) The current status of resistance to alpha-cypermethrin, ivermectin, and amitraz of the cattle tick (*Rhipicephalus microplus*) in Ecuador. PLoS One 12(4):e0174652
Isman MB (2006) Botanical insecticides, deterrents, and repellents in modern agriculture and an increasingly regulated world. Annu Rev Entomol 51:45–66
Isman MB (2015) A renaissance for botanical insecticides? Pest Manag Sci 71(12):1587–1590
Isman MB, Grieneisen ML (2014) Botanical insecticide research: many publications, limited useful data. Trends Plant Sci 19(3):140–145
Isman MB, Seffrin R (2014) Natural insecticides from the Annonaceae: a unique example for developing biopesticides. In: Singh D (ed) Advances in plant biopesticides. Springer, New Delhi, pp 21–33
Jardim ICSF, Andrade JA, Queiroz SCN (2009) Resíduos de agrotóxicos em alimentos: uma preocupação ambiental global: um enfoque às maçãs. Quim Nova 32:996–1012
Khot LR, Sankaran S, Maja JM et al (2012) Applications of nanomaterial in agricultural production and crop protection: a review. Crop Prot 35:64–70
Kostyukovsky M, Rafaeli A, Gileadi C et al (2002) Activation of octopaminergic receptors by essential oil constituents isolated from aromatic plants: possible mode of action against insect pests. Pest Manag Sci 58(11):1101–1106
Koul O, Isman MB, Ketkar CM (1990) Properties and uses of neem, *Azadirachta indica*. Can J Bot 68(1):1–11
Luiz AL, Perlatti B, Marques FA et al (2017) Efficacy of botanical extracts from Brazilian savannah against *Diabrotica speciosa* and associated bacteria. Ecol Res 32(3):435–444
Lundin O, Rundlöf M, Smith HG et al (2015) Neonicotinoid insecticides and their impacts on bees: a systematic review of research approaches and identification of knowledge gaps. PLoS One 10(8):e0136928

Massarolli A, Pereira MJB, Foerster LA (2017) *Annona crassiflora* Mart. (Annonaceae): effect of crude extract of seeds on larvae of soybean looper *Chrysodeixis includens* (Lepidoptera: Noctuidae). Bragantia 76(3):398–405

Menezes ELA (2005) Inseticidas botânicos: seus princípios ativos, modo de ação e uso agrícola. Embrapa Agrobiologia, Seropédica

Monnerat R, Martins E, Macedo C et al (2015a) Evidence of field-evolved resistance of *Spodoptera frugiperda* to Bt cornexpressing Cry1F in Brazil that is still sensitive tomodified Bt toxins. PLoS One 10(4):e0119544

Monnerat R, Martins E, Queiroz P et al (2015b) Insect resistance to Bt toxins in Brazil and Latin America Bt Resistance: characterization and strategies for GM crops producing *Bacillus thuringiensis*. Toxins 3:26–35

Muñoz E, Lamilla C, Marin JC et al (2013) Antifeedant, insect growth regulatory and insecticidal effects of *Calceolaria talcana* (Calceolariaceae) on *Drosophila melanogaster* and *Spodoptera frugiperda*. Ind Crop Prod 42:137–144

Napal GND, Buffa LM, Nolli LC et al (2015) Screening of native plants from central Argentina against the leaf-cutting ant *Acromyrmex lundi* (Guérin) and its symbiotic fungus. Ind Crop Prod 76:275–280

Ntalli NG, Caboni P (2012) Botanical nematicides: a review. J Agric Food Chem 60(4):9929–9940

Oliveira AP, Santana AS, Santana EDR et al (2017) Nanoformulation prototype of the essential oil of *Lippia sidoides* and thymol to population management of *Sitophilus zeamais* (Coleoptera: Curculionidae). Ind Crop Prod 107:198–205

Palmquist K, Salatas J, Fairbrother A (2012) Pyrethroid insecticides: use, environmental fate, and ecotoxicology. In: Perveen F (ed) Insecticides: advances in integrated pest management. InTech, New York, pp 251–278

Parreira DS, Alcántara-de la Cruz R, Leite GLD, Ramalho FS, Zanuncio JC, Serrão JE (2018) Quantifying the harmful potential of ten essential oils on immature *Trichogramma pretiosum* stages. Chemosphere 199:670–675

Pavela R (2016) History, presence and perspective of plant insecticides and farm products for protection against insects: a review. Plant Prot Sci 52:229–241

Pelletier SW (1983) The nature and definition of an alkaloid. In: Pelletier SW (ed) Alkaloids: chemical and biological perspectives. Wiley, New York, pp 1–31

Previero CA, Lima Júnior BC, Florencio LK et al (2010) Receitas de plantas com propriedades inseticidas no controle de pragas. CEULP/ULBRA, Palmas

Priestley CM, Williamson EM, Wafford KA et al (2003) Thymol, a constituent of thyme essential oil, is a positive allosteric modulator of human GABA(A) receptors and a homo-oligomeric GABA receptor from *Drosophila melanogaster*. Br J Pharmacol 140(8):1363–1372

Rais DS, Sato ME, Silva MZ (2013) Detecção e monitoramento da resistência do tripes *Frankliniella occidentalis* ao inseticida espinosade. Bragantia 72(1):35–40

Ribeiro LP, Ansante TF, Vendramim JD (2016) Effect of ethanolic extract from *Annona mucosa* seeds on development and feeding behavior of *Spodoptera frugiperda*. Bragantia 75(3):322–330

Sattelle DB, Cordova D, Cheek TR (2008) Insect ryanodine receptors: molecular targets for novel pest control chemicals. Invertebr Neurosci 8(3):107–119

Shukla YM (2009) Plant secondary metabolites. New India Publishing, Delhi

Silva MA, Bezerra-Silva GCD, Vendramim JD et al (2012) Inhibition of oviposition by neem extract: a behavioral perspective for the control of the Mediterranean fruit fly (Diptera: Tephritidae). Fla Entomol 95(2):333–337

Silva JE, Assis CPO, Ribeiro LMS et al (2016) Field-evolved resistance and cross-resistance of Brazilian *Tuta absoluta* (Lepidoptera: Gelechiidae) populations to diamide insecticides. J Econ Entomol 10:2190–2195

Simões CMO, Schenkel EP, Mello JCP et al (2017) Farmacognosia: do produto natural ao medicamento. Artmed, Porto Alegre

Sousa MF, Silva LV, Brito MD et al (2012) Tipos de controle alternativo de pragas e doenças nos cultivos orgânicos no estado de Alagoas, Brasil. Rev Bras Agroecol 7(1):132–138

Souza CM, Baldin ELL, Ribeiro LP et al (2017) Lethal and growth inhibitory activities of neotropical Annonaceae-derived extracts, commercial formulation, and an isolated acetogenin against *Helicoverpa armigera*. J Pestic Sci 90(2):701–709

Stoll A, Squeo FA (2012) Latin American plant sciences: from early naturalists to modern science. Plant Ecol Divers 5(2):147–151

Tavares WS, Cruz I, Petacci F et al (2009) Potential use of Asteraceae extracts to control *Spodoptera frugiperda* (Lepidoptera: Noctuidae) and selectivity to their parasitoids *Trichogramma pretiosum* (Hymenoptera: Trichogrammatidae) and *Telenomus remus* (Hymenoptera: Scelionidae). Ind Crop Prod 30(3):384–388

Turchen LM, Piton LP, Dall'Oglio EL et al (2016) Toxicity of *Piper aduncum* (Piperaceae) essential oil against *Euschistus heros* (F.) (Hemiptera: Pentatomidae) and non-effect on egg parasitoids. Neotrop Entomol 45(5):604–611

Verbel JO, Ballestas IT, Gallardo KC et al (2013) Essential oils applied to the food act as repellents toward *Tribolium castaneum*. J Stored Prod Res 55:145–147

Volpe HX, Fazolin M, Garcia RB et al (2016) Efficacy of essential oil of *Piper aduncum* against nymphs and adults of *Diaphorina citri*. Pest Manag Sci 72(6):1242–1249

War AR, Paulraj MG, Ahmad T et al (2012) Mechanisms of plant defense against insect herbivores. Plant Signal Behav 7(1):1306–1320

Werdin-González JO, Laumann RA, Silveira S et al (2013) Lethal and sublethal effects of four essential oils on the egg parasitoids *Trissolcus basalis*. Chemosphere 92(5):608–615

Wink M (1993) Production and application of phytochemicals from an agricultural perspective. In: Van Beek TA, Breteler H (eds) Phytochemistry and agriculture. Clarendon Press, Oxford, pp 171–213

Yazdani D, Tan YH, Zainal AMA et al (2011) A review on bioactive compounds isolated from plants against plant pathogenic fungi. J Med Plant Res 5(16):6584–6589

Yu SJ (2014) The toxicology and biochemistry of insecticides. CRC Press, London

Chapter 39
Effects of Genetically Modified Plants with Bt Toxins on Natural Enemies

Camila S. F. Souza, Rosangela C. Marucci, Daniela Resende Chaves, and Simone Martins Mendes

39.1 Introduction

Protection transgenic strategy against pest insects depends on the transference and the expression of defense genes for the target species. Delta-endotoxins of the bacterium *Bacillus thuringiensis* Berliner (Bt) are among the best-known and studied cases on resistance induction. The bacterium occurs naturally in the soil and builds protein crystals during the stationary or sporulation phase (Carneiro et al. 2009). Synthetized protein insecticides include crystalline protein, popularly known as delta-toxins (including toxins Cry and Cyt), and vegetative protein (Vip). Each Cry and Vip toxin is highly specific and reveals a narrow range of target insects (Peralta and Palma 2017). Vip proteins are a class of secreted insecticidal proteins that were first described in 1996 and which do not share sequence homology (Gomis-Cebolla et al. 2017).

The use of Bt plants has been a fundamental tool for pest management of crops such as corn, soybean, and cotton in countries such as Brazil and Argentina, considering a total of ten countries that approved cultivation in LA (International Service for the Acquisition of Agri-biotech Applications 2017). Easy logistics for crop cultivation, in addition to compatibility with other strategies of IPM, are characteristics that confirm the high percentage of cultivation of Bt plants in these countries.

The main activity related to Bt toxic proteins occurs by inserting the toxin in the insect through ingestion. Lithic pores are formed in the microvillosities of the

C. S. F. Souza · R. C. Marucci
Federal University of Lavras (UFLA), Department of Entomology, Lavras, MG, Brazil

D. R. Chaves
Consultor in Risk Assessment, Sete Lagoas, MG, Brazil

S. M. Mendes (✉)
Lab de Entomologia, Embrapa Milho e Sorgo, Sete Lagoas, MG, Brazil
e-mail: simone.mendes@embrapa.br

B. Souza et al. (eds.), *Natural Enemies of Insect Pests in Neotropical Agroecosystems*, https://doi.org/10.1007/978-3-030-24733-1_39

midgut membrane of the insects (Aronson and Shai 2001; Bravo et al. 2007). Cell lyse occurs, causing the selective permeability of the cations and the rupture of the epithelium of the middle intestine, releasing cell contents. Deep septicemia follows (De Maagd et al. 2001; Bravo et al. 2007).

Genetically modified plants (GMP) resistant to pest insects have basically codifying genes with Cry and Vip proteins for the Lepidoptera pest group and for the root pest Coleoptera. The commercial introduction of the plants demands the evaluation of possible impacts of this type of technology on the environments. Undesirable effects on nontarget organisms are among the probable impacts. A nontarget organism is any specimen present in the GMP crop environments and is not targeted for toxic protein control. Arthropods, plants, mammals, fish and birds, and others, featuring different ecological functions, such as scavengers, biological control agents, pollinizers, and even pest insects are included (Fontes et al. 2003; Capalbo et al. 2009).

The main advantages in the use of Bt plants are (a) decrease of liabilities caused by pests and the number of applications of insecticides, favoring natural enemies which help in control and which contribute toward the delay of the resistance evolution to insecticide proteins; (b) improvement of income in the production system, using plants resistant to insects with the best efficiency (Lourenção and Fernandes 2013); and (c) facility in the logistics of crop treatments, especially in great areas. On the other hand, the main disadvantages for the culture of Bt plants comprise the fast selection of the resistance of target pests to toxic proteins and the possibility of activity on nontarget organisms (Miranda et al. 2017; Tabashnik and Carrière 2017).

Great concern exists on the potential effects of crops with Bt toxins in the environment, specifically on nontarget organisms. Safety evaluation is mandatory in most countries where plants with Bt toxins are authorized (Romeis et al. 2011). Consequently, monitoring and in-depth studies on the association of these organisms with Bt plants are necessary (Costa 2011). Several studies have pinpointed possible toxic effects on proteins Bt on nontarget species, including other herbivores, scavengers, predators, and parasitoids (Hilbeck et al. 1998; Birch et al. 1999; Losey et al. 1999; Schuler et al. 1999). However, most researches have tested the effect of these proteins on the species under nonnatural conditions, without taking into account important characteristics of ecological interactivities and the true exposure level of sensitive stages in natural conditions (Dale et al. 2002). More in-depth and multi-variegated studies, which take into account exposure conditions similar to field situations, may greatly minimize the deleterious effects of Bt crops on nontarget species.

Brazil and other 130 countries signed the Cartagena Protocol on Biosafety. The protocol ensures an adequate protection level in the transference, handling, and use of genetically modified organisms (GMOs). In Latin America, Colombia, Costa Rica, Peru, and Mexico signed the protocol, whereas Argentina, the third largest producer of GMPs in the world, failed to do so. One should also underscore that Mexico is the country of origin of corn or, rather, the product with the greatest GM area worldwide and where the release of the commercial planting of GMPs is polemical, mainly due to the possibility of genic flow (Capalbo et al. 2009). In 2016, the USA still ranked first in countries with GM plantations, with 72.9 million hectares, followed by Brazil (49.1), Argentina (23.8), Canada (11.6), and India (10.8) (International Service for the Acquisition of Agri-biotech Applications 2016).

GM cultivars of cotton, corn, soybean, and sugarcane resistant to pests and tolerant to herbicides are available in Brazil. Since the commercial release of pest-insect-resistant GM plants, several studies have been undertaken on the effect of Bt toxic proteins on biodiversity, physiology, and environmental service provided by natural enemies (Yang et al. 2017) due to their activities in trophic interactions and to their ecological and economic importance.

This chapter deals with the impacts of plants with Bt toxins on natural enemies.

39.2 How May One Assess the Effect of Plants with Bt Toxins on Natural Enemies?

From the agricultural point of view, natural enemies are organisms, which control pests and diseases in plants. Predators, parasitoids (entomophagous insects), and entomopathogens may be included in the case of pest-arthropod control. In fact, these natural enemies conserve biological control since they maintain greater stability within agricultural ecosystems and provide, directly or indirectly, environmental, health, and economic advantages to farmers and to society (Östman et al. 2003; Bianchi et al. 2006; Zhang et al. 2007; Stallman and James 2015).

Sustainable agricultural production is linked to the conservation of the biodiversity of the agroecosystem through advantageous organisms for phytosanitary protection (Talebi et al. 2008; Tavares et al. 2015). Figueiredo et al. (2006) reported that the lack of natural enemies caused an increase in damages by *Spodoptera frugiperda* (J. E. Smith) (Noctuidae), one of the target insects of GM plants. Since damages cause more than 50% liabilities in grain yield and 48% losses in dry matter, the relevance of biological control in the population decrease of pest insects is obvious.

The identification of the ecological functions of organisms within the system ranks first to evaluate the effects of plants with Bt toxins on natural enemies. The adverse effects that must be analyzed should not merely deal with biological aspects but also with possible direct and indirect negative ones too, such as (a) pest control by parasitism or predation; (b) Bt gene inserted within the GM plant; (c) expression levels which may vary according to the development stage of the plant; (d) exposure mode of the natural enemy; and (e) site where the plant is cultivated (Capalbo et al. 2009). Evaluations should consider each case one by one, since impacts may be different according to the exposed organism. Furthermore, risks for the same organism within different environments and with great variations may not be the same (Pires et al. 2003).

Natural enemies may be affected by cultures with Bt toxins when there is a decrease in quantity, quality, or nutritional adequacy. Several natural enemies require not only preys or hosts but also parts of the plant for their nutritional needs. It is one of the direct exposure modes to toxic protein in the plant through pollen or nectar intake. In-depth information on the several manners through which natural enemies are exposed to toxins may be an asset within a more precise evaluation of the potential deleterious effects of GM cultures (Andow et al. 2006; Hilbeck and Schmidt 2006; Schrijver et al. 2016).

Risk assessment is usually done by stages. The adverse effects on the natural enemy are first evaluated in the laboratory. Later, assessment moves toward semi-field conditions. In this situation, conditions are characterized by greater ecological complexity. The field stage is the final evaluation site with regard to the consequences of adverse effects on the control agent (Romeis et al. 2011). The evaluation scheme described above may be considered oversimplified since laboratory tests may be insufficient or nonconclusive, especially in cases in which the effects of high toxin doses on natural enemies are not detected. Field assays may also forward divergent results due to the natural behavior of predation or parasitism and to the effects on the trophic chain (Lundgren et al. 2009; Mendes et al. 2012a, b).

However, evolutions in the assessments of impacts of GMOs on natural enemies are only perceived overtime. Romeis et al. (2014) have shown that effects perceived in laboratory conditions may not be translated into environmental damages. The authors argued that the development of conceptual models on the manner a plant with Bt toxins may damage a nontarget species or its ecological role also helps in the formulation of testable risk hypotheses.

According to Shelton et al. (2016), the most relevant issue of these studies is false positive or false negative, particularly for natural enemies. This is due to the fact that the potential effects of the quality of the host or the prey should be taken into account. Indirect effects on predators or parasitoids are the result of feed in susceptible hosts and preys that ingest Cry proteins and become less adequate as feed for the natural enemy.

39.3 Direct and Indirect Effects with Bt Toxins on Natural Enemies

Studies on the direct effects of GMPs on natural enemies have shown that it is highly important to evaluate how the organism will be exposed to toxic protein. It is also greatly relevant to have information on intra-guild interactions in the culture and on the most vulnerable organisms exposed to proteins (Romeis et al. 2013). Natural enemies, such as parasitoids and predators, may be indirectly affected by hosts and preys that feed on plants with Bt toxin or directly affected when they feed on parts of plants, such as pollen and nectar (Lundgren et al. 2009). Several studies have demonstrated that cultures with Bt toxins have negative effects on important natural enemies, especially parasitoids. However, most studies do not register any effect of these cultures on the organisms (Tian et al. 2015). Furthermore, studies demonstrating some negative effects of Bt protein on natural enemies evidence a lower impact with regard to organo-synthetic insecticides, especially the broad-spectrum ones. Consequently, Bt crops may contribute particularly toward biodiversity if integrated to alternative control methods (Fontes et al. 2003). According to Romeis et al. (2006), most researches showing Bt crop risks on natural enemies refer to their low abundance on the field. The authors attributed such results to the low number of available preys and hosts. This is due to the fact that most is controlled by toxin proteins in the plant.

Besides the concern on the impact on natural enemies, one of the main issues in Bt crops is the selection of target insects resistant to toxic proteins. Resistant insects that survive toxic proteins may be preys or hosts of bad quality, and, above all, they may alter the behavior of predation and parasitism. Several research studies on this subject were performed in Brazil and in other countries. Mendes et al. (2012a) evaluated the effect of protein Bt in maize on *Orius insidiosus* (Say) (Anthocoridae) in Sete Lagoas, State of Minas Gerais, Brazil, and concluded that *S. frugiperda* became more prone to predation and an easy target for the predator. The authors did not detect any changes in the biology of the predator. This fact makes feasible biological control through this Hemiptera. However, results were different in the case of *Podisus nigrispinus* (Dallas) (Pentatomidae) since the predator preferred *S. frugiperda* larvae to feeding on conventional corn. Consequently, predator behavior of each species should be taken into account since *P. nigrispinus* prefers large-size worms retrieved from feeding on conventional meals (Leite et al. 2014).

Assays with different proteins in tri-trophic studies involving resistant and sensitive herbivores revealed lack of negative effects for natural enemies (Tian et al. 2012; Shelton et al. 2016). Furthermore, natural enemies may delay the selection of target resistance of insects to Bt crops, since these agents prey or parasite resistant insects and decrease the population's resistance alleles (Liu et al. 2014).

A Brazilian survey performed to evaluate the diversity of insects in maize with different Bt proteins in the State of Minas Gerais failed to identify a negative effect on the diversity, richness, and abundance of entomofauna in the plantations. The survey showed 30 herbivores and 33 natural enemies in Bt and non-Bt maize, indiscriminately, with insects identified by species or family (Resende et al. 2016). Another survey on the diversity and abundance of nontarget arthropods in a soybean plantation with Cry1Ac and Cry1F in three Brazilian states (Goiás, Paraná, and Minas Gerais) also revealed that proteins did not influence the arthropod community (Marques et al. 2018).

Analysis of bioaccumulation is another method to detect the effect of plants with Bt toxins on natural enemies. Toxic proteins may be transferred in the trophic chain, passing from the plant to the herbivore and subsequently to the predator or parasitoid (third trophic level). This effect occurred in predators *Harmonia axyridis* (Pallas) (Coccinellidae) and *Cycloneda sanguinea* (Linnaeus) (Coccinellidae), which sequestered protein Cry1F and Cry1Ac of *Myzus persicae* (Sulzer) (Aphididae) fed on plants with the proteins (Paula and Andow 2016). Although proteins Cry have a greater molecular weight when compared to some metabolites of plants sequestered by lepidopterans, the protein Cry1Ac was absorbed by

Chlosyne lacinia (Geyer) (Nymphalidae) through the transovarian method, a pest insect in sunflower crops, and a nontarget on cotton crops, in Brasília, Distrito Federal, Brazil (Paula et al. 2014).

Understanding the above interactions may be of great help in the comprehension of the mechanisms involved in the selection of resistance by target species and in the implications that the protein passage from one trophic level to another may cause in trophic networks.

39.4 Final Remarks

There is generally no incompatibility in the use of GM cultures with Bt toxins with natural enemies. They are actually complementary and safe control strategies for farmers and consumers. Since Brazil is a country characterized by a vast biological diversity, the preservation of richness and abundance of species is crucial. Consequently, further researches are required to understand better the effects of these toxic proteins in natural enemies, a better form in reconciling biological control and plants with Bt toxins and the general arrangement of the agroecosystem for the preservation and sustainability of the agricultural environment.

International Service for the Acquisition of Agri-biotech Applications (2017) Global status of commercialized biotech/GM crops in 2017: biotech crop adoption surges as economic benefits accumulate in 22 years: brief n. 53. Isaa, Ithaca.

References

Andow DA, Lövei GL, Arpaia S (2006) Ecological risk assessment for Bt crops. Nat Biotechnol 24:749–751

Aronson AI, Shai Y (2001) Why *Bacillus thuringiensis* insecticidal toxins are so effective: unique features of their mode of action. FEMS Microbiol Lett 195:1–8

Bianchi F, Booij CJH, Tscharntke T (2006) Sustainable pest regulation in agricultural landscapes: a review on landscape composition, biodiversity and natural pest control. Proceedings B 273:1595–1715

Birch A, Geoghegan IE, Majerus MEN, Mcnicol JW, Hackett CA, Gatehouse AMR, Gatehouse J (1999) Tri-trophic interactions involving pest aphids, predatory 2-spot ladybirds and transgenic potatoes expressing snowdrop lectin for aphid resistance. Mol Breed 5:75–83

Bravo A, Gill SS, Soberón M (2007) Mode of action of *Bacillus thuringiensis* Cry and Cyt toxins and their potential for insect control. Toxicon 49:423–435

Capalbo DMF, Dusi NA, Pires CS, Paula DP, Arantes OMN, Melo IS (2009) OGM e biossegurança ambiental. In: Figueiredo A (ed) Biossegurança de OGM: uma visão integrada. Publit, Rio de Janeiro, pp 190–219

Carneiro MAC, Souza ED, Reis EF, Pereira HS, Azevedo WR (2009) Atributos físicos, químicos e biológicos de solo de cerrado sob diferentes sistemas de uso e manejo. Rev Bras Ciênc Solo 33:147–157

Costa MLM (2011) Caracterização molecular de genes cyt de cepas de *Bacillus thuringiensis* e avaliação da toxicidade de proteínas inseticidas contra *Spodoptera frugiperda* (J.E. Smith) (Lepidoptera: Noctuidae). Dissertation, Universidade Federal de Lavras

Dale PJ, Clarke B, Fontes EM (2002) Potential for the environmental impact of transgenic crops. Nat Biotechnol 20:567–574

De Maagd RA, Bravo A, Crickmore N (2001) How *Bacillus thuringiensis* has evolved specific toxins to colonize the insect world. Trends Genet 17:193–199

Figueiredo MLC, Dias AMPM, Cruz I (2006) Associação entre inimigos naturais e *Spodoptera frugiperda* (J.E. Smith, 1797) (Lepidoptera: Noctuidae) na cultura do milho. Rev Bras Milho Sorgo 5:340–350

Fontes EMG, Pires CSS, Suji ER (2003) O impacto de plantas geneticamente modificadas resistentes a insetos sobre a biodiversidade. In: Pires CSS, Fontes EMG, Suji ER (eds) Impacto ecológico de plantas geneticamente modificadas: o algodão resistente a insetos como estudo de caso. Embrapa, Brasília, pp 65–83

Gomis-Cebolla J, De Escudero IR, Vera-Velasco NM, Hernández-Martínez P, Hernández-Rodríguez CS, Ceballos T, Palma L, Escriche B, Caballero P, Ferré J (2017) Insecticidal spectrum and mode of action of the *Bacillus thuringiensis* Vip3Ca insecticidal protein. J Invertebr Pathol 142:60–67

Hilbeck A, Schmidt JEU (2006) Another view on Bt proteins: how specific are they and what else might they do? Biopestic Int 2:1–50

Hilbeck A, Baumgartner M, Fried PM, Bigler F (1998) Effects of transgenic *Bacillus thuringiensis* corn-fed prey on mortality and development time of immature *Chrysoperla carnea* (Neuroptera: Chrysopidae). Environ Entomol 27:480–487

International Service for the Acquisition of Agri-biotech Applications (2016) Global status of commercialized biotech/GM crops: brief n. 46. Manila

International Service for the Acquisition of Agri-biotech Applications (2017) Global status of commercialized biotech/GM crops in 2017: biotech crop adoption surges as economic benefits accumulate in 22 years: brief n. 53. Isaa, Ithaca

Leite NA, Mendes SM, Santos CA, Pereira EJG (2014) Does Cry1Ab maize interfere in the biology and behavioural traits of *Podisus nigrispinus*? Bull Insectol 67:265–271

Liu X, Chen M, Collins HL, Onstad DW, Roush RT, Zhang Q, Earle ED, Shelton AM (2014) Natural enemies delay insect resistance to bt crops. PLoS One 9:e90366

Losey JE, Rayor LS, Carter ME (1999) Transgenic pollen harms monarch larvae. Nature 399:214–215

Lourenção ALF, Fernandes MG (2013) Avaliação do milho Bt Cry1Ab e Cry1F no controle de *Spodoptera frugiperda* (J. E. Smith, 1797) (Lepidoptera: Noctuidae) em condições de campo. Científica 4:164–188

Lundgren JG, Gassmann AJ, Bernal J, Duan JJ, Ruberson J (2009) Ecological compatibility of GM crops and biological control. Crop Prot 28:1017–1030

Marques LH, Santos AC, Castro BA, Storer NP, Babcock JM, Lepping MD, Sa V, Moscardini VF, Rule DM, Fernandes OA (2018) Impact of transgenic soybean expressing Cry1Ac and Cry1F proteins on the non-target arthropod community associated with soybean in Brazil. PLoS One 13:e0191567

Mendes SM, Brasil KGB, Waquil MS, Marucci RC, Waquil JM (2012a) Biological and behavioral aspects of predator, *Orius insidiosus* (SAY, 1832) in Bt and non-Bt maize. Biosci J 28:753–761

Mendes SM, Resende DC, Leite NA, Oliveira FS, Santos CA, Barbosa TAN (2012b) Avaliação de variáveis comportamentais como metodologia para estudo de organismos não alvo em milho Bt (Circular Técnica; vol 21) Embrapa Milho e Sorgo, Sete Lagoas

Miranda JE, Mendes SM, Hirose E (2017) Refúgio como estratégia de manejo da resistência. A Granja 73:40–42

Östman Ö, Ekbom B, Bengtsson J (2003) Yield increase attributable to aphid predation by ground-living polyphagous natural enemies in spring barley in Sweden. Ecol Econ 45:149–158

Paula DP, Andow DA (2016) Uptake and bioaccumulation of Cry toxins by an aphidophagous predator. Environ Pollut 209:164–168

Paula DP, Andow DA, Timbó RV, Sujii ER, Pires CSS, Fontes EMG (2014) Uptake and transfer of a Bt toxin by a Lepidoptera to its eggs and effects on its offspring. PLoS One 9:e95422

Peralta C, Palma L (2017) Is the insect world overcoming the efficacy of *Bacillus thuringiensis*? Toxins 9:1–5

Pires CSS, Suji ER, Fontes EMG (2003) Avaliação de risco de plantas geneticamente modificadas resistentes a insetos sobre inimigos naturais. In: Pires CSS, Fontes EMG, Suji ER (eds) Impacto ecológico de plantas geneticamente modificadas: o algodão resistente a insetos como estudo de caso. Embrapa, Brasília, pp 85–115

Resende DC, Mendes SM, Marucci RC, Silva AC, Campanha MM, Waquil JM (2016) Does Bt maize cultivation affect the non-target insect community in the agro ecosystem? Rev Bras Entomol 60:82–93

Romeis J, Meissle M, Bigler F (2006) Transgenic crops expressing *Bacillus thuringiensis* toxins and biological control. Nat Biotechnol 24:63–71

Romeis J, Hellmich RL, Candolfi MP, Carstens K, De Schrijver A, Gatehouse AM, Herman RA, Huesing JE, Mclean MA, Raybould A, Shelton AM, Waggoner A (2011) Recommendations for the design of laboratory studies on non-target arthropods for risk assessment of genetically engineered plants. Transgenic Res 20:1–22

Romeis J, Raybould A, Bigler F, Candolfi MP, Hellmich RL, Huesing JE, Shelton AM (2013) Deriving criteria to select arthropod species for laboratory tests to assess the ecological risks from cultivating arthropod-resistant genetically engineered crops. Chemosphere 90:901–909

Romeis J, Meissle M, Álvarez-Alfagem F, Bigler F, Bohan DA, Devos Y, Malone LA, Pons X, Rauschen S (2014) Potential use of an arthropod database to support the non-target risk assessment and monitoring of transgenic plants. Transgenic Res 23:995–1013

Schrijver AD, Devos Y, De Clercq P, Gathmann A, Romeis J (2016) Quality of laboratory studies assessing effects of Bt-proteins on non-target organisms: minimal criteria for acceptability. Nat Biotechnol 25:395–411

Schuler TH, Potting RP, Denholm I, Poppy GM (1999) Parasitoid behaviour and Bt plants. Nature 400:825–829

Shelton AM, Romeis J, Naranjo SE, Tian JC, Hellmich RL (2016) Use of Bt-resistant caterpillars to assess the effect of Cry proteins on beneficial natural enemies. IOBC-WPRS Bull 114:51–55

Stallman HR, James HSJ (2015) Determinants affecting farmers' willingness to cooperate to control pests. Ecol Econ 117:182–192

Tabashnik EB, Carrière Y (2017) Surge in insect resistance to transgenic crops and prospects for sustainability. Nat Biotechnol 35:926–935

Talebi K, Kavousi A, Sabahi Q (2008) Impacts of pesticides on arthropod biological control agents. Pest Technol 2:87–97

Tavares J, Wang KH, Hooks CRR (2015) An evaluation of insectary plants for management of insect pests in a hydroponic cropping system. Biol Control 91:1–9

Tian J, Collins HL, Romeis J, Naranjo SE, Hellmich RL, Shelton AM (2012) Using field-evolved resistance to Cry1F maize in a lepidopteran pest to demonstrate no adverse effects of Cry1F on one of its major predators. Transgenic Res 21:1303–1310

Tian J, Yao J, Long LP, Romeis J, Shelton AM (2015) Bt crops benefit natural enemies to control non-target pests. Sci Rep 5:1–6

Yang H, Peng Y, Tian J, Wang J, Hu J, Song Q, Wang Z (2017) Review: biosafety assessment of Bt rice and other Bt crops using spiders as example for non-target arthropods in China. Plant Cell Rep 36:505–517

Zhang W, Ricketts TH, Kremen C, Carney K, Swinton SM (2007) Ecosystem services and disservices to agriculture. Ecol Econ 64:253–260

Chapter 40
Use of Silicon as Resistance Factor for Plants Against Insect Pests

Jair Campos Moraes, Franscinely Aparecida Assis, and Gleice Aparecida Assis

40.1 Introduction

Currently, extensive monoculture cropping combined with a need for more yield has resulted in the increased use of phytosanitary products – mainly synthetic insecticides – to manage insect pests. With its expansive agricultural sector producing annual, perennial, and semi-perennial crops, along with its managed forests and pastures, Brazil led the global ranking for the use of phytosanitary products in 2015. The sales of these products in Brazil accounted for 18.5% of the total global sales. All of Latin America, including Brazil, consumed 28% of insecticides for crop protection (Sindicato Nacional da Indústria de Produtos Para Defesa Vegetal – SINDIVEG 2016).

The improper use of crop protection products, particularly insecticides, is a concern for the environment and the consumer's health, due to the possible accumulation of residues from these substances in soil, water, and food. This might occur when the dose recommendations and indications prescribed by competent authorities are ignored. Therefore, these substances should be used in a sustainable manner to manage agricultural pests. Currently, there has been more research on the use of silicon for the management of insects (Nascimento et al. 2018), mites (Christalcatalani et al. 2017), fungi (Uchôa et al. 2014), and bacteria (Alves et al. 2015).

Despite not being an essential element for plant development, silicon acts as a beneficial substance (Guntzer et al. 2012). In addition to minimizing stress, silicon

J. C. Moraes (✉)
Federal University of Lavras (UFLA), Entomology Department, Lavras, Minas Gerais, Brazil

F. A. Assis
University Center of Goiatuba (UNICERRADO), Goiatuba, Goiás, Brazil

G. A. Assis
Federal University of Uberlândia (UFU), Monte Carmelo Campus, Institute of Agrarian Sciences (ICIAG), Monte Carmelo, Minas Gerais, Brazil

B. Souza et al. (eds.), *Natural Enemies of Insect Pests in Neotropical Agroecosystems*, https://doi.org/10.1007/978-3-030-24733-1_40

has become an important ally to manage agricultural pests worldwide, mainly in Latin America. Silicon in plants acts as an inducer of resistance, decreasing the injuries and/or damage caused by pests. Silicon-induced resistance in plants brings about cellular, physiological, and/or morphological changes in plants and, as a consequence, provides resistance to the injuries caused by herbivores (Pinto et al. 2012). Furthermore, silicon acts in the biology of insect pests, both chewing (Assis et al. 2015; Nascimento et al. 2018) and sucking insects (Camargo et al. 2008; Pereira et al. 2010), without causing deleterious effects to the natural enemies (Antunes et al. 2010).

Silicon can be used in conventional agriculture either alone or combined with synthetic or organic insecticides, such as diatomaceous earth (85% to 95% silicon dioxide: SiO_2), as approved by the Normative Instruction SDA/SDC nº 2, July 12, 2013 (Brasil 2013). These norms provide measures to control primary and secondary pests that affect large crops, fruit, vegetable, and ornamental cultures. Although there are many reports on its use in agriculture, silicon is generally used not for pest management but for soil correction and plant nutrition, with plant protection as a complementary function.

Currently, many sources of silicon have been explored to manage chewing and sucking insects and mites. The results are satisfactory, but there are gaps that are yet to be filled, as silicon can affect other types of insects in different ways. Therefore, it is necessary to understand the reason why plants do not respond to the application of silicon to control insect pests (Keeping and Kvedaras 2008).

In this chapter, we discuss the modes of action of silicon in soil and plants, as well as their role in the induction of resistance to insect pests. Furthermore, we consider the most relevant studies conducted in Latin America on contributions of silicon to the control of chewing and sucking insects. Lastly, we review the main sources of silicon used in experiments conducted in laboratories, greenhouses, and in the field.

40.2 Silicon in Soil and Plants

Silicon is the most common element in soil after oxygen, accounting for 28% of the Earth's crust (Richmond and Sussman 2003). The concentration of monosilicic acid (H_4SiO_4) in the soil ranges from 0.1 to 0.6 mM, and it is the most common form of silicon available for plants (Gunnarsson and Arnórsson 2000).

Silicon is absorbed by plant roots in the form of H_4SiO_4 and is then transported in the same form to the aerial parts via the xylem. In leaves, transpiration causes H_4SiO_4 to concentrate and polymerize as silica (SiO_2), which is deposited in the plant tissues. This happens initially in relatively young tissues and, subsequently, in senescent cells (Wiese et al. 2007). In roots, the apoplastic transfer of silicon from the soil to the symplast is mediated by specific transporters. Furthermore, in plants that naturally accumulate silicon, such as rice, corn, and barley, the transporters involved in the absorption and transport of silicon have been identified (Ma and Yamaji 2008).

Deposition of silica in the members of Poaceae, such as sugarcane, creates a mechanical barrier that increases the resistance of plants to herbivorous insects and to the penetration of larvae (Vilela et al. 2014). In corn plants, deposition of silicon in the outer epidermal wall results in the formation of a cuticle-silica double layer, toughening the plant leaves and, consequently, wearing down the jaws of leaf-eating insects (Goussain et al. 2002).

Many studies have also found that silicon increased the biosynthesis of cell wall components, such as lignin and polyphenols, in plants. This, in turn, affected the mechanical and chemical properties of plants, resulting in cell wall extensibility and improvements to defense mechanism (Hossain et al. 2007).

The amount of silicon found in terrestrial plants ranges from 0.1% to 10% dry mass (Liang et al. 2015). Intraspecific genotypic variation in silicon concentration has been observed; however, this was not as significant as interspecific variation (Hodson et al. 2005). Plants differ in their ability to absorb and accumulate silicon and can be classified, based on the percentage of SiO_2 in the dry mass of aerial parts, as (1) accumulators, which are generally species of Poaceae, containing more than 4% SiO_2; (2) intermediate accumulators, such as cereals, sugarcane, and a few dicots, with SiO_2 levels ranging from 2% to 4%; and (3) non-accumulators, including most dicots, such as beans, containing less than 2% SiO_2 (Ma and Takahashi 2002; Hodson et al. 2005).

Among the crops that respond to silicon application are the members of Poaceae, such as rice, corn, and wheat, which absorb silicon actively and accumulate it in large quantities; some dicots, such as cotton and soybean plants; and vegetables, such as cucurbits and tomatoes, that use specific carriers to accumulate silicon (Liang et al. 2015).

The benefits of using silicon in agriculture are evident when plants are exposed to biotic and abiotic stresses (Guntzer et al. 2012). Silicon can minimize the effects of salinity (Castellanos et al. 2016), heavy metal toxicity (Silva et al. 2015), and water deficiency (Moro et al. 2015). Among the attributes with respect to biotic stresses, protection against various diseases has been reported in a wide range of host plants, such as melon mildew (Dallagnol et al. 2015); bacterial wilt in sweet peppers (Alves et al. 2015); and anthracnose in sorghum (Santos et al. 2014). In addition to its protective role against pathogens, silicon can also increase the resistance level of plants against the damage caused by insect pests. This is mainly through two defense mechanisms, mechanical defense and biochemical defense, which will be addressed in the next section.

40.3 Role of Silicon in Induced Resistance Against Insect Pests in Plants

As mentioned above, silicon induces resistance in plants against insect pests through two major defense mechanisms: mechanical defense and biochemical defense. In mechanical defense, silicon is deposited on the cell walls of plant leaves and stems as a phytolith. This can act as a direct or indirect anti-herbivory defense mechanism by increasing the hardness and abrasiveness of the tissues and forming a mechanical barrier, causing damages to insect mouth parts. Mechanical defense mechanisms reduce the palatability and digestibility of leaves, which in turn reduce or prevent herbivory, and potentially decrease the effects of phytophagous insects (Massey and Hartley 2009; Reynolds et al. 2009, 2016).

On the other hand, silicon absorption also induces resistance by causing biochemical changes in plants, reducing the damage caused by insect pests. In wheat, silicate fertilization negatively affected the preference and growth rate of the aphid *Schizaphis graminum* (Rondani) (Aphididae), by increasing the activity of the oxidative enzymes peroxidase and polyphenol oxidase. These enzymes are involved in the process of resistance induction in plants (Gomes et al. 2005).

In addition to enzymes, silicon can also act as precursors of phenolic compounds (Mendonça et al. 2013), such as tannins (Gomes et al. 2008) and lignins (Fawe et al. 1998). These substances can also induce resistance in plants. Silicon is capable of inducing resistance in plants by nonfood preference mechanisms (Nascimento et al. 2014), oviposition (Peixoto et al. 2011), and antibiosis (Assis et al. 2015; Alvarenga et al. 2017; Nascimento et al. 2018). In the following sections, we present the practical evidence of silicon-induced resistance in plants to chewing and sucking insects.

40.4 Practical Evidence of Silicon-Induced Resistance in Plants to Chewing Insects

Most of the studies on silicon-induced resistance against herbivores in Latin American countries are concentrated mainly in Brazil. This might be because of its tropical climate and extensive areas of monoculture, which favors the occurrence of insect pests. This has contributed to the search for alternatives to chemical pesticides.

The effects of silicon can be observed in alterations to the biology (antibiosis) of chewing insects. Silicon also affects the insect's preference to feeding habits and/or oviposition, reducing the damage caused by both young and adult insects. In Brazil, Goussain et al. (2002) assessed the effect of application of sodium silicate to corn crops on the development of *Spodoptera frugiperda* (J E Smith) (Noctuidae). The results revealed that silicon can deposit on leaf cell walls and form a mechanical barrier, increasing the rigidity of the leaf tissues and wearing down the jaws of the larvae feeding on them. This research was considered a pioneering effort in Latin America to combat the insect pest *S. frugiperda*. Furthermore, an increase in the mortality rate of the second and sixth instar larvae was also observed. This was due to the difficulty for the larvae to scrape through the rigid leaf tissues of plants treated with silicon.

In line with this research, other studies have provided evidence of silicon as a resistance inducer due to its effect on the insects' biology. The females larvae of

S. frugiperda that were fed with corn plants treated with silicic acid solution (1% $SiO_2 \cdot XH_2O$) in the soil, at a dosage equivalent to 2 SiO_2 ha^{-1}, were found to have relatively low fertility. Fertility is one of the biological parameters most affected by the use of silicon. This deleterious effect on oviposition might be due to the accumulation of silicon in plants, which activated and increased the production of defense metabolites (Alvarenga et al. 2017).

In rice plants, the application of silicon (silicic acid) by drenching and foliar application, triggered biological alterations during the immature and adult phases of *S. frugiperda*. Consequently, a reduction in the leaf area consumed affected the larval and pupal weights of the insects. There was a reduction in the longevity of the adults, and number and viability of eggs laid on the first and second days of oviposition. The application of silicon to rice plants contributed to an increase in the silicon content in the plants when compared with the control. Thus, silicon content in plants had negative correlations with the foliar area consumed, weight of larvae, and longevity of adults, demonstrating the efficiency of silicate treatment inducing resistance in rice plants (Nascimento et al. 2018).

In the moth *Tuta absoluta* (Meyrick) (Gelechiidae), the different applications of silicon to tomato leaves caused biological alterations to insects, increasing the duration of their pupal phase. In addition, it decreased the larval and pupal survival rate and pupal weight and contributed to an increase in the duration of the larval phase (Santos et al. 2012).

The biology of *Chlosyne lacinia saundersii* (Doubleday & Hewitson) (Nymphalidae) was affected by the reduction in the weight of larvae 10 and 15 days after being fed with the leaves of sunflower plants treated with silicon. This was due to the increased accumulation of silicon in plants that were treated with silicon, resulting in a negative correlation between the silicon content in plants and biomass of larvae. The reduced biomass was due to chemical causes, as there was no difference in the leaf area consumed by the larvae, despite the accumulation of silicon in the plants (Assis et al. 2015).

There are several reports in relation to the preferences of insect pests. These reports indicate that silicon may affect the host selection by insects for feeding and oviposition. A study reported that 48 and 72 h after the release of the second instar larvae of *S. frugiperda*, rice plants treated with silicon by drenching or foliar application were less preferred for feeding when compared with the control plants. This indicated that silicon provided the plants with greater protection against insect pests, probably due to the difficulty the larvae had while feeding, which prompted them to seek other leaves not treated with silicon (Nascimento et al. 2014).

Diatomaceous earth, a powdered sediment of fossilized diatom algae, has silicon dioxide as its main component and has been used as a source of silicon for pest management. Silicon dioxide affects the water balance of insects, causing an abrasive effect on the integument, resulting in their death by dehydration (Lorini et al. 2003). In host plant preference test with potato plants, the use of diatomaceous earth, in either sprinkled or pulverized form, reduced the feeding rate of *Diabrotica speciosa* (Germar) (Chrysomelidae) and increased its mortality. This confirmed the insecticidal action of diatomaceous earth on chrysomelids (Assis et al. 2011).

The insecticidal effect of diatomaceous earth was also observed in field conditions, where there was a reduction in the number of damaged leaflets and holes on the leaves caused by *D. speciosa* in potato plants. Further, relatively fewer tubers were damaged when the plants were treated with diatomaceous earth. This shows that the plants treated with this compound are less preferred by insects for feeding (Assis et al. 2012).

Silicon can also be used with insecticides, such as insect growth regulators, to provide a relatively fast and efficient response to control insect pests, such as *S. frugiperda*. The host plant preference tests have revealed that using silicon in this manner reduces leaf consumption by insects and increases their mortality. Silicon can provide mechanical resistance to the leaves, making it difficult for insects to feed on them and making insects more susceptible to insecticides (Neri et al. 2005, 2009).

Regarding preference for oviposition, the application of silicon to tomato plants reduced the preference of *T. absoluta* for oviposition due to the formation of a layer of silica on the leaf epidermis. This might have contributed to reducing oviposition on the host plant (Santos et al. 2012).

40.5 Practical Evidence of Silicon-Induced Resistance to Sucking Insects in Plants

The application of silicon affected the feeding and oviposition preferences of sucking insects, as well as their biology. The application of 1% silicic acid to wheat crops via soil reduced the duration of the reproductive period of the green aphid *S. graminum*. The reduction was from 20 days in the control treatment to 9 days in silicon-treated plants. Furthermore, there was a reduction of 60%, 19%, and 34%, respectively, in the number of nymphs, growth rate, and longevity of the insect pest 15 days after the emergence of plants. The absorption and translocation of silicon were efficient during the early vegetative phase of wheat (Costa et al. 2009). The feeding behavior of *S. graminum* was also affected by silicon. Silicon decreased the number of honeydew drops secreted by aphids, indicating a reduction in the sap intake by aphids (Goussain et al. 2005). When pests feed on silicon-treated plants, they are affected by defense-related factors in the phloem that cease the ingestion earlier than that of the insects that feed on plants without silicon (Costa et al. 2009).

Calcium silicate on the leaves of cucumber crops negatively affected the population of *Bemisia tabaci* (Gennadius) biotype B (Aleyrodidae). In addition, it reduced oviposition from 487 to 223 eggs and the number of nymphs from 425 to 144 and increased the biological cycle from 19.2 to 24.8 days. It also increased the mortality of nymphs from 13.6% to 47.9% (Correa et al. 2005). In chrysanthemum crops, the use of 1% silicic acid via foliar application or drenching reduced the number of nymphs of *B. tabaci* from 4.8 to 2.8. It also contributed to a decrease in egg viability from 93% to 71.2% with foliar application (Melo et al. 2015).

In sugarcane crops, the use of potassium silicate increased the mortality of spittlebug nymphs of *Mahanarva fimbriolata* (Stål) (Cercopidae). Further, there were an increase in the duration of the nymphal phase and a decrease in the longevity of males and females (Korndörfer et al. 2011). Similar results were observed in peanut crops with the foliar application of silicon, which reduced the number of adults and nymphs of scraper-sucker insects (thrips) *Enneothrips flavens* (Moulton) (Thrypidae) (Dalastra et al. 2011). In *Pinus taeda* plants, the application of silicon negatively influenced the reproductive capacity of the giant aphid *Cinara atlantica* (Wilson) (Aphididae) (Camargo et al. 2008).

As silicon affects the feeding behavior and reproduction of sucking insects, it is necessary to make a detailed analysis of the stylet pathway from the cuticle to the phloem using the electrical penetration graphs (EPG) technique (Tjallingii 1978). This technique has already been used to investigate the transmission of viruses (Bonani et al. 2010), the resistance of host plant to aphids and whiteflies (Pereira et al. 2010; Yin et al. 2010), and the behavior of insect vectors and insecticides in plants. On the other hand, Moraes et al. (2013) demonstrated that the application of silicon did not interfere with the insertion of the *S. graminum* stylet into the leaf tissue of wheat plants, without altering the initial test behavior. Thus, the application of silicon did not prevent aphids from feeding normally and, therefore, did not constitute a mechanical barrier when deposited on the leaf (Goussain et al. 2005). In wheat crops, the application of silicon to control *S. graminum* reduced the number of nymphs and honeydew drops and the duration of phloem ingestion phase (during which the pest maintains its stylet in the phloem). Silicon might have acted as a defense-inducing substance in affected plants, reducing the feeding and, consequently, the fecundity of pests (Pereira et al. 2010). It was also observed that silicon did not affect the number of brief cell punctures during the stylet pathway phase. Therefore, the study concluded that there was no hardening of the cell wall due to the application of silicon.

A study on the white mealybug insect *Planococcus citri* (Risso) (Pseudococcidae) and coffee crops revealed that the insects easily reached the phloem both in silicon-treated and control plants. This showed the absence of a mechanical barrier to penetrating mouthparts. Moreover, the dietary behavior of the pest was not affected by the application of calcium silicate via soil (Santa-Cecília et al. 2014).

The barrier provided by silicon deposition on the epidermal cells is not the only defense mechanism against insect attack or penetration of fungal hyphae. Silicon also acts on the host tissue, influencing chemical signals between the host and the pathogen. This results in a rapid activation of plant defense mechanisms by increasing the production of enzymes, such as polyphenol oxidase, peroxidase, and phenylalanine ammonia lyase (Gomes et al. 2005). Furthermore, in soybean crops, the use of silicon also increased the production of lignin, thus interfering with the growth and development of insect pests (Moraes et al. 2009).

40.6 Sources of Silicon to Induce Resistance to Insect Pests in Plants

There are reports on silicon-induced resistance to chewing and sucking insects in crops of agricultural importance. Silicon from various sources have been used, in both commercial and research applications; the details are presented in Table 40.1.

Table 40.1 Silicon sources used in Latin America to induce insect pest resistance in some crops

Silicon sources	Crops	Insect pests	References
Silicic acid (100% de H_4SiO_4)	Sunflower (*Helianthus annuus*)	*Chlosyne lacinia saundersii* (Nymphalidae)	Antunes et al. (2010), Assis et al. (2013, 2015)
	Chrysanthemum (*Dendranthema grandiflorum*)	*Bemisia tabaci* Biótipo B (Aleyrodidae)	Melo et al. (2015)
	Sugarcane (*Saccharum* spp.)	*Diatraea saccharalis* (Crambidae)	Vilela et al. (2014)
	Rice (*Oryza sativa*)	*Spodoptera frugiperda* (Noctuidae)	Nascimento et al. (2014, 2018)
	Bean (*Phaseolus vulgaris*)	*B. tabaci* Biótipo B	Peixoto et al. (2011)
	Potato (*Solanum tuberosum*)	*Diabrotica speciosa* (Chrysomelidae) and *Myzus persicae* (Aphididae)	Silva et al. (2010)
	Pinus (*Pinus taeda*)	*Cinara atlantica* (Aphididae)	Camargo et al. (2008)
	Wheat (*Triticum aestivum*)	*Schizaphis graminum* (Aphididae)	Costa et al. (2011) Pereira et al. (2010)
	Corn (*Zea mays*)	*S. frugiperda*	Alvarenga et al. (2017)
	Soybean (*Glycine max*)	*B. tabaci* Biótipo B	Moraes et al. (2009), Ferreira et al. (2011)
Supa sílica® (10% de SiO_2)	Peanut *Arachis hypogaea*	*Enneothrips flavens* (Thripidae)	Dalastra et al. (2011)
Diatomaceous earth (Insecto® – 86.7% de SiO_2)	Potato (*S. tuberosum*)	*D. speciosa*	Assis et al. (2011, 2012)
Diatomaceous earth (KeepDry® – 86% de SiO_2)	Tomato and corn	*Spodoptera eridania* (Noctuidae)	Constanski et al. (2016)
Potassium silicate	Sugarcane (*Saccharum officinarum*)	*Mahanarva fimbriolata* (Cercopidae)	Korndörfer et al. (2011)
	Cocoa (*Theobroma cacao*)	*Toxoptera aurantii* (Aphididae)	Pinto et al. (2012)

40.7 Final Considerations

Over the years, a lot of progress has been made in relation to silicon-induced resistance to insect pests in plants. Reports have proved the effectiveness of silicon against chewing and sucking insects in laboratory, greenhouse, and field experiments. Moreover, silicon is a versatile compound, as it can be used in both organic and conventional farming. These facts prove that silicon can be used as an alternative to chemical control and is feasible to be used in any integrated pest management program.

However, there are gaps, such as target organisms and dosage, period, and means of application, which are yet to be filled. The main reason for the relatively low use of silicon in the management of insect pests is the lack of availability and transfer of technology to the farmers.

References

Alvarenga R, Moraes JC, Auad AM et al (2017) Induction of resistance of corn plants to *Spodoptera frugiperda* (J E Smith, 1797) (Lepidoptera: Noctuidae) by application of silicon and gibberellic acid. Bull Entomol Res 107(4):527–533

Alves AO, Santos MMB, Souza LJN et al (2015) Use of silicon for reducing the severity of bacterial wilt of sweet pepper. J Plant Pathol 97(3):419–429

Antunes CS, Moraes JC, Antonio A et al (2010) Influência da aplicação de silício na ocorrência de lagartas (Lepidoptera) e de seus inimigos naturais chaves em milho (*Zea mays* L.) e em girassol (*Helianthus annuus* L.). Biosci J 26(4):619–625

Assis FA, Moraes JC, Nascimento AM et al (2011) Efeitos da terra diatomácea sobre *Diabrotica speciosa* (Germar, 1824) (Coleoptera: Chrysomelidae) em batata inglesa. Cienc Agrotec 35(3):482–486

Assis FA, Moraes JC, Silveira LCP et al (2012) Inducers of resistance in potato and its effects on defoliators and predatory insects. Rev Colomb Entomol 38(1):30–34

Assis FA, Moraes JC, Auad AM et al (2013) The effects of foliar spray application of silicon on plant damage levels and components of larval biology of the pest butterfly *Chlosyne lacinia saundersii* (Nymphalidae). Int J PestManage 59(2):128–134

Assis FA, Moraes JC, Assis GA et al (2015) Induction of caterpillar resistance in sunflower using silicon and acibenzolar-s-methyl. J Agr Sci Tech-Iran 17(3):543–550

Bonani JP, Fereres A, Garzo E et al (2010) Characterization of eletrical penetration graphs of the Asian citrus psyllid, *Diaphorina citri*, in sweet orange seedlings. Entomol Exp Appl 134(1):35–49

Brasil. Ministério da Agricultura, Pecuária e Abastecimento, Secretaria de Defesa Agropecuária (2013) Instrução Normativa Conjunta n° 2, de 12 de julho de 2013. Brasília, DF. http://www.agricultura.gov.br/assuntos/insumos-agropecuarios/insumos-agricolas/agrotoxicos/legislacao/arquivos-de-legislacao/inc-02-2013-especificacao-de-referencia. Accessed 6 Dec. 2017

Camargo JMM, Moraes JC, Oliveira EB et al (2008) Efeito da aplicação do silício em plantas de *Pinus taeda* L., sobre a biologia e morfologia de *Cinara atlantica* (Wilson) (Hemiptera: Aphididae). Cienc Agrotec 32(6):1767–1774

Castellanos CIS, Rosa MP, Deuner C et al (2016) Aplicação ao solo de cinza de casca de arroz como fonte de silício: efeito na qualidade de sementes de trigo produzidas sob stresse salino. Rev CiêncAgrár 39(1):95–104

Christalcatalani G, Vieira MR, Camargos LS et al (2017) Effects of potassium silicate application on papaya plants in the two spotted spider mite population. Rev Bras Frutic 39(2):1–9

Constanski KC, Zorzetti J, Santoro PH et al (2016) Inert powders alone or in combination with neem oil for controlling *Spodoptera eridania* and *Spodoptera frugiperda* (Lepidoptera: Noctuidae) larvae. Semina: Ciênc Agrár 37(4):1801–1810

Correa RSB, Moraes JC, Auad AM et al (2005) Silicon and acibenzolar-s-methyl as resistance inducers in cucumber, against the whitefly *Bemisia tabaci* (Gennadius) (Hemiptera: Aleyrodidae) biotype B. Neotrop Entomol 34(3):429–433

Costa RR, Moraes JC, Costa RR (2009) Interação silício-imidacloprid no comportamento biológico e alimentar de *Schizaphis graminum* (Rond.) (Hemiptera: Aphididae) em plantas de trigo. Cienc Agrotec 33(2):455–460

Costa RR, Moraes JC, Dacosta RR (2011) Feeding behaviour of the greenbug *Schizaphis graminum* on wheat plants treated with imidacloprid and/or silicon. J Appl Entomol 135(1/2):115–120

Dalastra C, Campos AR, Fernandes FM et al (2011) Silício como indutor de resistência no controle do tripes do prateamento *Enneothrips flavens* Moulton, 1941 (Thysanoptera: Thripidae) e seus reflexos na produtividade do amendoinzeiro. Cienc Agrotec 35(3):531–538

Dallagnol LJ, Rodrigues FA, Pascholati SF et al (2015) Comparison of root and foliar applications of potassium silicate in potentiating post-infection defences of melon against powdery mildew. Plant Pathol 64(5):1085–1093

Fawe A, Abou-Zaid M, Menzies JG et al (1998) Silicon-mediated accumulation of flavonoid phytoalexins in cucumber. Phytopathology 88(5):396–401

Ferreira RS, Moraes JC, Antunes CS (2011) Silicon influence on resistance induction against *Bemisia tabaci* Biotype B (Genn.) (Hemiptera: Aleyrodidae) and on vegetative development in two soybean cultivars. Neotrop Entomol 40(4):495–500

Gomes FB, Moraes JC, Santos CD et al (2005) Resistance induction in wheat plants by silicon and aphids. Sci Agr 62(6):547–551

Gomes FB, Moraes JC, Santos CD et al (2008) Uso de silício como indutor de resistência em batata a *Myzus persicae* (Sulzer) (Hemiptera: Aphididae). Neotrop Entomol 37(2):185–190

Goussain MM, Moraes JC, Carvalho JG et al (2002) Efeito da aplicação de silício em plantas de milho no desenvolvimento biológico da lagarta do cartucho *Spodoptera frugiperda* (J E Smith) (Lepidoptera: Noctuidae). Neotrop Entomol 31(2):305–310

Goussain MM, Moraes JC, Prado E (2005) Effect of silicon applied to wheat plants on the biology and probing behaviour of the green bug *Schizaphis graminum* (Rond.) (Hemiptera: Aphididae). Neotrop Entomol 34(5):807–813

Gunnarsson I, Arnórsson S (2000) Amorphous silica solubility and the thermodynamic properties of H_4SiO_4 degrees in the range of 0 degrees to 350 degrees C at P-sat. Geochim Cosmochim Ac 64(13):2295–2307

Guntzer F, Keller C, Meunier JD (2012) Benefits of plant silicon for crops: a review. Agron Sustain Dev 32(1):201–213

Hodson MJ, White PJ, Mead A et al (2005) Phylogenetic variation in the silicon composition of plants. Ann Bot 96(6):1027–1046

Hossain MT, Soga K, Wakabayashi K et al (2007) Modification of chemical properties of cell walls by silicon and its role in regulation of the cell wall extensibility in oat leaves. J Plant Physiol 164(4):385–393

Keeping MG, Kvedaras OL (2008) Silicon as a plant defence against insect herbivory: response to massey, ennos and Hartley. J Anim Ecol 77(3):631–633

Korndörfer AP, Vendramim JD, Grisoto E (2011) Induction of plant resistance to the spittlebug *Mahanarva fimbriolata* Stål (Hemiptera: Cercopidae) in sugarcane by silicon application. Neotrop Entomol 40(3):387–392

Liang Y, Nikolic M, Bélanger R et al (2015) Silicon in agriculture: from theory to practice. Springer, Dordrecht

Lorini I, Moras A, Beckel H (2003) Tratamentos de sementes armazenadas com pós inertes à base de terra de diatomáceas. Embrapa Trigo, Passo Fundo. (Embrapa Trigo. Comunicado Técnico Online, 113)

Ma JF, Takahashi E (2002) Soil, fertilizer, and plant silicon research in Japan. Elsevier Science, Amsterdam

Ma JF, Yamaji N (2008) Functions and transport of silicon in plants. Cell Mol Life Sci 65(19):3049–3057

Massey FP, Hartley SE (2009) Physical defenses wear you down: progressive and irreversible impacts of silica on insect herbivores. J Anim Ecol 78(1):281–291

Melo BA, Moraes JC, Carvalho LM (2015) Resistance induction in chrysanthemum due to silicon application in the management of whitefly *Bemisia tabaci* Biotype B (Hemiptera: Aleyrodidae). Rev Ciênc Agroamb 13(2):1–8

Mendonça AO, Tavares LC, Brunes AP et al (2013) Acúmulo de silício e compostos fenólicos na parte aérea de plantas de trigo após a adubação silicatada. Biosci J 29(5):1154–1162

Moraes JC, Ferreira RS, Costa RR (2009) Indutores de resistência à mosca-branca *Bemisia tabaci* biótipo B (Genn., 1889) (Hemiptera: Aleyrodidae) em soja. Cienc Agrotec 33(5):1260–1264

Moraes JC, Goussain MM, Vilela M et al (2013) Comportamento de prova do pulgão-verde *Schizaphis graminum* (Rond.) (Hemiptera: Aphididae) em plantas de trigo tratadas com silício e dimetoato. Biosci J 29(1):83–90

Moro AL, Broetto F, Moro E (2015) Relação hídrica e teor de clorofila em dois cultivares de arroz submetido à deficiência hídrica e adubação silicatada. Irriga 20(3):570–586

Nascimento AM, Assis FA, Moraes JC et al (2014) Não preferência a *Spodoptera frugiperda* (Lepidoptera: Noctuidae) induzida em arroz pela aplicação de silício. Rev Bras Ciênc Agrár 9(2):215–218

Nascimento AM, Assis FA, Moraes JC et al (2018) Silicon application promotes rice growth and negatively affects development of *Spodoptera frugiperda* (J E Smith). J Appl Entomol 142(1/2):241–249

Neri DKP, Moraes JC, Gavino MA (2005) Interação silício com inseticida regulador de crescimento no manejo da lagarta-do-cartucho *Spodoptera frugiperda* (J E Smith, 1797) (Lepidoptera: Noctuidae) em milho. Cienc Agrotec 29(6):1167–1174

Neri DKP, Gomes FB, Moraes JC et al (2009) Influência do silício na suscetibilidade de *Spodoptera frugiperda* (J E Smith) (Lepidoptera: Noctuidae) ao inseticida lufenuron e no desenvolvimento de plantas de milho. Cienc Rural 39(6):1633–1638

Peixoto ML, Moraes JC, Silva AA et al (2011) Efeito do silício na preferência para oviposição de *Bemisia tabaci* biótipo B (Genn.) (Hemiptera: Aleyrodidae) em plantas de feijão (*Phaseolus vulgaris* L.). Cienc Agrotec 35(3):478–481

Pereira RRC, Moraes JC, Prado E et al (2010) Resistance inducing agents on the biology and probing behaviour of the greenbug in wheat. Sci Agr 67(4):430–434

Pinto DG, Aguilar MAG, Souza CAS et al (2012) Alterações fisiológicas após aplicação de silício em cacau e sua influência na preferência por pulgões. Ceres 59(3):360–367

Reynolds OL, Keeping MG, Meyer JH (2009) Silicon-augmented resistance of plants to herbivorous insects: a review. Ann Appl Biol 155(2):171–186

Reynolds OL, Padula MP, Zeng R et al (2016) Silicon: potential to promote direct and indirect effects on plant defense against arthropod pests in agriculture. Front Plant Sci 7:1–13

Richmond KE, Sussman M (2003) Got silicon?: the non-essential beneficial plant nutrient. Curr Opin Plant Biol 6(3):268–272

Santa-Cecília LVC, Prado E, Moraes JC (2014) Avaliação do silício no comportamento alimentar da cochonilha-branca [*Planococcus citri* (Risso) (Pseudococcidae)] em cafeeiro. Coffee Sci 9(1):10–13

Santos MC, Junqueira AMR, Sá VGM et al (2012) Efeito do silício em aspectos comportamentais e na história de vida de *Tuta absoluta* (Meyrick) (Lepidoptera: Gelechiidae). RBAS 2(1):76–88

Santos GR, Rodrigues AC, Bonifacio A et al (2014) Severidade de antracnose em folhas de sorgo submetido a doses crescentes de silício. Rev Cienc Agron 45(2):403–408

Silva VF, Moraes JC, Melo BA (2010) Influence of silicon on the development, productivity and infestation by insect pests in potato crops. Cienc Agrotec 34(6):1465–1469

Silva AJ, Nascimento CW, Gouveia-Neto AS et al (2015) Effects of silicon on alleviating arsenic toxicity in maize plants. Rev Bras Cienc Solo 39(1):289–296

Sindicato Nacional da Indústria de Produtos para Defesa Vegetal (2016) Consumo de produtos fitossanitários no Brasil. http://www.sindiveg.org.br. Accessed 6 Dec. 2016

Tjallingii WF (1978) Eletronic recording of penetration behaviour by aphids. Entomol Exp Appl 24(3):521–530

Uchôa CN, Pozza EA, Uchôa KSA et al (2014) Acibenzolar-S-Metil e silício como indutores de resistência à Sigatoka-negra em bananeira cultivar Grand Naine (AAA). Rev Agrarian 7(24):189–196

Vilela M, Moraes JC, Alves E et al (2014) Induced resistance to *Diatraea saccharalis* (Lepidoptera: Crambidae) via silicon application in sugarcane. Rev Colomb Entomol 40(1):44–48

Wiese H, Nikolic M, Romheld V (2007) Silicon in plant nutrition: effects on zinc, manganese and boron leaf concentration and compartmentation. In: Sattelmacher B, Horst WJ (eds) The apoplast of higher plants: compartment of storage, transport and reactions. Springer Netherlands, Dordrecht, pp 33–47

Yin HD, Wang XY, Huang KXCH et al (2010) Impacts of transgenic Bt cotton on the stylet penetration behaviors of *Bemisia tabaci* biotype B: evidence from laboratory experiments. Insect Sci 17(4):344–352

Chapter 41
Use of Semiochemical-Based Strategies to Enhance Biological Control

Maria Fernanda Gomes Villalba Peñaflor

41.1 Introduction

Semiochemicals (*semeîon* means signal in Greek) comprise substances that mediate interactions among organisms. A single or mixture of chemicals emitted by an organism is detected by a receiver that often responds modifying its behavior or physiology. If the organisms involved in the interaction belong to the same species, the chemicals are named pheromones, whereas those involved in interspecific interactions are allelochemicals.

Insects are among the most specialized organisms to make use of semiochemicals in the environment in order to locate mates, food resources, and oviposition sites, as well as communicate among them. Their antennae contain a myriad of trichoid sensillum hairs and basiconic sensilla that house olfactory sensory neurons, odor-binding proteins, receptors, and enzymes that all together are responsible for odor chemoreception (Leal 2013). In the case of parasitoids and arthropod predators, they must have specialized chemoreceptors for recognizing odors associated with their host or prey in order to localize them (Wang et al. 2015). These semiochemicals can constitute odors derived from host/prey bodies and their subproducts, such as silk and feces (Mattiacci and Dicke 1995). Even though some odors are not derived from the target stage of the host/prey, as sex and aggregation pheromones, which are chemicals involved in the insect adult communication, natural enemies can also explore them to find the host/prey community (Reddy et al. 2002; Jumean et al. 2009). In this way, eavesdropping the intraspecific communication can facilitate finding host/prey stages by natural enemies (Leal et al. 1995).

Odors from host plants have shown to be important signals for natural enemies in host/prey location. In the 1990s, researchers found that herbivore-damaged plants emit an attractive volatile blend (named herbivore-induced plant volatiles, HIPVs)

M. F. G. V. Peñaflor (✉)
Federal University of Lavras (UFLA), Department of Entomology, Lavras, MG, Brazil
e-mail: fernanda.penaflor@ufla.br

B. Souza et al. (eds.), *Natural Enemies of Insect Pests in Neotropical Agroecosystems*, https://doi.org/10.1007/978-3-030-24733-1_41

to herbivore natural enemies (Turlings et al. 1991; Dicke et al. 1990a). Since then, a large body of literature has accumulated showing the intricacies of tritrophic interactions mediated by plant chemicals (Turlings and Erb 2018, Wäschke et al. 2013; Mumm and Dicke 2010).

In general, host/prey odors are less detectable than those emitted by plants for two main reasons: (i) plant biomass is larger than host/prey, and therefore plant volatiles are more abundant than odors emitted by insects; and (ii) host/prey have been under selective pressure of being attacked by natural enemies, and their chemicals can be less volatiles; on the other hand, plants emitting more detectable signals to herbivore natural enemies have benefited in terms of fitness. Despite being less detectable, host/prey odors are more reliable signals to natural enemies than HIPVs as they indicate more precisely the herbivore's identity. Therefore, HIPVs seem to play an important role on recruiting natural enemies at long distances, while host/prey derived odors at short distances (Vet and Dicke 1992). In host selection process (Vinson 1976, 1984), HIPVs must be important cues at the first steps, such as host habitat location and host location by parasitoids, and host chemicals must guide them to locate hosts in the plant as well as help in host recognition and acceptance.

Given the attractiveness of host/prey odors (kairomones) and HIPVs (allelochemicals) to natural enemies, several studies have addressed their potential to be incorporated in tactics to enhance the efficacy of biological control by manipulating the abundance of natural enemies in crops or optimizing their ability to locate their host/prey (Khan et al. 2008; Kaplan 2012; Peñaflor and Bento 2013). In this chapter, I reviewed the main semiochemical-based practices that have been suggested to improve biological control and critically discussed their feasibility. It is important to note that I addressed strategies to manipulate natural enemy behavior exploring semiochemicals in the field. There are ways to improve the ability of natural enemies to detect important semiochemicals associated with the pests by "training" (or conditioning) them in mass rearings before field releases (Giunti et al. 2015), but these were not addressed here. Also, given the vast literature on natural enemy behavior to host/prey and plant odors, I gave special attention to practical studies conducted in greenhouses or field conditions over those performed under controlled conditions, and highlight those that were conducted in Latin America. So far, many of the tactics have yet to be tested in Latin American crops.

41.2 Semiochemical-Based Strategies

Over the last three last decades, advances in analytical chemistry and synthesis have allowed to test and employ numerous synthetic semiochemicals in pest management tactics in the field (Mori and Tashiro 2004). However, recruiting biological control agents and enhancing their foraging efficiency with the use of semiochemicals is not as straightforward as attracting insect adults to sex pheromone-baited lures in tactics for pest monitoring and control.

Most of the studied strategies to recruit, retain, or optimize natural enemy foraging efficiency explored semiochemicals derived from host/prey or plant odors. Although natural enemy sex or aggregation pheromones to recruit and retain them in crops have been far less studied (Aldrich and Cantelo 1999), it is important to discuss a case of success to incite future works.

Use of semiochemicals to manipulate natural enemies in crops was first suggested in the 1970s (Lewis et al. 1972, 1975; Gross et al. 1975; Nordlund et al. 1977). Most of these works used host odors, obtained from extracts or their synthetics (e.g., sex and aggregation pheromones, odors from moth scales), as attractants and/or arrestants of parasitoids in agricultural crops (Beevers et al. 1981; Lewis et al. 1982).

Since the 1990s, when the first studies on the role of HIPVs in host search by natural enemies came out, the focus on developing semiochemical-based tactics to improve biological control has shifted to HIPVs (Dicke et al. 1990b; Whitfield 2001). Since the mechanisms underlying plant defenses triggered by arthropod herbivory or oviposition have been unraveled (Farmer et al. 1992; Mithofer and Boland 2008; Köllner et al. 2008; Erb et al. 2012; Reymond 2013), novel tactics have been developed and explored by manipulating the host plant-induced defenses. Besides deploying synthetic lures in the field, knowledge on plant-induced response at the molecular, genomic, and biochemical levels has allowed to elicit the emission of HIPVs by treating plants with elicitors or genetically engineering plants to emit a specific attractive volatile to natural enemies, for example. Alternatively, classical plant breeding with wild relatives of crops can help to strengthen plant-induced defenses.

These semiochemical-based strategies have been employed in the field at various degrees of effectiveness, and their advantages and drawbacks to biological control efficacy are discussed below.

41.3 Natural Enemy Pheromones

Arthropod predators and parasitoids make use of chemical communication to locate mates (Pompanon et al. 1997; Ruther et al. 2000). Use of their pheromones would be useful to increase and/or retain their population levels in crops and, consequently, levels of parasitism and predation. However, unlike lepidopteran, coleopteran, and hemipteran pests, not much effort has been invested in identifying and synthesizing sex or aggregation pheromones of natural enemies.

In the literature, to the best of our knowledge, there is only one study in which natural enemy pheromone was used as a strategy to increment biological control. Pheromone-baited traps using synthetic aggregation pheromone of the spined soldier bug *Podisus maculiventris* (Say) (Pentatomidae), an abundant predator in North America, were used to harvest male and female adults, thereafter released in potato plants to control the Colorado potato beetle, *Leptinotarsa decemlineata* (Say) (Chrysomelidae) (Aldrich and Cantelo 1999). The authors also tested another method that comprised in transferring the harvested spined soldier bug adults to porous nursery cages, that allowed nymphs to escape, but not adults, located peripherally around plots. Pheromone dispensers at the center of plots likely promoted the dispersion of the spined soldier bug nymphs, which are also attracted to aggregation pheromone (Sant'Ana et al. 1997). Both methods proved to be successful at controlling the Colorado potato beetle in potato as population levels reduced, reflecting in increased yields. Importantly, these strategies likely reduce costs of rearing the spined soldier bug, particularly for small growers that cannot afford to purchase predators. The pheromone is available commercially (Soldier Bug Attractor™, Sterling International, USA) and is recommended for controlling pests in gardens.

41.4 Host/Prey-Derived Odors

There are numerous studies reporting the attraction or arrestment of natural enemies to chemicals emitted from the host/prey target and nontarget stages as well as their by-products (Rutledge 1996; Afsheen et al. 2008; Hatano et al. 2008). More recently, it has been shown that volatiles emitted from host-associated bacteria are also important for parasitoids to locate hosts (Leroy et al. 2011). These allelochemicals are called kairomones as they evoke a behavioral response that is adaptively favorable to the receiver individual, but negative, in terms of fitness, to the emitter individual (Dicke and Sabelis 1988).

Kairomones used by natural enemies for host/prey finding and recognition originate from caterpillar silk thread (Thibout 2005), frass (Mattiacci and Dicke 1995), honeydew (Pascal et al. 2014), defensive secretion (Mattiacci et al. 1993), sex or aggregation pheromone (Hardie et al. 1994; Mizutani 2006), alarm (Abassi et al. 2000), oviposition-marking (Onodera et al. 2002), anti-aphrodisiac pheromones (Fatouros et al. 2005), exuviae (Battaglia et al. 2000), wing scales, cuticular hydrocarbons (Usha Rani et al. 2006), egg cement (Bin et al. 1993), and chemical footprints (Wölfling and Rostás 2009). Despite this wide variety of origin sources, only some of them have potential to be explored in semiochemical-based tactics to recruit natural enemies because of their volatility/detectability. For example, compounds responsible for the kairomonal effect of accessory gland secretion to egg parasitoids are proteins (i.e., high molecular weight) and only detectable in contact, likely playing a role on host recognition (Nordlund et al. 1987). Besides this limitation, far more studies focused on identifying the origin source of kairomones, while only a few examined the biologically active compounds responsible for the attraction or arrestment behavior (Afsheen et al. 2008).

Host/prey pheromones in turn are among the most promising kairomones for being employed in strategies to recruit natural enemies given their volatility/detectability at long distances and their availability in synthetic versions. Moreover, in general, natural enemies that prey/parasitize different development stages exploit host/prey pheromones. However, based on the number of studies, egg parasitoids seem to rely more on host pheromone than larval, pupal, or adult parasitoids (Afsheen et al. 2008). As insect eggs are inconspicuous and emit low amounts of odors, egg parasitoids may have overcome the difficulty to locate host eggs by recognizing other host cues that, although are not directly associated with host eggs, are more abundant and detectable at long distances and may guide them to the host habitat.

Among the pioneer studies to use kairomones as attractants of natural enemies, Lewis et al. (1982) showed a 1.7-fold increase in parasitized host eggs (*Trichogramma* spp.) in cotton plots treated with synthetic sex pheromone-baited dispensers than in the control plots. Follow-up studies using similar approach with host pheromone-baited dispensers found that egg parasitism was enhanced in pheromone-treated plots in the field (Nakashima et al. 2016).

Innovative strategy using host pheromone-lured baits associated with nonviable refrigerated host eggs for parasitoid reproduction not only recruits egg parasitoids from the vicinities but also retains and augment egg parasitoid population levels in the field (Alim and Lim 2011). Moreover, the association of the pheromone-baited lure with host eggs in this strategy may reinforce the natural enemy behavioral response to the kairomone, what is important for maintaining its foraging ability.

Parasitism of hemipterans has also been successfully manipulated with sex pheromone-baited lures in the field (Hardie et al. 1994). A more recent study has shown that synthetic aphid sex pheromone lures led to augmented parasitism levels (Aphidiidae) without recruiting hyperparasitoids or promoting intraguild predation in alfalfa fields (Nakashima et al. 2016).

In contrast to sex pheromones, tactics using aphid alarm pheromone to recruit aphid parasitoids have shown to be dubious at controlling aphids. Besides dispensing synthetic lures in the field, emission of (*E*)-β-farnesene, the most common component in aphid alarm pheromone (Pickett and Griffiths 1980), by crop plants has also been suggested not only to prevent aphids from colonizing but also to attract aphid natural enemies (Crock et al. 1997). Although parasitoid wasps were attracted to the transformed wheat line under laboratory conditions, parasitism levels did not differ between untransformed and transformed-wheat plots (Bruce et al. 2015). Overall, the inconsistent results across studies seem to reflect the varying effect of (*E*)-β-farnesene according to natural enemy identity, previous experience, and compound concentration (Vosteen et al. 2016).

Host pheromone-based tactic can be unsuccessful when populations of the target natural enemy is not present in the surrounding areas. Therefore, the use of pheromone-baited dispensers integrated with releases of parasitoids results in greater levels of parasitism than using augmented biological control and semiochemical-based tactic separately (Mansour et al. 2010).

41.5 Herbivore-Induced Plant Volatiles

Unlike host/prey pheromones, herbivore-induced plant volatiles (HIPVs) mediating the attraction of natural enemies are classified as synomones. The nomenclature refers to allelochemicals mediating interactions in which both emitter and receiver individuals reciprocally benefit, in terms of fitness.

As extensively described in the literature, natural enemies usually orient to HIPV blends emitted by plants damaged by their hosts/prey (Mumm and Dicke 2010; Turlings and Erb 2018). However, this positive chemotaxis by HIPVs seems to be more variable than natural enemy response to host/prey odors, as it depends on the coevolutionary history among species (De Moraes et al. 1998), experience (Allison and Hare 2009), and preimaginal conditioning (Takemoto et al. 2012). Moreover, composition of HIPV blend is complex and highly variable according to abiotic (light, temperature, water stress, humidity, nutrition) and biotic factors (co-occurrence of herbivores and diseases, plant genotype, tissue/organ upon attack) (Gouinguené and Turlings 2002; Holopainen and Gershenzon 2010). Nevertheless, natural enemy's ability to learn and therefore expand the repertoire to novel HIPV compositions is likely an important strategy to overcome the variability of HIPV blend in a given environment (McCormick 2016).

Despite the variability in the blend composition, HIPVs are common to several plant species, and their synthetics are usually available for purchase. Therefore, similar to the kairomone-based strategy above mentioned, lures containing single or multiple plant volatile components can also be employed to attract natural enemies in the field (James 2003, 2005). Nevertheless, as plant volatiles are also important cues for herbivores, depending on the lure, herbivores can be recruited to HIPV-baited plots (Halitschke et al. 2007).

Another drawback of this strategy refers to the natural enemy ability for associative learning. Attraction to HIPV lures in the absence of the host/prey can be detrimental to the natural enemy foraging as HIPVs become unreliable cues of the host/prey, resulting in reduced biological control efficiency (Kaplan 2012). This consequence on the natural enemy foraging efficiency is difficult to detect in the field.

Study in Brazilian soybean fields showed that synthetic (*E*)-2-hexenal dispensers recruited stink bug egg parasitoids (Scelionidae) to treated plots, but parasitism rates were similar compared to control plots (Vieira et al. 2013), indicating that release of an attractive semiochemical to a given natural enemy not necessarily enhances the parasitoid ability.

With the aim to prevent these negative effects, the strategy was adapted by combining HIPV lures with an alternative food source as "reward" (Simpson et al. 2011). The strategy, named "attract and reward," seemed to be promising, but the "attract" (HIPV lure) and "reward" elements (e.g., flowering plants with suitable pollen and nectar) do not interact, and the expected synergistic effects between them have not been evidenced (Orre Gordon et al. 2013).

Among the most promising HIPV for recruiting natural enemies in ways that translate into increased parasitism and predation in the field is the methyl salicylate (MeSA), which is derived from the esterification of the phytohormone salicylic acid (SA). Studies have shown positive effects with a number of parasitoids and predators in a variety of crop plants (Rodriguez-Saona et al. 2011). Because of its activity, MeSA-baited lures are commercially available for gardens (PredaLure® Beneficial Insect Lure, AgBio Inc.). Nevertheless, there are some debates around the relevance of MeSA dispensers in pest management tactics, including the attraction of natural enemies (Snoeren et al. 2010; Rowen et al. 2017; Turlings and Erb 2018).

In Brazil, use of MeSA dispensers in common bean fields recruited higher abundance not only of predators (Anthocoridae and Syrphidae) but also of a leaf beetle (Chrysomelidae) (Salamanca et al. 2018). Although carnivore mites were not collected, which are well-known for being oriented to MeSA (De Boer and Dicke 2004), MeSA-treated plots had lower densities of pest spider mites (Tetranychidae). Also, aphid predation was greater in MeSA-treated plots. Despite of that, treatment with MeSA did not result in increased common bean yield.

The same rationale as synthetic HIPV lures to attract natural enemies has been used in transformed plants emitting a given attractive HIPV continuously. Comparing these two HIPV-based strategies, the only clear benefit of transformed HIPV-emitter plants is the cost savings in installing dispensers over crop fields. The genetic transformation can change the plant physiology and other plant-insect interactions, resulting in undesirable effects on yield (Robert et al. 2013). Moreover, the tactic presents the same drawbacks as mentioned above for HIPV-baited lures.

In this context, enhancing plant herbivore-induced signaling is more promising than tactics that present HIPVs continuously in the field, but plant transformation in ways that expression of attractive signals is inducible will be a turning point (Turlings and Erb 2018). Nowadays, strengthening plant induced responses can be accomplished by crop breeding (Tamiru et al. 2015) and use of defense elicitors (Stout et al. 2002).

Domestication and plant breeding programs have selected crop genotypes to optimize yield and agricultural management in detriment of plant defensive traits. As a result, crop plants present a weak defense system, such as reduced chemical induced defenses, and wild relatives represent interesting sources for breeding programs to restore at some extent the plant's ability to defend against insect pests (Chen et al. 2015). Indeed, there is much variation on the composition of HIPVs across domestication gradient (i.e., wild genotypes, landraces and domesticated cultivars), reflecting on reduced parasitism in domesticated cultivars compared to wild relatives in the field (Tamiru et al. 2011; Li et al. 2018). The classic study from Rasmann et al. (2005) showed that roots of North American corn lines have lost the ability to emit an important volatile signal for entomopathogenic nematodes, the (*E*)-β-caryophyllene, while European lines did not. The signal was restored in North American lines by inserting a (*E*)-β-caryophyllene synthase gene from oregano (Köllner et al. 2008). Even though authors showed satisfying results on the attraction of entomopathogenic nematodes to the transformed plant, the constitutive emission of (*E*)-β-caryophyllene incurred in physiological and ecological consequences undesirable in the agricultural context (Robert et al. 2013).

The second tactic that can reverse, to a certain extent, the flawed constitutive defensive system of crop plants consists of plant treatment with elicitors, which are chemical substances that activate plant induced defenses against pathogens or pests (Sobhy et al. 2014). The elicitors that induce resistance against pests promote augmented levels of direct (toxic, deterrent or repellant metabolites to herbivores) and indirect defenses (e.g., emission of attractive HIPVs to natural enemies). The best studied elicitors for field approaches are the derivatives and mimics of jasmonic acid and salicylic acid (Smith et al. 2009). Application of those elicitors usually provokes changes on HIPV profile similar to that emitted by herbivore-damaged plants, being attractive to natural enemies, but not always translating into greater parasitism and predation rates (Thaler 2002).

Treatment of soybean plants with cis-jasmone, a volatile compound derived from jasmonic acid pathway, promotes changes in plant volatile emission that attracts egg parasitoids (Moraes et al. 2009). Subsequent study showed that cis-jasmone-treated plants recruited a greater diversity and abundance of parasitoid wasps (Scelionidae) of stink bug eggs in soybean fields in Brazil, but parasitism rates were unaltered (Vieira et al. 2013). This result suggests that treatment with cis-jasmone attracts egg parasitoids, but do not enhance their host searching efficiency. Similarly, treatment of corn fields in Mexico with another jasmonic acid derivative, methyl jasmonate (MeJA), showed inconsistent results in parasitism rates of the fall armyworm *Spodoptera frugiperda* (J.E. Smith) (Noctuidae) as there was much variation among replicates (von Mérey et al. 2012).

Unlike the studies conducted in Latin America with elicitors, there have been successful cases with the use of jasmonates to enhance field parasitism and predation in temperate regions (Thaler 1999; Lou et al. 2005). However, because insects respond differently to plants treated with jasmonic acid (Thaler 2002), this variability on the effects of elicitor-treated plants on biological control efficacy is context dependent.

Other substances do not trigger plant-induced response, but they prime plants for defense (i.e., plants express defenses faster and more intense once damage occurs). Some plant volatiles, such as cis-jasmone, indole, and some green-leaf volatiles (GLVs) (C6-volatile compounds released immediately after mechanical damage), play a role on plant-plant signaling priming neighbor receiver plants (Frost et al. 2008; Matthes et al. 2010; Erb et al. 2015). In the field, primed plants would exhibit a stronger and faster induced response, what can translate into increased biological control efficiency. Therefore, use of dispensers releasing important plant-plant signals (e.g., GLVs and cis-jasmone) or strategically positioned transformed plants emitting them continuously can function not only as attractants of natural enemy populations but also as priming agents of crop plants.

So far, field studies did not find encouraging results with respect to the priming effect of HIPVs dispersers to increase biological control (Kalaivani et al. 2016; Rowen et al. 2017). For example, in Mexican corn fields, dispensers releasing a blend of GLVs that primes neighbor corn plants did not alter parasitism rates of the fall armyworm. Indeed, GLV-primed plants had more herbivores and were more damaged by insects than non-primed plants (von Mérey et al. 2011).

Plant strengtheners or stimulants, as plant-plant signaling volatiles, can prime or enhance plant defenses, besides other benefits in plant growth. For example, plant or seed inoculation with plant growth-promoting bacteria can augment control by natural enemies (Pangesti et al. 2015). This effect caused by the plant-microbial interaction can result from changes in plant size/vigor, altered primary and secondary metabolism and microbial volatiles that prime plants (Rasmann et al. 2017). Because plant-microbial interactions can alter plant chemical profile in ways that more herbivores are attracted (Babikova et al. 2014), the real potential of this tactic still needs to be addressed. By contrast, silicon (Si) amendment as integrated pest management (IPM) has shown more consistent results. Recent studies indicate that Si amendment primes grasses defenses, and increases biological control efficiency in Si-treated plots likely because of increased emissions of HIPVs (Ye et al. 2013; Liu et al. 2017). Further information about the effects of Si amendment on plant resistance are given in Chap. 27, Part 5.

41.6 Final Considerations

It is clear that semiochemicals represent important tools to be explored in the IPM context, especially because they recruit, retain, and increase searching efficiency of natural enemies in agricultural systems. However, as we are aware, chemical control can disrupt the integration of biological control with semiochemical-based strategies, and therefore, rational use of selective insecticides can prevent this problem (see Chap. 37, Part V).

Tactics employing host/prey pheromones are more practical because of the commercial availability of these chemicals, but further studies may verify the compatibility of using pheromone lures to monitor pest population and, at the same time, attract and retain their natural enemies in areas.

Despite the extensive literature on herbivore-induced plant responses and the ecological interactions mediated by plant signaling, exploring HIPVs as semiochemicals to recruit and enhance searching efficiency of natural enemies is still challenging. Lately, studies have focused more on providing reliable signals of host/prey presence to natural enemies than recruiting them to semiochemical-treated crops. But strategies using elicitors can imply in energy costs for plant physiological processes what can lead to reduced energy allocation to produce flowers, fruits, and seeds. In this sense, using priming agents and plant strengtheners seem to be the best tactic so far. Still, we face unpredictable effects while exploring HIPVs in the field because plant-insect interactions is highly species-specific and context dependent.

Finally, there is a perspective of integrating multiple semiochemical-based strategies that together may boost plant indirect defenses. Initial study under laboratory conditions has shown that herbivore aggregation pheromone exposure induced in cotton the emission of attractive volatiles to parasitic wasps (Magalhães et al. 2019). If crop plants are generally able to detect their herbivore pheromones and have not lost this ability across domestication and breeding processes, we can develop IPM strategies using pheromone lures for multiple functions in the field, like pest monitoring, natural enemy recruitment, and priming of plant-induced defenses. Undoubtedly chemical ecology is a fruitful research area for developing IPM tactics with the aim to make agriculture more sustainable. Given the worldwide importance of Latin America in food production, semiochemicals-based strategies should be more explored in tropical conditions, where I expect different outcomes due to the richer biodiversity of species and interactions. Moreover, the predominance of extensive monocultures in agricultural landscape of most Latin America likely limits the efficacy of semiochemical-based strategies (Kaplan and Lewis 2015).

References

Afsheen S, Wang X, Li R et al (2008) Differential attraction of parasitoids in relation to specificity of kairomones from herbivores and their by-products. Insect Sci 15(5):381–397

Abassi SA, Birkett MA, Pettersson J et al (2000) Response of the seven-spot ladybird to an aphid alarm pheromone and an alarm pheromone inhibitor is mediated by paired olfactory cells. J Chem Ecol 26:1765–1771

Aldrich JR, Cantelo WW (1999) Suppression of Colorado potato beetle infestation by pheromone-mediated augmentation of the predatory spined soldier bug, *Podisus maculiventris* (Say) (Heteroptera: Pentatomidae). Agric For Entomol 1:209–217

Allison JD, Hare JD (2009) Learned and naïve natural enemy responses and the interpretation of volatile organic compounds as cues or signals. New Phytol 184:768–782

Alim MA, Lim, UT (2011) Refrigerated eggs of *Riptortus pedestris* (Hemiptera: Alydidae) added to aggregation pheromone traps increase field parasitism in soybean. J Econ Entomol 104:1833–1839

Babikova Z, Gilbert L, Bruce T et al (2014) Arbuscular mycorrhizal fungi and aphids interact by changing host plant quality and volatile emission. Funct Ecol 28:375–385

Battaglia D, Poppy G, Powell W et al (2000) Physical and chemical cues influencing the oviposition behaviour of *Aphidius ervi*. Entomol Exp Appl 94:219–227

Beevers M, Lewis WJ, Gross HR, Nordlund DA (1981) Kairomones and their use for management of entomophagous insects: X. Laboratory Studies on Manipulation of Host-Finding Behavior of *Trichogramma pretiosum* Riley with a Kairomone Extracted from *Heliothis zea* (Boddie) Moth Scales. J Chem Ecol 7(3):635–648

Bin F, Vinson SB, Strand MR et al (1993) Source of an egg kairomone for *Trissolcus basalis*, a parasitoid of *Nezara viridula*. Physiol Entomol 18:7–15

Bruce TJA, Aradottir GI, Smart LE et al (2015) The first crop plant genetically engineered to release an insect pheromone for defence. Sci Rep 5:11183

Chen YH, Gols R, Benrey B (2015) Crop domestication and its impact on naturally selected trophic interactions. Annu Rev Entomol 60:35–58

Crock J, Wildung M, Croteau R (1997) Isolation and bacterial expression of a sesquiterpene synthase cDNA clone from peppermint (*Mentha* x *piperita*, L.) that produces the aphid alarm pheromone (*E*)-beta-farnesene. Proc Natl Acad Sci USA 94:12833–12838

De Boer JG, Dicke M (2004) The role of methyl salicylate in prey searching behavior of the predatory mite *Phytoseiulus persimilis*. J Chem Ecol 30:255–271

De Moraes CM, Lewis WJ, Pare PW et al (1998) Herbivore-infested plants selectively attract parasitoids. Nature 393:570–573

Dicke M, Van Beek TA, Posthumus MA et al (1990a) Isolation and identification of volatile kairomone that affects acarine predator prey interactions Involvement of host plant in its production. J Chem Ecol 16:381–396

Dicke M, Sabelis MW, Takabayashi J et al (1990b) Plant strategies of manipulating predator-prey interactions through allelochemicals: prospects for application in pest control. J Chem Ecol 16:3091–3118

Dicke M, Sabelis MW (1988) Infochemical terminology: based on cost-benefit analysis rather than origin of compounds? Func Ecol 2:131–139

Erb M, Meldau S, Howe GA (2012) Role of phytohormones in insect-specific plant reactions. Trends Plant Sci 17:250–259

Erb M, Veyrat N, Robert CAM et al (2015) Indole is an essential herbivore-induced volatile priming signal in maize. Nat Commun 6:6273

Farmer EE, Johnson RR, Ryan CA (1992) Regulation of expression of proteinase inhibitor genes by methyl jasmonate and jasmonic Acid. Plant Physiol 98:995–1002

Fatouros NE, Huigens ME, van Loon JJA et al (2005) Butterfly anti-aphrodisiac lures parasitic wasps. Nature 433:704–704

Frost CJ, Mescher MC, Dervinis C et al (2008) Priming defense genes and metabolites in hybrid poplar by the green leaf volatile *cis* -3-hexenyl acetate. New Phytol 180:722–734
Giunti G, Canale A, Messing RH et al (2015) Parasitoid learning: current knowledge and implications for biological control. Biol Control 90:208–219
Gouinguené SP, Turlings TCJ (2002) The effects of abiotic factors on induced volatile emissions in corn plants. Plant Physiol 129:1296–1307
Gross HR, Lewis WJ, Jones RL, Nordlund DA (1975) Kairomones and their use for management of entomophagous insects. III. Stimulation of *Trichograrnma achaeae*, *T. pretiosum*, and *Microplitis croceipes* with host-seeking stimuli at time of release to improve their efficiency. J Chem Ecol 1(4):431–438
Halitschke R, Stenberg JA, Kessler D et al (2007) Shared signals –'alarm calls' from plants increase apparency to herbivores and their enemies in nature. Ecol Lett 1(1):24–34
Hardie J, Hick AJ, Höller C et al (1994) The responses of *Praon* spp. parasitoids to aphid sex pheromone components in the field. Entomol Exp Appl 71:95–99
Hatano E, Kunert G, Michaud JP, Weisser WW (2008) Chemical cues mediating aphid location by natural enemies. Eur J Entomol 105:797–806
Holopainen JK, Gershenzon J (2010) Multiple stress factors and the emission of plant VOCs. Trends Plant Sci 15:176–184
James DG (2003) Synthetic herbivore-induced plant volatiles as field attractants for beneficial insects. Environ Entomol 32:977–982
James DG (2005) Further field evaluation of synthetic herbivore-induced plan volatiles as attractants for beneficial insects. J Chem Ecol 31:481–495
Jumean Z, Jones E, Gries G (2009) Does aggregation behavior of codling moth larvae, *Cydia pomonella*, increase the risk of parasitism by *Mastrus ridibundus*? Biol Control 49:254–258
Kalaivani K, Kalaiselvi MM, Senthil-Nathan S (2016) Effect of methyl salicylate (MeSA), an elicitor on growth, physiology and pathology of resistant and susceptible rice varieties. Sci Rep 6:34498
Kaplan I (2012) Attracting carnivorous arthropods with plant volatiles: The future of biocontrol or playing with fire? Biol Control 60:77–89
Kaplan I, Lewis D (2015) What happens when crops are turned on? Simulating constitutive volatiles for tritrophic pest suppression across an agricultural landscape. Pest Manag Sci 71:139–150
Khan ZR, James DG, Midega CAO, Pickett JA (2008) Chemical ecology and conservation biological control. Biol Control 45:210–224
Köllner TG, Held M, Lenk C et al (2008) A maize (*E*)-beta-caryophyllene synthase implicated in indirect defense responses against herbivores is not expressed in most American maize varieties. Plant Cell 20:482–494
Leal WS (2013) Odorant reception in insects: roles of receptors, binding proteins, and degrading enzymes. Annu Rev Entomol 58:373–391
Leal WS, Higuchi H, Mizutani N et al (1995) Multifunctional communication in *Riptortus clavatus* (Heteroptera: Alydidae): conspecific nymphs and egg parasitoid *Ooencyrtus nezarae* use the same adult attractant pheromone as chemical cue. J Chem Ecol 21:973–985
Leroy PD, Sabri A, Heuskin S et al (2011) Microorganisms from aphid honeydew attract and enhance the efficacy of natural enemies. Nat Commun 2:348
Lewis WJ, Jones RL, Nordlund DA, Gross HR (1975) Kairomones and their use for management of entomophagous insects: II. Mechanisms causing increase in rate of parasitization by *Trichogramma* spp. J Chem Ecol 1:349–360
Lewis WJ, Jones RL, Sparks AN (1972) A host-seeking stimulant for the egg parasite *Trichogramma evanescens*: its source and a demonstration of its laboratory and field activity. Ann Entomol Soc Am 65:1087–1089
Lewis WJ, Nordlund DA, Gueldner RC et al (1982) Kairomones and their use for management of entomophagous insects. XIII. Kairomonal Activity for *Trichogramma* spp. of Abdominal Tips, Excretion, and a Synthetic Sex Pheromone Blend of *Heliothis zea* (Boddie) Moths. J Chem Ecol 8(10):1323–1331

Li X, Garvey M, Kaplan I et al (2018) Domestication of tomato has reduced the attraction of herbivore natural enemies to pest-damaged plants. Agric For Entomol 20:390–401

Liu J, Zhu J, Zhang P et al (2017) Silicon supplementation alters the composition of herbivore induced plant volatiles and enhances attraction of parasitoids to infested rice plants. Front Plant Sci 8:1–8

Lou Y-G, Du M-H, Turlings TCJ et al (2005) Exogenous application of jasmonic acid induces volatile emissions in rice and enhances parasitism of *Nilaparvata lugens* eggs by the parasitoid *Anagrus nilaparvatae*. J Chem Ecol 31:1985–2002

McCormick AC (2016) Can plant-natural enemy communication withstand disruption by biotic and abiotic factors? Ecol Evol 6:8569–8582

Magalhães DM, Fidelis IAS, Borges M et al (2019) *Anthonomus grandis* aggregation pheromone induces cotton indirect defence and attracts the parasitic wasp *Bracon vulgaris*. J Exp Bot 70(6):1891–1901. https://doi.org/10.1093/jxb/erz040

Matthes MC, Bruce TJA, Ton J et al (2010) The transcriptome of cis-jasmone-induced resistance in *Arabidopsis thaliana* and its role in indirect defence. Planta 232:1163–1180

Mattiacci L, Dicke M (1995) The parasitoid *Cotesia glomerata* (Hymenoptera: Braconidae) discriminates between first and fifth larval instars of its host *Pieris brassicae*, on the basis of contact cues from frass, silk, and herbivore-damaged leaf tissue. J Insect Behav 8:485–498

Mattiacci L, Vinson SB, Williams HJ et al (1993) A long-range attractant kairomone for egg parasitoid *Trissolcus basalis*, isolated from defensive secretion of its host, *Nezara viridula*. J Chem Ecol 19:1167–1181

Mithofer A, Boland W (2008) Recognition of herbivory-associated molecular patterns. Plant Physiol 146:825–831

Mizutani N (2006) Pheromones of male stink bugs and their attractiveness to their parasitoids. Japanese J Appl Entomol Zool 50:87–99

Moraes MCB, Laumann RA, Pareja M et al (2009) Attraction of the stink bug egg parasitoid *Telenomus podisi* to defence signals from soybean activated by treatment with *cis* -jasmone. Entomol Exp Appl 131:178–188

Mori K, Tashiro T (2004) Useful reactions in modern pheromone synthesis. Curr Org Synth 1:11–29

Mumm R, Dicke M (2010) Variation in natural plant products and the attraction of bodyguards involved in indirect plant defense the present review is one in the special series of reviews on animal–plant interactions. Can J Zool 88:628–667

Mansour R, Suma P, Mazzeo G, Russo A et al. (2010). Using a kairomone-based attracting system to enhance biological control of mealybugs (Hemiptera: Pseudococcidae) by *Anagyrus* sp. near pseudococci (Hymenoptera: Encyrtidae) in Sicilian vineyards. J Entomol Acarol Res 42:161–170

Nakashima Y, Ida TY, Powell W et al (2016) Field evaluation of synthetic aphid sex pheromone in enhancing suppression of aphid abundance by their natural enemies. BioControl 61:485–496

Nordlund DA, Lewis WJ, Todd JW, Chalfant RB (1977) Kairomones and their use for management of entomophagous insects. J Chem Ecol 3:513–518

Nordlund DA, Strand MR, Lewis WJ, Vinson SB (1987) Role of kairomones from host accessory gland secretion in host recognition by *Telenomus remus* and *Trichogramma pretiosum*, with partial characterization. Entomol Exp Appl 44:37–43

Onodera J, Matsuyama S, Suzuki T, Fujii K (2002) Host-recognizing kairomones for parasitic wasp, *Anisopteromalus calandrae*, from larvae of azuki bean weevil, *Callosobruchus chinensis*. J Chem Ecol 28:1209–1220

Orre Gordon GUS, Wratten SD, Jonsson M et al (2013) 'Attract and reward': combining a herbivore-induced plant volatile with floral resource supplementation – Multi-trophic level effects. Biol Control 64:106–115

Pangesti N, Weldegergis BT, Langendorf B et al (2015) Rhizobacterial colonization of roots modulates plant volatile emission and enhances the attraction of a parasitoid wasp to host-infested plants. Oecologia 178:1169–1180

Pascal LD, Raki A, Sabrine A et al (2014) Aphid honeydew: an arrestant and a contact kairomone for *Episyrphus balteatus* (Diptera: Syrphidae) larvae and adults. Eur J Entomol 111:237–242
Peñaflor MFGV, Bento JMS (2013) Herbivore-induced plant volatiles to enhance biological control in agriculture. Neotrop Entomol 42:331–343
Pickett JA, Griffiths DC (1980) Composition of aphid alarm pheromones. J Chem Ecol 6:349–360
Pompanon F, De Schepper B, Mourer Y et al (1997) Evidence for a substrate-borne sex pheromone in the parasitoid wasp *Trichogramma brassicae*. J Chem Ecol 23:1349–1360
Rasmann S, Bennett A, Biere A et al (2017) Root symbionts: powerful drivers of plant above- and belowground indirect defenses. Insect Sci 24:947–960
Rasmann S, Köllner TG, Degenhardt J et al (2005) Recruitment of entomopathogenic nematodes by insect-damaged maize roots. Nature 434:732–737
Reddy GVP, Holopainen JK, Guerrero A (2002) Olfactory responses of *Plutella xylostella* natural enemies to host pheromone, larval frass, and green leaf cabbage volatiles. J Chem Ecol 28:131–143
Reymond P (2013) Perception, signaling and molecular basis of oviposition-mediated plant responses. Planta 238:247–258
Robert CAM, Erb M, Hiltpold I et al (2013) Genetically engineered maize plants reveal distinct costs and benefits of constitutive volatile emissions in the field. Plant Biotechnol J 11:628–639
Rodriguez-Saona C, Kaplan I, Braasch J et al (2011) Field responses of predaceous arthropods to methyl salicylate: a meta-analysis and case study in cranberries. Biol Control 59:294–303
Rowen E, Gutensohn M, Dudareva N, Kaplan I (2017) Carnivore attractant or plant elicitor? Multifunctional roles of methyl salicylate lures in tomato defense. J Chem Ecol 43:573–585
Ruther J, Homann M, Steidle JLM (2000) Female-derived sex pheromone mediates courtship behaviour in the parasitoid *Lariophagus distinguendus*. Entomol Exp Appl 96:265–274
Rutledge CE (1996) A survey of identified kairomones and synomones used by insect parasitoids to locate and accept their hosts. Chemoecology 7:121–131
Salamanca J, Souza B, Rodriguez-Saona C (2018) Cascading effects of combining synthetic herbivore-induced plant volatiles with companion plants to manipulate natural enemies in an agro-ecosystem. Pest Manag Sci 74:2133–2145
Sant'Ana J, Bruni R, Abdul-Baki AA, Aldrich JR (1997) Pheromone-induced movement of nymphs of the predator, *Podisus maculiventris* (Heteroptera: Pentatomidae). Biol Control 10:123–128
Simpson M, Gurr GM, Simmons AT et al (2011) Field evaluation of the 'attract and reward' biological control approach in vineyards. Ann Appl Biol 159:69–78
Smith JL, De Moraes CM, Mescher MC (2009) Jasmonate- and salicylate-mediated plant defense responses to insect herbivores, pathogens and parasitic plants. Pest Manag Sci 65:497–503
Snoeren TAL, Mumm R, Poelman EH et al (2010) The herbivore-induced plant volatile methyl salicylate negatively affects attraction of the parasitoid *Diadegma semiclausum*. J Chem Ecol 36:479–489
Sobhy IS, Erb M, Lou Y, Turlings TCJ (2014) The prospect of applying chemical elicitors and plant strengtheners to enhance the biological control of crop pests. Philos Trans R Soc B Biol Sci 369:20120283
Stout MJ, Zehnder GW, Baur ME (2002) Potential for the use of elicitors of plant resistance in arthropod management programs. Arch Insect Biochem Physiol 51:222–235
Takemoto H, Powell W, Pickett J et al (2012) Two-step learning involved in acquiring olfactory preferences for plant volatiles by parasitic wasps. Anim Behav 83:1491–1496
Tamiru A, Bruce TJ, Woodcock CM et al (2011) Maize landraces recruit egg and larval parasitoids in response to egg deposition by a herbivore. Ecol Lett 14:1075–1083
Tamiru A, Khan ZR, Bruce TJ (2015) New directions for improving crop resistance to insects by breeding for egg induced defence. Curr Opin Insect Sci 9:51–55
Thaler JS (2002) Effect of jasmonate-induced plant responses on the natural enemies of herbivores. J Anim Ecol 71:141–150
Thaler JS (1999) Jasmonate-inducible plant defences cause increased parasitism of herbivores. Nature 399:686–688

Thibout E (2005) Role of caterpillar silk thread in location of host pupae by the parasitoid *Diadromus pulchellus*. J Insect Behav 18:817–826
Turlings TCJ, Erb M (2018) Tritrophic interactions mediated by herbivore-induced plant volatiles: mechanisms, ecological relevance, and application potential. Annu Rev Entomol 63:433–452
Turlings TCJ, Tumlinson JH, Heath RR et al (1991) Isolation and identification of allelochemicals that attract the larval parasitoid, *Cotesia marginiventris* (Cresson), to the microhabitat of one of its hosts. J Chem Ecol 17:2235–2251
Usha Rani P, Indu Kumari S, Sriramakrishna T, Ratna Sudhakar T (2006) Kairomones extracted from rice yellow stem borer and their influence on egg parasitization by *Trichogramma japonicum* Ashmead. J Chem Ecol 33:59–73
von Mérey G, Veyrat N, Mahuku G et al (2011) Dispensing synthetic green leaf volatiles in maize fields increases the release of sesquiterpenes by the plants, but has little effect on the attraction of pest and beneficial insects. Phytochemistry 72:1838–1847
von Mérey GE, Veyrat N, De Lange E et al (2012) Minor effects of two elicitors of insect and pathogen resistance on volatile emissions and parasitism of *Spodoptera frugiperda* in Mexican maize fields. Biol Control 60:7–15
Vet MEL, Dicke M (1992) Ecology of infochemical use by natural enemies in a tritrophic context. Annu Rev Entomol 37:141–172
Vieira CR, Moraes MCB, Borges M et al (2013) cis-Jasmone indirect action on egg parasitoids (Hymenoptera: Scelionidae) and its application in biological control of soybean stink bugs (Hemiptera: Pentatomidae). Biol Control 64:75–82
Vinson SB (1984) Parasitoid-host relationship. In: Chemical ecology of insects. Springer US, Boston, MA, pp 205–233
Vinson SB (1976) Host selection by insect parasitoids. Annu Rev Entomol 21:109–133
Vosteen I, Weisser WW, Kunert G (2016) Is there any evidence that aphid alarm pheromones work as prey and host finding kairomones for natural enemies? Ecol Entomol 41:1–12
Wang S-N, Peng Y, Lu Z-Y et al (2015) Identification and expression analysis of putative chemosensory receptor genes in *Microplitis mediator* by antennal transcriptome screening. Int J Biol Sci 11:737–751
Wäschke N, Meiners T, Rostás M (2013) Foraging strategies of parasitoids in complex chemical environments. In: Chemical Ecology of Insect Parasitoids. Wiley-Blackwell, Chichester, UK, pp 37–63
Whitfield J (2001) Making crops cry for help. Nature 410:736–737
Wölfling M, Rostás M (2009) Parasitoids use chemical footprints to track down caterpillars. Commun Integr Biol 2:353–355
Ye M, Song Y, Long J et al (2013) Priming of jasmonate-mediated antiherbivore defense responses in rice by silicon. Proc Natl Acad Sci 110:E3631–E3639

Chapter 42
Aphid-Tending Ants and Their Effects on Natural Enemies Used in the Biological Control

Ernesto Oliveira Canedo-Júnior, Ângelo Barbosa Monteiro, Antônio César Medeiros de Queiroz, and Graziele Santiago Silva

42.1 Introduction

Survival of species is based on acquiring resources that are often limited in the nature, such as sufficient quantity and quality food. Individuals within a population must compete for resources in order to survive, which depends upon their adaptations and ability to respond to natural changes in the environment, as well as the adopting of strategies to reduce the efficiency of their natural enemies, also known as biological control agents or organisms that search for feeding alternatives from the existing prey population within the ecosystem. However, the optimization process in obtaining of resources and natural enemies escaping affords greater competitive advantage with regard to other species found within the same ecological context (Begon et al. 2007).

Plants, which are described as sessile organisms, are not able to escape from high exploitation conditions. They deploy diverse defense strategies against natural enemies in order to survive, grow, and reproduce under a certain level of herbivory (Boiça-Jr et al. 2018). Plants may also experience phenological changes, show interspecific differences in leaf morphology to reduce recognition by herbivores, develop mechanisms to prevent invasive actions on essential plant parts, and influence the nutritional quality of plant tissues by means of production of secondary compounds or physical structures (Raven et al. 2007). Plants may also benefit indirectly from herbivores by producing attractive metabolites, which favor the perma-

E. O. Canedo-Júnior (✉)
Departamento de Educação e Ciências Humanas, Universidade do Estado de Minas Gerais - Unidade Poços de Caldas, Poços de Caldas, Minas Gerais, Brazil

Â.B. Monteiro (Deceased)

A. C. M. de Queiroz · G. S. Silva
Laboratório de Ecologia de Formigas, Department of Biology - Setor Ecologia, Universidade Federal de Lavras, Lavras, Minas Gerais, Brazil

B. Souza et al. (eds.), *Natural Enemies of Insect Pests in Neotropical Agroecosystems*, https://doi.org/10.1007/978-3-030-24733-1_42

nence of predators in the area (Altieri et al. 2003). Interactions between ants and extrafloral nectaries, food corpuscles, elaiosomes, domacies, and nesting sites have been reported (Hölldobler and Wilson 1990). The pequi tree (*Caryocar brasiliensis*, Caryocaraceae), for example, is a tree native to Brazilian Cerrado in the Caryocaraceae family, commonly consumed by various herbivore species (Lepidoptera, Hemiptera, Diptera, Hymenoptera). This tree has extrafloral nectaries that show attractive effect to more than 30 ant species, which defends the tree against herbivores, reducing significantly the infestation by these herbivorous insects on the tree (Rodrigues et al. 2008).

At the current state of understanding of mutualistic relationships, it is found that interactions between plants and predators depend on the ecological context where interactions take place, such as biotic (e.g., genetic variation among individuals within a population and population variations) and abiotic variations in the habitat. Herbivores are able to develop a specialization for a specie or group of plants throughout species evolution. When plants produce chemical defense compounds, these herbivores may excrete such compounds or detoxify themselves, or even store them within tissues for own defense against predators (Chown and Nicolson 2004). This may be the worst scenario for plants, where the investment should be concentrated on defense mechanisms and attractive compounds derived from plants to reduce herbivore damage and increase the effectiveness of the control strategy. The control of herbivore population growth depends on the stable equilibrium of three trophic levels, as well as the ecological context, since natural enemies will attempt to avoid herbivores with high degree of toxicity (Price et al. 1980; Schowalter 2006).

The specialization degree of a predator may also influence the dynamic of plant-herbivore interactions. In this context, ant-attractive compounds produced by plants are usually subjected to selection by specialist predators. Although a review is provided of the current state of understanding of ants as generalist predators, focusing on the reduction of population density of herbivores, ants are still not fully considered great for biological control of herbivores (Chamberlain and Holland 2009; Rosumek et al. 2009). Generalist predators do not show both numeric and functional responses to prey. They commonly produce lesser progeny than herbivores and show satiation associated with predation rates (below the threshold due to the time spent in the search and apprehension of prey) as well as easy change of food preference motivated by the availability and access of alternative prey in the habitat (Cohen et al. 1993). In addition, herbivores are also able to use ants for own protection. In this relation, also known as trophobiosis, ants are induced by herbivores to attend them, offering honeydew and protecting them from natural enemies (e.g., Hemiptera: Sternorrhyncha, Auchenorrhyncha, and some Lepidoptera worms) (Fiedler and Maschwitz 1988; Delabie 2001; Zhang et al. 2012).

Species are all usually found to be under biological control conditions, and population outbreaks are considered exceptions in natural conditions. However, agricultural systems are also found to be simplified environments, which propitiate favorable conditions for herbivore population outbreaks (Odum and Barret 2007). These outbreaks are likely to occur, especially in monoculture production systems, generating significant losses in crops, making such organisms to be considered agricultural pests (Gallo et al. 2002). In this context, the control of such populations is required in order to minimize economic damages caused by the action of these organisms. Thus, the use of techniques of mass introduction or conservation of natural enemies of herbivore – pests in agricultural systems – has been growing in the last few decades as part of integrated pest management programs (Gravena 1992; Berti-Filho and Macedo 2010).

The use of natural enemies in biological control of herbivore – pests in agroecosystems – faces various challenges, such as selection of high-performance natural enemies in the control of pests, biological characterization for breeding purpose, field release of predators, and many others (Panizzi 1990). Besides, the trophobiotic relation between some ants and herbivore – pests, especially aphids – contributes negatively to the implementation of control programs, which are not yet well studied. Aphids are highlighted by showing high reproductive rates, by being polyphagous since they can raise successful colonies on a wide range of crops, and by being vectors of many plant pathogenic microorganisms responsible for crop losses worldwide (van-Emedem and Harrington 2007).

Therefore, this chapter approaches the issue of how the diversity of ants found in interaction with aphids may affect negatively the action of natural enemies in biological control programs and how such thematic relationships has been discussed in the Latin American context.

42.2 How Does the Ant-Aphid Interaction Take Place?

Ants have shown high diversity and specialization to a wide range of environments, which is reflected in diverse ant traits, such as the diversity of feeding habits. Generalist ants, for example, may feed on many species or individuals of the same species, as well as on alive or dead things, derived from animal, vegetal, and fungal origin (Baccaro et al. 2015). Aphids, however, are essentially phytophagous and feed upon sap of diverse plant organs. Aphids need to ingest large amount of phloem sap to appease their requirements, once it is found to be relatively rich in water and carbohydrates and naturally poor in nitrogenous compounds. They have specialized intestines to filter phloem sap, then select required nutrients, and expel the excess as waste droplets through anal siphon (Dixon 1998).

Excreta are known as honeydew, mainly rich in carbohydrates, nutrients found to be essential in the diet of ants. The honeydew usually falls onto plant leaves or on the ground and serves as trail for ants, which follow the drops to aphid colonies, searching for food (Reznikova and Novgorodova 1998; Detrain et al. 2010). In addition to the accidental encounter with honeydew droplets, ants may be attracted by volatile semiochemicals expelled by plants infested by aphids (Córdova-Yamauchi et al. 1998). After the encounter, depending on the ant specie and nutritional requirements of the colony, ants may switch to alternative sources, keeping aphids as sporadic sources of food or returning to the ant-nest to recruit another worker ants and lead them to attend aphids, either by collecting aphids' honeydew and by removing exuviae or for own protection against natural enemies. However, ants do not use some aphid species for the protection strategy, since they only use myrmecophilous species (Fiedler and Maschwitz 1988; Delabie 2001; Stadler and Dixon 2008).

Ant-aphid interactions may affect certain aphid attributes, such as behavioral pattern, reduction of mobility, delay on dispersion cycles, and morphological changes in the anal siphon, which makes easy the honeydew collection by ants (Novgorodova 2002; Tegelar et al. 2012). Aphids may produce honeydew with the most attractive chemical composition for ants, seeking to keep the ants loyal (Völkl et al. 1999; Stadler and Dixon 2005). Such changes occur because the larger number of ants in the habitat and the long permanence time in the colony mean a likely reduction in deaths by predation or parasitism (Buckley 1987). The species of ant and aphid involved in the interaction, the quantity and quality of honeydew produced by aphids, and the size of aphid colonies interfere in the compulsoriness of interactions and, then, in the protection effectiveness of aphids by ants against a predatory (Sakata 1995; Flatt and Weinsser 2000).

42.3 What Is the Diversity of Aphid-Tending Ants in Agricultural Systems and Where Do They Come From?

The first thing that comes in mind when thinking about ants in Latin-America agroecosystems, especially in Brazil, are possible damages and prejudices potentially caused by ants of the genus *Atta* and *Acromyrmex*, also known as leaf-cutting ants. However, there is high diversity of ants foraging and nesting in agricultural environments. This diversity goes beyond leaf-cutting ants and becomes larger as agricultural and natural environments show high diversity of resources and species (Perfecto and Snelling 1995; Armbrecht et al. 2005). In most of cases, ants have reduced colonization capacity as function of changes caused by the soil use (e.g., decrease in plant cover, introduction of exotic species, and soil disturbance for agricultural use) (Queiroz et al. 2017). In addition, the colonization of ants in agricultural environments is also associated with the quantity and proximity of areas with surrounding natural formations (Roth et al. 1994). The main genera of ants related to myrmecophilous insects, members of the order Hemiptera found in Brazil, for example, are *Camponotus*, *Pheidole*, *Linepithema*, *Solenopsis*, and *Crematogaster*, some of them showing high diversity worldwide (Resende et al. 2006; Penteado et al. 2012; Guindani et al. 2017; Canedo-Jr et al. 2017a). Insect species of these genera occur with lesser or greater frequency in natural environments and are commonly affected by changes in the soil use and management practices.

42.4 How Do Aphid-Tending Ants Affect Negatively the Effectiveness of Natural Enemies in Biological Control Programs?

Ants have shown high diversity to a wide range of environments, which is reflected within interactions of ant with aphids and with other pest insects in agroecosystems, as explained above. These ants are, in most of cases, generalist predators, and they are adapted to open environments and with high disturbance degree (Gibb and Hochuli 2003). They have an intense foraging habit and are highly aggressive in the protection of food sources, recruiting greater number of workers and soldiers ants, which are responsible for building and maintaining the nest, protecting the colony from other ants, and, most importantly, feeding the colony (Lach et al. 2010; Canedo-Jr et al. 2017a).

Aphids are sources of available and accessible healthy food for ants, and they provide food for long periods of time. This is the reason why ants protect aphids from other competitor ants, predators, and parasitoids, which address the role of natural enemies (Herbert and Horn 2008; Stewart-Jones et al. 2008; Powell and Silverman 2010; Dao et al. 2013) (Table 42.1). Once established the ant-aphid interaction, ants attack ferociously any organism that approaches the ant-attended aphids colony, either herbivore or natural enemy. Sometimes, ants may specifically respond to alert pheromones released from aphids under attack and increase the sense of protection and aggressiveness against enemies of aphids (Nault et al. 1976; Verheggen et al. 2012).

This protection model between species may complicate the understanding and success of biological control programs of aphids, since this control strategy uses natural enemies to control populations of aphids. Aphid-tending ants may negatively affect the natural enemies that approach the colony of ant-attended aphids by biting them, injecting venom, or sprinkling formic acid, either individually or group

Table 42.1 Effect of aphid-tending ants on aphid predators (larvae and adults) and parasitoids

		Predators of aphids (larvae)	Predators of aphids (adults)	Parasitoids of aphids
Tending ants	*Direct effects*[a]	Mutilation or death	Mutilation or death	Mutilation or death
		Repulsion from the host plant	Repulsion from the host plant	Repulsion from the host plant
		Immobility behavior		Removal or predation of parasitized aphids
	Indirect effects[b]	Avoidance on preying attended aphids	Avoidance on preying attended aphids	Avoidance on ovipositing on attended aphids

[a]*Direct effects*. Ants attack predators or parasitoids of attended aphid directly by biting them, stinging, or sprinkling of acid. Ants may conduct individual or group attacks involving a number of workers and soldier ants

[b]*Indirect effects*. These refer to the presence of ants per se, which can be detected by aphid predators or parasitoids by means of visual and/or chemical signals emitted in the environment by ants

attacks. These attack strategies result in the repulsion of natural enemies from plants, mainly winged enemies such as ladybugs, earwigs, predatory hemiptera, and spiders (Rodrigues et al. 2005). Laboratory studies report that *Camponotus rufipes* Fabricius (Formicidae) attending *Aphis craccivora* Koch (Aphididae) show an extremely aggressive behavior toward adults of *Hippodamia convergens* Guerin-Meneville (Coccinellidae). Findings show that when ladybugs prey on ant-attended aphids in the same host plant, ants attempt to attack them actively by biting or injecting an acid. In most of cases, ladybugs move away or fly off from the host plant to their escape; however, they are captured and preyed upon by ants in some cases (personal statement from Ernesto O. Canedo-Júnior, based on daily laboratory observations).

In case of predatory larvae, which have less escaping capacity when compared to adult insects (e.g., larvae of Coccinellidae, Chrysopidae, and Syrphidae), the attack of ants results in the predation of these natural enemies in most cases. Therefore, ants protect the colony and the food source, and they use larvae as source of proteins and lipids to the colony (Vinson and Scarborough 1989).

The negative effects of ants on natural enemies may also be considered behavioral indirect effect, once natural enemies may avoid preying upon aphid-tending ants (Canedo-Jr et al. 2017a), or, in case of parasitoids, they may avoid ovipositing on aphid-tending ants. This parasitoids avoidance process occurs because some tending ant species can recognize parasitized aphids, removing them from the host plant or by predating them and then killing parasitoid larvae (Rodrigues et al. 2005). In case of larvae, members of the family Syrphidae, for example, ants do not recognize and attack larvae by standing immobile and then reducing the predation efficiency (Bächtold and Del-Claro 2013).

42.5 What Is the Current State of Understanding of the Effects of Aphid-Tending Ants on Natural Enemies of Pests in Latin America?

A review is provided of the current state of understanding of effects of aphid-tending ants or other myrmecophilous insects on natural enemies of pests in agroecosystems (Nechols and Seibert 1985; Vinson and Scarborough 1989; Itioka and Inoue 1996; Herbert and Horn 2008; Stewart-Jones et al. 2008; Piñol et al. 2009; Powell and Silverman 2010; Dao et al. 2013; Nagy et al. 2015) However, none of these studies was developed in Latin America.

The effect of ants on natural enemies used in biological control programs is not widely studied at present. Silva and Perfecto (2013) investigated the effects of aphid-tending ants on distribution and occurrence of two aphid predators, fly larvae (Syrphidae) and ladybug larvae (Coccinellidae), in a cocoa plantation in Northeast of Brazil. They found that tending ants attacked larvae members of the family Syrphidae. However, they did not show any aggressive behavior on ladybug larvae.

Thus, they stated that the aggressive behavior of ants on predatory larvae may show a positive effect on the spatial overlapping and coexistence of two species of predatory larvae.

Canedo-Jr et al. (2017b) tested the effect of aphid-tending ants, *A. craccivora*, on the permanence of larvae of chrysopid predator species in host plants of a trial consisting of plants of common bean (*Phaseolus vulgaris*). They established an experiment on presence or exclusion of ants and compared the number of remaining larvae in host plants, about 24 hours after releasing. Two releases of chrysopid larvae were performed, 1 and 2 months after aphid infestation in the plantation. They found that ants show negative effect on predatory larvae. However, this effect might not be observed due to the turnover of ant species over time, which describes the ant-aphid interaction dynamic in the agroecosystems.

Although the dynamic of assemblies of aphid-tending ants over time is found to be an important factor for the understanding of interactions between ants and aphids, and their effects on aphid natural enemies, there are still few studies developed in this context. Canedo-Jr et al. (unpublished manuscript) assessed the temporal dynamics of assemblies of aphid-tending ants over 10 weeks. They found that the richness of ant species increased over time. In addition, the ant species composition suffered a progressive turnover, i.e., the composition of assemblies of aphid-tending ant in the first data collection week was very different from the last data collection week, which suggests putative relative differences to effects of ants on natural enemies, depending on the moment these natural enemies were released in the agroecosystem.

Studies developed in Latin America only describe the effect of aphid-tending ants on aphid natural enemies, based on field observation and with no data analysis or quantification of effects under study. Rodrigues et al. (2005) found that ants attacked parasitoids in chrysanthemum grown in protected environment, then preventing their oviposition. Oliveira et al. (2006) found that the majority of ants studied in the organic integrated crop management system interacted with spittlebugs and aphids, then affecting the action of natural enemies. Leite et al. (2006) studied tending ants in the biological control system, focused in protecting aphid populations from natural enemies on seedlings of pequi trees. Rodrigues et al. (2010) stated that the presence of tending ants in organic tangerine (*Citrus reticulata*) farming systems is a determining factor for low predation and parasitism efficiencies on the aphid *Toxoptera citricida* Kirkaldy (Aphididae).

42.6 Effect of Aphid-Tending Ants in the Biological Control: Final Considerations

The Latin American region is found to be a potential for the development of biological control programs for pests due to the existing diversity of organisms in the agroecosystem, especially arthropods, which can be used as predators. However, the establishment of this control program in this region is challenging (van-Lenteren and Bueno 2003), although there is an extensive cultivated area in the agroecosystem. The intensive use of agrochemicals in the crops production system, weak dissemination of information on the biological control method in the region, and lack of public political support to assist farmers in the implementation of this control method are the main challenging factors (Parra 2014; Halfeld-Vieira et al. 2016).

Aphid-tending ants contribute negatively to the success of biological control programs, as described in this chapter. In addition, at the current state of understanding of this control method, there are still few studies focused on interactions between ants and aphids and their effects on aphid natural enemies and other pests, members of the order Hemiptera. The Latin American region, which comprises the majority of area of the Neotropical region, contains more than 4100 ant species already described (Antwiki 2019), where most of them attend aphids and other myrmecophilous Hemiptera. Studies developed in this area focus on the description of ant species interacting with insects, members of the order Hemiptera; the understanding of the dynamic of assemblies of aphid-tending ants in these systems; the understanding of interactions between ants and aphids, and their effects in the effectiveness of biological control methods; and design of applied methods that can weaken the interaction between ants and hemiptera-pests. Therefore, this science knowledge area is not yet well studied in Latin America, so that further researches are needed. In this context, two essential base ideas are described as follows: (i) development of accessible technologies comprising nutritional food baits that reduce ants' interest on honeydew and reduce the occurrence of interactions between ants and aphids and their effects in the biological control programs and (ii) test of the effect and quantification of different plantations designs and its environmental heterogeneity on the interaction. These ideas may lead the development of the theory related to this study area and generate research products with direct return to farmers involved in biological control programs.

References

Altieri MA, Silva EN, Nicholls CI (2003) O papel da biodiversidade no manejo de pragas. Holos, Ribeirão Preto

Antwiki (2019) Neotropical region species list. http://www.antwiki.org/wiki/Neotropical_Region_Species_List. Accessed 30 Jan 2019

Armbrecht I, Rivera L, Perfecto I (2005) Reduced diversity and complexity in the leaf-litter ant assemblage of colombian coffee plantations. Conserv Biol 19(3):897–907

Baccaro FB, Feitosa RM, Fernandez F et al (2015) Guia para gêneros de formigas do Brasil. INPA, Petrópolis

Bächtold A, Del-Claro K (2013) Predatory behavior of *Pseudodorus clavatus* (Diptera, Syrphidae) on aphids tended by ants. Rev Bras Entomol 57(4):437–439

Begon M, Towsend CR, Harper JL (2007) Ecologia: de indivíduos a ecossistemas, 4a edn. Artmed, Porto Alegre

Berti-Filho E, Macedo LPM (2010) Fundamentos de controle biológico de insetos-praga. IFRN, Natal

Boiça-Júnior AL, Freitas CA, Freitas MM et al (2018) Estratégia de defesa de plantas a insetos. In: Castilho RC, Truzi CC, CPG P (eds) Tópicos em entomologia agrícola XI. Multipress, Jaboticabal, pp 71–93
Buckley RC (1987) Interactions involving plants, homoptera, and ants. Ann Rev Ecol Syst 18(1):111–135
Canedo-Júnior EO, Santiago GS, Ribas CR et al (2017a) The effect size of aphid-tending ants in an agricultural tri-trophic system. J Appl Entomol 142(3):349–358
Canedo-Júnior EO, Santiago GS, Zurlo LF et al (2017b) Isolated and community contexts produce distinct responses by host plants to the presence of ant-aphid interaction: plant productivity and seed viability. PLoS One 12(1):e0170915. https://doi.org/10.1371/journal.pone.0170915
Chamberlain SA, Holland JN (2009) Quantitative synthesis of context dependency in ant–plant protection mutualisms. Ecology 90(9):2384–2392
Chown SL, Nicolson SW (2004) Insect physiological ecology mechanisms and patterns. Oxford University Press, New York
Cohen JE, Pimm SL, Yodzis P et al (1993) Body sizes of animal predators and animal prey in food webs. J Anim Ecol 62(1):67–78
Córdova-Yamauchi L, Gianoli E, Quiroz A et al (1998) The ant, *Linepithema humile* (Hymenoptera: Formicidade:Dolichoderinae) is sensitive to semiochemical involved in the spacing behavior in the bird cherry-oat aphid *Rhopalosiphum pali* (Sternorrhyncha: Aphididae). Eur J Entomol 95:501–508
Dao HT, Meats A, Beattie GAC et al (2013) Ant-coccid mutualism in citrus canopies and its effect on natural enemies of red scale, Aonidiella aurantii (Maskell) (Hemiptera: Diaspididae). Bull Entomol Res 104(2):1–6
Delabie JHC (2001) Trophobiosis between formicidae and hemiptera (Sternorrhyncha and Auchenorrhyncha): an overview. Neotrop Entomol 30(4):501–516
Detrain C, Verheggen FJ, Diez L et al (2010) Aphid–ant mutualism: how honeydew sugars influence the behaviour of ant scouts. Physiol Entomol 35(2):168–174
Dixon AFG (1998) Aphid ecology, 2a edn. Springer Science, New York
Fiedler K, Maschwitz U (1988) Functional analysis of the myrmecophilous relationships between ants (Hymenoptera: Formicidae) and lycaenids (Lepidoptera: Lycaenidae). Oecologia 75(2):204–206
Flatt T, Weinsser WW (2000) The effects of mutualistic ants on aphid life history traits. Ecology 81(12):3522–3529
Gallo D, Nakano O, Silveira-Neto S et al (2002) Entomologia agrícola. FEALQ, Piracicaba
Gibb H, Hochuli DF (2003) Colonisation by a dominant ant facilitated by anthropogenic disturbance: effects on ant assemblage composition, biomass and resource use. Oikos 103(3):469–478
Gravena S (1992) Controle biológico no manejo integrado de pragas. Pesq Agropec Bras 27(13):281–299
Guindani AN, Nondillo A, Wolff VRS et al (2017) Interação mutualística entre cochonilhas e formigas em videira. RICA 2(4):6–11
Halfeld-Vieira BA, Marinho-Prado JS, Nechet KL et al (2016) Defensivos agrícolas naturais: uso e perspectivas. Embrapa, Brasília
Herbert JJ, Horn DJ (2008) Effect of ant attendance by *Monomorium minimum* (Buckley) (Hymenoptera: Formicidae) on predation and parasitism of the soybean aphid *Aphis glycines* Matsumura (Hemiptera: Aphididae). Environ Entomol 37(5):1258–1263
Hölldobler B, Wilson EO (1990) The ants. Harvad University Press, Cambrige
Itioka T, Inoue T (1996) The consequences of ant-attendance to the biological control of the red wax scale insect ceroplastes rubens by anicetus beneficus. J Appl Ecol 33(3):609–618
Lach L, Parr CL, Abbott KL (2010) Ant ecology. Oxford University Press, New York
Leite GLD, Veloso RVS, Redoan AC et al (2006) Artrópodes associados a mudas de pequizeiro. Arq Inst Biol 73(3):365–370
Nagy C, Cross JV, Markó V (2015) Can artificial nectaries outcompete aphids in ant-aphid mutualism? Applying artificial sugar sources for ants to support better biological control of rosy apple aphid, Dysaphis plantaginea Passerini in apple orchards. Crop Prot 77:127–138

Nault LR, Montgomery ME, Bowers WS (1976) Ant-aphid association: role of aphid alarm pheromone. Science 25(192):1349–1351

Nechols JR, Seibert TF (1985) Biological control of the spherical mealybug, *Nipaecoccus vastator* (Homoptera: Pseudococcidae): assessment by ant exclusion. Environ Entomol 14:45–47

Novgorodova TA (2002) Study of adaptations of aphids (Homoptera, Aphidinea) to ants: comparative analysis of myrmecophilous and nonmyrmecophilous species. Entomol Rev 82(5):569–576

Odum EP, Barret GW (2007) Fundamentos de ecologia, 5a edn. Thomson Pioneira, São Paulo

Oliveira IR, Andrade LNT, Nunes MUC et al (2006) Pragas e inimigos naturais presentes nas folhas das plantas de feijão-caupi e milho-verde em cultivo consorciado e com sistema orgânico de produção. Embrapa, Aracajú

Panizzi AR (1990) Ecologia nutricional de insetos e suas implicações no manejo de pragas. Manole, Brasilia

Parra JRP (2014) Biological control in Brazil: an overview. Sci Agric 71(5):345–355

Penteado SRC, Lazzari SMN, Reis-Filho W et al (2012) Associações entre *Cinara atlantica*, seus inimigos naturais e formigas. Pesqui Florest Bras 32(71):309–320

Perfecto I, Snelling R (1995) Biodiversity and the transformation of a tropical agroecosystem: ants in coffee plantations. Ecol Appl 5(4):1084–1097

Piñol J, Espadaler X, Cañellas N et al (2009) Effects of the concurrent exclusion of ants and earwigs on aphid abundance in an organic citrus grove. BioControl 54(4):515–527

Powell BE, Silverman J (2010) Impact of *Linepithema humile* and *Tapinoma sessile* (Hymenoptera: Formicidae) on three natural enemies of *Aphis gossypii* (Hemiptera: Aphididae). Biol Control 54(3):285–291

Price PW, Bouton CE, Gross P et al (1980) Interactions among three trophic levels: influence of plants on interactions between insect herbivores and natural enemies. Ann Rev Ecol Syst 11(1):41–65

Queiroz ACM, Rabello AM, Braga DL et al (2017) Cerrado vegetation types determine how land use impacts ant biodiversity. Biodivers Conserv 26:1–18

Raven PH, Evert RF, Eichhorn SE (2007) Biologia vegetal, 7a edn. Guanabara Koogan, Rio de Janeiro

Resende ALS, Silva EE, Silva VB et al (2006) Primeiro Registro de *Lipaphis pseudobrassicae* Davis (Hemiptera: Aphididae) e sua associação com insetos predadores, parasitóides e formigas em couve (Cruciferae) no Brasil. Neotrop Entomol 35(4):551–555

Reznikova ZI, Novgorodova TA (1998) The importance of individual and social experience for interaction between ants and symbiotic aphids. Dokl Biol Sci 359:173–175

Rodrigues CA, Araújo MS, Cabral PID et al (2008) Comunidade de formigas arborícolas associadas ao pequizeiro (*Caryocar brasiliense*) em fragmento de Cerrado Goiano. Pesqui Florest Bras 57:3944

Rodrigues SMM, Bueno VHP, Sampaio MV (2005) Efeito da liberação inoculativa sazonal de *Lysiphlebus testaceipes* (Hym.: Aphidiidae) na população de *Aphis gossypii* (Hem.:Aphididae) em cultivo de crisântemo em casa de vegetação comercial. Bol San Veg Plagas 31:199–207

Rodrigues WC, Spolidoro MV, Zinger K et al (2010) Dinâmica populacional de pulgão preto dos citros (Sternorrhyncha) em cultivo orgânico de tangerina (*Citrus reticulata* Blanco) em Seropédica, RJ. EntomoBrasilis 3(2):38–44

Rosumek FB, Silveira FA, Neves FDS et al (2009) Ants on plants: a meta-analysis of the role of ants as plant biotic defenses. Oecologia 160(3):537–549

Roth DS, Perfecto I, Rathcke B (1994) The effects of management systems on ground-foraging ant diversity in Costa Rica. Ecol Appl 4(3):423–436

Sakata H (1995) Density-dependent predation of the ant *Lasius niger* (Hymenoptera: Formicidae) on two attended aphids *Lachnus tropicalis* and *Myzocallis kuricola* (Homoptera: Aphididae). Res Popul Ecol 37(2):159–164

Schowalter TD (2006) Insect ecology: an ecosystem approach. Academic Press, Oxford

Silva EN, Perfecto I (2013) Coexistence of aphid predators in cacao plants: does ant-aphid mutualism play a role? Sociobiology 60(3):259–265

Stadler B, Dixon AFG (2005) Ecology and evolution of aphid-ant interactions. Ann Rev Ecol Evol Syst 36:345–372
Stadler B, Dixon T (2008) Mutualism: ants and their insect partners. Cambridge Press, Cambridge
Stewart-Jones A, Pope TW, Fitzgerald JD et al (2008) The effect of ant attendance on the success of rosy apple aphid populations, natural enemy abundance and apple damage in orchards. Agric For Entomol 10(1):37–43
Tegelar K, Hagman M, Glinwood R et al (2012) Ant – aphid mutualism: the influence of ants on the aphid summer cycle. Oikos 121(1):61–66
Van-Emedem HF, Harrington R (2007) Aphid as a crop pests. CABI, Oxford
Van-Lenteren JC, Bueno VHP (2003) Augmentative biological control of arthropods in Latin America. BioControl 48(2):123–139
Verheggen FJ, Diez L, Sablon L et al (2012) Aphid alarm pheromone as a cue for ants to locate aphid partners. PLoS One 7(8):e41841
Vinson SB, Scarborough TA (1989) Impact of the imported fire ant on laboratory populations of cotton aphid (*Aphis gossypii*) predators. Fla Entomol 72(1):107–111
Völkl W, Woodring J, Fischer M et al (1999) Ant-aphid mutualisms: the impact of honeydew production and honeydew sugar composition on ant preferences. Oecologia 118(4):483–491
Zhang S, Shang Y, Ma K (2012) The ecological effects of the ant-hemipteran mutualism: a meta-analysis. Basic Appl Ecol 13(12):116–124

Index

B. Souza et al. (eds.), *Natural Enemies of Insect Pests in Neotropical Agroecosystems*, https://doi.org/10.1007/978-3-030-24733-1

W

Y

Z

FSC
www.fsc.org
MIX
Papier | Fördert
gute Waldnutzung
FSC® C083411

Kolibri 360 GmbH
Ferdinand-Jühlke-Straße 7
99095 Erfurt, Deutschland
produktsicherheit@kolibri360.de